PRIMATES OF WESTERN UGANDA

Developments in Primatology: Progress and Prospects

Series Editor:
Russell H. Tuttle
University of Chicago, Chicago, Illinois

This peer-reviewed book series melds the facts of organic diversity with the continuity of the evolutionary process. The volumes in this series exemplify the diversity of theoretical perspectives and methodological approaches currently employed by primatologists and physical anthropologists. Specific coverage includes: primate behavior in natural habitats and captive settings; primate ecology and conservation; functional morphology and developmental biology of primates; primate systematics; genetic and phenotypic differences among living primates; and paleoprimatology.

ANTHROPOID ORIGINS: NEW VISIONS

Edited by Callum F. Ross and Richard F. Kay

MODERN MORPHOMETRICS IN PHYSICAL ANTHROPLOGY

Edited by Dennis E. Slice

BEHAVIORAL FLEXIBILITY IN PRIMATES: CAUSES AND CONSEQUENCES

By Clara B. Jones

NURSERY REARING OF NONHUMAN PRIMATES IN THE 21ST CENTURY

Edited by Gene P. Sackett, Gerald C. Ruppenthal and Kate Elias

NEW PERSPECTIVES IN THE STUDY OF MESOAMERICAN PRIMATES: DISTRIBUTION, ECOLOGY, BEHAVIOR, AND CONSERVATION

Edited by Paul Garber, Alejandro Estrada, Mary Pavelka and LeAndra Luecke

HUMAN ORIGINS AND ENVIRONMENTAL BACKGROUNDS

Edited by Hidemi Ishida, Martin Pickford, Naomichi Ogihara and Masato Nakatsukasa

PRIMATE BIOGEOGRAPHY

Edited by Shawn M. Lehman and John Fleagle

REPRODUCTION AND FITNESS IN BABOONS: BEHAVIORAL, ECOLOGICAL, AND LIFE HISTORY PERSPECTIVES

Edited By Larissa Swedell and Steven R. Leigh

PRIMATE ORIGINS: ADAPTATIONS AND EVOLUTION

Edited by Matthew J. Ravosa and Marian Dagosto

RINGAILED LEMUR BIOLOGY*: LEMUR CATTA IN MADAGASCAR*

Edited by Alison Jolly, Robert W. Sussman, Naoki Koyama and Hantanirina Rasamimanana

PRIMATES OF WESTERN UGANDA

Edited by Nicholas E. Newton-Fisher, Hugh Notman, James D. Paterson and Vernon Reynolds

PRIMATES OF WESTERN UGANDA

Edited by

Nicholas E. Newton-Fisher
Department of Anthropology, University of Kent, Canterbury, Kent, United Kingdom

Hugh Notman
Department of Anthropology, University of Calgary, 2500 University Drive NW, Calgary, Alberta, Canada

James D. Paterson
Emeritus, Department of Anthropology, University of Calgary, Canada

Vernon Reynolds
Emeritus, School of Anthropology, University of Oxford, 51 Banbury Road, Oxford, United Kingdom

Springer

Editors:
Nicholas E. Newton-Fisher
Department of Anthropology,
University of Kent,
Canterbury, Kent, United Kingdom

Hugh Notman
Department of Anthropology,
University of Calgary,
2500 University Drive NW,
Calgary, Alberta, Canada

James D. Paterson
Emeritus, Department of Anthropology,
University of Calgary, Canada

Vernon Reynolds
Emeritus, School of Anthropology,
University of Oxford,
51 Banbury Road, Oxford,
United Kingdom

Cover illustration: (clockwise from top left): blue monkey *Cercopithecus mitis* (© T. Furuichi), mountain gorilla *Gorilla beringei* (© J. Rothman), chimpanzee *Pan troglodytes schweinfurthii/marungensis* (© N.E. Newton-Fisher), red colobus *Piliocolobus tephrosceles* (© A.J. Plumptre).

Library of Congress Control Number: 2006920916

ISBN-10: 0-387-32342-2 e-ISBN 0-387-33505-6
ISBN-13: 978-0387-32342-8

Printed on acid-free paper.

Printed in the United States of America. (TB/EB)

9 8 7 6 5 4 3 2 1

springer.com

CONTENTS

Preface ix
List of Contributors xiii

SECTION 1 TAXONOMY 1

1. Taxonomy and Biogeography of the Primates of Western Uganda 3
Colin Groves

SECTION 2 ECOLOGY 21

2. Factors Influencing Variation in the Population Densities of *Colobus guereza* Within Selectively Logged Forest at the Budongo Reserve: The Importance of Lianas During a Subsistence Diet 23
Graham Anthony Preece

3. How Does the Golden Monkey of the Virungas Cope in a Fruit-Scarce Environment? 45
Dennis Twinomugisha, Colin A. Chapman, Michael J. Lawes, Cedric O'Driscoll Worman, and Lisa M. Danish

4. The Diet of Olive Baboons (*Papio anubis*) in Budongo Forest Reserve, Uganda 61
Adam Alberts Okecha and Nicholas E. Newton-Fisher

5. Aspects of Diet, Foraging, and Seed Predation in Ugandan Forest Baboons 75
James D. Paterson

6. The Interaction of Hormones with Ecological Factors in Male Budongo Forest Chimpanzees 93
Sally B. Seraphin, Patricia L. Whitten, and Vernon Reynolds

7. **The Role of Diet in Self-Medication Among Chimpanzees in the Sonso and Kanyawara Communities, Uganda** 105
Paula Pebsworth, Sabrina Krief, and Michael A. Huffman

8. **Geophagy in Chimpanzees (*Pan troglodytes schweinfurthii*) of the Budongo Forest Reserve, Uganda: A Multidisciplinary Study** 135
Mnason Tweheyo, Vernon Reynolds, Michael A. Huffman, Paula Pebsworth, Shunji Goto, William C. Mahaney, Michael W. Milner, Anthony Waddell, Randy Dirszowsky, and Ronald G. V. Hancock

9. **Nutritional Aspects of the Diet of Wild Gorillas: How Do Bwindi Gorillas Compare?** 153
Jessica M. Rothman, Alice N. Pell, J. Bosco Nkurunungi, and Ellen S. Dierenfeld

10. **The Parasites of the Gorillas in Bwindi Impenetrable National Park, Uganda** 171
Jessica M. Rothman, Dwight D. Bowman, Gladys Kalema-Zikusoka, and John Bosco Nkurunungi

11. **Preliminary GIS Analysis of Range Use by Sympatric Mountain Gorillas and Chimpanzees in Bwindi Impenetrable National Park, Uganda** 193
J. Bosco Nkurunungi and Craig B. Stanford

SECTION 3 BEHAVIOR AND PHYSIOLOGY **207**

12. **Comparison of Sex Differences in Gregariousness in Fission–Fusion Species: Reducing Bias by Standardizing for Party Size** 209
Melissa Emery Thompson and Richard W. Wrangham

13. **Urinary Estrone Conjugates and Reproductive Parameters in Kibale (Kanyawara) and Budongo (Sonso) Chimpanzees** 227
Melissa Emery Thompson, Richard W. Wrangham, and Vernon Reynolds

14. **Frequent Copulations by Females and High Promiscuity in Chimpanzees in the Kalinzu Forest, Uganda** 247
Chie Hashimoto and Takeshi Furuichi

15. **Effects of Injury on the Locomotion of Free-Living Chimpanzees in the Budongo Forest Reserve, Uganda** 259
Julie Munn

16. **Effect of Snare Injuries on the Fig-Feeding Behavior of Chimpanzees of the Budongo Forest, Uganda: Behavioral Adaptations and Long-Term Implications** 281
Emma J. Stokes and Richard W. Byrne

17. **Deciphering Junglespeak: An Update on Chimpanzee Vocalizations, Referential Communication, and Their Relationship to Language** 299
Hugh Notman

18. **Instrumental Leaf Use by Chimpanzees of the Budongo Forest (Sonso Community)** 313
Duane Quiatt

SECTION 4 CONSERVATION **327**

19. **A Survey of Prosimians in the National Parks and Forest Reserves of Uganda** 329
Lesley Ambrose

20. **The Diets, Preferences, and Overlap of the Primate Community in the Budongo Forest Reserve, Uganda: Effects of Logging on Primate Diets** 345
Andrew J. Plumptre

21. **Behavioral Patterns of Colobus in Logged and Unlogged Forests: The Conservation Value of Harvested Forests** 373
Colin A. Chapman, Michael D. Wasserman, and Thomas R. Gillespie

22. **Threats to, and Protection of, the Chimpanzees of the Budongo Forest Reserve** 391
Vernon Reynolds

23. **Gorillas Living on the Edge: Literally and Figuratively** 405
Michele L. Goldsmith, Joel Glick, and Evarist Ngabirano

24. Local Ecological Perceptions of Chimpanzees and Forest Resources: A Case Study Near Budongo Forest, Uganda 423
Cristy Watkins

References **439**
Index **509**

PREFACE

From galagos to gorillas, the primates of western Uganda comprise a very diverse collection of species. Western Uganda has a long history of primatological research extending back to the publications of the Uganda Virus Institute in 1947 and even to the notable first encounters of Count Beringe with mountain gorillas in 1913. Many forested areas of Uganda (Figure i) have been the focus of research continuously since 1970, and thus western Uganda has a central place in primatology that it maintains to the present day. In this book, we present a series of new, unpublished scientific accounts of a selection of the species in the region, each chapter focusing on one or more particular characteristics of the species concerned. The book falls naturally into four sections. First, we introduce the primates of western Uganda, with a chapter on their taxonomy. We have left authors to follow the taxonomic terminology with which they are most comfortable, but present this first chapter to reflect recent developments in the understanding of taxonomic relationships among the Ugandan primates. Second, we present a section with an ecological focus, followed by a collection of chapters on behavior and physiology. Finally the focus shifts to conservation.

Chimpanzees and gorillas have always attracted a lot of interest both among the general public and among researchers; consequently, this interest is reflected in the present volume. This book contains 12 chapters on chimpanzees, 4 chapters on gorillas, 6 chapters on monkeys and prosimians, and 1 general chapter on taxonomy. The chimpanzee bias is unfortunate in a book on "primates," but is an unavoidable reflection on the recent research that has been conducted in western Uganda. Studies of other primate species have been made in the past, but this volume is not a compilation of past studies; every study is new and is here published for the first time.

If there is a message emerging from this book, it is that western Uganda provides an excellent area for future studies of primates. In the modern setting of population increase, depletion of primate habitats, hunting with snares and

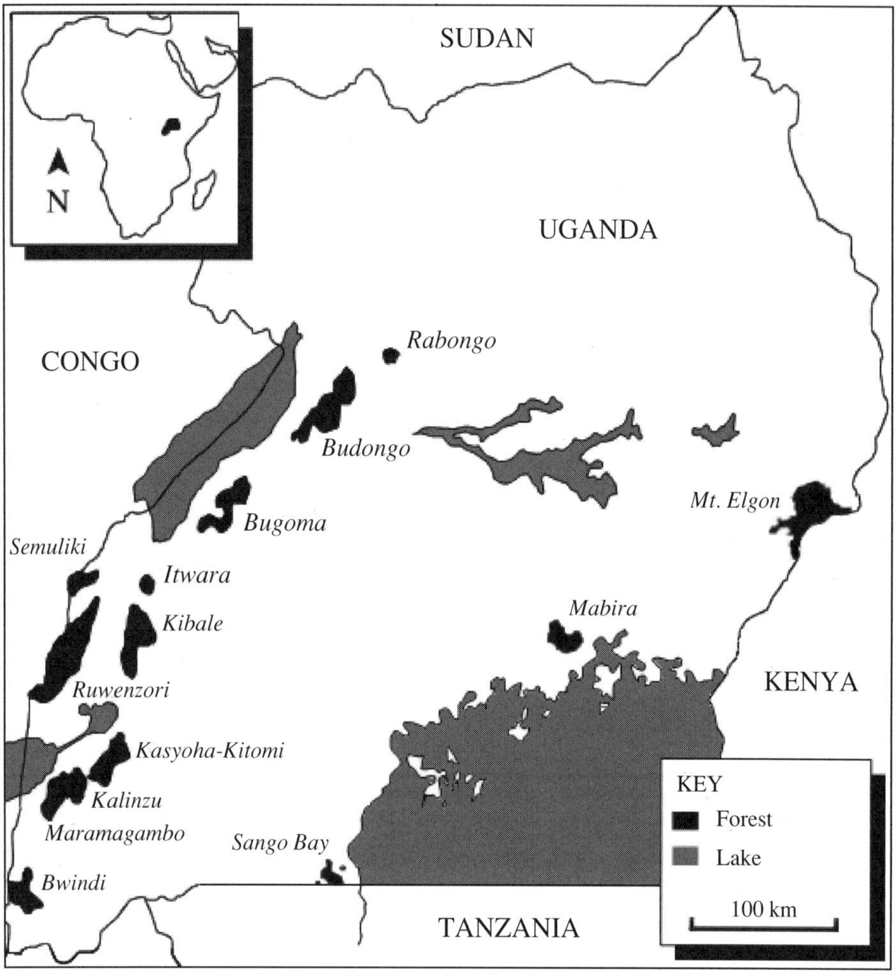

Figure i. Map of Uganda showing the major forested areas.

traps, and other challenges to the forests and their natural inhabitants, scientific projects can provide havens of security for wildlife. Scientific research needs to be more widely recognized as a primary mechanism for conservation, and its importance needs to be spelled out to wildlife NGOs and governmental authorities. All those who work with primates, whether in research or in conservation, must assist in the effort to bring their work to the attention of the Uganda National Forest Authority and the Uganda Wildlife Authority, in particular, as well as to local authorities. It is thus with great pleasure that we publish here the work of a number of Ugandan primatologists. They will be the stewards of Uganda's primates in the years to come.

Finally, a word of thanks. The idea of putting together a book about the primates of western Uganda first arose after a very successful meeting of the Primate Society of Great Britain held in London on December 5, 2001. The meeting attracted a large audience and it was clear that there was a lot of interest in the primates of western Uganda, nowadays sometimes referred to as the Albertine Rift, although technically most of the study areas here are outside of the Rift itself. We are therefore very grateful to the Primate Society of Great Britain for making possible the meeting, which led to the present book.

Nicholas E. Newton-Fisher
Hugh Notman
James D. Paterson
Vernon Reynolds

LIST OF CONTRIBUTORS

Lesley Ambrose, Nocturnal Primate Research Group, School of Social Sciences and Law, Oxford Brookes University, Oxford OX3 QBP, United Kingdom.

Dwight D. Bowman, Department of Veterinary Microbiology & Immunology, College of Veterinary Medicine, Cornell University, Ithaca, NY 14853, USA.

Richard W. Byrne, Scottish Primate Research Group, University of St. Andrews, School of Psychology, St. Andrews, Fife KY16 9JU, Scotland.

Colin A. Chapman, Department of Zoology, University of Florida, Gainesville, FL 32611, and Wildlife Conservation Society, 2300 Southern Boulevard, Bronx, NY 10460, USA.

Lisa M. Danish, Department of Zoology, University of Florida, Gainesville, FL 32611, USA.

Ellen S. Dierenfeld, Department of Animal Science, Cornell University, Ithaca, NY 14853, and St. Louis Zoo, One Government Drive, St. Louis, MO 63110, USA.

Randy Dirszowsky, Geomorphology and Pedology Laboratory, York University, 4700 Keele Street, North York, Ontario, M3J 1P3, and Department of Chemistry and Chemical Engineering, University of Toronto, Toronto, Ontario, Canada, K7K 7B4.

Melissa Emery Thompson, Department of Anthropology, Harvard University, 11 Divinity Avenue, Cambridge, MA 02138, USA.

Takeshi Furuichi, Faculty of International Studies, Meiji-Gakuin University, Totsuka, Yokohama, Kanagawa 244-8539, Japan.

Thomas R. Gillespie, Department of Zoology, University of Florida, Gainesville, FL 32611, USA.

Michele L. Goldsmith, Emerson College, 120 Boylston Street, Boston, MA 02116, USA.

Shunji Goto, Primate Research Institute, Kyoto University, Kanrin 41-2, Inuyama, Aichi 484-8506, Japan.

Colin Groves, School of Archaeology and Anthropology, Australian National University, Canberra, ACT 0200, Australia.

Ronald G.V. Hancock, Department of Chemical Engineering and Applied Chemistry, University of Toronto, Toronto, Canada, M5S 1A4.

Chie Hashimoto, Primate Research Institute, Kyoto University, Inuyama, Aichi 484-8506, Japan.

Michael A. Huffman, Primate Research Institute, Kyoto University, Kanrin 41-2, Inuyama, Aichi 484-8506, Japan.

Gladys Kalema-Zikusoka, Conservation Through Public Health, P.O. Box 4483, Kampala, Uganda.

Sabrina Krief, Eco-anthropologie et Ethnobiologie – Chimie des Substances Naturelles, Muséum National d'Histoire Naturelle, 43 rue Buffon, 75231 Paris, France.

Michael J. Lawes, Forest Biodiversity Programme, School of Botany and Zoology, University of Natal, Post Bag X01, Scottsville, 3209, South Africa.

William C. Mahaney, Geomorphology and Pedology Laboratory, York University, 4700 Keele Street, North York, Ontario, Canada, M3J 1P3.

Michael W. Milner, Geomorphology and Pedology Laboratory, York University, 4700 Keele Street, North York, Ontario, Canada, M3J 1P3.

Julie Munn, Australian National University, Canberra, ACT 0200, Australia.

Nicholas E. Newton-Fisher, Department of Anthropology, University of Kent, Canterbury, Kent, CT2 7NR, United Kingdom.

J. Bosco Nkurunungi, Department of Zoology, Makerere University, Kampala, Uganda.

Hugh Notman, Department of Anthropology, University of Calgary, 2500 University Drive NW, Calgary, Alberta, Canada T2N 1N4.

Cedric O'Driscoll Worman, Department of Zoology, University of Florida, Gainesville, FL 32611, USA.

Adam Alberts Okecha, Faculty of Forestry and Nature Conservation, Makerere University, Kampala, Uganda.

James D. Paterson, Emeritus, Department of Anthropology, University of Calgary, 2500 University Drive NW, Canada T2N 1N4.

Paula Pebsworth, Primate Research Institute, Kyoto University, Kanrin 41-2, Inuyama, Aichi 484-8506, Japan.

Alice N. Pell, Department of Animal Science, Cornell University, Ithaca, NY 14853, USA.

Andrew J. Plumptre, Wildlife Conservation Society, Plot 39 Acacia Avenue, P.O. Box 7487, Kampala, Uganda.

Graham Anthony Preece, University of Wales, Bangor.

Duane Quiatt, Department of Anthropology, University of Colorado, Denver, P.O. Box 173364, Denver, CO 80217, USA

Vernon Reynolds, School of Anthropology, University of Oxford, 51 Banbury Road, Oxford OX2 6PE, United Kingdom.

Jessica M. Rothman, Department of Animal Science, Cornell University, Ithaca, NY 14853, USA.

Sally B. Seraphin, Department of Anthropology, Emory University, 207 Anthropology Building, 1557 Dickey Drive, Atlanta, GA 30322, USA.

Craig B. Stanford, Jane Goodall Research Center, Departments of Anthropology and Biological Sciences, University of Southern California, Los Angeles, CA 90089, USA.

Emma J. Stokes, Wildlife Conservation Society, Nouabalé-Ndoki Project, BP 14537, Brazzaville, Congo.

Dennis Twinomugisha, Institute of Environment and Natural Resources, Makerere University, Kampala, Uganda.

Mnason Tweheyo, Department of Forest Biology and Ecosystems Management, Makerere University, Kampala, Uganda.

Anthony Waddell, Geomorphology and Pedology Laboratory, York University, 4700 Keele Street, North York, Ontario, Canada, M3J 1P3.

Michael D. Wasserman, Department of Zoology, University of Florida, Gainesville, FL 32611, USA.

Cristy Watkins, School of Natural Resources and Environment, University of Michigan.

Patricia L. Whitten, Department of Anthropology, Emory University, 207 Anthropology Building, 1557 Dickey Drive, Atlanta, GA 30322, USA.

Richard W. Wrangham, Department of Anthropology, Harvard University, 11 Divinity Avenue, Cambridge, MA 02138, USA.

SECTION ONE

Taxonomy

Figure ii. Red colobus *Piliocolobus tephrosceles* (above, photo: A.J. Plumptre) and chimpanzee *Pan troglodytes schweinfurthii/marungensis* (below, photo: N.E. Newton-Fisher).

CHAPTER ONE

Taxonomy and Biogeography of the Primates of Western Uganda

Colin Groves

INTRODUCTION

In this brief survey, I will list the species (and subspecies) of nonhuman primates that have been recorded from western Uganda, with an outline of their known ranges within Uganda and in neighboring countries, and use this to try to assess the biogeographic affinities of the region as a whole.

I adopt the following definitions, paraphrased after Groves (2001):

- A species is a population (or group of populations), distinguished by the possession of one or more consistent (fixed, absolute) heritable differences from other such populations;
- A subspecies is a geographic segment of a species, distinguished by the possession at high frequencies, but not as much as 100%, of one or more heritable differences from other such segments;

Colin Groves • School of Archaeology and Anthropology, Australian National University, Canberra, ACT 0200, Australia.

Primates of Western Uganda, edited by Nicholas E. Newton-Fisher, Hugh Notman, James D. Paterson, and Vernon Reynolds. Springer, New York, 2006.

- A genus is a monophyletic group of species (or a single species), which separated from other such groups earlier than the Miocene–Pliocene boundary;
- A family is a monophyletic group of genera (or a single genus), which separated from other such groups before the Oligocene–Miocene boundary.
- Species-groups, tribes, subfamilies, and superfamilies are best taken as further convenience categories, set up to assort an otherwise unwieldy group of families or genera into monophyletic clusters.

The number of taxa is still uncertain. In some cases, especially in that of the *Cercopithecus mitis* group, there are problems that accurate field records, by experienced observers, can easily untangle. In other cases, especially involving the genus *Galagoides*, solution of the problems will take more specialized fieldwork. In no case is further collecting necessary, which is not to say that pick-up specimens (including bones from owl pellets and from the ground below eagles' nests) should be ignored.

In what follows, the taxonomic outlines presented in Groves (2001) will be followed, except where indicated. I may be forgiven for reiterating the guiding principle laid down in that book: "I hope that the classifications in the book will be taken as a starting point, not as a solution to problems. I hope that others will take up some of the propositions and test them" (Groves, 2001:viii).

STREPSIRRHINI: LORISIFORMES

Galagidae Gray, 1825

Jenkins (1987) reinstated the original spellings "Loridae" and "Galagonidae" for the lorises and bushbabies, and was supported in this change by Groves (2001). The previously better known forms of the names have recently, however, been officially sanctioned (International Commission on Zoological Nomenclature, 2002). This means that all names of superfamilies, families, subfamilies, and tribes based on these two names take the stem form Loris- and Galag-, respectively. Also, the higher-category name should preferably be Lorisiformes, not Loriformes, although ranks above the family-group are not actually covered by the rules of nomenclature.

How many genera there are in the Galagidae is unclear. Groves (2001) recognized three, while noting that one of them, *Galago*, was probably not monophyletic; Grubb *et al.* (2003) provisionally increased the number to five genera, separating *Galagoides* from *Galago* but noting that it, too, may not be monophyletic.

Galago *E. Geoffroy St. Hilaire, 1796*

Two species live in western Uganda, one in tree-savannah and thorn-bush, the other in rainforest.

Lesser or Senegal bushbaby, Galago senegalensis *E. Geoffroy St. Hilaire, 1796.* These are the common small bushbabies of the nonforested regions. There are two quite distinct subspecies in western Uganda (Groves, 2001):

- *Galago senegalensis senegalensis* E. Geoffroy, 1796: from the northern part, grey in color with creamy yellow limbs, a grey-brown tail, and yellow-white underparts.
- *Galago senegalensis sotikae* Hollister, 1920: from Ankole and south into Tanzania, more brownish grey and noticeably larger in size (mean skull length 48.3 mm [$n = 4$] as against 43.6 mm [$n = 21$]).

Eastern needle-clawed bushbaby, or Spectacled galago, Galago matschiei *Lorenz von Liburnau, 1917.* A rainforest species, whose main distribution is the eastern Democratic Republic of Congo (DRC). In Uganda, according to Kingdon (1971), it lives mainly in the southwest, north perhaps to the Victoria Nile, and is characteristic of medium-altitude *Parinari excelsa* forest and along forest margins. It is dark in color, with large yellow eyes ringed with black, and has sharp needle-pointed nails. Vocalizations are distinctive (Bearder *et al.*, 1995).

Galagoides *A. Smith, 1833*

The genus may or may not include the "rolling callers" and "incremental callers" listed in Grubb *et al.* (2003)—the species that are central to the genus being the "crescendo callers." While Bearder *et al.* (1995) recognized only two crescendo-calling species, their taxonomic diversity is probably far greater than this (Groves, 2001; Grubb *et al.*, 2003), and the two species of Bearder *et al.*

(1995) probably actually represent species-groups. The two groups are widely sympatric across the Central and West African rainforest belt.

Thomas's bushbaby, Galagoides thomasi *(Elliot, 1907).* The common small rainforest bushbaby of western Uganda, it extends at least from Bwindi in the south to the forests east of Lake Albert in the north, and in DRC at least to Idjwi island in Lake Kivu. Vocalizations resembling it, but not necessarily identical, have been recorded by Bearder *et al.* (1995) as far west as Cameroon, Gabon, and Bioko island.

This is the largest of the Central African crescendo callers, the skull length being always above 40 mm. It is larger and more blackish grey than others, with a more conspicuous median facial white stripe.

Demidoff's bushbaby, Galagoides demidoff *Fischer von Waldheim, 1806.* On the evidence of vocalizations, Bearder *et al.* (1995) recorded *G. demidoff* in the Ugandan sector of Semliki forest. Vocalizations attributed by them to this species are rather uniform across the West and Central African rainforests, but physical phenotypic diversity is considerable (Groves, 2001), so that it is probable that there are several species in the group rather than one.

Otolemur *Coquerel, 1859*

The generic distinctiveness of the greater galagos or thick-tailed bushbabies has been acknowledged for many years. They live in forested regions, but not strictly rainforest. The number of species is still unclear; two are generally recognized, but Kingdon (1997) added a third, *Otolemur argentatus,* citing several differences, in particular the structure of the penis. Groves (2001:105) noted that externally the Lake Victoria *Otolemur* "can barely be distinguished" from those from Angola, and that the name *monteiri* (given to the Angolan form) has 50 years' priority over the name *argentatus* (given to the Lake Victoria form).

Silvery Greater galago, Otolemur monteiri *(Bartlett, 1863).* This is a large, usually silvery-white, bushbaby, with dark hands and feet, and is creamy yellow on the midline of the underparts. The tail is often nearly white. Melanistic individuals are common.

The Uganda subspecies is *Otolemur monteiri argentatus* (Lönnberg, 1913). It reaches Uganda only in the far southwest, on the borders of Tanzania and Rwanda.

Lorisidae Gray, 1821

Perodicticus *Bennett, 1831*

Grubb *et al.* (2003) predict that this genus will prove to contain several valid species, but until further research is done only one can be recognized.

Potto, Perodicticus potto *(Müller, 1776).* This is a rainforest species, extending from DRC through Uganda as far east as the Kakamega forest in Kenya. The "eastern subspecies" is designated *Perodicticus potto ibeanus* Thomas, 1910; as noted above, this may turn out to be a distinct species. The names *arrhenii* and *nebulosus* were given to pottos from eastern DRC (Masisi and Ukaika, respectively), near the border with western Uganda.

HAPLORRHINI: SIMIIFORMES

Cercopithecoidea Gray, 1821: Cercopithecidae Gray, 1821

Both subfamilies of Cercopithecidae occur in western Uganda.

Cercopithecinae Gray, 1821

Groves (2001) divided this subfamily into two tribes, whose validity and content have recently been confirmed by molecular studies (Tosi *et al.*, 2003).

Cercopithecini Gray, 1821. The number of genera in this family has recently been thrown into question by the finding of Tosi *et al.* (2004) that the genus *Cercopithecus*, as traditionally recognized, is nonmonophyletic. It turns out that the *Cercopithecus lhoesti* species-group is part of a clade containing *Chlorocebus* (vervets) and *Erythrocebus* (patas). The postcranial skeleton of the *Cercopithecus lhoesti* species-group shows similarities to that of patas monkeys and, to a lesser extent, to that of vervets (Gebo & Sargis, 1994), but the skull does not (Verheyen, 1962).

The question is how to recognize this taxonomically? Tosi *et al.* (2004) list the options: transferring both the lhoesti group and the patas monkey to the genus *Chlorocebus*, or splitting up the clade into three different genera—*Chlorocebus* for vervets only, *Erythrocebus* for patas, and *Allochrocebus* for the lhoesti group. They indicate that they marginally prefer the first of these two options.

Preliminary determinations indicate that the divisions between the three subclades are deep, probably late Miocene (T. Disotell, personal communication). If this is so, the time criterion recommended by Goodman *et al.* (1998) is amply met, and even more so the modified one of Groves (2001). Therefore, it is appropriate to recognize three separate genera among the terrestrial cercopithecins:

Chlorocebus *Gray, 1870.* These are predominately wooded savannah monkeys, not rainforest. There are several species in the genus, although they may well not be precisely as delimited by Groves (2001), and a full revision is needed. Western Uganda has two species that are parapatric and hybridize.

Tantalus monkey Chlorocebus tantalus *(Ogilby, 1841).* This species extends into Uganda from Sudan, and extends as far south as the latitude of Entebbe, and west into the DRC (Mawambi in the north, Rutshuru plains in the south), east into Kenya. The dorsum is grizzled olive brown, the limbs grey with blackish digits, the underparts white, the tail greyer with a white tip and a white basal tuft, the scrotum sky blue surrounded by a long orange tuft. There is a black line from eye to temple, separating the sinuous, tapered brow-band from the long, stiff whitish yellow, black-tipped cheek whiskers. The subspecies in Uganda is *Chlorocebus tantalus budgetti* (Pocock, 1907).

Vervet monkey Chlorocebus pygerythrus *(F. Cuvier, 1821).* This is the southwestern Ugandan species, whence it extends into Rwanda and round the southern and eastern shores of Lake Victoria to Entebbe. The dorsum, limbs, and tail are fawn to orange yellow, the hands, feet, and tail-tip darker, the underparts whitish often infused with reddish, the tail base red but without tufts, the scrotum turquoise blue. There is no black line, so the white face ring is complete; the cheek whiskers are shorter and speckled. The Ugandan subspecies is *Chlorocebus pygerythrus rufoviridis* (I. Geoffroy St. Hilaire, 1843).

Where this species meets the last, there are hybrids along with the parent forms. The hybrid zone has been described in some detail by Dandelot (1959), who maps it as extending approximately from the Kazinga Channel southeast to the Tanzanian border.

Erythrocebus Trouessart, 1897. There is a single species in this genus:

Patas monkey, Erythrocebus patas *(Schreber, 1774).* The patas monkey in Uganda is found only north of the latitude of Lake Albert, and north of the Victoria Nile.

Allochrocebus *Elliot, 1913.* As explained above, what has hitherto been called the *Cercopithecus lhoesti* group is here recognized as a distinct genus. They are robustly built monkeys, with distinct terrestrial adaptations in the postcranial skeleton (Gebo & Sargis, 1994).

L'Hoest's monkey, Allochrocebus lhoesti *(Sclater, 1899).* This mainly terrestrial monkey is known from several rainforest areas, predominately in montane forest regions: Bwindi, Kalinzu, and Kibale. It is black with an orange, speckled dorsal saddle, and has bushy white cheek whiskers.

Cercopithecus *Linnaeus, 1758.* With the expulsion of the lhoesti group, the genus *Cercopithecus* becomes a homogeneous group of agile, generally brightly colored, short-faced arboreal rainforest monkeys. Four of the seven species-groups have representatives in western Uganda. These groups are the *cephus*, *mitis*, *mona*, and *neglectus* groups; three of these are taxonomically simple, but the *C. mitis* group is diverse and somewhat controversial.

Red-tailed monkey, Cercopithecus ascanius *(Audebert, 1799).* This species, widespread in Central Africa, extends into western Uganda as far north as Budongo, to the shores of Lake Victoria (including Buvuma Island) and into Kenya (Kakamega Forest). It is distinguished by its white heart-shaped nose spot, white ear tufts and red tail. The subspecies in Uganda is *Cercopithecus ascanius schmidti* Matschie, 1892, which is so strikingly—and, apparently, consistently—different from other members of the species that it should probably be ranked as a distinct species.

Blue monkey, Cercopithecus mitis *Wolf, 1822.* Blue monkeys, like Redtails, are found in all the western forests of Uganda and across it into Kenya, where the range is much wider than the Redtail and it extends to the Rift Valley (and is represented east of it by a related species, *Cercopithecus albogularis*), and north into Ethiopia. In Uganda it reaches higher altitudes than the Redtail.

The subspecies in western Uganda is *Cercopithecus mitis stuhlmanni* Matschie, 1893, which is not closely related to *Cercopithecus mitis mitis* (Grubb *et al.*, 2003), and may not in fact be conspecific. Outside Uganda it occurs in the Ituri and Semliki districts of DRC to west of Lake Kivu, and west as far as the Lualaba. In Uganda, it is found in all the western forests from Budongo south to about Lake George, including Kibale (where it is rare), Semliki, and the Rwenzoris. It was recorded by Kingdon (1971) in Bugoma, but appears no longer to occur there. It is characterized by the dark speckled blue-grey color, with dark (sometimes partly black) legs, paler underparts, and black crown, which contrasts strongly with a light speckled grey frontal diadem.

Silver monkey, Cercopithecus doggetti *Pocock, 1907.* This monkey replaces the Blue monkey, to which it is very closely related, in the southwest, from the Virungas and Maramagambo and Bwindi forests, east to Sango Bay; south of the border it extends into Rwanda, Burundi, and the mountains bordering Lake Tanganyika. Darker animals, presumed to be hybrids with *Cercopithecus mitis stuhlmanni*, are known from some areas of the eastern DRC, in the Rift mountains on the western side of Lakes Edward, Kivu, and Tanganyika as far as 4° S. It is light silvery grey-brown, with black feet and arms, dark grey legs, long grizzled cheek whiskers, and black crown corresponding with a pale buff-speckled frontal diadem. In some respects it is a paler (more silvery) version of a Blue monkey, but the respective distributions of the two are complex, and need elucidation by careful field observations.

Golden monkey, Cercopithecus kandti *Matschie, 1905.* This is the third presumptive species of the *C. mitis* group to occur in Uganda. It is bright golden colored, with black limbs and tail, and black crown contrasting with the golden cheeks and frontal diadem. In the field, females appear distinctly more brightly colored than males (Twinomugisha *et al.*, 2003).

The distribution centers on the high altitude forests of the eastern Virungas (not the Hagenia forests of the western Virungas), but is also claimed to occur in Bwindi and in the Nyungwe forest of southern Rwanda. In all these areas, *Cercopithecus doggetti* has also been recorded, but Twinomugisha *et al.* (2003) did not encounter it in Mgahinga although they cite an unpublished thesis by Werikhe (1997) that recorded few *C. doggetti* there in the past. There are also grey-olive monkeys that may be hybrids between them; to this color type the name *schoutedeni* has been given. There are two explanatory models for this situation: *Cercopithecus kandti* is a high altitude species whose range has been progressively invaded by *Cercopithecus doggetti* until it is now restricted to the highest altitudes; or it is not a separate taxon at all, but simply a high-altitude morph of *Cercopithecus doggetti*.

I prefer the first explanation, because the pelage pattern is not simply a golden variant of that of *Cercopithecus doggetti*, and because of the existence of the "*schoutedeni*" intermediates. In fact, the latter may constitute a different taxon yet again, as they are reputed to form the entire population on Idjwi and Shushu islands in Lake Kivu. This may be a case where mtDNA and Y chromosome DNA could help to untangle potential ancestral strands.

Dent's mona, Cercopithecus denti *Thomas, 1907.* This representative of the *Cercopithecus mona* group is widespread in eastern DRC and enters Ugandan territory only in the Semliki forest, where, according to Kingdon (1971), it lives in high canopy mixed forest. It also occurs in Rwanda. It is dark brown with blackish limbs and white underparts sharply demarcated from the dark upper side. The head is yellowish, set off with black lateral crown stripes.

De Brazza's monkey, Cercopithecus neglectus *Schlegel, 1876.* A heavy-bodied, short-tailed largely terrestrial monkey favoring swamp forests. Its distribution in western Uganda is curiously restricted to the Lake Albert region; it is found again at Sango Bay, and then again on Mt. Elgon in the far east.

Papionini Burnett, 1828

Papio Erxleben 1777. Baboons are widespread in western Uganda; here and in the neighboring northeastern DRC they penetrate more deeply into rainforest than is usual for baboons. Of the five species, one occurs in Uganda:

Olive or Anubis baboon, Papio anubis *(Lesson, 1827).* A sort of "patchwork" morph, described as *Papio tesselatum*, of this species predominates in rainforest habitats. The light and dark bands of the hairs in this morph happen to coincide over wide areas to give the patches of dark and light colors.

Lophocebus *Palmer, 1903*

These arboreal mangabeys were formerly included with the true "white-eyelid" mangabeys in *Cercocebus*, but they are nowadays universally separated from them. *Lophocebus* are closely related to baboons, while *Cercocebus* are related to mandrills. Molecular clock estimates (Goodman *et al.*, 1997) suggest that *Mandrillus* separated from *Cercocebus* only in the Pliocene, and if this rather shallow time depth is corroborated then it would be appropriate to combine them into a single genus; it is unclear when *Lophocebus* separated from *Papio* and *Theropithecus*.

Gray-cheeked mangabey, Lophocebus albigena *(Gray, 1850).* Mangabeys are widespread from the DRC border to Sango Bay, and are especially common in swamp forests. They are absent from Budongo but occur in Kibale; in Bugoma, according to Kingdon (1971), they are "almost the only monkey."

The Uganda mangabey was included by Groves (1978) in the subspecies *Lophocebus albigena johnstoni* (Lydekker, 1900), which elsewhere was distributed in northern and eastern DRC; but, to judge by the skull, Ugandan mangabeys are noticeably smaller, and may rate a different subspecies, in which case the name *ugandae* Matschie, 1913 is applicable.

Colobinae Jerdon, 1867

Colobus monkeys, both red and black-and-white, are presently numerous in the forests of western Uganda. Their high biomass masks a disconcerting vulnerability; it is as well to remember that the only primate taxon that seems to have been exterminated during the 20th century was a red colobus (Oates *et al.*, 2000b).

Piliocolobus *Rochebrune, 1887*

Although all red colobus have been traditionally included in one single species, it is very clear that this is an oversimplification. There are several species, distinguished by characters of pelage and skull, vocalizations, and other features (Groves, 2001; Grubb *et al.*, 2003). The Ugandan red colobus, therefore, is not to be referred to as *Piliocolobus* (or *Procolobus*, or *Colobus*) *badius*; that name denotes a West African species.

Ugandan red colobus, Piliocolobus tephrosceles *(Elliot, 1907).* This species has long, loose glossy black dorsal pelage, with a red crown, light to white underparts, dull light grey forearms and legs, and light tufts at the tail base. There are also prominent tufts at the base of the ears, and a prominent dark red-brown forehead crest bordered by a black stripe that runs back to the temporal region.

The distribution is patchy; it is abundant in Kibale, but absent from Bugoma and the Kagombe–Matiri forest complex, and it does not reach Budongo. South of Uganda, the species occurs in suitable habitats along the whole eastern shore of Lake Tanganyika, to Lake Rukwa in far southern Tanzania. Ugandan examples tend to be more red-tinged, with a lighter grey-brown rump than of those from further south.

Central African red colobus, Piliocolobus foai *(Pousargues, 1899).* This diverse species, hard to define, includes a diversity of forms mostly from northern and eastern DRC. All have dark to black hands and feet, red crown, black browband, and light-colored cheeks. One population enters Uganda, where it occurs in the Semliki forest. The subspecies is *Piliocolobus foai semilikiensis* (Colyn, 1991). It is distinguished from *P. tephrosceles* by having legs blackish to reddish grey, and arms red-brown, with black hands (instead of all four limbs being dirty grey); throat reddish, and rest of underparts and inner aspects of limbs grey (instead of white or whitish); and pelage short and dull-colored (instead of long and shiny) (Colyn, 1991).

Colobus *Illiger, 1811*

Black-and-white colobus are represented in Uganda by two species, about whose distinctness there is no doubt. While, overall, *C. guereza* has a more northerly range and *C. angolensis* a more southerly one, they overlap in the Ituri forest.

Mantled guereza, Colobus guereza *Rüppell, 1835.* In western Uganda, this species in found west of the Nile, in all forested areas as far north as Budongo. The subspecies is *Colobus guereza occidentalis* Rochebrune, 1887, in which the white flank-veil or mantle is shorter than in other subspecies, and the white tail tuft occupies only a third of the length of the tail. It extends north into southernmost Sudan and west through northern DRC into Congo, Cameroon, and northeastern Gabon.

Angola colobus, Colobus angolensis *Sclater, 1860.* This species has a restricted distribution in Uganda. It occurs in the Semliki region, where it replaces *C. guereza* in the montane forests of the Rwenzoris (Kingdon, 1971); and in the forests of Mt. Kakuka, Maramagambo, Kaiso, Katera, and Sango Bay. The subspecies in Uganda is said to be *Colobus angolensis ruwenzorii* Thomas, 1901, which ranges south into Rwanda, Burundi, and northwestern Tanzania, and in DRC it occurs on Mt. Kahuzi and along the Ruzizi River. A different subspecies, *Colobus angolensis cottoni*, is widespread in the northern DRC, and extends at least to the borders of the Semliki valley according to Colyn (1991). It is very different in appearance from *C. a. ruwenzorii*, having much thinner white shoulder tufts ("epaulettes"), a longer grey terminal zone on the tail, and no white in the pubic region. The two possibly should be regarded as different species. *C. a. ruwenzorii* is closely related to an East African form, *C. a. palliatus.*

Hominoidea Gray, 1825: Hominidae Gray, 1825

There are two subfamilies in this family: Ponginae (of which the only living representative is the Asian orangutan) and Homininae, which includes *Gorilla*, *Pan*, and *Homo.*

Homininae Gray, 1825

Of the three genera, *Gorilla* is more divergent from *Pan* and *Homo* than these are from each other. A strong case has recently been made that *Pan* ought to be sunk into *Homo*, in effect making chimpanzees a kind of human being (Goodman *et al.*, 1998; Watson *et al.*, 2001; Wildman *et al.*, 2003).

Gorilla *I. Geoffroy St. Hilaire, 1853*

Groves (2001) divided the genus into two species: *Gorilla gorilla*, found in West Central Africa; *G. beringei*, found in Uganda, Rwanda, and DRC.

Eastern gorilla, Gorilla beringei *Matschie, 1903.* In Uganda, Eastern gorillas occur in Mgahinga (the Uganda sector of the Virunga Volcanoes) and in Bwindi. The Virunga population belongs to the subspecies *Gorilla beringei beringei*, the Mountain gorilla. According to Sarmiento *et al.* (1996a,b), the

Bwindi gorillas differ in several respects from those in Virunga. The possibility that they are different is intriguing, and can be resolved only by larger samples, especially from Bwindi. The problem must be approached by assessing whether there are heritable differences between them, such that most (presumably not all) individuals can be correctly sorted; ecological differences are not strictly relevant, except as possible explanations for why they may differ.

Pan *Oken, 1816*

Common chimpanzee, Pan troglodytes *Blumenbach, 1775.* This species is no longer "common," but is declining throughout its range. Uganda is guardian to one of the relatively few populations that is both substantial and stable. Chimpanzees occur in all the western forests, north as far as Budongo (A recent survey reported a small number of chimpanzees further north, in the Otzi Forest Reserve; see Plumptre *et al.*, 2003).

Hitherto, all common chimpanzees from Uganda, Sudan, Rwanda, Burundi, Tanzania, and the northern and eastern parts of DRC have been regarded as belonging to a single subspecies, *Pan troglodytes schweinfurthii* (Giglioli, 1872). Recently (Groves, 2005), I compared measurements of different geographic samples of skulls, and concluded that not one but two subspecies are represented, as follows:

- *Pan troglodytes schweinfurthii* (Giglioli, 1872): northern and eastern DRC into southernmost Sudan, east to the Ituri region and Lakes Edward and Albert, and southeast to Lake Kivu. Large size, with relatively long and wide upper face, and wide braincase.
- *Pan troglodytes marungensis* (Noack, 1887): Rutshuru district to western Uganda south to Rwanda, Burundi and northwestern Tanzania; Kivu district south through Maniema and the Itombwe Mountains to Marungu. Small size, with relatively broad muzzle and fairly long palate.

The Ugandan subspecies would therefore be *marungensis*, not *schweinfurthii*. I am confident that this would apply to those from the southern part of western Uganda, but my Ugandan sample included mainly skulls from the far southwest (bordering Rutshuru and Rwenzori). I had only a single skull from Budongo and one from Toro, so their allocation to the newly resurrected subspecies *marungensis* must be provisional only.

BIOGEOGRAPHY OF UGANDA'S PRIMATES

The primates and other fauna of the western Ugandan forests are generally regarded as "overspills" from the northeastern rainforests of DRC, the "East-Central Primate Zone" of Grubb (2001). Within this zone, Grubb distinguishes two Centers of Endemism: an Ubangui-Uele and a Kivu centre. Some discussion of the East-Central Primate Zone, placed in the context of African mammal zoogeography in general, seems necessary to put Uganda's primates into some perspective.

The African Forest Biome is divided into five faunistic regions: Western, West Central, East Central, South Central, and Eastern (Grubb, 1978). These have substantially different subsets of the general forest mammal fauna, with different duikers, dwarf antelopes, squirrels, genets, and so on, as well as primates. The East Central region is divided from the South Central by the great bend (Cuvette Centrale) of the Congo/Lualaba; to the west, it may be divided from the West Central by the Oubangui (Ubangui), but where detailed distributions are known the divider actually turns out, in some cases at least, to be a much more easterly tributary of the Congo, the Itimbiri, which marks the boundary between *Cercopithecus mitis* and *Cercopithecus nictitans*, and between *Cercopithecus denti* and *Cercopithecus pogonias* (Colyn, 1991). It is plausible that in the Middle Pleistocene the Oubangui flowed west into Lake Chad, and it was the Itimbiri that carried most of flow that is now carried by the Uele, the major tributary of the Oubangui; it would then be only rather recently that the Oubangui was captured by the Congo.

There is no doubt, certainly, that the area of highest diversity of primates in the East-Central region is the Kivu area. Here are found not only apparently widespread (perhaps only because poorly studied) species like *Galagoides demidoff*, *Galagoides thomasi*, and *Perodicticus potto*, but all the taxa that are characteristic of the region, and wholly or partly endemic to it: *Galago matschiei*, *Gorilla beringei*, *Lophocebus albigena*, *Cercocebus agilis*, *Allochrocebus lhoesti*, *Cercopithecus (ascanius) schmidti*, *Cercopithecus denti*, *Cercopithecus hamlyni*, *Cercopithecus (mitis) stuhlmanni*, *Piliocolobus foai*, and the *Colobus cottoni/cordieri/prigoginei* group. These have spread varying distances to the east; *Cercocebus agilis* and *Cercopithecus hamlyni* have not reached Ugandan territory at all, *Cercopithecus denti* and *Piliocolobus foai* (and *Galagoides demidoff*) enter Uganda only in the Semliki valley, while the others all have substantial Ugandan populations.

A number of authors, notably Colyn (1991), have noted that there is an odd hiatus within the range of two species of colobus in the Kivu region. The southern boundary of *Colobus angolensis cottoni* is the Lindi River; there are no black-and-white colobus in the Kivu lowlands until the Ulindi is reached, and south of this *C. a cordieri* occurs. The southern boundary of *Piliocolobus foai ellioti* is the Maiko River, south of which there is again a hiatus (though smaller than that within *C. angolensis*), until the Ulindi, south of which *P. f. foai* is found (there is one questionable record of red colobus north of the Ulindi, south of the Lowa). It is especially interesting that the hiatus in the range of red colobus is precisely the area where *Gorilla beringei graueri* extends into the lowlands, toward the Lualaba, and what makes this even more intriguing is that almost the entire western (West Central region) range of gorillas also fills a hiatus within the range of red colobus.

In western Uganda there is a strong north–south division in primate faunas, but it is uneven (Table 1). The Savannah Biome has markedly distinct Northern and Southern savannah mammalian faunas, and some groups of primates have a northern and a southern vicar, but the patas monkey is northern only. Forest primates are mainly southern in distribution; this is not—or not entirely—due to the way the forests themselves change from true rainforest in the south to a drier semievergreen formation to the north, because in two cases there are specifically northern forest representatives: *Colobus guereza*, replacing *Colobus angolensis*, and *Cercopithecus mitis stuhlmanni*, replacing (and, apparently, hybridizing with) *Cercopithecus doggetti*. The colobus is not part of the East Central fauna, but a (mainly forest-edge) species widespread along

Table 1. The north–south division in the primate fauna of western Uganda

Northern	Southern
Savannah species	
Galago senegalensis senegalensis	*Galago senegalensis sotikae*
Chlorocebus tantalus	*Chlorocebus pygerythrus*
Erythrocebus patas	–
Forest species	
–	*Otolemur monteiri*
Cercopithecus (mitis) stuhlmanni	*Cercopithecus doggetti*
–	*Piliocolobus tephrosceles*
Colobus guereza	*Colobus angolensis*
–	*Gorilla beringei*
–	*?Pan troglodytes marungensis*

the margins of several faunal regions. Only the blue monkey is a genuine East Central taxon.

Superimposed on this is a montane element. The Rwenzoris have no endemics, but appear in some instances to be the conduit for entry of species into Uganda, and at least three species—*Galagoides demidoff*, *Piliocolobus foai*, and *Cercopithecus denti*—got no further. *Colobus angolensis* got no further along this route, but evidently found a second, more successful, southerly route of entry. On the other hand, the Virungas developed two endemics, *Gorilla beringei beringei* and *Cercopithecus kandti*; the former was able to spread into Bwindi when there was a forest connection, but the latter was unable to do so, blocked by the presence of a closely related species, *C. doggetti*.

Three species of primates in the western Uganda forests are not part of the East Central faunal region at all. These are *Piliocolobus tephrosceles*, which does not occur in the DRC at all; *Cercopithecus doggetti*, which perhaps occurs there in the form of hybrids with *Cercopithecus mitis stuhlmanni*, and their recombinants; and *Colobus angolensis ruwenzorii*, which is restricted to the Kahuzi highlands and the Ruzizi River. The ranges of all three extend well south through Rwanda, Burundi, and western Tanzania (*Piliocolobus tephrosceles* extends south to Lake Rukwa). The two colobus—but perhaps not the Silver Monkey—actually belong to Eastern Region sections of their respective genera.

The forests of western Uganda have undergone alternate episodes of contraction and expansion over the past few hundred thousand years. The periods of contraction seem to have been accompanied by lowering of vegetation zones in mountainous areas, most lately during the Last Glacial Maximum (LGM), peaking at about 18,000 bp (Hamilton, 1988). The drying was intense; for example, Lake Victoria apparently dried out completely during the LGM, filled at 14.5 ka during an ultra-wet phase, and, after a reversal in the refilling trend corresponding to the Younger Dryas, finally established a continuous outflow into the Nile at 11.2 ka. The Rift lakes probably did not actually dry out, but certainly rose and fell, and around Lake Albert the evidence of drying, with arid grasslands, was patent (Beuning *et al.*, 1997). These two phenomena have presumably had considerable effects on primate ranges in recent times.

A much older geological phenomenon that would have affected primate distributions was uplift along the Western Rift, which went on throughout the Pleistocene (Andrews & van Couvering, 1975). Some of the cichlid fishes of Lake Victoria show affinities not with those of the Nile, as one would predict from present-day river patterns, but with those of the Congo (Seehausen *et al.*,

2002), indicating a former connection. The lake itself is said to be only about 400,000 years old, formed by blockage and back-ponding of rivers that had once flowed from East Africa into the Congo, and now obtained their headwaters in the Rift mountains. This may relate to the renewed Middle Pleistocene uplift of the Rwenzoris, which, as described by Krafft (1990), was the event that split Lakes Albert and Edward into separate basins.

Vulcanism in the Virunga region goes back to the Late Oligocene, but the formation of the present chain is much more recent: the oldest of the group, Mikeno and Sabinyo, formed between the Late Pliocene and Early Pleistocene, while Muhavura, Mgahinga, Visoke, and Karissimbi were formed as recently as the Late Pleistocene, or even later (Krafft, 1990:10). The Toro-Ankole volcano region has a not dissimilar history; Kiyambogo has an age of 450 ka, and the latest volcanic activity in the region is only 6.5 ka.

These dramatic changes imply a history of continual depopulation and repopulation of western Uganda by primates. Prior to the uplift of the Rwenzoris and Virungas, and the onset of Toro-Ankole vulcanism, continuity between present-day Uganda and the forests of the Congo basin was complete; gradually, this continuity was rendered intermittent. Meanwhile climatic fluctuations saw forests periodically withdrawn from Uganda, then reestablished across the southwest reaching as far as Lake Victoria and severing the northern from the southern savannahs. With each spread of the forests, the East Central fauna spread as far as it could, whether not quite to Uganda, just across the Uganda border (in the Semliki region), or far into Ugandan territory. At the same time, continuity was reestablished with the forests of the Eastern region along the eastern edge of Lake Tanganyika, and two taxa of colobus and one of Cercopithecus (see above) spread north into Uganda. The spread of the two colobus may be what prevented *Piliocolobus foai* and *Colobus angolensis cottoni* from spreading further into Uganda than the Semliki forest; the spread of *Cercopithecus doggetti*, and the further spread of *Colobus angolensis*, however, were checked by *Cercopithecus* (*mitis*) *stuhlmanni* and *Colobus guereza* coming the other way, having apparently entered Uganda around the northern end of Lake Albert.

Most recently, human deforestation of southwestern Uganda has left "orphan" populations of several forest species clinging to the shores of Lake Victoria, isolated from the western forests, and has brought northern and southern savannah populations of *Chlorocebus* and of *Galago senegalensis* into contact, resulting (in the former case at least) in a belt of hybridization.

In summary, the primate fauna of the forests of western Uganda is a unique blend of East Central and Eastern and of montane and lowland forest faunas. Only here do they meet, interact, and overlay. A multitude of questions arise as a consequence. What determines which species shall exist where? Is it chance, or is it competition? Why have some species not spread further than the Semliki? Why have the species from the south apparently spread faster than their sister species from the west? What blocks the further northward spread of red colobus? Or gorillas? Is it lack of time; is it human interference; or is it unsuitable ecological setting? There is a rich research field here, for Ugandan and expatriate primatologists alike.

SECTION TWO

Ecology

Figure iii. Adult black and white colobus *Colobus guereza occidentalis* (above, photos T. Furuichi) and juvenile gorilla *Gorilla beringei* (below, photo, J. Rothman).

CHAPTER TWO

Factors Influencing Variation in the Population Densities of *Colobus guereza* Within Selectively Logged Forest at the Budongo Forest Reserve

The Importance of Lianas During a Subsistence Diet

Graham Anthony Preece

INTRODUCTION

Populations of *Colobus guereza* typically show a positive response to habitat modification, attaining some of their highest population densities within the most heavily or recently logged forest areas. The response has been viewed as idiosyncratic for colobines and may be linked to the ability of *C. guereza* to subsist on mature leaves for protracted periods during times of annual food scarcity. Specifically, explanations have focused on *C. guereza*'s proclivity to

---

Graham Anthony Preece • University of Wales, Bangor.

Primates of Western Uganda, edited by Nicholas E. Newton-Fisher, Hugh Notman, James D. Paterson, and Vernon Reynolds. Springer, New York, 2006.

exploit the leaves of lianas and colonizing tree species, which represent a major source of high quality mature leaves and are typically found in greater abundance within more open or disturbed forest sites (Clutton-Brock, 1975; Oates, 1977). An important corollary of these findings is the assumption that it is not the preferred food items that sets the carrying capacity for *C. guereza* but the abundance and/or quality of subsistence foods.

On the basis of the data collected at the Budongo forest, Western Uganda, together with data from previous studies, this study tested the prediction that variation in the population densities of *C. guereza occidentalis* will be positively related to the abundance of lianas. In addition to examining factors that may be significant in determining the habitat preferences of *C. guereza*, this chapter also discusses possible physiological and social adaptations that may facilitate population increases under conditions created by habitat modification, specifically those associated with logging.

Guereza and Their Response to Habitat Modification

Selective logging opens up the canopy, creating the conditions for the regeneration of colonizing species of plants and the potential for increased productivity of vegetation of a high nutritional quality (Ganzhorn, 1995). These changes may benefit guereza, as they appear to be adapted to exploit this type of vegetation, primarily selecting resources from colonizing species of midstorey trees. However, increases in the population densities of *C. guereza* do not appear to be related to food availability. For example, at the Kibale forest, Struhsaker (1997) recorded the highest levels of population densities within the most heavily logged areas even though the abundance of both *Celtis gomphophylla* [syn. *C. durandii*] and *Markhamia platycalyx* (which comprised >60% of the diet) were the same or even lower within these areas (see also Dunbar & Dunbar, 1974; Clutton-Brock, 1975; Skorupa, 1986). It is possible that populations are maintained below carrying capacity in moist forest habitats or that factors other than food availability are limiting guereza populations. Alternatively, a few key, but not heavily used, resources may have a disproportionate effect on guereza population densities (see "nutrients as constraints" model; Pulliam, 1975).

In the Budongo Forest, Plumptre and Reynolds (1994) found no evidence to support a link between differences in forest type (including colonizing forest) and the abundance of *C. guereza*. Nevertheless, as the density of *C. guereza* showed a significant positive correlation with logging date, it was concluded that guereza populations actually benefited from logging itself, rather than from any differences in forest composition. This would suggest that factors intrinsic to logging or habitat modification benefits guereza groups, although a causal link for this relationship remains undefined. If the population changes associated with logging can be attributed to food availability, then by inference, logged sites must provide *C. guereza* with an additional food resource or resource of greater quality (Waterman *et al.*, 1988; Oates *et al.,* 1990; Ganzhorn, 1992; Struhsaker, 1997; Chapman *et al.*, 2000). As *C. guereza*'s successful colonization of more open forest has been linked to their capacity to subsist on mature leaves (Clutton-Brock, 1975; Oates, 1977), increases in the population densities of *C. guereza* should be mirrored by a proportionate increase in the abundance and dispersion of plants with mature leaves that are of a high nutritional quality.

Colobine Feeding Strategies and Their Implications for Habitat Preferences

Guereza's usual pattern of feeding is to rely on relatively few plant species. Their diet is monotonous, favoring young leaves, and subsisting on mature leaves and fruit during periods of annual scarcity (but see Fashing, 2001a). An ability to cope with a monotonous diet has been viewed as an adaptation to maximize the benefits of a more abundant and/or evenly distributed resource (Clutton-Brock, 1975; Oates, 1977). However, as more abundant resources are generally low in quality (Wrangham, 1980), the cost in absolute terms is a diet low in energy intake constrained by the need to select foliage with a favorable protein-to-fiber ratio and possibly low in toxins. Oates (1977) asserts that these traits characterize adaptations to more marginal forest habitats subject to pronounced seasonal shortfalls in high quality food resources. Certainly, guereza occupy a range of different habitats and are found at very high population densities (>300 individuals per km^2; Rose, 1978; Dunbar, 1987; Krueger *et al.*, 1998) within dry and gallery forests, suggesting they are well adapted to cope in marginal forest environments.

By contrast most other African colobines, including related species of black-and-white colobus, rely on a wider diversity of food resources (Table 1). While this may be associated with a strategy to maintain food quality (Dasilva, 1992), diversifying the diet may also provide a means of dealing with a range of different toxins at suitably low levels (Oates, 1977). However, relying on a high quality diet may restrict distributions to areas with less (across species) synchrony in fruit

Table 1. Long-term studies of colobines having detailed information on dietary composition

Species	ML	YL	LB	UL	FR	WH	SD	FL	OT	Habitat	PC	NS	References
Colobus guereza	12.4	57.7	4.0	2.5	13.6	12.1	1.0	2.1	7.7	Kibale, Uganda	69	43	Oates, 1997
Colobus guereza	4.6	21.7	2.5	19.3	44.0	42.6	1.4	0.7	8.5	Kakemega, Kenya (T)[a]	43.7	>28	Fashing, 2001a
Colobus guereza	8.6	19.1	4.1	2.5	33.1	32.1	1.0	0.2	12.7	Kakemega, Kenya (O)	39.5	>28	Fashing, 2001a
Colobus guereza	3.8	26.2	3.5	24.2	24.6	2.6	22.0	2.9	14.5	Ituri, DRC	–	31	Bocian in Fashing, 2001a
Colobus angolensis	2.4	23.5	2.4	22.0	27.5	5.4	22.1	7.2	14.0	Ituri, DRC	–	37	Bocian in Fashing, 2001a
Colobus satanas	18.1	19.7	0.8	0.0	53.2	0.0	53.2	3.3	4.8	Douala, Cameroon	–	84	McKey, 1978
Colobus satanas	3	23	–	0.0	64.0	4	60	5	4	Lope, Gabon	–	65	Harrison, 1986
Colobus polycomos	26.4	29.7	–	1.6	35.0	3.2	31.8	2.7	3.1	Tiwai, Sierra Leone	26	56	Dasilva, 1989
Procolobus badius	5.4	24.0	17.5	0.0	35.9	17.4	18.5	8.7	8.5	Fathala, Senegal	–	39	Gatinot in Oates, 1994
Procolobus badius	11.8	26.2	8.7	0.1	41.6	38.7	2.9	8.7	2.9	Abuko, Gambia	–	89	Starin in, Oates, 1994
Procolobus badius	20.2	31.7	– –	– –	31.2	5.9	25.3	16.1	0.8	Tiwai, Sierra Leone	21	51	Davies in Oates, 1994
Procolobus badius	23.1[b]	50.6	–	10.4	5.7	4.9	0.8	11.9	2.9	Kibale, Uganda	30.1	≥55	Strusaker & Oates, 1979
Procolobus badius	44.1	34.8[c]	–	–	11.4	–	–	6.8	2.9	Gombe, Tanzania	40	>58	Clutton-Brock, 1975
Procolobus badius	11.5	36.0	16.4	0.9	25.0	24.1	0.9	6.2	4.0	Tana, Kenya	–	22	Marsh, 1981[d]

Procolobus badius	7.3	32.1	14.6	7.3[e]	31.7	–	–	10.6	2.3	Jozani, Zanzibar (I)	50.8	63	Mturi, 1993[d]
Procolobus badius	6.3	31.8	21.6	5.6[e]	31.2	–	–	5.4	1.3	Jozani, Zanzibar (II)	41.0	62	Mturi, 1993[d]
Presbytis comata	5.6	59.1	–	–	14.2	13.5	0.7	7.0	12.2	West Java	–	74	Ruhiyat, 1983
Presbytis melalophos	11	24	–	–	56	48	8	6	2	Kuala Lompat, Malaysia	22	137	Curtin, 1976
Presbytis obscura	22	36	–	–	35	32	3	7	–	Kuala Lompat, Malaysia	31	87	Curtin, 1976
Trachypithecus johnii	26.8	25.5	5.7	4.2	25.1	–	–	9.3	3.4	Kakachi, India	44.2	>107	Oates *et al.*, 1980
Trachypithecus pileatus	42.0	10.9	4.9	0.0	33.7	24.4	9.3	7.0	1.5	Madhupur,	>64	35	Stanford, 1991
Trachypithecus senex	40	20	–	–	28	–	–	12	–	Polonnaruwa, Sri Lanka	70	–	Hladik, 1977
Semnopithecus entellus	21	27	–	–	45	–	–	7	–	Polonnaruwa, Sri Lanka	30	–	Hladik, 1977
Semnopithecus entellus	34.9	11.5	2.8	2.4	24.4	–	–	9.5	14.5	Kanha, India	33.5	53	Newton, 1992
Semnopithecus entellus	–	–	–	64.2	15.1	–	–	6.3	4.9	Ramnagar, Nepal	54	69	Koenig *et al.*, 1998

ML = mature leaves; YL = young leaves; LB = Leaf buds; UL = unclassified leaves; FR = total fruit (fruit + seeds); WH = whole fruit; SD = seeds; FL = flowers; OT = other; PC = % contribution of top three species to diet; NS = number of species in diet; DRC = Democratic Republic of Congo.

[a] Letters within parentheses denote group identity where more than one group was studied.

[b] Mainly the petioles of mature leaves eaten and not mature leaf blades, which averaged only 2.3% of the diet over a 12-month period.

[c] New leaves and "shoots."

[d] Total exceeds 100%.

[e] leaf stalk.

and/or young leaf production and therefore a greater year-round availability of high quality food (Clutton-Brock, 1975). These traits more likely characterize adaptations to mature rain forest environments, in which seasonal variation in plant production is less pronounced and where the ability to deal with toxicity develops from adapting to an environment more strongly associated with plant secondary compounds (Struhsaker & Oates, 1979). In many respects, this may represent a more typical pattern of dietary adaptation, certainly among the majority of African (and Southeast Asian) colobines that are largely restricted to moist lowland forest habitats (Oates & Davies, 1994).

Although colobine populations may benefit—or at least be buffered against food shortages—from increases in productivity following low intensity selective logging (Skorupa, 1986; Thomas, 1991; Struhsaker, 1997; Chapman *et al.*, 2000), the overall trend is for a general decline in population densities following habitat disturbance (Yeager & Kool, 2000) (Table 2). For example, although the population densities of certain *Presbytis* species may increase after disturbance, their response is largely dependent on the nature or quality of subsequent regeneration (see Johns, 1988; Plumptre & Reynolds, 1994; Chapman *et al.*, 2000). In addition, density-dependent population regulation can set animal densities below any limit set by food availability, so populations may be buffered against changes in resource availability—at least in the short term (Krebs & Davies, 1993; Yeager & Kool., 2000). In the long term, however, post logging, variation in the abundance and regeneration of high quality resources can be expected to have a concomitant effect on the primate populations that rely on these resources (Terborgh, 1992; Cowlishaw & Dunbar, 2000). The differential response of colobines to changes in resource availability may be related to niche separation, and understanding the accordant behavioral, morphological, and physiological adaptations will provide an insight into habitat preferences.

Relationship Between Colobine Feeding Ecology and Plant Chemistry

Colobines have evolved an enlarged fore-stomach that facilitates microbial digestion of fibrous plant material by retaining food in a fermenting chamber to allow for the microbial breakdown of complex structural polysaccharides such as cellulose (Oates & Davies, 1994). This allows the colobines to subsist on relatively indigestible foliage often not available to monogastric primate species. However, this predisposes the colobines to a diet high in fiber and potentially

Table 2. Colobine responses to habitat disturbance

Study site	Habitat (time since logging)	Species	Population density (groups/km^2)		Direction		Source
			Old forest	Disturbed forest	+/−	*p*	
Borneo, Silabukan/ Bakapit	Hill dipterocarp forest— selectively logged (19 years)	*Presbytis aygula/* *Presbytis hosei* *Presbytis rubicunda*	3.6 0.9	1.3 1.3	− +		Payne & Davies in Johns & Skorupa, 1987
Borneo, Bole Kecil/ Malabuk	Hill dipterocarp forest— selectively logged (15 years)	*P. rubicunda* *P. aygula/P. hosei*	4.5 0.7	1.1 0.0	− −		Payne & Davies in Johns & Skorupa, 1987
West Java	Montane forest—montane secondary forest	*Presbytis comata*†	11.5	35	+		Ruhiyat, 1983
Malaysia, Sungai Tekam	Hill dipterocarp— selectively logged (5–6 years)	*Presbytis melalophos* *P. obscura*	3.4 0.5	3.4 0.8	ns +		Johns in Johns & Skorupa, 1987
Malaysia, Lesong	Lowland diptorecarp—selectively logged (5 years)	*P. melalophos* *P. obscura*	6.5 3.8	3.9 5.0	− +		Marsh & Wilson, 1981
Ghana, Bia	Lowland mixed forest—selectively logged (<2 years)	*Colobus verus* *Colobus polycomos* *Procolobus badius*	1.1 1.7	0.8 1.6	− −		Martin & Aisbey in Johns & Skorupa, 1987
Gabon, Lopé	Lowland Primary forest—selectively logged	*Colobus satanas*	0.75	0.89	+	ns	Brugière, 1998

(*Cont*).

Table 2. (*Continued*).

			Population density (groups/km²)		Direction		
Study site	Habitat (time since logging)	Species	Old forest	Disturbed forest	+/−	*p*	Source
Democratic Republic of Congo, Ituri	Mature mixed forest—secondary forest	*Colobus angolensis*[a]	1.06–2.66	0.0–0.99	−		Thomas, 1991
		P. badius[a]	1.19–3.80	0.0–1.7	−		
		Colobus guereza[a]	0.0	0.0–2.55	+		
Uganda, Kibale	Moist evergreen forest—selectively logged (12 years)	*P. badius*	5.46	3.08	−	**	Chapman *et al.*, 2000
		C. guereza	0.89	4.81	+	***	
Uganda, Kibale	Moist evergreen forest—selectively logged (28 years)	*P. badius*	5.50	4.43	−	ns	Chapman *et al.*, 2000
		C. guereza	2.00	9.12	+	***	
Uganda, Budongo	Moist semideciduous forest—selectively logged	*C. guereza*[b]	27.0	44.2	+	***	Plumptre & Reynolds, 1994

"Direction" indicates if the population increased after disturbance (+), or decreased after disturbance (−), and (p) if the change is significant.
[a] Range of estimates from seven census locations.
[b] Individual density per km^2.
*** $p < 0.001$ (Z tests).
** $p < 0.01$ (Z tests).

high in plant secondary compounds (Oates & Davies, 1994). This places a number of constraints on their choice of food items as fiber can inhibit the uptake of protein, while plant secondary compounds can act as digestion inhibitors (such as tannins) or toxins (such as alkaloids). Numerous studies have highlighted the importance of plant chemistry in determining food selection in primate species (Milton, 1980; Waterman & Choo, 1981; Wrangham & Waterman, 1981; Ganzhorn, 1992). In colobines, a positive relationship has been shown to exist between mature leaf digestibility and colobine biomass (McKey, 1978; Davies, 1994; Oates *et al.*, 1990). More specifically, Chapman *et al.* (2002) provided evidence for a positive relationship between colobine biomass and the protein-to-fiber ratio of mature leaves. They conclude that, because of the constraints associated with food quality, colobine monkeys may live close to a protein deficit that sets the threshold for colobine biomass. This will be more acutely felt during times of annual food scarcity, when high quality food resources are at a premium and colobines need to rely more heavily on mature leaves. These factors provide evidence to support the speculation that the major determinant of colobine biomass is the availability and quality of food resources (Davies, 1994), or, more specifically, that the density of plants providing digestible mature leaves at times of annual food scarcity sets the carrying capacity for colobines.

Throughout periods of annual food scarcity, the mature leaves of lianas form an important resource in the diet of many colobine species, producing leaves that are often of a far higher quality than those produced by trees (*Colobus satanas*: McKey *et al.*, 1981; *Presbytis rubicunda*: Davies, 1991; *Semnopithecus entellus* : Koenig *et al.*, 1998). Given that they are positively correlated both qualitatively and quantitatively with levels of disturbance (Putz, 1983; Hegarty, 1991; Hegarty & Caballé, 1991; Dewalt *et al.*, 2000; Gerwing & Farias, 2000), lianas may represent a key resource variable and could offer a habitat-wide explanation for variation in *C. guereza* densities. Black-and white-colobus (*C. polycomos*) at Tiwai, Sierra Leone, have been described as liana "specialists" (Davies *et al.*, 1999) feeding "almost exclusively" on the mature leaves of lianas when preferred food items such as young leaves and seeds were not available (Dasilva, 1992). Oates (1977) observed *C. guereza* "feeding heavily" on the foliage of lianas and has subsequently suggested they may play an important role in the diets of *C. guereza*, especially in more disturbed forest habitats. This chapter aims to examine, on the basis of the data collected from the Budongo forest, if liana availability can be considered a proximate cause of variation in *C. guereza* densities and an important factor in determining the carrying capacity of the environment.

METHODOLOGY

Procedure

During 10 weeks of fieldwork, 50 census walks were conducted following five 1-km transect lines to provide measures of Colobus density within a single forest block (Nyakafunjo). The observations were applied to the transect width estimation method of Whitesides *et al.* (1988). In addition, data from previous census work (Plumptre *et al.*, unpublished) was used to provide a measure of Colobus density across six individual compartments all logged at different time intervals (Table 3). Within the logged forest area, a total of fifty 0.04-ha plots were enumerated within each location censused for *Colobus*, to provide measures of liana abundance and forest structural attributes (basal area and canopy tree density). A further ten 0.04-ha plots were located within the unlogged nature reserve for comparison.

Analysis

Correlation coefficients were calculated between measures of the basal area of lianas across different size classes and (i) the density of *C. guereza* across several forest compartments all logged at different time intervals, and (ii) measures of *C. guereza* densities at individual locations within a single forest block

Table 3. Logging history of each study compartment

	Logging history of each study compartment			
Compartment: Block	Area (ha)	Date logged	Volume of timber removed (m^3/ha)	Arboricide treatment
N15: Reserve	777	Unlogged	None	None
B1: Biiso	582	1935	19.9	1958
		1983–1986	21.5	None
B4: Biiso	748	1941–1942	34.8	1957
N1: Nyakafunjo	412	1945	58.7	1962–1963
N2: Nyakafunjo	630	1945–1947	46.2	1955–1956
N3: Nyakafunjo	620	1947–1952	80.0	1960–1961
W21: Waibiira	1116	1963–1964	36.1	1963–1964

Source: Plumptre and Reynolds (1994) and Plumptre (1996). Compartments B1, B4, and W21 have also been subject to low levels of pitsawing.

(Nyakafunjo) of uniform habitat type. Calculations were made across different size classes because, unlike trees, increases in liana leaf biomass are not directly proportional to the size of stem cross-sectional area, with larger increases found the larger the diameter increment (Putz, 1983).

RESULTS

Ecological Correlates of *C. guereza* Density with Those of Forest Structural Attributes

There was no significant relationship found between the total basal area of trees and the density of *C. guereza*, or the total basal area of lianas and the density of *C. guereza* either at the inter- or intrasite level. No significant relationships were found between the abundance of canopy trees (>20 cm Diameter at Breast Height (DBH)) or the number of trees carrying lianas and the density of *C. guereza* either at the inter- or intrasite level. There were no significant relationships found between measures of liana basal area across different size classes and the group or individual density of *Colobus*. The relationship between *Colobus* (individual) density and the basal area of lianas more than 3 cm was positive but not significant. Correlation coefficients between *C. guereza* density and measures of forest structural attributes are shown in Table 4.

Intrasite Variation in the Density of *C. guereza* Groups

Estimates of *C. guereza* density from across different sites located within a single forest block (Nyakafunjo) show a significant and positive relationship between

Table 4. Spearman correlations between *C. guereza* density and the forest attributes (a) across study compartments (intersite) and (b) within the Nyakafunjo block (intrasite)

		Liana basal area (cm^2/ha)				Forest structural Characteristics		
		Size class (cm)						
	Population characteristics	1–2	>3	>4	>5	Canopy tree BA	Midtree BA	Trees with lianas
Intersite[a] density	Groups/km^2	−0.77	0.33	0.5	0.5	0.71	0.34	−0.25
	Individuals/km^2	−0.14	0.81	0.74	0.68	−0.09	−0.09	0.31
Intrasite[b] density	Groups/km^2	−0.7	0.8	1*	1*	0.17	−0.33	0.08
	Individuals/km^2	−0.6	0.6	0.9	0.9	0.08	−0.42	0.0

[a] r_S value equal to or greater than 0.866 ($n = 6$) is significant ($^* p < 0.05$).
[b] r_S value equal to 1 ($n = 5$) is significant ($^* p < 0.05$).

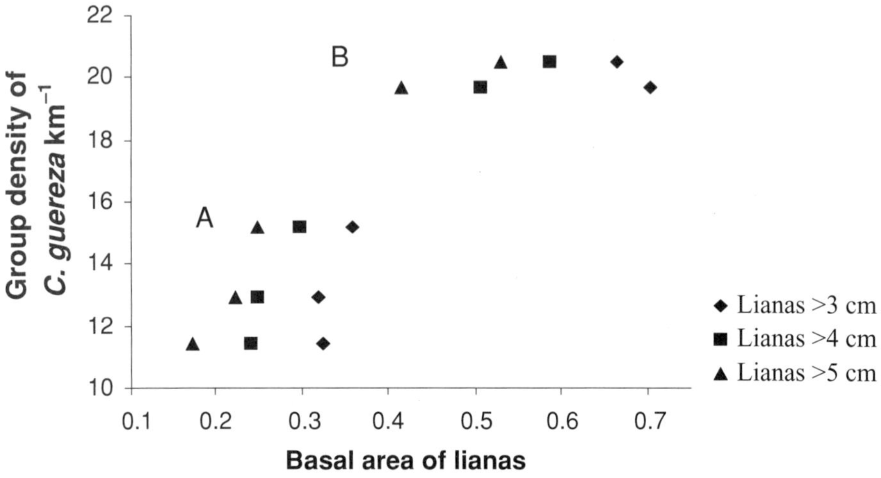

Figure 1. The relationship between intrasite variation in *C. guereza* group density and the basal area of lianas by size class.

the group density of *C. guereza* and the basal area of lianas across different size classes. The results also show that guereza densities are positively related to lianas in the larger size classes (all lianas ≥4 cm in diameter).

Furthermore, the population densities of guereza groups within those areas of high liana biomass were significantly different ($U = 92.5$, $n_1 = 20$, $n_2 = 30$, $p < 0.001$) from those in which liana biomass was low (indicated by two clusters A/B in Figure 1).

DISCUSSION

Originally I hypothesized that the abundance of lianas may set the carrying capacity for *C. guereza*. There was no direct evidence to support this speculation, as the density of *C. guereza* did not covary with liana biomass across sites. However, the positive correlation of liana biomass with intrasite variation in *Colobus* density provides compelling evidence that lianas play an important role in the diet of *C. guereza* at Budongo. This is further supported by the findings that indicate a significant increase in the density of *C. guereza* beyond a certain threshold in liana biomass (Figure 1). These results also indicate that it is the larger-diameter lianas (≥4 cm) that covary with the density of *C. guereza* at the intrasite level. This may be due to the higher leaf biomass relative to cross-sectional area the larger the size of liana. However, the abundance of lianas (≥4 cm) does not differ significantly between logged and unlogged forest habitats. This suggests that the abundance of lianas is not a causal factor eventuating the usual pattern of increase in guereza population densities under the conditions created by selective logging. However, Johns (1986) has shown that the quality and dispersion of food resources are key factors in determining their suitability to primates. In the absence of young leaves, lianas may fit these requirements being both high in quality and more evenly dispersed in logged forest areas (discussed below).

If lianas represent a major source of high quality mature leaves that *C. guereza* can subsist on during periods of annual food scarcity, then the strongest relationship between liana biomass and the density of *C. guereza* can be expected at those periods of the year coincident with low levels of young leaf production. This study was conducted over a period of low, young leaf availability concurrent with the fruiting of *C. gomphophylla*, which is a preferred food item for *C. guereza* at Budongo (Plumptre *et al.*, 1997). The evidence to suggest the leaves of lianas make a positive and significant contribution to the diet of *C. guereza* therefore needs to be viewed in the wider context of their feeding ecology and the consumption of a mixed fruit–leaf diet (discussed below).

The Importance of Lianas as a Key Resource Variable

Within disturbed forest habitats, elevated light levels significantly increase the potential for carbon gain in liana leaves. Higher photosynthetic rates result in rapid leaf development and an increase in leaf turnover rate (Hegarty, 1990). This may be linked to a greater abundance of leaves of higher nutritional quality (rather than an absolute increase in abundance) in more open, logged forest areas. The leaves of lianas may therefore be preferentially selected within logged forest habitats owing to the high quality of leaf material under the conditions associated with logging. Nevertheless, within these habitats, choice of leaf material should be strongly influenced by its distribution. Wrangham (1980) proposes that during a "subsistence diet," leaves should be selected from ubiquitous resources that occur in large (defensible) scattered clumps that can easily satiate the needs of the group, or, if small in size (such as lianas), be widely dispersed (see also van Schaik, 1989).

Previous research has shown that leaves are not selected on the basis of abundance; rather, the most preferred leaves are often taken from "rare" tree or liana species (*C. satanas*: McKey, 1978; *P. johnii*: Oates *et al.*, 1980; *C. polycomos*: Dasilva, 1992; *S. entellus*: Koenig *et al.*, 1998; *C. guereza*: Fashing,

2001a). Notably, these findings have demonstrated that mature leaves can show significant variation in phytochemical quality, leading to a high degree of spatial variability in the distribution of preferred (as defined by selection ratios) fallback resources. In addition a number of studies affirm a link between a need to match patterns of ranging with both the quality and distribution of a subsistence diet. For example, McKey (1978, at Douala Edea) found that the distance traveled by *C. satanas* "significantly" increased during periods of annual food scarcity, when mature leaves were predominant in the diet. Bocian (1997; cited in Fashing, 2001a) reports a similar association for *C. angolensis* in the Ituri forest.

While it is true that mature leaves should be available in large abundant clumps, potential sources of high quality mature leaves from lianas and pioneer and midstorey trees may be patchily distributed throughout the forest community. As such, patterns of ranging may be less influenced by the availability of the most abundant food items and more by the distribution of a few high quality subsistence resources. The index of dispersion for Budongo shows that the lianas within the logged forest habitat (Nyakafunjo block) are clumped in distribution ($IP = 0.501$, $p < 0.05$, standardized Morisita index) although they occur frequently within plots (80% of plots had ≥ 3 lianas ≥ 4 cm in diameter). In previous studies, the movements of guereza have not been closely associated with the abundance of primary food items. However, the relationship between guereza distribution (revealed through discrete or subtle shifts in their density) and a resource that is potentially high in quality, but patchily distributed, could be linked to patterns of ranging in guereza at this site, although this assumption requires more rigorous testing.

Colobine Physiology and a Mixed Fruit–Leaf Diet

Whole fruit typically represents an important element in the diet of *C. guereza*, often accounting for more than 50% of monthly feeding records (Dunbar & Dunbar, 1974; Oates, 1977; Fashing, 2001a). As fruit is low in protein and high in fiber, its selection by *C. guereza* is more likely determined by the content of available sugars rather than protein. However, as there is thought to be an associated risk of acidosis when consuming large volumes of fruit pulp, its selection can be viewed as unusual—in contrast to previous findings concerning colobine digestive physiology (see Waterman & Kool, 1994). In contrast, related species of black-and-white colobus typically show a greater reliance on or even preference for seeds in their diet (*C. satanas*, Douala Edea: McKey, 1978;

C. satanas, Lopé: Harrison, 1986; *C. angolensis*, Salonga: Maisels *et al.*, 1994; *C. polycomos*, Tiwai Island: Davies *et al.*, 1999). Seeds represent a source of food that is both high in calorie and protein and low in fiber, making it a highly digestible food item and a "source of readily available energy" (McKey *et al.*, 1981). However, seeds are potentially high in toxins, and guereza are possibly choosing for low toxicity as well as quality because detoxification "will make demands upon the time budget" (Dasilva, 1992) and increase the requirements for energy and nutrients (Foley, 1992). Longer handling times may also mean that seeds are a more contestable food item than are fruits (Korstjens *et al.*, 2002). This has implications for energy expenditure.

It should be noted that fruit selection could simply be related to the sites in which guereza have been extensively studied rather than a direct preference for fruit over seeds, that is, good quality habitats such as Kibale, where few leguminous species are found and where leaves contain high concentrations of protein (but see Bocian, 1997). Nevertheless, throughout periods of annual food scarcity, the usual pattern of guereza food choice is to rely heavily on whole fruit and mature leaves. Understanding what drives this pattern of food choice may help explicate key determinants of their habitat preference.

It is now widely accepted that colobines have evolved to optimize the digestion of a mixed fruit–leaf or seed–leaf diet and are not adapted to cope with a completely folivorous diet (Cork, 1996). Dasilva (1992) found that *C. polycomos* were possibly energy limited when mature leaves formed the "dominant part of their diet" (see also *S. entellus*: Hladik, 1977). Dasilva's findings suggest that an important criterion for the selection of mature leaves may also be their calorific value as well as protein content (at least at bottleneck periods) and alternatives such as fruit or seeds should be selected to compensate for the energy deficiencies of a highly folivorous diet (Cork, 1994). However, colobines are also limited by the amount of food they can process as their strategy is to maximize absorption, which is facilitated by long transit times (Lambert, 1998). This places a limit on the amount of fleshy fruit they can consume, as large volumes need to be passed through the digestive system quickly in order to meet their daily protein and energy requirements (Milton, 1981). Because of these limitations, leaves of a high nutritional quality should be selected to balance the requirements for both protein and energy (when fruit is in short supply) in a mixed fruit–leaf diet. For *C. guereza*, an ability to select the most digestible leaf items may be particularly important owing to their heavy (seasonal) reliance on low quality mature leaves. As a resource high in nutritional quality, liana leaves may complement the feeding on fruit to offset the protein and/or energy constraints associated with an extremely low quality diet.

A major influence on food choice will be the interaction between the environment and the physiological constraints on the absorption and processing of food items. An optimal diet should include a sufficient combination of food items that maximizes nutritional requirements, while minimizing the consumption of plant secondary compounds and fiber that are inimical to digestion. Maintaining a balance between the ratio of fruits to leaves in the diet may help to ensure an assimilation equilibrium (or nutritional threshold) by offsetting absorption and processing constraints. The evidence to date shows that the most important criterion in mature leaf selection is the protein-to-fiber ratio (Chapman *et al.*, 2000; Wasserman & Chapman, 2003), although leaves may also be selected on the basis of their calorific value (Dasilva, 1992; Koenig *et al.*, 1998). Fruits with a relatively high pH and a low level of organic acids may also be more suited to colobine digestion, helping to maintain an alkaline environment and thereby mitigating the risk of acidosis. Likewise, fruits high in complex structural carbohydrates may better suit the colobine digestive strategy that maximizes the extraction of nutrients through long processing times.

Coping With a Low Energy High Fiber Diet: Is Energy Limiting in Guereza Groups?

Although food choice may help balance the requirements for energy (as well as protein) in the diet, behavioral (rather than physiological) adaptations may be the main mechanism by which colobines can compensate for a low energy diet (Dasilva, 1992). Increased resting time and thermoregulatory behavior (such as early morning sunbathing; Clutton-Brock, 1975) can therefore be viewed as a response to a low energy intake associated with a low quality folivorous diet. It is therefore telling that *C. guereza* typically spend a greater proportion of their time resting in comparison to other colobine species (typically >50%). For example, red colobus typically spend more time traveling and feeding and less time resting than guereza groups (Table 5). Measures of monthly dietary diversity are typically higher for red colobus and it is generally considered that their diet is more consistently of a higher quality (Struhsaker & Oates, 1979).

This is a strong indication that there is a trade-off linking factors of food quality and its distribution with the expenditure of time and energy in procuring and handling food resources (see also Milton, 1980). The constraints of

Table 5. Comparison of annual activity budgets and mean index of species diversity of red colobus and guereza colobus

	Species					
Activity %	*Procolobus badius*			*Colobus guereza*		
Rest	35[a]	55[b]	51[c]	57[d]	>63[e]	59[f]
Feed	47	37	32	20	>25	20
Travel	9	5	8	5	2	–
Social	–	–	–	6	8	8
Other	8	3	9	11	1	>13
Diversity index, *H′*	2.65	–	1.93	1.72	1.61/1.72*	–

All samples across a 12-month period, except for the present study sample.
[a] Kibale, Uganda: (Stuhsaker, 1975).
[b] Tiwai island, Sierra Leone (Davies, cited in Oates, 1994).
[c] Tana river, Kenya (Marsh, 1978; cited in Mturi, 1993).
[d] Kibale, Uganda (Oates, 1977).
[e] Kakamega, Kenya (Fashing, 2001a), average for two groups.
[f] Budongo forest, this study.
* Mean values for two separate groups (T/O), (Fashing 2001a).

the diet will necessarily dictate the most efficient use of time and energy. Increased resting time should benefit guereza in two interdependent ways, first by conserving energy and second by extending the period of time for food to pass through the gut, thus maximizing the extraction of energy and nutrients from foodstuffs ingested.

Fashing (2001a) suggests that feeding heavily on abundant fruit resources helped reduce daily energy expenditure in guereza groups (at Kakamega, Kenya), by minimizing the distance traveled between resources. Struhsaker and Oates (1979) associate the monotony of the guereza diet with their heavy reliance on common tree species such as *C. gomphophylla* (see also *P. senex*: Hladik, 1977). Dietary decisions may therefore help minimize energy expenditure, either as a consequence or function of monotony—by reducing both search and travel times.

A defining characteristic of colobines is their ability to utilize mature leaves as a fallback resource and in all situations they are highly selective of the leaves they "choose" to eat. By contrast, suitable fruit and seeds are often taken from the most widely available resources (*C. satanas*, Legumonosae, Harrison, 1986; *P. badius, C. angolensis*, Legumonosae, Maisels *et al.*, 1994; *P. badius, C. polycomos*, Legumonosae, Davies *et al.*, 1999; *C. guereza*, Moraceae fruit, Fashing, 2001a).

Interestingly, Whitehead (1986) found that feeding in infant howler monkeys (*Allouatta palliata*) was "strongly" related to their mother's feeding behavior in the case of leaves but not fruit. These findings suggest that the constraints on the choice of fruit are less stringent than for leaves, possibly due to a reduced risk of ingesting harmful toxins or less variability in quality (but see Ganzhorn, 1988; Simmen & Sabatier, 1996). If colobines in general are not energy limited and leaves are the principal source of protein (but see McKey *et al.*, 1981), then perhaps the most readily available fruit resources will suffice to balance their energy needs (but see Dasilva, 1992). For instance, across a range of different habitats within and around the Kibale forest, Wasserman and Chapman (2003) found "no correlation between energy content . . . and foraging effort" in *C. guereza*. Furthermore, none of the study groups were found to experience an energy deficit and they did not select foods high in energy content. This suggests that energy is not a limiting factor in the diet of *C. guereza*, at least at Kibale, where foraging effort was more closely related to the protein-to-fiber ratio in mature leaves.

CONCLUSIONS AND IMPLICATIONS

A Synthesis of Dietary Strategies and Feeding Ecology in Colobines

Although colobines have been broadly classified as folivores, they show wide variation in their feeding behavior, relying on flowers, fruits, shoots, and seeds to varying degrees (Clutton-Brock, 1975; Hladik, 1977; McKey, 1978; Harrison, 1986; Dunbar, 1987; Stanford, 1991; Dasilva, 1992; Mitchell, 1994; Oates, 1994; Fashing, 2001a). While some dietary flexibility may be attributable to ecological factors, intergeneric variation in diet has been linked to the adaptation of specific dietary strategies that are more related to phylogenetic inheritance (Clutton-Brock, 1974; Stanford, 1990; Fashing, 2001a). For example, the langurs of the Indian subcontinent, which inhabit some extreme seasonal environments, have a tendency to include more mature leaves and less fruit in their diet than the langurs of Southeast Asia (see Koenig & Borries, 2001). This suggests that an important adaptation for coping in environments with greater seasonal variation in plant production is an increased capacity to process mature foliage. Other studies suggest that colobines may vary in their capacity to break down digestion inhibitors and are constrained by their relative ability to extract nutrients from the environment (Hladik, 1977; Stanford, 1990).

Colobines have also been differentiated in terms of how their body size is related to dietary quality. For example, Yeager and Kool (2000) infer that the larger stomach size of *Trachypithecus* in comparison to *Presbytis* species could be related to an increased capacity to process foliage. Evidence would seem to support this supposition as *Trachypithecus* have been shown to include more mature leaves in their diet (Yeager & Kool, 2000).

However, the fermenting chambers in colobines do not scale linearly with body size, suggesting that simple allometric relationships are a crude measure for differentiating between digestive capabilities (Chivers & Hladik, 1980). A more precise means for understanding the relationship between digestive physiology and feeding ecology is to factor in a measure of the rate at which food passes through the gut, that is, transit times (Penry, 1993; Lambert, 1998). *C. guereza* have long food transit times relative to their body mass (Table 6) and one of the longest actual food transit times recorded for any primate, suggesting an exceptional ability to cope with difficult-to-digest plant material (Watkins *et al.*, 1985). This ability may be manifest in the capacity of *C. guereza* to thrive on a small number of tree species and subsist on mature leaves throughout periods of seasonal shortage.

Onderdonk and Chapman (2000) found that within a number of small forest fragments surrounding the Kibale forest, guereza "maintained" its monotonous diet even though the actual plant species that comprised the diet differed across sites. This suggests that monotony is a key dietary adaptation that is modulated in response to ecological conditions. For example monotony may help to de-

Table 6. A comparison of transit time and mean retention time (MRT) from transit markers in colobine species

Species	Transit time (h)	MRT (h)	Body weight (kg)[a]	Reference
Nasalus larvatus	14	49	15.0/9.0	Dierenfild *et al.*, 1992
Colobus guereza	38	–	13.5/9.2	Kay & Davies, 1994
Trachypithecus abscuris[b]	15.6, 9.3 1	55.4, 62.1	7.0/5.0	Caton, 1999
Trachypithecus cristatus[b]	10.1, 5.0	59.1, 59.9	6.6/5.7[c]	Caton, 1999

[a] Body weights are given in the first instance for the males and then females.
[b] Times are for both solute and particle markers respectively.
[c] Mean body weight (from Oates *et al.*, 1994) rather than actual body weight of individuals used in the study.

velop a gut microflora that is specific to the guereza diet and therefore highly efficient at digestion. This form of physiological acclimation is known to occur in other mammals that use microbial symbionts in fermentation, for example, green turtles, *Chelonia mydas* (Penry, 1993). However, the dietary flexibility described by Wasserman and Chapman could equally be related to an ability to expand the resource base safely and incorporate novel foods into their diet (e.g., *Macaca fuscata*: Fedigan, 1991). Alternatively, when food quality varies widely but the rate of food intake does not then the capacity for gastrointestinal modulation (pliant throughput rate) may be particularly important in dealing with seasonal fluctuations in diet composition, as it influences the absorption of available nutrients and energy from foodstuffs ingested (Penry, 1993). Studies have shown that the ratio of activity to rest varies in *C. guereza* dependent on the habitat, time of day, and season (Oates, 1977; Wasserman & Chapman, 2003), and these behavioral changes may help modulate digestion in response to the diet. As yet, this hypothesis remains untested and the degree of modulation within, and variability between, species is little understood.

The adaptations to a particular diet and the association with the spatial and temporal availability of food resources will have a profound effect on social organization in primates (Johns, 1986; Yeager & Kool, 2000). An interesting feature of guereza ecology is their rigid pattern of social organization. Although the density of guereza populations may vary widely between a range of different habitats, their social organization remains fairly inelastic, with little variation in group size and structure (Marler, 1969; Dunbar & Dunbar, 1974; Suzuki, 1979; Oates, 1994; Fashing, 2001a). Moreover, the uni-male pattern typical of *C. guereza* is unusual among other black-and-white colobus monkeys, in which multi-male groups form a common and sometimes "characteristic" pattern of social organization (Oates *et al.*, 2000). Clutton-Brock (1974) suggests that variation in social organization could represent one aspect of evolutionary divergence that may be the consequence of long-term differences in the response to similar ecological pressures, such as food availability (see also Sussman, 1977; Frumhoff, 1995; Barton, 2000). Thus, as low quality resources are generally more evenly and heavily distributed throughout the forest, *C. guereza* may benefit from living in small social groups and defending (core) territories because the returns per unit area will be high. However, the tight margins of the guereza diet may place a cap on group size as the energy costs of contesting over low quality food resources will also be high ("ecological constraints" model: Gillespie & Chapman, 2001).

ACKNOWLEDGMENTS

I especially thank Prof. Vernon Reynolds for providing me with the opportunity to conduct research at the Budongo Forest Project and for inviting me to contribute to this volume. Many thanks to Dr. Andrew Plumptre, who offered his advice and generously provided access to data for the purposes of the study. I wish to express special thanks to Dr. John Hall, Bangor University, for this advice and help throughout the study. I would like to thank field researchers Andama Kennedy and Muhumuza Geresomu, who contributed their observations and advise, and to all the staff at the Budongo Forest Project. I am grateful for permission to work in Budongo, which was granted by the Ugandan Forest Department and the National Council for Science and Technology. Special thanks to Dr. Angela Pitts for helpful comments and encouragement throughout.

CHAPTER THREE

How Does the Golden Monkey of the Virungas Cope in a Fruit-Scarce Environment?

Dennis Twinomugisha, Colin A. Chapman, Michael J. Lawes, Cedric O'Driscoll Worman, and Lisa M. Danish

INTRODUCTION

Understanding the processes determining the density and distribution of species is one of the primary goals of ecology (Boutin, 1990). The importance of this information has increased with the need to develop informed management plans for endangered or threatened species. With respect to primates, these theoretical issues are critical because the tropical forests they occupy are undergoing rapid

Dennis Twinomugisha • Institute of Environment and Natural Resources, Makerere University, Kampala, Uganda **Colin A. Chapman** • Department of Anthropology & McGill School of Environment, McGill University, Montreal, Quebec, Canada H3A 2T7; Wildlife Conservation Society, 2300 Southern Boulevard, Bronx, NY 10460 **Michael J. Lawes** • School of Biological and Conservation Sciences, Forest Biodiversity Research Unit, University of KwaZulu-Natal, P/Bag X01, Scotsville, 3209, South Africa **Cedric O'Driscoll Worman and Lisa M. Danish** • Department of Zoology, University of Florida, Gainesville, FL 32611.

Primates of Western Uganda, edited by Nicholas E. Newton-Fisher, Hugh Notman, James D. Paterson, and Vernon Reynolds. Springer, New York, 2006.

anthropogenic transformation and modification. For example, countries with primate populations are cumulatively losing approximately 125,000 km^2 of forest annually (Chapman & Peres, 2001). Other populations are being affected by forest degradation (logging and fire) and hunting. However, predicting the responses of particular species has often proved difficult.

The blue monkeys (*Cercopithecus mitis*) of Uganda fit this generalization in that it has proven difficult to predict how they will respond to disturbance or to natural variation in forest structure. For example, blue monkey abundance was severely affected by logging at Kibale National Park, Uganda: 15 years after logging, areas had 20–30% fewer blue monkeys than unlogged areas (Skorupa, 1988) and this trend continues to this day (Chapman *et al.*, 2000). In contrast, in Budongo Forest Reserve, Uganda, blue monkeys are 3.7 times more abundant in logged areas than in unlogged areas (Plumptre & Reynolds, 1994). Similarly, within Kibale National Park, blue monkeys are common in the north of the park but their numbers gradually decline toward the south (Chapman & Lambert, 2000). There is no corresponding change in forest structure that explains this gradual decline. Thus, it appears that predicting responses of blue monkeys to disturbance or understanding responses to natural changes in the environment are difficult.

The blue monkey has been characterized as a species capable of occupying a variety of habitat types and forest conditions (Lawes, 1991). This forest species has an extremely wide distribution, extending from the forests of southern Sudan to the Eastern Cape Province in South Africa (Lawes, 1990). It occurs in forests from sea level to over 3000 m. Given this wide distribution, it is surprising that blue monkeys are also one of the most recently derived species within the Cercopithecini (Leakey, 1988; Lernould, 1988; Ruvolo, 1988). Their wide distribution, recent origin, and tendency to generate subspecies are generally attributed to high dispersal ability and their capacity for survival in the fragmented forests that existed at the end of the last glacial period centered on 18,000 BP (Lawes, 1990).

The mechanisms that facilitate blue monkeys' having such a wide distribution are poorly understood. However, a number of studies have suggested that blue monkeys have a broad diet (Rudran, 1978a; Struhsaker, 1978; Gautier-Hion, 1988; Butynski, 1990; Lawes *et al.*, 1990; Chapman *et al.*, 2002). This flexibility allows some populations to turn to a diet with a large leaf component (Beeson, 1989), while others to insects (Rudran, 1978a; Butynski, 1990) or flowers (Schlichte, 1978), when preferred fruit is not available (Lawes, 1991). Their flexible diet appears to have a morphological basis: blue monkeys have a significantly larger hindgut (caecum and colon) and the surface area of the small intestine is greater than in vervet monkeys (*Cercopithecus aethiops*; Bruorton & Perrin, 1991) and specialized symbiotic gut microflora (Bruorton *et al.*, 1991). Thus, blue monkeys may have the ability to include a larger amount of fibrous leaf material in their diets than other guenons (Lawes, 1991). A high level of folivory, at least on a seasonal basis, has been noted by a number of studies (Rudran, 1978a; Schlichte, 1978; Beeson, 1989; Lawes, 1991). This should allow blue monkeys to do well in marginal or disturbed habitats, and facilitate their survival in areas where there are seasonal shortages of preferred foods. Although it is generally agreed that blue monkeys have adapted to a broad diet, little is known about the dietary requirements of the species and how populations, in what might be thought of as marginal habitats, are able to meet their nutritional requirements. For example, how do populations occupying areas where fruit is scarce obtain an adequate supply of sugars? Because fruit is often a major energy source sustaining primate populations, the density of fruit-eating primates has been suggested to be limited by the lowest seasonal level of fruit availability (Janson & Emmons, 1990; Janson & Chapman, 1999).

This study compares the nutritional ecology of the golden monkey (*Cercopithecus mitis kandti*) of Mgahinga Gorilla National Park, Uganda, to that of the blue monkey (*C. mitis stuhlmanni*) of Kibale National Park, Uganda, approximately 200 km away. These are two very closely related subspecies, with the golden monkey being isolated in the high elevation forests. Interbreeding between subspecies of *C. mitis* has been described (Kingdon, 1971). Mgahinga is a high elevation site (>3000 m) where fruiting trees are extremely rare and are represented by only a few species (Schaller, 1963, 1964; Kalina, 1991). In contrast, Kibale is a midelevation forest (~1500 m) with a relatively diverse and abundant fruiting tree community (Chapman *et al.*, 1997). We describe the diets of each of these populations and then consider the nutritional quality of the foods eaten with respect to protein, fiber, lipids, sugars, and a series of secondary compounds.

METHODOLOGY

Study Areas

Mgahinga Gorilla National Park (MGNP), Uganda (33 km^2) encompasses the slopes of three volcanoes (Mgahinga, 3474 m; Muhabura, 4127 m; and Sabinyo, 3634 m) and is part of the greater Virunga Conservation Area, which covers 434 km^2 (Figure 1). The park lies in the Albertine rift region, which

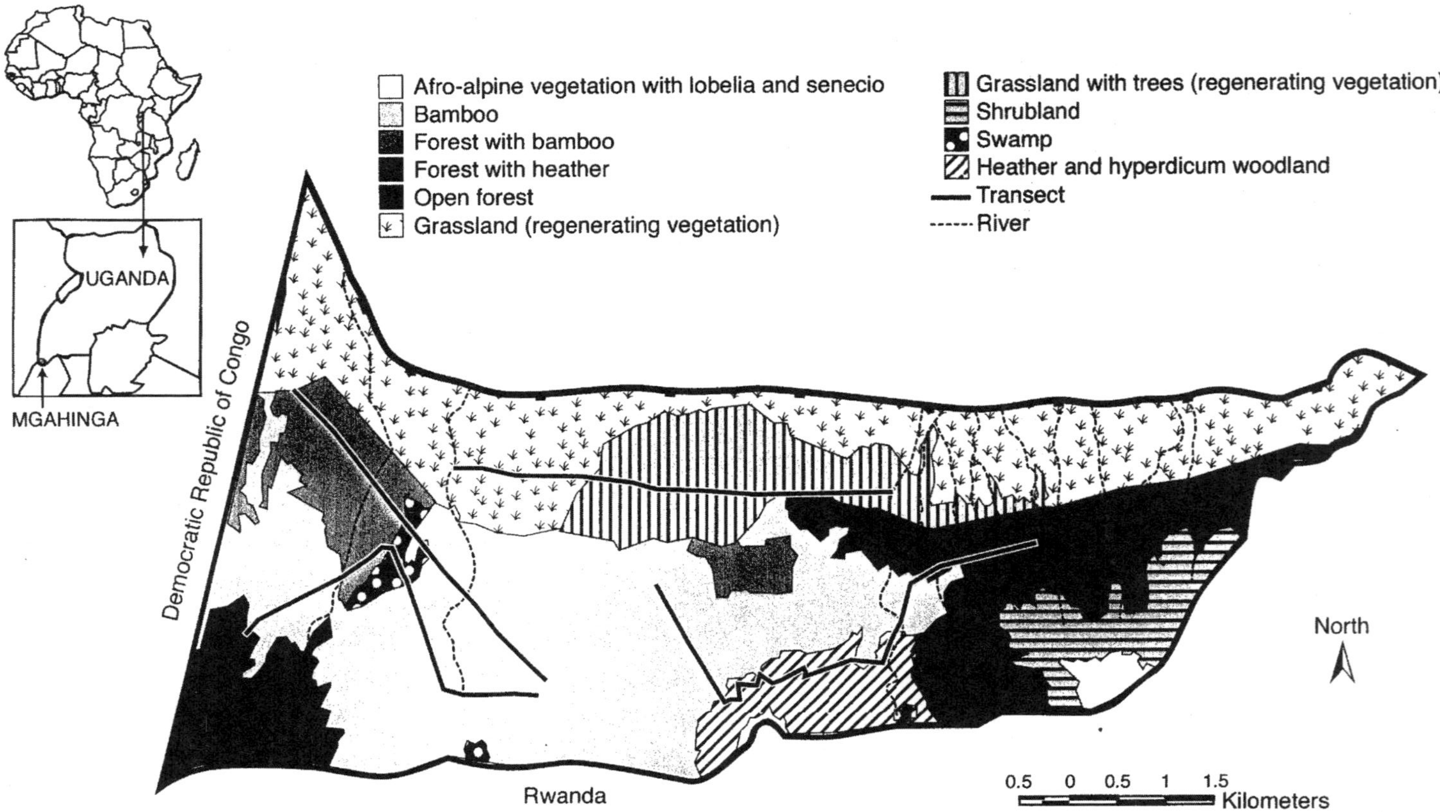

Figure 1. A map of Mgahinga Gorilla National Park, Uganda, illustrating its location within Uganda and the major vegetation types found within the park.

is characterized by a high degree of avian and mammalian endemism (Bibby *et al.*, 1992) owing to its proximity to a glacial forest refugium (Hamilton, 1988). The vegetation types of the park are diverse and are broadly classified into three belts and several zones within the belts (Figure 1). The vegetation belts are alpine, subalpine (ericaceous), and montane forest (Schaller, 1963). The alpine belt is prominent at the highest altitude. The subalpine belt is composed of moorland, montane grassland, and ericaceous zones—the moorland zone being transitional between the ericaceous zone and alpine belt. The ericaceous zone is characterized by the species *Philipa johnstonii*, *Erica arborea*, and *Hypericum revolutum*, which are often densely laden with *Usnea* sp. lichens (Kalina, 1991). The montane forest belt is the most extensive vegetation type, encompassing 40% of the park and is characterized by low tree species diversity (22 species in 2.2 ha, Twinomugisha, 1999). Within this forest belt the most extensive vegetation zone is bamboo (*Arundinaria alpina*). The remaining 33% of the park is covered by grassland and wooded grassland, and was previously under cultivation (Figure 1). The area was declared a National Park in 1991. However, since being gazetted in 1930, it has undergone a number of changes of name, status, size, and management. These changes have had effects on the conservation of the area in terms of habitat degradation and poaching.

Kibale National Park is located in western Uganda near the base of the Rwenzori Mountains (Struhsaker, 1997; Chapman & Lambert, 2000). Kibale is a midaltitude moist evergreen forest that is more diverse than Mgahinga (68 tree species in 4.8 ha; Chapman *et al.*, 1997). The study was conducted at Kanyawara (compartment K-30, ~1500 m elevation). The forest here is considered *Parinari* forest by foresters because of the spreading crowns of *Parinari excelsa*, which can be distinguished on aerial photographs. Canopy codominants include other important timber trees such as *Olea welwitschii*, *Aningeria altissima*, *Strombosia scheffleri*, and *Newtonia buchananii* (Osmaston, 1959; Chapman *et al.*, 1997). Kanyawara receives approximately 1741 mm of rainfall annually (1990–2002), which peaks during two rainy seasons, although rainfall is well dispersed throughout the year, falling on an average of 166 days per year.

Observation of Study Groups

The diet of golden monkeys was quantified during two periods. During the first period (January to September 1998), two already partially habituated groups of golden monkeys were further habituated during the first 2 months. Starting in March 1998, systematic instantaneous scan samples of feeding behavior were conducted during day-long follows for three consecutive days each month for 7 months. Four scan samples, each lasting 5 min, with 10-min intervals between scans, were conducted each hour on as many individuals as possible. Individuals were observed for 5–10 s and the food item eaten was recorded. During a single 5-min sample period, a feeding observation by any individual on a particular food item was scored only once unless the same individual fed on different parts of the same food plant. Group 1 (Ntebeko group) was followed for 19 days in total, during which 69 h of observations were made. Group 2 (Gatalabana group) was followed for 17 days (85 h). Feeding observations were also recorded opportunistically. Secondary indications (e.g., discarded fruit) were also used, as were interviews of rangers about the foods that they had observed the golden monkey eating. During a second period (January to August 2003) the same methods were used to observe another group for a total of 57 days (485 h). On average, 7 days of observations were conducted each month (range = 3–11 days per month).

Comparative data from Kibale were obtained from Rudran (1978a,b) and Butynski (1990), who collected data using a similar instantaneous scan sample procedure. Butynski (1990) studied five groups of blue monkeys in two subpopulations over a 6-year period (1978–1984). Rudran (1978a,b) studied two groups of blue monkeys between November 1972 and October 1974.

When there are appreciable differences in mean values, variation can be evaluated using the coefficient of variation (CV; Sokal & Rohlf, 1981). We use the CV to evaluate variation in time devoted to different plant parts. The CV was calculated as the standard deviation of the foraging effort devoted to a specific plant part divided by the mean. This value is multiplied by 100 to express the standard deviation as a percentage of the mean.

Plant Collections and Nutritional Analyses

Samples for nutritional analyses were obtained using a tree-pruning pole to cut down limbs, typically from the middle of the tree's canopy. The trees used were located in the same general areas as the groups foraged (with the exception of Group 33, Butynski 1990), but were not necessarily the same tree that the group fed in. No collections were made from trees growing in unusual situations, such as tree fall gaps or forest edges (except for species typically only found in such habitats, such as *Prunus africana* on edges; see Chapman *et al.*, 2003, for a discussion of sources of variation in nutritional values created by method of collection). Only those food items selected by the animals were collected. For example, if the animals ate leaf petioles, the length of petiole typically consumed was recorded. In Kibale, plant samples were collected at a time when blue monkeys and redtail monkeys (*Cercopithecus ascanius*) were known to be eating these items. Sample collections for the Kibale blue monkeys were part of our long-term studies of the primates of Kibale, and occurred when we were concentrating observations on redtail monkeys (Rode & Chapman, unpublished data).

Samples were dried in the field using a dehydrator that circulated warm air past the samples (the majority of the samples), by using a lightbulb to heat a box containing a series of racks, or by sun drying. All samples were dried at temperatures below 50°C. For samples dried in an oven, the heat setting was at its lowest (37°C). Dried samples were sealed in plastic bags and taken to the University of Florida for analysis.

Dried samples were ground to pass through a 1-mm mesh screen in a Wiley mill (stainless steel). Dry matter mass was determined by drying a portion of each sample overnight at 105°C. Samples were analyzed in duplicate, and replicates were considered acceptable if the relative error was less than 2%. This 2% criterion was applied to dry matter, organic matter, fiber, protein, and saponins.

The protein (nitrogen) content of the plant parts was assessed using Kjeldahl procedures (Horwitz, 1970). Samples were digested using a modification of the aluminum block digestion procedure of Gallaher *et al.* (1975). The digestion mix contained 1.5 g of 9:1 K_2SO_4:$CuSO_4$, and digestion was conducted for at least 4 h at 375 C° using 6 ml of H_2SO_4 and 2 ml of H_2O_2. The nitrogen in the digestate was determined by semiautomated colorimetry (Hambleton, 1977). Measuring total nitrogen provides an estimate of crude protein and traditionally the N content multiplied by 6.25, a conversion factor that has been used as an index of protein levels. A better conversion factor for tropical foliage may be approximately 4.3 (Conklin-Brittain *et al.*, 1999) or 4.4 (Milton and Dintzis, 1981). The 4.3/4.4 conversion factors probably underestimate nitrogen, while the 6.25 overestimates available protein, but it does not necessarily overestimate nitrogen (Conklin-Brittain, *et al.*, 1999). We used a conversion factor of 4.3.

Fiber (Acid Detergent Fiber [ADF]) was measured using the methods outlined by van Soest (1963) and modified by Goering and van Soest (1970) and

Robertson and van Soest (1980). ADF is a measure of cell wall cellulose and lignin. It has been found to have a strong negative correlation with food selection by some primates (Glander, 1982; Oates *et al.*, 1990). However, ADF is somewhat fermentable, while lignin is not (van Soest, 1982).

The primary components of plant carbohydrates that are easily digestible by mammalian enzymes were quantified using a method that requires differential extractions (80% ethanol) and digestions with colorimetric analysis of filtrates (Hall *et al.*, 1999). This procedure allows an assessment of organic acids and simple sugars (mono- and oligosaccharides). For ease of discussion, we refer to this as an evaluation of sugar content.

Many alkaloids are bitter tasting and perhaps play a role as a feeding deterrent (Harborne, 1993; Roberts and Wink, 1998); however, it has not been demonstrated that primates avoid foods high in alkaloids (Waterman, 1993). The presence of alkaloids was tested using a spot test with Dragendorff's reagent (Waterman, 1993). Dragendorff's reagent is known to sometimes produce false positive results (Waterman, 1993).

Saponins are surfactants and have a "soaplike" foam-forming property in aqueous solutions, hence their name. These compounds are bitter tasting and are found in over 70 plant families. Saponins have been documented to cause bloat in ruminants and have been implicated in diet selection of cattle, but do not influence red colobus diet selection (Chapman & Chapman, 2002). They also have the ability to irritate the digestive tract, and can serve as a steroid hormone precursor (Phillips-Conroy, 1986; Francis *et al.*, 2002). The quantity of saponins present in a 0.25-g sample was indexed using the Froth Test (Fong *et al.*, unpublished guide) using 60 and 1800 s criteria. This relative measure involves shaking the sample in a set fashion and measuring the height of the foam after 60 and 1800 s.

Cyanogenic glycosides are capable of releasing toxic hydrogen cyanide, but their role in deterring herbivory is questionable (Seigler, 1991; Jones, 1998). The presence or absence of hydrogen cyanide was determined by the Feigl–Anger test (Feigl & Anger, 1966; Glander *et al.*, 1989).

To compare the quality of the diet of *C. mitis* at Kibale and Mgahinga, we contrasted the nutritional characteristics in the 10 most frequently eaten foods for two groups of blue monkeys in Kibale (Rudran, 1978a) and the group of golden monkeys studied in 2003. We analyzed 73% of the total possible 210 nutrient–plant combinations (10 species/parts from each of the three populations and seven nutrient/secondary compounds). The majority of the nutrients/secondary compounds that were not analyzed were from species or parts reported to be eaten by Rudran (1978a,b), but were not observed being eaten during our study and thus were not collected. Percentages were arc-sine square root transformed for correlations between foraging effort and nutritional characteristics. Differences between groups/populations were analyzed taking a univariate (i.e., one-way analysis of variance [ANOVA], contrasting specific nutrients one at a time) and multivariate approaches (i.e., MANOVA contrasting all nutrients in the same analysis). The multivariate approach is somewhat limited, because if one nutrient out of the five continuous nutrients considered (protein, fiber, sugars, lipids, and saponins) could not be measured because of the lack of sample, that species/part had to be dropped from the analysis. Differences in nutrients of the major foods are also illustrated graphically.

RESULTS

Although there are few fruiting tree species in Mgahinga, fruit was a major component of the diet of some golden monkey groups (Table 1). In general, golden monkeys ate fruit less frequently (average 26.3%) than blue monkeys (35.3%); however, fruit eating among blue monkeys at Kibale was highly variable (15–30.1%) and some groups ate less fruit than the average golden monkey group. Two golden monkey groups fed more frequently on young leaves than blue monkey groups from Kibale, but a third golden monkey group used young leaves less frequently than any group from Kibale.

In terms of the plant parts eaten, the diet of the golden monkeys varied over time and among groups (Table 1). For example, the frequency with which young leaves (including bamboo) were eaten varied among groups from 11.3 to 58.6%, while the use of insects varied from 8.0 to 30.5% (Table 1). Blue monkey diets from Kibale were less variable than those of the golden monkey. The average coefficient of variation for the major plant parts (fruits, young leaves, flowers, and insects) was 31.2% for blue monkeys from Kibale ($n = 6$ groups), 46.4% for all studies of *C. m. stuhlmanni* ($n = 11$), and 68.3% for golden monkeys ($n = 3$).

Bamboo (*Arundinaria alpina*) was particularly important in the diet of the golden monkeys and they fed on bamboo leaves, culms, and shoots. The group observed in the 2003 field season ate bamboo for an average of 52.4% of their foraging time and in 1 month bamboo foraging constituted 61.7% of their

Table 1. The percentage of foraging effort devoted to different plant parts by diet of different subspecies of *Cercopithecus mitis*

Species	FR	ML[a]	YL	FL	Insect	SD	PT	ST	Other	Source	Study site, country
C. mitis stuhlmanni	54.6	2.3	16.6	3.7	16.8	2.5		1.1	2.4	Cords, 1986	Kakamega Forest, Kenya
C. mitis stuhlmanni	55.8		21.8	4.55	8.6	1.5				Fairgrieve, 1995	Budongo, Uganda
C. mitis stuhlmanni	44.9		29	6.2	9.7	5.9				Fairgrieve, 1995	Budongo, Uganda
C. mitis stuhlmanni	37		14	20	11					Schlichte, 1978	Lake Kivu, DRC
C. mitis stuhlmanni	42.7		19.1	11.7	19.8					Rudran, 1978a,b[b]	Kanyawara, Kibale, Uganda
C. mitis stuhlmanni	30.1		22.8	9.8	35.9					Butynski, 1990	Ngogo, Kibale, Uganda
C. mitis stuhlmani	22.2		34.3	7.4	35.4				0.7	Butynski, 1990	Kanyawara, Kibale, Uganda
C. mitis stuhlmani	28.9		22.4	2.9	45.4				0.4	Butynski, 1990	Kanyawara, Kibale, Uganda
C. mitis stuhlmani	22.1		33.3	7.8	35.1				1.7	Butynski, 1990	Kanyawara, Kibale, Uganda
C. mitis stuhlmani	15		35.4	7.4	41.8				0.4	Butynski, 1990	Kanyawara, Kibale, Uganda
Average[c]	35.3	–	24.9	8.2	25.9	–	–	–	–		
C. mitis erythrarchus	51.7	14.0[d]	11.8[e]	13.4	5.8			0.4	2.8	Lawes, 1991	Cape Vidal, South Africa
C. mitis erythrarchus	59.4		23.9	5.5	<5					Scorer, 1980, in Lawes, 1991	Cyprus, South Africa
Average	55.6	–	17.9	9.45	5.8	–	–	–	–		
C. mitis doggetti	47.4		6.2	6.2	24.9	9.3			6.2	Kaplin, 2001	Nyungwe, Rwanda
C. mitis nyasae	24.2		51.9	17.9	0.3					Beeson, 1989	Zomba, Malawi
C. mitis labiatus	91.1		3	2.1					3.8	Lawes *et al.*, 1990	Ngoye Forest, South Africa
C. mitis kandti	31.1	0.24	47.4	1.38	10.5		0.1	5.8	3.5	This study, Time 1 Group N	Mgahinga, Uganda
C. mitis kandti	36.7	0	11.3	14.0	30.5		0.5	7	0.02	This study, Time 1 Group G	Mgahinga, Uganda
C. mitis kandti	11.0	0	58.6	21.9	8.0		0.08	0.3	0	This study, Time 2	Mgahinga, Uganda
Average	26.3	–	39.1	12.4	16.4	–	–	–	–		

Methods of determining foraging effort vary among studies. FR = fruit; ML = mature leaves; YL = young leaves; FL = flowers; SD = seeds; PT = pith; ST = stems; DRC = Democratic Republic of Congo.

[a] Many studies provide only the total amount of leaves eaten and do not separate young versus old leaves; in this case no information is presented for mature leaves.

[b] Group 1 from February 73 to January 74.

[c] Averages only calculated for parts that were consistently recorded among studies.

[d] Dry leaves included with mature leaves.

[e] Leave buds included with young leaves.

foraging time (see also Aveling, 1984; Kingdon, 1971). No one plant species was this important in the diet of blue monkeys in Kibale: the most frequently eaten plant (*Ficus exasperata*) constituted only 15.1% of a group's foraging effort.

The golden monkey fed on few food sources. In 2003 the golden monkey group fed on between 3 and 12 species of plants in any given month. Over a period of 8 months, only 16 plant species were eaten. Four plant species were added from opportunistic observations. The golden monkey is known to feed on a total of 33 plant species. In contrast, Rudran (1978a) reported that over an annual cycle the blue monkeys at Kibale (group 1) fed on 59 plant species and 101 specific food items.

Comparing the diets of blue monkeys of Kibale and the golden monkeys of Mgahinga to other published accounts of *C. mitis* diets confirms previous assessments of the dietary flexibility of this species (Lawes, 1991). The foraging effort devoted to fruits was as high as 91.1% and as low as 11.0% (Table 1). Similarly the foraging effort devoted to eating leaves was as high as 58.6% and as low as 3%.

Conducting univariate analysis of variance considering each nutritional character one at a time revealed that the diets of golden and blue monkey groups did not differ in terms of any of the continuous nutritional variables (protein, fiber, lipids, sugars, or saponins; $p > 0.1$). Considering this question from a multivariate perspective we contrasted the nutritional content of the most frequently eaten food items among groups using a MANOVA and this analysis revealed no overall effect (Wilks' $\lambda = 0.134$, $F = 1.388$, $p = 0.328$). These patterns were graphically illustrated by producing a three-dimensional plot showing the position of the top 10 foods in relation to their protein, fiber, and lipid contents (Figure 2). This figure illustrates little structuring of the different populations/groups. However, note that bamboo has the highest protein level for the Mgahinga group and is somewhat separated from other foods.

No group had a food item in their top 10 most frequently eaten foods that had cyanogenic glycocides. Of the top 10 most frequently eaten foods by blue monkeys in Kibale, 40% of the species examined tested positive for alkaloids in one group, 60% tested positive in a second group. In the top 10 foods in the diet of the golden monkey group, 50% of the species examined tested positive for alkaloids.

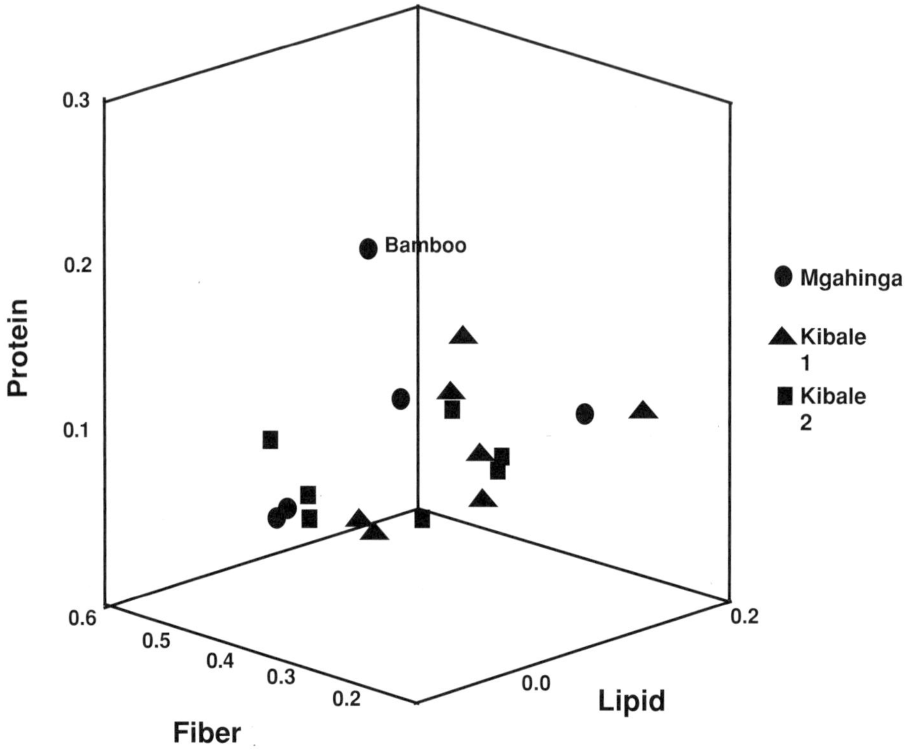

Figure 2. A plot of the nutrient composition of foods eaten by the golden monkey from Mgahinga National Park, Uganda, and two groups of blue monkeys from Kibale National Park, Uganda.

Correlations between foraging effort (the number of point samples observed feeding on an item/all feeding point samples) and nutritional components of the foods suggest that one of the Kibale groups tended to avoid food high in fiber ($r = -0.790$, $p = 0.001$). No other correlation between foraging effort of Kibale blue monkeys and nutritional components of the foods were found. For the Mgahinga group there were no correlations between any of the nutritional components and foraging effort.

Ultimately the quality of an animal's diet affects fecundity and fitness. The adult female-to-infant ratio was contrasted among different subspecies of *C. mitis* (Table 2). Golden monkey groups had a lower infant-to-adult female ratio than blue monkey groups (*C. m. stuhlmanni*), as well as *C. m. erythrarchus* and *C. m. labiatus* groups, suggesting that fewer infants are born into golden monkey groups (Table 3).

Table 2. The age/sex composition of different groups of different subspecies of *Cercopithecus mitis*

Species	Total	AM	AF	Imm.	Inf.	Unk.	Source
C. m. stuhlmanni	45	1	17	18	9	0	Cords, 1986
C. m. stuhlmanni	35	1	19	6	9	0	
C. m. stuhlmanni	34	1	17	11	5	0	
C. m. stuhlmanni	28	1	9	12	3	3	
C. m. stuhlmanni	21	1	9	7	4	0	
Average Kakamega	32.6	1	14.2	10.8	6	0.6	
C. m. stuhlmanni	24	1	11	8	4	0	Rudran, 1978a,b
C. m. stuhlmanni	13	1	5	6	1	0	
C. m. stuhlmanni	27	1	12	11	3	0	
C. m. stuhlmanni	13	1	4	7	1	0	
C. m. stuhlmanni	27	2[a]	9	12	4	0	
C. m. stuhlmanni	17	1	8	7	1	0	Butynski, 1990
C. m. stuhlmanni	11	1	6	3	1	0	
C. m. stuhlmanni	24	1	18	5	0	0	
C. m. stuhlmanni	19	1	8	8	2	0	
C. m. stuhlmanni	18	1	12	3	2	0	
Average Kibale	19.3	1.1	9.3	7	1.9	0	
Average all *C. m. stulhmanni*	23.7	1.1	10.9	8.3	3.3	1.2	
C. m. erythrarchus	26	2	9	13	2	0	McLeod, 2000
C. m. erythrarchus	22	1	7	8	6	0	
C. m. erythrarchus	22	1	8	13	0	0	
Average	23.3	1.3	8	11.3	2.7	0	
C. m labiatus	16	1	6	6	3	0	Lawes *et al.*, 1990
C. m. labiatus	21	1	8	8	4	0	
Average	18.50	1	7	7	3.5	0	
C. m. kandti	41	5[a,b]	11	17	3	4	This study
C. m. kandti	38	1	14	21	2	0	This study
C. m. kandti	41	1	14	26	0	0	This study
Average *C. m. kandti*	40	2.3	13	21.3	1.7	1.3	

AM = adult males; AF = adult females; Imm. = immatures; Inf. = infants; Unk. = unknown.
[a]Sometimes two to five males are seen within 25 m of one another and males who spend a majority of their time away from social groups may join a group during the mating season (Cords, 1986).
[b]Three males, which have been in the group at least for 14 months, were now feeding separately from the group. The remaining subdominant male feeds with the group, but seems to remain on the peripheral.

Table 3. Comparison of group composition data for subspecies of *C. mitis*

Species	Infant-to-adult female ratio
C. m. stuhlmanni (Kibale)	0.2043
C. m. stuhlmanni (Kakamega)	0.4225
C. m. stuhlmanni (all)	0.2988
C. m. labiatus	0.5000
C. m. erythrarchus	0.3333
C. m. kandti	0.1282

DISCUSSION

The density and species richness of fruiting trees at Mgahinga is very low. Despite this fact, the golden monkeys appear to cope well. Some groups focused their feeding effort on a few fruiting species and trees, while other groups relied heavily on flowers and leaves and all groups obtained their protein from bamboo. The overall nutritional characteristics of the foods used by the Mgahinga animals were not different from those foods used by the Kibale groups. This suggests that golden monkeys can substitute nonfruit foods for fruits and still obtain a balanced diet.

In general, fruits are known to provide an easily assimilated source of sugars and energy, but have been suggested to supply inadequate amounts of protein (Gaulin, 1979). This may explain why some populations of *C. mitis* appear to select foods based on their protein content (Beeson, 1989; Lawes, 1991). However, none of the three groups studied here selected foods high in protein (but see discussion of bamboo below). On the other hand, protein is likely readily available from some easily digestible insects, or less readily digestible young leaves (Lawes, 1991). Golden monkeys at Mgahinga consistently fed on bamboo. Bamboo has a relatively high protein content (22% of dry matter), but it is a very poor source of sugars (just trace amounts). Bamboo was eaten in every month of the year and is probably an important source of protein and vital to the survival of golden monkeys in these mountain forests. The importance of bamboo is suggested by the fact that in Mgahinga there were higher sighting rates and densities of golden monkeys in the bamboo zone and in forests with bamboo vegetation types (Twinomugisha *et al.*, 2003).

The golden monkeys obtained their sugars from the few fruits that were available, from flowers, and from the leaves of *Nuxia congesta*, which had higher levels of sugars (19%) than many fruits eaten by the blue monkeys in Kibale. *Hypericum revolutum* flowers were a particularly important source of sugars (29% of dry weight) and were available and eaten year round. There was no evidence of avoidance of secondary compounds and it may be that plant diet selection is little affected by secondary compounds in these monkeys. Wrangham *et al.* (1998) documented that three cercopithecines (*C. mitis*, *C. ascanius*, and *Lophocebus albigena*) had higher absolute intake levels of secondary compounds than chimpanzees (*Pan troglodytes*), suggesting a high tolerance. Only one of our study groups at Kibale appeared to select foods that were low in fiber, suggesting that fiber is an antifeedant for these animals. In contrast, Conklin-Brittain *et al.* (1998) demonstrated that three species of cercopithecines (*C. mitis*, *C. ascanius*, and *Lophocebus albigena*) had a constant level of intake of the different fiber fractions throughout the year, suggesting that even when they could have avoided eating foods with high fiber content they did not do so. However, if the foods typically eaten by these animals were not high in fiber there may have been no need to avoid such foods. Thus, the role of fiber in cercopithecine diet selection warrants further consideration.

This study confirms the suggestion from a number of studies that *C. mitis* has a very flexible dietary strategy (Rudran, 1978a; Struhsaker, 1978; Gautier-Hion, 1988; Beeson, 1989; Butynski, 1990; Lawes *et al.*, 1990; Lawes, 1991), and cautions against evaluating habitat suitability on the basis of only the availability of different types of foods (e.g., the scarcity of fruit) and without assessing the nutritional value of foods. Golden monkeys appear to be able to obtain an adequate diet by balancing the nutrients they need from a few plant species that are available year-round. Thus they derive their protein from bamboo and their sugars from fruits, flower, and leaves. In addition, the golden monkey group fed on between 3 and 12 plant species in any given month and only 16 plant species were recorded in the diet over the entire study period.

This is not to suggest that the diet of the golden monkey at Mgahinga was optimal in any sense, but merely adequate, and there is evidence to suggest that adopting this diet may have a reproductive cost. Nutrition can affect the age at which a female becomes sexually mature, the ovulatory cycle, the length of time it takes to conceive, interbirth intervals, birth rates, and infant survival (Koenig, 2000). As a result, the relative reproductive success of a population provides information regarding the quality of the population's diet that may integrate a long time frame. Golden monkey groups had a lower infant-to-adult female ratio than any of the other blue monkey subspecies for which data exists, suggesting that fewer infants are born into these groups (Table 3). While we found no differences in the nutritional quality of the populations' diets, the data on infant-to-adult female ratio may still indicate that golden monkeys are under greater nutritional stress than other subspecies. Thus, further investigation of the cause of the lower infant-to-adult female ratio is warranted, and this should include nutritional elements not evaluated here (e.g., minerals).

Cords (1986) reported that pregnant and lactating female blue monkeys eat 63–83% less fruit than other females and 1.2–3 times as many insects and suggested that this represents the added protein needs associated with childbirth and rearing. The fact that the relative densities of golden monkeys in Mgahinga is highest in the bamboo zone and in forests with bamboo vegetation types (Twinomugisha *et al.*, 2003), that bamboo is a major food item, and that it provides a significant proportion of the group's protein suggests that bamboo is a critical resource for these animals. As a result, efforts should be increased to stop the illegal extraction of bamboo from the national park, and permission for the extraction of bamboo in community-based conservation development projects (Ugandan Wildlife Authority, 1996) should be critically evaluated.

ACKNOWLEDGMENTS

This study was funded by the Mgahinga Bwindi Impenetrable Conservation Trust (MBIFCT), the Wildlife Conservation Society, the Leakey Foundation, the Compton Foundation and the National Science Foundation (grant number SBR-9617664, SBR-990899). The first Mgahinga component was co-coordinated by the Institute of Tropical Forest Conservation (ITFC), which also provided the field equipment and facilities. We are grateful to S. Cavel, C. Lalobo Oryem, and R. Malenky for their constant encouragement. This work also benefited from comments, suggestions, and reading material provided by D. Pomeroy, G. Basuta, D. Baranga, A. Plumptre, E. Williamson, and B. Kaplin. We are very thankful to the MGNP rangers, particularly E. Sahane, M. Shwekyerera, E. Bizimana, B. Byamugisha, G. Baziyaka, and E. Mugisha, and park authorities under the leadership of Mr. Ignatius Achoka and Chris Oryem for their unwavering support during fieldwork. We thank Nancy-Lou Conklin and Richard Wrangham for providing nutritional data for Pancovia and Warburgia. Mike Wasserman provided vital assistance with the nutritional analysis. Lauren Chapman, Nancy-Lou Conklin, and Karyn Rode provided helpful comments on the manuscript.

CHAPTER FOUR

The Diet of Olive Baboons (*Papio anubis*) in the Budongo Forest Reserve, Uganda

Adam Alberts Okecha and Nicholas E. Newton-Fisher

INTRODUCTION

Baboons (genus *Papio*) are large-bodied, semi-terrestrial monkeys that occupy a diversity of habitats. Across populations, they show wide variation in dietary composition and in their foraging behavior (Whiten *et al.*, 1991). Early studies concluded that baboons were generalist feeders (De Vore & Hall, 1965; Rowell, 1966; Ransom, 1971; Harding, 1981), with Rowell (1966) saying of Ugandan baboons: "a list of plants eaten would probably be approximate to the botanical species for the area," but it is now clear that baboons selectively exploit their environment (Hamilton *et al.*, 1978; Norton *et al.*, 1987; Whiten *et al.*, 1991).

Adam Alberts Okecha • Faculty of Forestry and Nature Conservation, Makerere University, Kampala, Uganda **Nicholas E. Newton-Fisher** • Department of Anthropology, University of Kent, Canterbury, Kent, United Kingdom.

Primates of Western Uganda, edited by Nicholas E. Newton-Fisher, Hugh Notman, James D. Paterson, and Vernon Reynolds. Springer, New York, 2006.

Such selectivity would be predicted from a consideration of the wide variation in nutritional content and digestibility of potential food items, together with the need for foraging baboons to satisfy nutritional and energetic demands. The baboon foraging adaptation, in general terms, may be to selectively exploit a wide array of plant foods to satisfy energetic and nutritional needs when faced with a shifting mosaic of possibilities (Norton *et al.*, 1987).

The variation in diet between baboon populations has been used to investigate the influence of ecology on both diet and foraging behavior (Hill & Dunbar, 2002), but few such quantitative data have been published for forest-dwelling baboons (cf. Paterson, Chapter 5, this volume). These data are likely to be important, as savannah and forest habitats will present different ecological conditions and resources to foraging baboons. Tropical forests typically show less dramatic seasonality in comparison with savannah habitats, and so forest-living baboons should experience a wider and potentially more consistent resource base, which may influence both dietary composition and feeding time.

Furthermore, such data will be useful for interspecific comparisons. Forest-living baboons may be significant competitors with sympatric chimpanzees (*Pan troglodytes*) despite contrasts in foraging strategies. While baboons follow a selective–generalist strategy, chimpanzees typically show a reliance on ripe fruit (Wrangham, 1986). Baboons may have the advantage in scramble competition where the two species target fruit of the same plant species, as they are better able to digest unripe fruit (Wrangham *et al.* 1998; Lambert, 2005). There are few data available to test whether such competition occurs. Peters and O'Brien (1981) reported a dietary overlap between baboons and chimpanzees of only 18.3%, but a much greater degree of overlap (around 50%) was found recently in a study of sympatric baboons (*Papio cynocephalus*) and chimpanzees (*Pan t. schweinfurthii*) in the Mahale Mountains National Park, Tanzania (Matsumoto-Oda & Kasagual, 2000).

Here, as a preliminary step in investigating potential feeding competition and to broaden the knowledge of baboon foraging strategies, we describe the diet of forest-living olive baboons (*Papio anubis*) from the Sonso region of the Budongo Forest Reserve, where they live sympatrically with chimpanzees (*P. t. schweinfurthii*). The diet of these chimpanzees has been well studied (Newton-Fisher, 1999b; Tweheyo & Obua, 2001), but comparable data for baboons have not been collected.

METHODOLOGY

Study Site and Subjects

This study was conducted in the Budongo Forest Reserve, Uganda, in the region (compartment N3) of the largely defunct Sonso sawmill and the Budongo Forest Project (BFP) field station. Budongo forest is classified as moist, medium-altitude, and semi-deciduous (Eggeling, 1947; Langdale-Brown *et al.*, 1964; Howard *et al.*, 1991), and is one of the most botanically diverse forest reserves in Uganda (Synnott, 1985; Howard, 1991).

The olive baboons that were the subjects of this study were members of the Sonso Sawmill troop. This troop consisted of nine individuals at the beginning of the study, increasing to 10 with the birth of an infant, and represented an immigrant population. Baboons were not present in this area in the early 1970s, but were resident from the early 1990s (V. Reynolds, personal communication). It seems likely that baboons moved into the sawmill clearing after the cessation of the intensive sawing operations, and came from a population of olive baboons living beyond the edge of the forest (Paterson, Chapter 5, this volume). These baboons were partially habituated at the start of the study; they were commonly sighted within the site of the BFP field station, coming mainly to raid garbage.

Data Collection

Data were collected between September 1999 and February 2000, for a total of 20 weeks, divided evenly between the wet and dry season. Scan and *ad libitum* sampling (Altmann, 1974) was conducted from dawn to dusk or until poor visibility set in. Altogether, 7722 feeding records were taken for both seasons, representing 2621 and 5301 feeding records for rainy and dry seasons, respectively. Lower sampling in the wet season was the result of data collection being interrupted by heavy rain and poorer levels of habituation of the study subjects.

RESULTS

The Sonso baboons spent 47% of the time that they spent feeding consuming fruits and seeds (34% on fruits, 13% on seeds), 17% of this time eating leaves, and 17% feeding on food items raided from garbage. The remaining 19% of feeding time was spent on stems, bark seedlings, roots, and items obtained through agricultural crop raiding.

A total of 51 identified wild plant species were recorded as providing food items for baboons during the 6-month study period (Table 1). The cumulative number of species eaten approached an asymptote after about 12 weeks of study (Figure 1), suggesting that these 51 species represent the bulk of the diet of this baboon population. Observations were made in both the wettest and driest seasons, and so any large seasonal shifts in diet should be included in these data, although undoubtedly more species would be recorded with longer observation time and at other times of the year. Twenty-one species were regularly and consistently consumed, with each accounting for 0.5% or more of the time spent feeding (Table 2).

Diet by Food Item (Figure 2)

Leaves

The bulk of foliage was provided by *Broussonetia papyrifera* (paper mulberry), accounting for an average of 13% of total feeding records. Other species that contributed appreciably to the amount of recorded feeding on leaves were *Amaranthus* sp. (herb), *Desmodium intotuene* (creeper), and *Pannicum maximum* (grass).

Fruit

During the wet season, *Celtis durandii* constituted the highest proportion of fruit records (7%), while those of *Raphia farinifera* constituted the highest records (7%) during the dry season. *C. durandii* is one of the most common main canopy trees in the study area (Plumptre, 1996; Newton-Fisher, 1997). It produces large quantities of fruit from September through October and occasionally into December. *R. farinifera*, a palm tree, produces long strands of nuts. The mesocarp of a ripe fresh nut is scraped from the central nut and eaten by the baboons. The remaining nut is eaten only after it has become hydrated and begun to sprout (below). Other species that contribute appreciably to fruit diet are *Lantana camara*, *B. papyrifera*, *Physalis peruviana*, *Psidium guajava*, *Ficus sur*, *Ficus exasperata*, *Cleitopholis patens*, *Tetrapleura tetraptera*, and *Margaritaria discoideus*. *Lantana camara*, *B. papyrifera*, and *P. guajava* are

Table 1. Food items in the diet of the olive baboons from the Sonso sawmill troop, Budongo Forest, Uganda

Species	Leaves	Fruit	Seed	Stem	Bark	Root	Seedling
Acalypha sp.	✓						
Aframomum sp.	✓	✓	✓	✓		✓	
Albizia coriaria					✓		
Albizia zygia			✓		✓		
Amaranthus sp.	✓		✓				
Aneiloma euquietiale						✓	
Brachiaria brizantha			✓				
Broussonetia papyrifera	✓	✓					
Caloncoba schweinfurthii		✓					
Canna indica		✓		✓			
Celtis durandii		✓					
Citropsis sp.		✓					
Cleitopholis patens		✓					
Climbers (unknown)		✓				✓	
Coffea robusta		✓					
Coffea sp.		✓					
Commelina sp.	✓						
Cordia melinii		✓					
Croton macrostachys		✓					
Curcunis arculeatus		✓					
Datura sp.		✓					
Desmodium intotuene	✓					✓	
Eleusine jaegeri			✓				
Erythrophleum suaveolens		✓					
Euphorbia teke	✓						
Ficus exasperata		✓				✓	
Ficus polita		✓					
Ficus sansibarica		✓					
Ficus saussureana		✓					
Ficus sp.						✓	
Ficus sur		✓					
Ficus varifolia		✓				✓	
Iodes africana		✓					
Khaya anthotheca					✓		
Lantana camara		✓					
Leptapsis sp.	✓						
Marantochloa sp	✓			✓		✓	
Margaritaria discoideus		✓					
Myrianthus holstii		✓					
Palisota mannii	✓	✓					
Pannicum maximum	✓						

(*Cont.*)

Table 1. *(Continued)*

Species	Leaves	Fruit	Seed	Stem	Bark	Root	Seedling
Physalis peruviana		✓					
Pollia condensate		✓					
Prunus africana		✓					
Psidium guajava		✓					
Raphia farinifera		✓	✓	✓			✓
Ricinodendron heudelotii			✓				
Sarcophrynium schweinfurthianium	✓				✓		
Teclea nobilis		✓					
Tetrapleura tetraptera		✓					
Unknown	✓	✓					

Data are from scan sampling and ad libitum observations. A ✓ indicates that the baboons were observed to feed on the item.

exotics. *F. sur* was the primary fig species consumed between September and November 1999, while *F. exasperata* became the main fig species between December 1999 and February 2000.

Seeds

Two main species contributed to the bulk of seed diet: *Ricinodendron heudelotii* and *R. farinifera*. *R. heudelotii* produces a fruit containing two hard-shelled

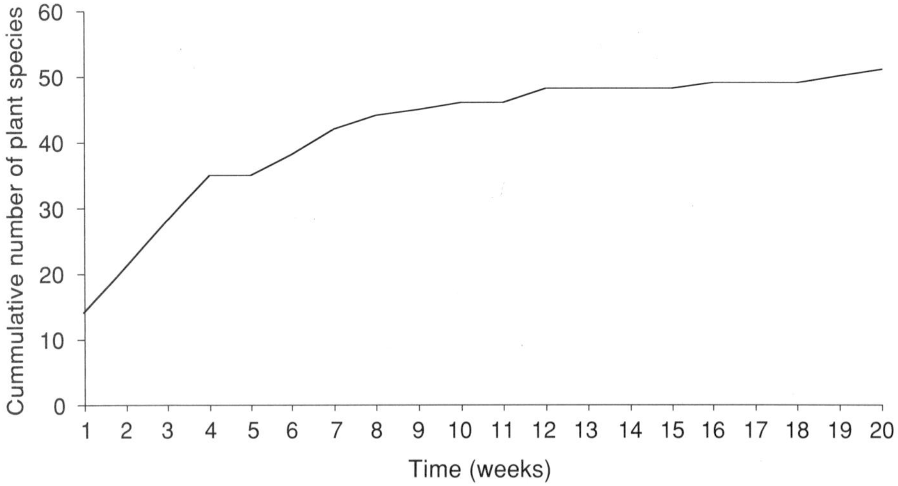

Figure 1. Cumulative number of species on which baboons were observed to feed, plotted by number of weeks of observation.

Table 2. Species accounting for 0.5% or more of time spent feeding

Species	Percentage
Broussonetia papyrifera	17.65
Raphia farinifera	12.49
Ricinodendron heudelotii	8.52
Celtis durandii	5.78
Lantana camara	4.17
Ficus exasperata	4.11
Physalis peruviana	3.42
Sarcophrynium schweinfurthianium	2.74
Psidium guajava	2.54
Amaranthus sp.	2.07
Ficus sur	1.92
Desmodium intotuene	1.73
Pannicum maximum	1.49
Pennisetum purpureum	1.39
Aframomum sp.	1.17
Khaya anthotheca	1.09
Cleitopholis patens	1.07
Eleusine jaegeri	0.96
Marantochloa sp.	0.75
Tetrapleura tetraptera	0.71
Palisota mannii	0.60

nuts. The baboons were not seen eating the flesh of this fruit, nor climbing up trees to obtain them; they made use of the fruit after it had fallen and rotted sufficiently and the nut had been disengaged. They usually cracked the hard nut with their molar teeth and ate the endosperm, spitting out the shell. *R. farinifera* seeds were only eaten when hydrated and sprouting, the baboons uprooting the seedling, consuming the meristem or pith, cracking the nut, and eating the remainder of the embryo. Other species that provided seeds to the diet were *Amaranthus* sp., *Albizia zygia*, *Eleussine jaegeri*, and *Brachiaria brizantha*. The last two of these are grasses, which are mainly available during the dry season (December 1999–February 2000).

Stems

These accounted for 7% of feeding records, of which *Sarcophrynium schweinfurthianium* (herb species found in swamps along River Sonso) contributed 2.7%. Of the tree species, only the stem of *R. farinifera* trees that had been

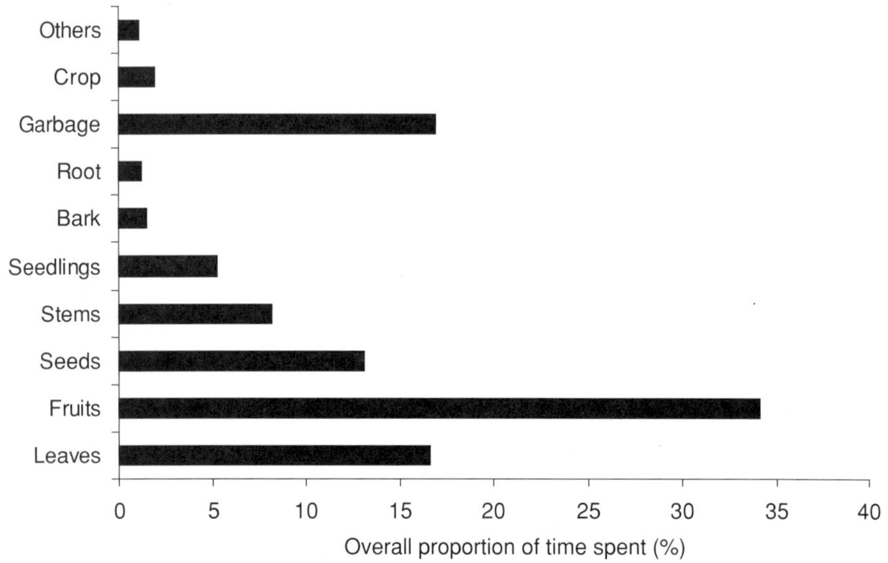

Figure 2. Breakdown of baboon diet by proportion of time spent feeding on each food item.

felled by local people to tap alcohol was consumed. The fermented stems were also consumed by other primates and bush pigs. Other species from which the baboons ate the stems were *Pennisetum purpureum*, *Afromomum*, and *Marantochloa*.

Bark

The bark of trees was occasionally eaten, in some cases to obtain gums. Bark alone accounted for only 1.6% of the total food. Overall, the bark of *Khaya anthotheca* was the most eaten (1.1%), followed by *F. exasperata* (0.44%). Other species that contributed bark to the diet were *Ficus varifolia* and *Albizia zygia*.

Roots

These accounted for an average of 1.2% of total food. Herb species formed the highest proportion of these items, including species such as *D. intotuene*, *Aframomum*, *Marantochloa*, and *Aneiloma aquietiale*. Many climbers were also uprooted and the root bark chewed. It was, however, difficult to identify climber species, and these were recorded as unidentified climber species. Of the tree species, the baboons were only recorded feeding on roots of *Ficus* seedlings.

Seedlings

Overall, seedlings accounted for about 5% of the food. Seedlings of *R. farinifera* contributed the whole of this proportion. The baboons uprooted the seedling, stripped off the outer part, and consumed the pith.

Garbage Raiding

Items raided from garbage accounted for approximately 17% of the overall food. It was difficult to make a concise list of the various items raided from garbage, although it included most of what people in the camp ate and/or any edible item dumped in the garbage pits, ranging from processed food items such as bread, margarine, and biscuits to locally grown food items such as cassava, potatoes, and fruit. Baboons spent as much time feeding on garbage as they did on the most consumed plant species: *B. papyrifera*. It was also noted that the intensity of garbage raiding increased appreciably during the dry season, possibly due to the decline in seasonal availability of plant items (see Figure 3).

Crop Raiding

This accounted for a very small proportion of the overall feeding recorded (about 2%). Among the primate species, baboons have been reported to be the most notorious crop raiders (Naughton-Treves, 1998). This small proportion can be attributed to the fact that crops were only grown in small gardens belonging to the houses of Budongo Sawmills Limited labor line, which were easily guarded and were not otherwise available to this baboon population.

Other

Baboons were also observed to feed on other items. These included insects (particularly termites and grasshoppers), birds (and their eggs), and mushrooms. On one occasion, a baboon named "Magezi" fed on chicks of domestic fowls. Mushrooms were encountered seasonally, and were their favorite when encountered.

Season Changes in Diet

Baboons feed for longer on fruit and stems in the wet season that in the dry season (fruits: $U = 22$, $p < 0.05$; stems: $U = 19$, $p < 0.05$), but there were no significant seasonal differences in feeding time for leaves, seeds, bark, roots, or seedlings. Diet in the dry season appeared to be broader than in the wet season (Figure 3).

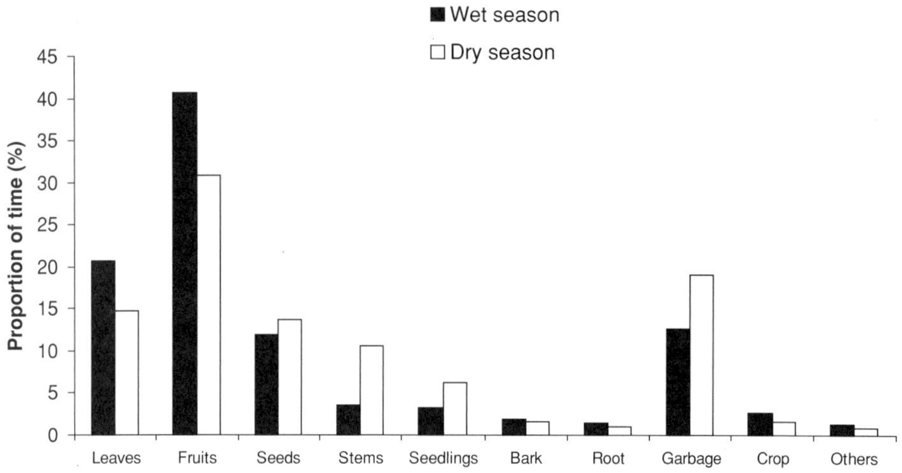

Figure 3. Comparison of baboon diets between wet and dry seasons.

DISCUSSION

The olive baboons of Sonso consumed a variety of food items and broadly appeared to follow a generalist feeding strategy: their diet included, among others, leaves, fruits, stems, seedlings, roots, bark, and animal items. Fruit appeared to be eaten in preference to leaves, stems, and seeds, whenever available, and the high proportion of fruit in the diet appears typical of most *Papio anubis* populations (Table 3).

The baboons fed on the seeds of only a few plant species, mainly *R. heudelotii* and *R. farinifera*. Seeds constitute a class of food item that imposes time and energy costs on would-be consumers. In contrast to fruit, seeds have been selected to avoid destruction during ingestion and gut passage: they may have hard rinds, sticky resins, irritant hairs, or poisonous secondary chemicals. Even if an animal can open such seeds it may not be worth the necessary time and

Table 3. Dietary profiles of olive baboon populations based on proportion of time spent feeding on different categories of food

Population	Fruit (%)	Leaves (%)	Roots (%)	Other (%)
Bole, Ethiopia	41	41	1	17
Budongo, Uganda	47	17	1	35
Chololo, Kenya	23	27	15	35
Gilgil, Kenya	10	53	27	10
Gombe, Tanzania	49	14	7	30
Masai Mara, Kenya	46	44	8	2
Shai Hills, Ghana	59	8	17	16

Sources: Data taken from Hill and Dunbar (2002:Table 1), where original sources are indicated, except for Budongo, for which data are from this study.

energy (Raemaekers *et al.*, 1980), particularly if easier food sources are available, which may be the case for the Sonso baboons.

These baboons showed considerable selectivity in the food items they consumed. This was true even for fruit and leaves, items that the baboons fed on the most. The animals were observed to concentrate on leaves and fruits of *B. papyrifera*; fruits of *C. durandii*, *Ficus* spp., and *R. farinifera*; and seeds of *R. heudelotii*. Many of the species on which the baboons fed were patchy in their distribution (Newton-Fisher, 1997; Newton-Fisher *et al.*, 2000). *B. papyrifera*, for example, was locally abundant but concentrated in areas bordering the old sawmill clearing, while *R. farinifera* is not particularly abundant and is confined to areas of swamp forest. *F. sur* and *R. heudelotii* also appear to be unevenly distributed in the Sonso region. Targeting patchily distributed species implies considerable selectivity of plant food items by the baboons.

The generalist aspect of the baboon foraging strategy is evident when compared with the feeding behavior of the sympatric chimpanzee population (Newton-Fisher, 1999b). The diet for both baboons and chimpanzees contained a similar number of plant species, which were targeted for at least 0.5% of the time spent feeding (19 species for baboons, 17 species for chimpanzees). But while these species accounted for 93% of chimpanzee feeding time, they represented only 75% of baboon feeding time. Furthermore, the baboons did not focus their efforts as noticeably as did the chimpanzees on the most commonly consumed plant species (Figure 4).

The data also suggest limited overlap in the diets of the two species. *F. sur*, the top-ranked dietary item for chimpanzees (23% of feeding time), accounted for only 1.9% of baboon feeding time while *R. farinifera* (fruit, seed, stems

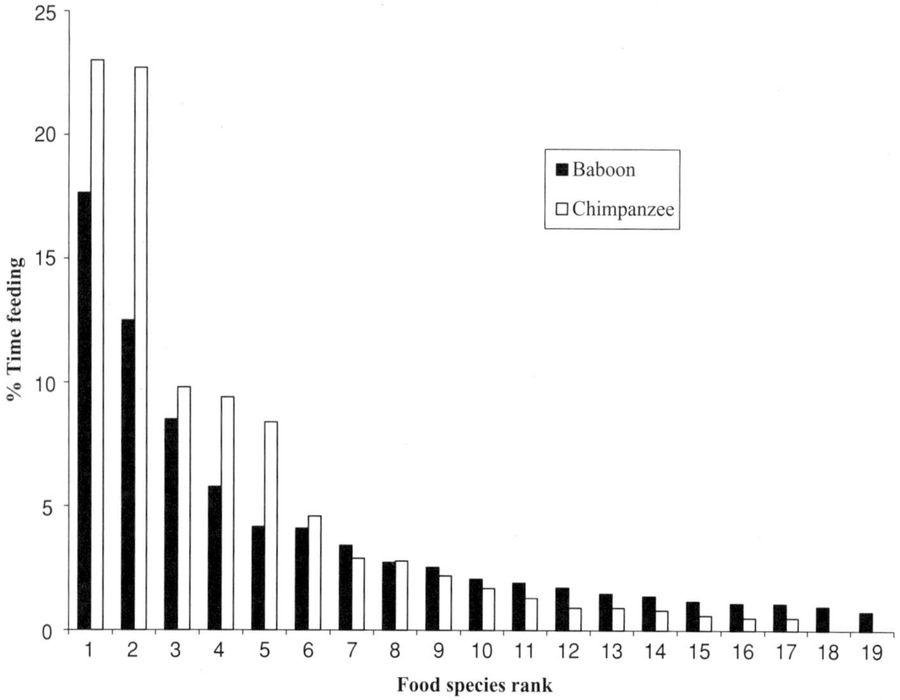

Figure 4. Comparison of top-ranked (by % feeding time) food species for sympatric baboon and chimpanzee populations. All species that accounted for 0.5% or more of feeding time are included.

and seedlings) for 12.5% of their feeding time, but chimpanzees fed on this species for only 0.6% of their feeding time (eating only the rotten wood). Only six of the species that were targeted by baboons for at least 0.5% of the time spent feeding were similarly targeted by chimpanzees. Both species spent considerable time feeding on the fruit and leaves of *B. papyrifera* (baboons: 17.7%; chimpanzees: 22.7%), as well as the fruit of *C. durandii* (baboons: 5.8%; chimpanzees: 8.4%) and that of *F. exasperata*, although baboons also targeted the roots of this species (baboons: 4.1%; chimpanzees: 2.2% of total feeding time targeting this species). These comparisons suggest that feeding competition between baboons and chimpanzees is unlikely to be important. Any scramble competition is most likely to occur with either *B. papyrifera* or *C. durandii*, but the first of these is a locally abundant, highly productive exotic and the second is common throughout the region (24% of the Sonso region is *Celtis*-dominated mixed forest; Newton-Fisher, 1997). Baboons spent little time feeding on the fruit of the fig species *F. sur* (1.9%) and *F. mucuso* (0%), which are important staple foods for the sympatric chimpanzees (Newton-Fisher, 1999b); this suggests that their ability to consume and digest unripe fruit does not lead to significant scramble competition between these two populations.

Further studies of baboon diet and foraging behavior are required. The determination of distribution and abundance of plant food resources were beyond the scope of this study but remain an important next step to better address the nature of the baboon foraging strategy. In addition, a study of the dietary quality of preferred food items is essential to investigate why the baboons select or concentrate on a few plant items in the face of a diversity of plant food resources.

CHAPTER FIVE

Aspects of Diet, Foraging, and Seed Predation in Ugandan Forest Baboons

James D. Paterson

INTRODUCTION

The common "olive" savannah baboon, *Papio cynocephalus anubis*, is a widely spread and locally common monkey in Africa. Baboons have a large number of characteristics in common with humans, they are omnivorous and often have similar taste preferences, but differ in prefering fruits at an immature state as well as many items that present a pronounced bitter or astringent taste to human perception. They live in large communal social groups, are hierarchically organized, and are opportunistic, seizing any opportunity to obtain food, especially the concentrated nutrients in domesticated species. Thus they can play the role of classic "pest" or "weed" primates. In most of their social and behavioral characteristics, they parallel humans, and on previous occasions (Paterson, 1973, 1976) I have argued that baboons and humans are the survivors of a coevolutionary radiation involving a predator–prey relationship between hominids and cynopithecini. The parallels in diet and behavior can be seen as

James D. Paterson • Emeritus, Department of Anthropology, University of Calgary, Canada.

Primates of Western Uganda, edited by Nicholas E. Newton-Fisher, Hugh Notman, James D. Paterson, and Vernon Reynolds. Springer, New York, 2006.

a consequence of humans and baboons competing for the same resources—a scramble competition—in much the same environments, combined with a pattern of predation by humans on other primates. The result is that much of baboon social structure and behavior can be considered as a response to competition and especially to predation by humans. Even today, humans eat baboons in many areas, and actively compete with them for access to resources.

Baboons have typically been studied in short-grass savanna environments because of the acknowledged difficulties of observing and tracking them in high grass or forest areas (Washburn & DeVore, 1961; DeVore & Washburn, 1963; Paterson, 1976). This chapter represents a rare study of baboons in a forest/forest-edge habitat. Baboons are noted as having extremely broad ranges of dietary input, but the patterns for forest-resident populations may be the most divergent from the averages. Harding (1981) stated in a comparative study that focused on savanna- and thicket-dwelling populations, that baboons had the most diverse diets of any primate species, with all nine of the dietary compartments assessed being utilized by them. The plant species listed in their dietary range is further expanded in forest habitats. Okecha and Newton-Fisher (Chapter 4, this volume) record 51 plant species consumed by the Sonso troop in 1999–2000, with few of the species recorded in other studies. My earlier study on Busingiro Hill (Paterson, 1973, 1976) recorded 31 plant species in the baboon diet, in addition to a very high level of geophagy.

METHODOLOGY

Subjects

The Sonso baboon troop, resident in the clearing of the Budongo Sawmills Limited, is undoubtedly a recent immigrant population, as they were not present in the area during 1970–1971 (personal observations). It is likely that the troop moved into the sawmill clearing after the cessation of intensive sawing operations in 1978. They appear to have been resident at least from 1990, as Vernon Reynolds noted their presence during his visit to Sonso (V. Reynolds, personal communication). The logical origin for the troop is as an offshoot of forest-edge populations in the Nyabyeya or Nyakafunjo area, which voluntarily moved, or was forced by population pressures, down the Royal Mile to the Sonso clearing.

The troop, in September and October 1996, comprised 12 individuals, with a very unbalanced sex ratio. There were three adult males, two adult females, one 2-year-old juvenile female, five juvenile males (one aged 4–5 years, one 3+ years, two 2+ years, and one 1+ year), and an infant-2, holding membership in the group. The unusual sex ratio (3:1, males to females) is likely the result of historical events. Baboons are male dispersers, with most males leaving the natal group as young adults and moving into another troop. Bercovitch (1992, 1994, 1995) noted, in the Amboseli population, that aging males also migrate, moving into a new troop where they are less likely to be persecuted by younger males who have recently arrived at high ranks. By November 1996, composition of the Sonso troop had changed to just 10 members with the departure or disappearance of one adult male, the 3-year-old juvenile male, and the infant-2, counterbalanced by birth of an infant male on October 9. It is possible the first two individuals moved together to another group in the Nyakafunjo area and that the infant-2 was the baboon body noted in passing by our research assistant (however, we were unable to relocate it). Since the Sonso troop is essentially at the end of a linear transfer corridor, the resident adult males likely emigrated from the Nyabyeya or Nyakafunjo area and might return along that corridor. The second demographic factor is that until shortly before this study, the troop must have had a third adult female. In light of the presence of three juveniles at the 2 to 2+ age level, a third female had to have been present until at least March 1995. The situation at the time—with one female considerably older than the other and only one juvenile female—placed the long-term survival of the troop in jeopardy, yet the troop remained viable through 2004 (J. Wallis, personal communication). Without the production of more female offspring, the troop has the potential to die out, leaving an empty range open to be filled by a new migration from the Nyabyeya or Nyakafunjo areas.

Data Collection

The study was conducted with several observing techniques: focal animal sampling; regular scan sampling; and, unfortunately for statistic analyses, the most useful turned out to be a regularized form of *ad libitum* sampling. Data was collected through focal animal sampling (Altmann, 1974), using 30-min samples of individual animal behavior, and through the use of half-hourly-interval sampling (scans) of activity collected by the assigned research assistant. Ecologically relevant data was gathered in several ways:

1. Sample plots of forest structure. The plots were 10 by 25 m in size, and two plots were placed in proximity to each other, to make up a "sample quadrat" for use in calculating a version of Morisita's Index of similarity (Krebs, 1999; C. Chapman, personal communication). Within each plot, each tree of more than 4 m in crown height was numbered with an aluminum tag, identified, and the diameter at breast height (DBH), the crown height, and the phenological state were recorded. Each was also mapped as to position within the plot. Sixteen plots yielding eight sample quadrats were positioned around the Sonso clearing in areas either utilized or traversed by the baboon troop during the study period. Each sample plot was assessed for crown density using a Lemmon Crown Densiometer (Lemmon, 1956, 1957) to evaluate overstory at each end and at the middle of the plot.
2. Biological sampling proceeded throughout the course of the study period; collection of food species for identification where these were unknown to the research assistants resulted in more than 20 species being added to the herbarium.
3. Fecal sampling for several purposes was also carried out. Samples identified and associated with specific individuals were regularly collected. Each sample was divided and treated in different ways. A set of samples was preserved in 95% ethanol for DNA analysis by Dr C. J. Jolly. A sample was completely dried on a glass slide before being placed in a dry vial and a sample was washed through a medium-small stainless steel strainer, the residue sun-dried on photocopy paper and stored in a dry vial. The net weight of dried samples averaged 11% of the wet weight, indicating that approximately 89% of the feces was water. Washed samples were examined under a microscope, all recognizable seeds and other materials extracted and evaluated for damage. Assessments of fiber content, of bark fragments, and two submillimeter-sized seeds were recorded. All other material was counted, either as intact or damaged. While this technique—similar to that presented by Overdorff and Strait in 1996—cannot assess the role of seed disperser, it clearly reveals the role of seed predator since neither damaged nor destroyed seeds can be classified as eligible for germination.

The data on the Busingiro Hill population (four groups ranging in size from 28 to 70 individuals) was collected January 1970 through April 1971. The species diet list is included for comparison with the Sonso population. Data was collected using an *ad hoc–ad libitum* method.

RESULTS

The primary observer expended a total of 302 fieldwork hours, that is actual time either with or searching for the troop. Of this time, 202 h (67%) were spent in actual proximity when observations were being recorded. The general difficulty of observing conditions, as noted earlier, which resulted in a much lower number of focal sample hours being recorded (39.5 h of focal data = 19.5% of the contact hours), falls into this category. The remaining contact time is recorded in *ad libitum* notes. The half-hourly-interval scans made by the research assistants were not added into these totals. A further 300 h were expended in preparation and preservation of botanical and fecal sample materials, and in setting and evaluating sample plots.

Diet Diversity

Baboons are noted as having extremely broad ranges of dietary input. Whiten *et al.* (1991) noted: "foraging profiles described even at the level of very gross food types show remarkable variations across the habitats: . . . 3 to 74% (fruit), 1 to 53% (subterranean items), and 8 to 53% (leaves)." At Sonso, study subjects were repeatedly recorded as ingesting more than 10 different species of food materials during the course of a half-hour focal sample, and it was not unusual to observe several different species being consumed within a minute. Yet on other occasions, intensive feeding on a single foodstuff would persist for several hours.

Baboons, like humans, are omnivores but their tastes differ in some respects from those of humans. They are much more likely to eat fruits at a very unripe state—inedible to most humans—and consume a great quantity of leafy material and pith from grasses and trees, which humans find tasteless or unpleasantly astringent. Table 1 presents the lists of observed plant food species at Sonso and at Busingiro Hill. There is relatively little overlap between the two lists, with only two tree species and two grass species appearing in both. This is due to the ecological difference between the two sites. Sonso is a clearing, mostly artificial, within a selectively logged, mixed forest habitat, while Busingiro Hill is a mosaic of high-grass–tree savanna and *Maesopsis* colonizing forest. The variance in the

Table 1. Species observed to be consumed at two study sites within the Budongo forest reserve

Sonso (1996)	Busingiro (1970–1971)
Trees	Trees
Albizia coriaria	*Albizia grandibracteata*[a]
Broussonetia papyrifera	*Albizia zygia*
Calconcoba schweinfurthi	*Annona chrysophylla*
Celtis durandii	*Calconcoba schweinfurthi*
Cleistopholis patens	*Cassia mannii*
Croton macrostachys	*Combretum molle*
Ficus exasperata	*Cordia ovalis*
Ficus polita	*Erythrina abysinica*
Ficus sur	*Ficus capensis*
Ficus variifolia	*Ficus igens*
Khaya anthotheca	*Ficus vallis-choudae*
Lychnodiscus cerospermus	*Grewia mollis*
Margaritaria discoideus	*Jacaranda ovalifolia*
Psidium guajava	*Lannea kerstingii*
Raphia farinifera	*Maesopsis eminii*
Ricinodendron heudelotii	*Mangifera indica*
	Psidium guajava
	Spathodea campanulata
	Strychnos innocua
Herbs and Shrubs	Herbs and Shrubs
Aframomum sp	*Centrosema pubescens*
Aneiloma aequietiale	*Commelina erecta*
Canna indica	*Dioscorea dumeforum*
Desmodium intotueme	*Indigoflora emarginella*
Desmodium repandum	*Oxytenanthera abysinica*
Desmodium canum	*Securinega virosa*
Lantana camara	
Marantochloa leucantha	
Pollia condensata	
Grasses	Grasses
Eleusine jaegeri	*Hyparrhenia cymbarica*
Pannicum maximum	*Hyparrhenia folipendula*
Pennisetum purpureum	*Paspalum conjugatum*
	Pannicum maximum
	Pennisetum purpureum

[a] Known to contain bioactive compounds (Krief, 2004; Krief *et al.*, 2004).

two lists serves to highlight the catholic and opportunistic dietary patterns of baboons.

The members of the Sonso troop also ate gums, bark, insects, and about once per month one of the adult males took a domestic chicken. The male we named "Studly" was the main predator on three observed occasions. (The first observed incident occurred within half an hour of our arrival at Budongo Forest

Project [BFP]!). They also obtain an indeterminate amount of human domestic foods, foraging for the young leaves of bean plants, sweet potato roots, cassava stems and tubers, Uganda yam (taro) roots, as well as discarded food waste such as matoke, bananas, rice, millet, posho, etc. The amount of feeding time devoted to, and calories derived from, human foodstuffs, garbage, and garden raiding at Sonso during this study was equivocal. By far the majority of their food and caloric intake was derived from native and exotic vegetation. The two most important species during the September–October period for the troop were the exotics *Broussonetia papyrifera* and *Lantana camara*. These provided more than 90% of the observed ingested material in some weeks. Inspection of fecal samples collected during September–October showed that they contained substantial quantities of seeds, especially those of *Ficus sur*, *Psidium guajava*, *L. camara*, and *B. papyrifera*. During the latter half of October, and in November, both behavioral observation and examination of fecal samples indicate a slow shift toward more fibrous foodstuffs, with several varieties of low growing creepers, especially *Desmodium intotueme*, *Desmodium repandum*, *Desmodium canum*, leaves of *Broussonetia*, piths of grasses of the *Eleussine* and *Pennisetum* genera, *Aframomum* spp, *Canna indica* (an exotic Asiatic lily), as well as the fruits of *Celtis durandii*, *Raphia farinifera*, and *Cleistopholis patens* becoming important components of their diet. In addition, some novel observations of gummivory combined with bark ingestion, pure bark feeding, and seed predation are reported below.

Seed Predation

This study has identified four species of trees that are subjected to seed predation by the baboons. *Ricinodendron heudelotii*, *C. patens*, *C. durandii*, and *R. farinifera* all suffer complete destruction of their embryos through the actions of the baboons. In light of the identification of these species being preyed on during this 3-month study, it may be expected that other species suffer seed predation at other seasons of the year. Each of the identified species is exploited in a distinctive fashion.

R. heudelotii produces a 3- by 1.5-cm cherry-like fruit with a green skin, with yellow-green flesh smelling of peaches, containing two hard-shelled nuts each about 1 cm in diameter. The baboons were not seen to eat the flesh of this fruit, or to climb into the trees to obtain them; they were seen only to make use of the fruit after it had fallen and rotted sufficiently that the flesh could be disengaged from the nut. They then crack the nut between their molar teeth, and use their tongue and incisors to extract the nutmeat, spitting out the remains of the nutshell. The nut is extremely hard, and the initial cracking produces a sound that is audible for 20–30 m.

C. patens produces a very similar fruit. It is almost exactly the same in design—a double-pitted cherry—but turns black on ripening and falling to the ground. The flesh is green, with a minimal odor like green feed (chopped unripe barley) or shredded cabbage. The nut case is distinctively rough textured and, judging from the sounds produced by feeding baboons, it is less hard than the *Ricinodendron* nut. Baboons again feed primarily on the ground from dropped fruit; only on one occasion were two juveniles observed to eat the unripe fruit still attached to the parent tree. Both the flesh and the nut are consumed; normally the flesh of a ripe fruit is ingested and then the endocarp is cracked to extract the nutmeat. Rotted fruits, those from which the flesh has separated, are not consumed.

C. durandii is a very common secondary forest species and the fruits are much favored by all primates as food. The flesh of this 5–6-mm-diameter fruit is yellow and bland, and the kernel is dark brown and rugose. Baboons consume them both from the trees and as pickups from the ground. In mastication, they crush the whole fruit and the shell fragments can be found in the feces. None of the *C. durandii* seeds were passed intact.

R. farinifera, unlike the previous species, is a palm. It produces long masts of fruits, and the baboons process these twice. The fresh, ripening fruit, about 5 cm in length and 4 in diameter, is selected from the fruit mast. Baboons use their incisors to pick off the outer scales, which are discarded, and the paste-like mesocarp is scraped from the central nut. This nut is too hard for any primate to crack and is immediately discarded. This mesocarp is only 2–3 mm in thickness and similar in texture to a fig paste. It appears to be sufficiently nutrient-rich to be worth the effort for an adult baboon to climb up to the fruit mast and feed. The second use of the *Raphia* nut occurs weeks or months later, once it has hydrated and begun to sprout. Baboons then pull up the seedling and consume the meristem or pith, and access the remainder of the embryo.

Seeds of exotic (introduced) species were also consumed and destroyed by the baboons. *C. indica* is an exotic lily once planted at the Sawmill manager's residence, and has spread extensively around the edges of the clearing. The ripening seedpods are opened and the bean-like seed consumed; residues of these seeds were never found in the fecal samples. *L. camara*, a berry-producing, spreading shrub 3–4 m high was strongly favored during September and October, although rarely did any of the grape-like seeds pass intact through the baboon digestion. *Lantana* seeds did however contribute a major proportion of the crushed shell material found in the fecal samples. Native tree species *Croton macrostachys*, *Margaritaria discoideus*, and on occasion, *Calconcoba schweinfurthi* were similarly fed upon without intact seed being identified in the samples.

Several species of plants demonstrated variable patterns of survival and predation in the seeds recovered from fecal samples. *B. papyrifera*, the paper mulberry—a translocated exotic—benefited most from the baboon feeding, with high numbers of the 1-mm seed surviving and relatively few being damaged (Figure 1A). An unidentified discoidal 1-mm seed (Figure 1B) was found consistently in samples throughout the study period but the relative proportions of intact and damaged seed reversed between September–October and November–December. *F. sur* was the major fig species consumed, but small amounts of fruit were ingested from *Ficus polita* during November (Figure 1C). Roughly 50% of fig seeds were damaged, and there was a clear "seasonal" effect of the mid-September fruiting of *F. sur* and the November crop of *F. polita*. Baboons were never seen to feed in the *Ficus mucuso* trees beloved of the chimpanzees during November. Indeed, on occasion they were seen to move through a large *F. mucuso* tree, avoiding both the masses of fruit and the 30+ chimpanzees feeding there.

Another exotic—*P. guajava*, the guava (Figure 1D)—was present both in observed feeding and in fecal samples. Guavas were taken at an extremely green state, mastication noise indicating both the hard fruit and the immature seed being crushed. The seeds are large and relatively hard, yet somewhat below 50% survive oral treatment and gut passage. A native ground plant common on the forest floor is *Pollia condensata*, which is used by local people as a "natural bead," and which they consider to be poisonous. Since intact *Pollia* seeds (Figure 1E) were generally more prevalent than damaged ones, the plant may derive some small benefit in seed dispersal from its association with baboons. The final component of Figure 1(F) shows the counts of identifiable insect exoskeleton parts, and minerals—crystals and stony fragments—that were regularly encountered in the samples. The former are deliberately consumed, as observations of bark removal and scale bug consumption were recorded on several occasions. In addition, consumption of arthropods was associated with *Khaya* bark ingestion. The crystal and mineral items were most likely accidental inclusions in soils and

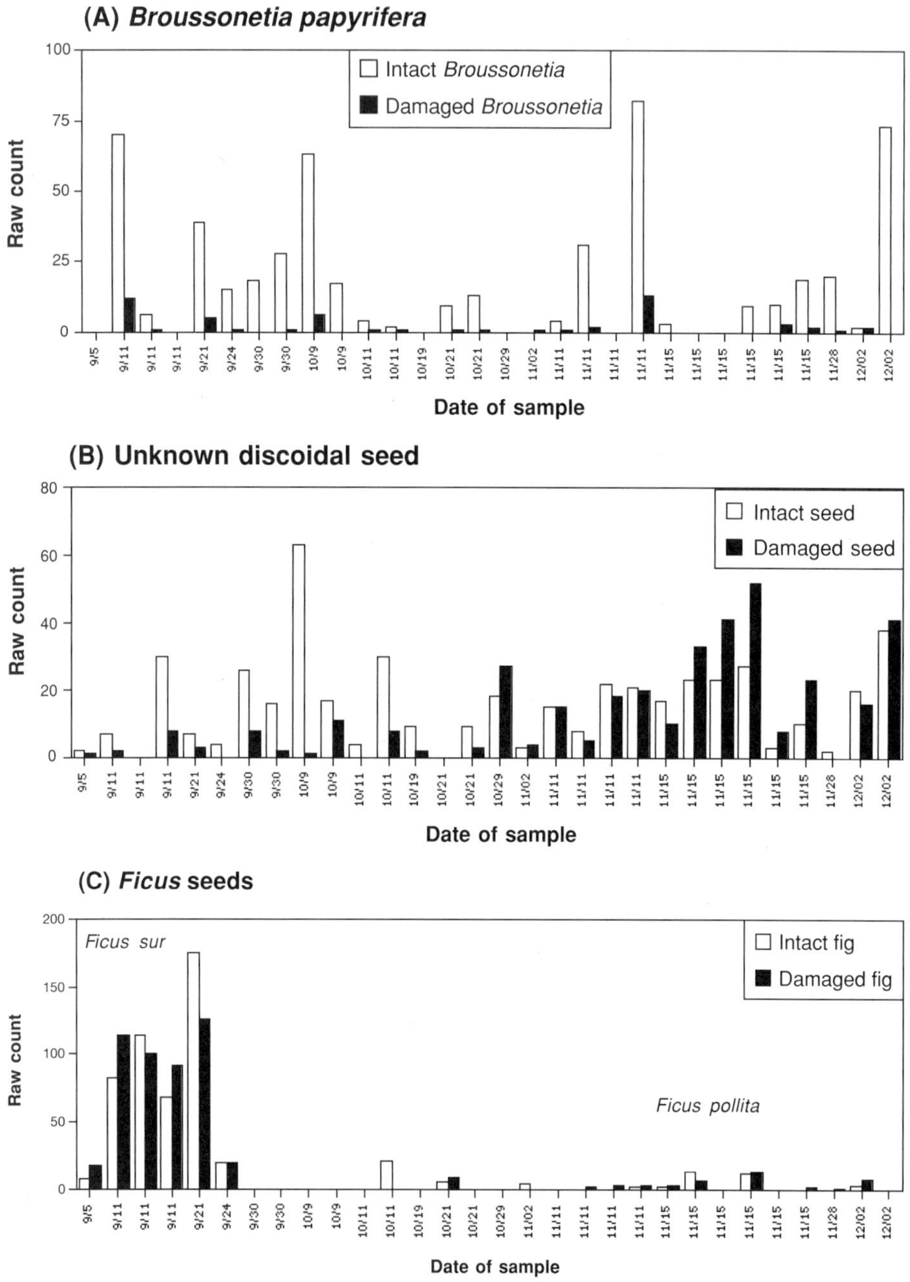

Figure 1. (A)–(F) Seed survival and destruction over 15 weeks. *x*-axis is date of sample collection; *y*-axis is raw frequency of seeds in 1-cm^3 sample.

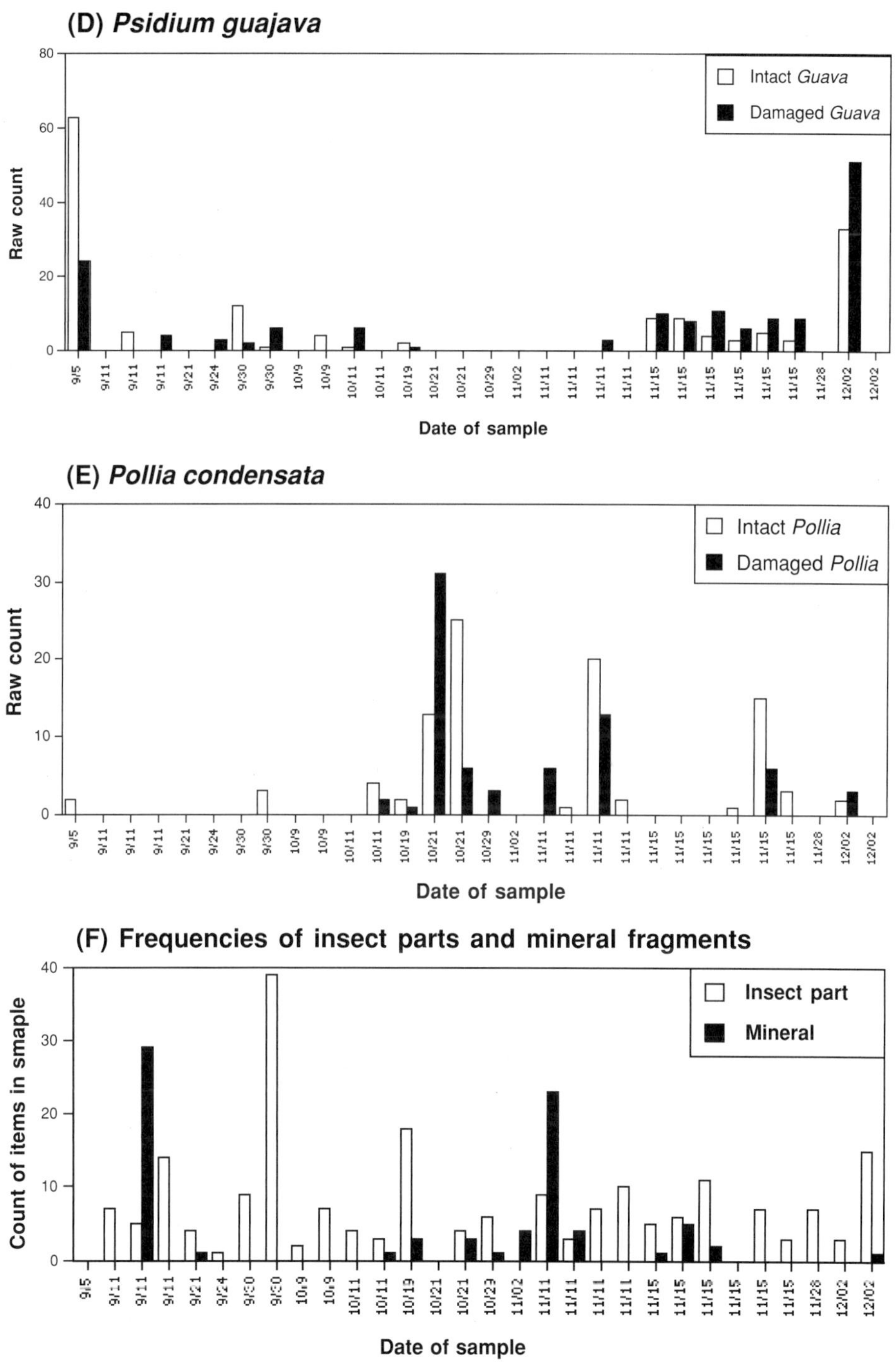

Figure 1. *(Continued)*

on plant materials rather than geophagy. No geophagy was observed at Sonso, unlike at Busingiro, where the brick used in the Busingiro House buildings and soils from collapsed termite mounds were regularly consumed by all members of the main study troop. X-ray diffraction analysis (courtesy of Mr. D. Harvey, Department of Geology and Geophysics) of a sample of the brick revealed that the major components were dehydrated clays "Gibbsite" –($Al(OH_3)$) and "Goethite" –($Fe_2O_3H_2O$), which dissociate in an acid solution to produce an alkaline slurry with some effects like the commercial "Kaopectate" (cf. Tweheyo *et al.*, Chapter 8, this volume).

Two rare unidentified seeds and several species of grasses are represented in Figure 2. These two unidentified seeds were occasional findings in samples

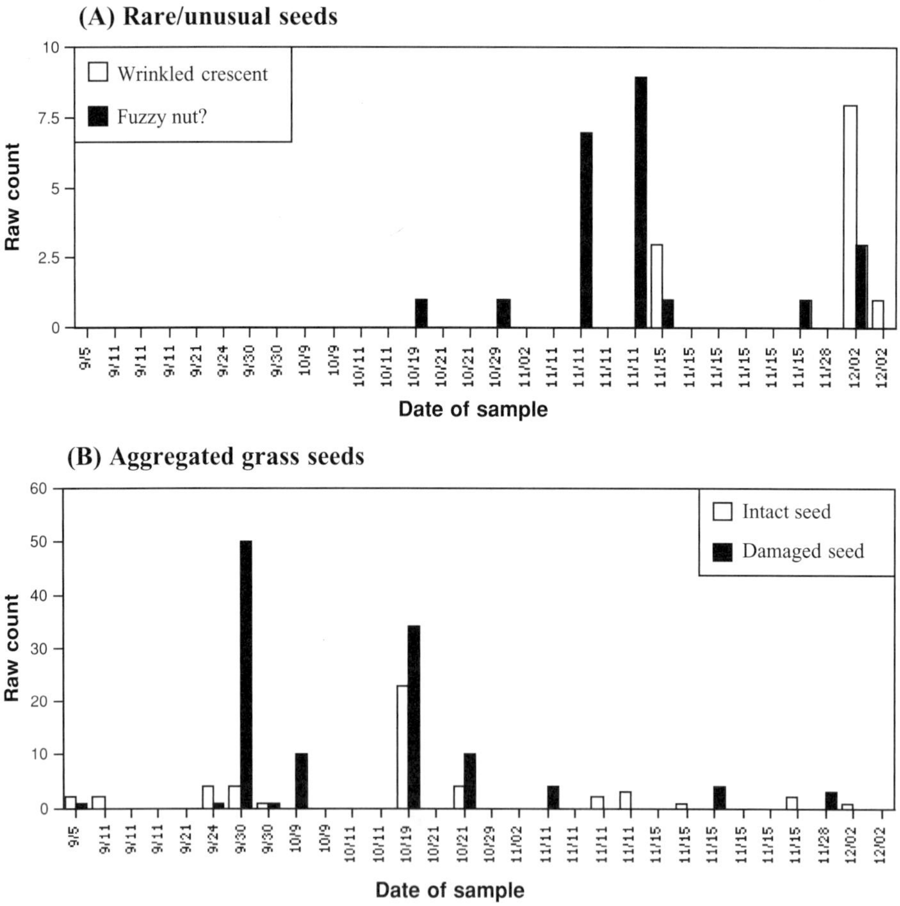

Figure 2. (A) Rare or unusual seeds noted in fecal samples. (B) Grass seeds of Eleusine, Pennisetum and Pannicum genera, intact seeds and remnants of damaged seed. x-axis is date of sample collection; y-axis is raw frequency of seeds in 1-cm^3 sample.

predominantly from November and early December. Grass seeds were frequently observed to be consumed during the study, with elevated intake during late September to late October, when *Pannicum maximum*, "Guinea grass," and *Brachiaria brizantha*, a minor component, were in seed. Relatively few grass seeds made it through the digestive path, and no intact seeds of *Pannicum* were encountered in the samples.

Pith, Root, and Bark Eating

The Sonso baboons were observed to feed on the structural components of a number of forest and grassland plants, specifically the internal, soft, and less fibrous material sometimes called "pith," the root structures either whole or in part of both trees and understory plants, and either the outer or inner bark of particular tree species. Table 2 presents a listing of the plants that are treated in this fashion by the Sonso baboons.

Pennisetum purpureum, known as Napier grass or elephant grass; *Marantochloa leucantha*, a common ground-level succulent; *R. farinifera*, the swamp-resident small nut palm; and *B. papyrifera*, the paper mulberry (an imported species), are the most frequently affected plants. As regards the other major grass species, *Pannicum maximum*, no pith consumption was observed. *Khaya* bark consumption is treated separately below, and cambium consumption is irregular in both time and space. Many *Ficus exasperata* and *Ficus variifolia* trees within the forest, but not all, show some evidence of bark gnawing along the buttresses of moderate-sized specimens. Some of this is known to be done by

Table 2. Pith, root, and bark ingested by Sonso baboons, and consequences for the plant involved

Type of plant	Species	Part	Consequences for plant
Asiatic lilly	*Canna indica*	Pith	Destruction of stem
Grass	*Eleusine jaegeri*	Pith	Destruction of stem
Grass	*Pennisetum purpureum*	Pith	Destruction of stem
Herb	*Aframomum* sp	Pith	Destruction
Herb	*Aneiloma aequietiale*	Root nodes	Destruction
Herb	*Marantochloa leucantha*	Pith of petiole	Destruction of leaf
Tree	*Broussonetia papyrifera*	Leaf stems	Pruning
Tree	*Ficus exasperata*	Cambium	Unknown
Tree	*Ficus variifolia*	Cambium	Unknown
Tree	*Khaya anthotheca*	Outer bark	Unknown
Tree	*Albizia coriaria*	Root pith	Destruction of seedling
Tree	*Lychnodiscus cerospermus*	Root bark	Destruction of seedling
Tree	*Raphia farinifera*	Pith	Destruction of seedling

the chimpanzees, who seem to favor the slight vinegar tang of *F. variifolia*, but on five occasions baboons were observed actively gnawing on fig buttresses—four *F. exasperata* and one *F. variifolia*. By a very substantial amount, the two common plants *P. purpureum*, and *B. papyrifera* are the most affected by pith feeding, some being consumed every observation day. For 5 days in late November, pithing of *Pennisetum* became an intensive activity, with at least an hour each day being devoted to it by the whole troop. The remainder of the species in Table 2 are rare observational occurrences, and *Raphia* was discussed earlier. Perhaps the most interesting is *Albizia coriaria* (one of the adult females uprooted a 1-m-tall seedling and nibbled the outer bark off of the entire root) which is used by the West Nile district human population as an effective fish poison. However, bioactivity of this species of *Albizia* is unknown.

While pith consumption was observed during a study centered around Busingiro Hill (approximately 9 km west of the Sonso site, in 1970–1971) it was seen only on grasses—*P. purpureum* and *Paspalum conjugatum*. No bark consumption was observed at that time. The ecological circumstances of the Sonso population are very different from that of the Busingiro Hill populations with respect to vegetation, animal, and avian communities, as well as temperatures and precipitation. Both locations have numerous "exotic" or imported species present, but they are different sets (see Table 1). At Sonso, the exotic *Broussonetia papyrifera* has become a major pest species on any disturbed soil and, owing to its pattern of rapid growth, vegetative colonizing as well as massive fruiting on a twice-a-year pattern, is becoming the dominant species around the clearing. Primates of all five resident species pay great attention to it as almost all parts are edible. Leaves, petioles, flowers, and fruits have been observed in the diets of all five species.

Khaya anthotheca and *Khaya grandifoliola* (often these two species are not discriminable under field conditions) are commercially important mahogany trees in Uganda and present an unsolved puzzle, which served as the focal point for a study in 1997. It has been recognized by the researchers at the BFP that all five of the resident primate species at least occasionally consume *Khaya* bark, and some blue monkeys consume very large amounts (Fairgrieve, 1995). But in light of the condensed tannin component of its makeup, 14–17 mg/g as reported in Plumptre *et al.* (1994), the question must be 'Why?' Condensed tannins are most frequently viewed as toxins rather than attractors, but recent work on the benefits of various tannins as cancer preventives in humans may change this perspective. One of the common characteristics of *Khaya* trees is the presence of delaminated areas on the midtrunk and lower main branches. These have been labeled as "scabs" in previous reports. Close observation of the activities of a 2-year-old female baboon, particularly because she is an intensive investigator of *Khaya* bark, indicates that scabs, when encountered, are pried open and the underside eaten, and the cavity is also nibbled on. Strings of resin from these scabs, whether clear, white, or tan, are avidly consumed. To humans, the resin is almost tasteless, but remnant bark smells somewhat like malt vinegar and is somewhat bitter to the tongue, yet a fresh slash through the bark yields a watermelon-like odor. The cambium is always avoided as it produces an instant anesthesia effect. Direct observation of openings in the outer bark produced by baboons record a rapid recovery, with the area being closed over with new outer bark in 12–14 days. Examination of a number of lower trunk scabs found them to be inhabited by nymphs of wood-boring beetles, solitary brown ants, wasp larvae, and scale bugs.

Biochemical analyses of the bark using thin layer chromatography (TLC) were undertaken during 1997–2001. In addition to finding condensed tannins of the anthocyanidin cluster, my students and I found limonoids, at least one other terpenoid compound and two coumarins—scoparone and scopoletin—in the bark. Extended TLC plates have suggested the presence of as many as eight major compounds, some having as many as four variants, differing only slightly in molecular weight. The published identifications of *Khaya* bark chemicals are predominantly from West African species of *Khaya* and populations of *K. grandifoliola*, and hence may differ from the Uganda specimens. Table 3 shows the identified compounds and their bioactive effects. Note that in spite of direct TLC comparisons with control samples of the three identified sterols, none of these were detected in the Sonso bark collection.

However, the actual energy content of the bark may also contribute to the reasons for its consumption. Bomb calorimetry was conducted on four samples of gum and eight bark samples. The mean energy content of the *Khaya* gum was 2922 cal/g and the bark was 3720 cal/g, but the largest part of the bark energy is unavailable to primate digestion as it is derived from cellulose and hemicelluloses. The sugar content found by Reynolds *et al.* (1994) accounts for 578 cal/g of dry bark, or 15.5% of the average yield.

At this point a tentative hypothesis might be that the tree's bark is prone to invasion by insects, that the insect invasion causes the bark to delaminate—forming the scab, the tree tries to compensate or "drown" the insects with an overproduction of resin, which in turn attracts the attention of the primates,

Table 3. Phytochemicals found in *Khaya* bark

Compound	Subgroups	Effects	Reference
Terpenoids	Limonoid[a] (methyl angolensate)	Molluscicide	Odyek *et al.*, 1990 Harborne, 1991 Adesina *et al.*, 1971
Sterols	Campesterol Stigmasterol β-sitosterol	Antipyretic Anti-inflammatory	Adesina, 1983
Phenols	Flavonoids Isoflavones Anthocyanidins[a]	Antioxidants	Markham, 1982 Harborne, 1991
Coumarins	Scoparone[a] Scopoletin[a] Umbelliferone Aesculetin	Anticonvulsants	Gupta *et al.*, 1980 Adesina, 1983

[a] Compounds confirmed in this study.

who consume both insects and resin—thus providing the tree with a pest control service. In addition, one can speculate that the primates receive a medicinal dosage of bioactive compounds and some energy from the sugars in the bark and gum.

Leaf, flower, and stem material is also recognized as an important component of the Sonso baboon diet, and is represented by approximately 10–15% of the washed fecal sample in the form of "fibrous wool" in every sample. This study was unable to identify the species of leaf fibers in these samples, and the focus in observation was upon seed ingestion. Table 4 lists the minor (small amounts ingested) species, their families and the plant parts consumed for those plants observed to be ingested, but for which no identification in the fecal samples could be made.

Forest Structure

The forest structure around the Sonso clearing as assessed in the sample plots measured in 1996 is heterogeneous. The Morisita Index of Similarity for the plots range from 0.16 to 0.99, with most index values around 0.5. The primary interpretation of the Morisita values is that the forest is diverse in species

Table 4. Leaf, fruit, and flower material observed to be consumed but unidentifiable in fecal samples

Species name	Family	Plant part consumed
Achyranthes aspera	Amaranthaceae	Flower spike
Aframomum sp.	Zingiberaceae	Root-bulb
Alternanthera nodiflora	Amaranthaceae	Leaf, flower
Alternanthera pungens	Amaranthaceae	Leaf, flower
Aneilema beniniense	Commelinaceae	Whole plant
Aneilema sp.	Commelinaceae	Flower
Bidens pilosa	Asteraceae	Young leaf, flower
Broussonetia papyrifera	Moraceae	Fruit, flower, bark, leaf-petiole, leaf
Casaeria engleri	Flacourtiaceae	Leaf
Cucumella engleri	Cucurbitaceae	Leaf, fruit
Cyathua cylindrica	Amaranthaceae	Leaf, flower
Datura sauvelens	Solanaceae	Young leaf
Desmodium repandum	Fabaceae	Young leaf
Desmodium canum	Fabaceae	Young leaf
Diplocyclos palmatus	Curcubitaceae	Fruit, flower (creeper)
Greenwayodendron sauveiolens	Annonaceae	Leaf
Gynura scandens	Asteraceae	Young leaf
Mallotus oppositifolius	Euphorbiaceae	Leaf
Melanthera scandeus	Asteraceae	Leaf (creeper)
Olyra latifolia	Poaceae	Leaf (grass)
Ouratea hiernii	Ochnaceae	Leaf
Physalis peruviana	Solanaceae	Leaf, fruit (creeper)
Piper umbellatum	Piperaceae	Flower, fruit
Rubus pinnatus	Rosaceae	Fruit
Solanum anguivi	Solanaceae	Fruit
Solanum welsitschii	Solanaceae	Fruit
Sorghum arundinaceum	Poaceae	Leaf, seed

composition, and that there is a moderate degree of patchiness or local concentrations for most species. The crown density measurements cluster between 90 and 96%, with only a few canopy cover measurements at the edge of clearings falling to the 60% level. The diversity and heterogeneity of the forest is consistent with its history of having been logged in the 1940s and 1960s. As in other studies, the diversity and productivity of logged forests is higher than that of unlogged forest (Chapman et al, 2000; Johns & Skorupa, 1987; Sheppard, 2000).

CONCLUSION

Baboon diets have been discussed extensively over the last half-century, but most of the focus has been on the variations and extent seen in savanna-dwelling populations. Forest-living populations have essentially been ignored owing to the great difficulties inherent in observation on the forest floor, combined with research foci on behavioral rather than ecological issues. In this study of the Sonso baboons, the range of food materials included in the species diet is expanded substantially, and a number of unusual processing sequences are noted in addition to the variations on plant parts utilized. The Sonso baboons incorporate a rather high proportion of barks, piths, and gums into their diet and, as noted in the results, appear to ingest a number of phytochemical compounds that might be bioactive. Whether or not these bioactive compounds should be categorized as "regular diet components," "diet supplements," or as "medicinal compounds" is indeterminable at present. More research is required.

Baboons in forest conditions dramatically expand the range and variety of food materials, thus confirming their opportunistic omnivore categorization. At the same time, the differences in their preferences for unripe fruits and a broad range of nutmeats, piths, barks, and grasses points up the evolutionary disruptive selection that has forced their diet to diverge from but remain in substantial overlap with that of humans.

ACKNOWLEDGMENTS

My thanks to the Uganda National Council for Research and Technology for research permits, and the President's Office for permission to work in Uganda; Andy Plumptre for initially inviting me to study the Sonso baboons and for providing his maps and tree species listing; Vernon Reynolds as principal investigator for information about the project and sharing his enthusiasm; Jeremy Lindsell for toleration and good humor in his takeover of an unusual administrative job; Sandy Paterson, Titia Achobo, Andama Kennedy, and Jacan Godfrey for their field assistance in observation; and Tinka John for aid in tree identification in the sample plots. The staff and technicians of the Makerere University Herbarium are also thanked for their expertise in identification of a number of obscure plants. All of the BFP staff are thanked for making our stays at Sonso comfortable. Also thanks to the University of Calgary Research Grants Committee for partial funding of this project.

CHAPTER SIX

The Interaction of Hormones with Ecological Factors in Male Budongo Forest Chimpanzees

Sally B. Seraphin, Patricia L. Whitten and Vernon Reynolds

INTRODUCTION

Behavioral endocrine studies have increasingly become a powerful tool for elucidating the ways in which specific hormones influence the expression of certain types of behavior. For instance, recent decades of research have implicated androgens in aggression, glucocorticoids in social dominance, gonadal steroids in courtship, and neuropeptides in social affiliation. However, no investigation of the behavioral endocrinology of wild animals is adequate without consideration of nonsocial and physiological–ecological, or ecophysiological, factors that may shape hormone secretions. While behavioral endocrine studies typically focus on the interaction between hormones and an organism's social environment and

Sally B. Seraphin and Patricia L. Whitten • Department of Anthropology, Emory University, 207 Anthropology Building, 1557 Dickey Drive, Atlanta, GA 30322 **Vernon Reynolds** • Emeritus, School of Anthropology, University of Oxford, 51 Banbury Road, Oxford, United Kingdom.

Primates of Western Uganda, edited by Nicholas E. Newton-Fisher, Hugh Notman, James D. Paterson, and Vernon Reynolds. Springer, New York, 2006.

behavior, ecophysiological analyses endeavor to reveal the mechanisms through which hormone secretions are patterned by temporal and spatial fluctuations of nonsocial features of the environment, such as flora and climate. By examining the relationships between androgens, glucocorticoids, and natural aspects of ecology, our field study addresses the extent to which chimpanzee (*Pan troglodytes*) hormone secretions are modulated by the nonsocial environmental factors of diet and forest temperature.

Hormones and Diet

One of the more significant ecophysiological factors influencing endocrine function in free-ranging animals is diet. Most notably, the plant estrogens or "phytoestrogens" found in fungi, berries, seeds, fruits, nuts, vegetables, and leaf matter have been linked to various reproductive functions in animals and humans, such as potentially influencing the synthesis, metabolism, and action of steroids along the hypothalamic–pituitary–gonadal axis (Whitten & Naftolin, 1998; Whitten & Patisaul, 2001). Macro and micro nutrients supplied through the diet can also modulate endocrine function in wild animals. Minerals, vitamins, fats, proteins, and sugars are of paramount importance for normal endocrine functions because they support overall health and vigor. Animals have developed neuroendocrine adaptations for achieving energy balance by physiologically monitoring their nutritional status. Specifically, neurons in the hindbrain can not only detect subtle nutritional deficits, but may also directly inhibit the secretion of gonadotrophins (luteinizing hormone [LH] and follicle-stimulating hormone [FSH]) through their projections to neurons in the hypothalamus that secrete corticotropin-releasing hormone, reducing sexual receptivity and culminating in nutritional infertility (Wade & Jones, 2004).

Because it is involved with controlling the excretion of salt in urine, the adrenal gland and the glucocorticoids it secretes play key roles in the regulation of sodium homeostasis (Lewicka *et al.*, 1998). In mammals, sodium excretion is additionally modulated by behavioral stressors that potentially increase plasma volume and raise blood pressure with chronic exposure through actions of the hypothalamic–pituitary–adrenal axis. A diet that is high in salt challenges the kidneys' ability to clear excess sodium. Combined with behavioral stressors that increase blood pressure, the additive effects of high salt consumption were demonstrated in one study of adult male baboons. Here, mean arterial pressure was not elevated over baseline after 5 months of high dietary salt (6 ± 5 mmHg) alone but was increased by an average of 17 (±3 SEM) mmHg after 5 months of a high salt diet plus the emotional stress of social conflict (Turkkan, 1994). In this study, "salt and stress resistant" or "salt and stress sensitive" animals could be further deciphered on the basis of their degree of pressure diuresis and natriuresis, urinary free cortisol (CORT), and a behavioral index of stress sensitivity (Turkkan, 1994).

As a part of the behavioral adaptations to acute stress, glucocorticoids are responsible for allocating blood sugars to expensive tissues, like the major skeletal muscle and the brain, enabling mind and body to focus on the stressor and execute the appropriate fight or flight response. That corticosterone (CCT) increases plasma glucose levels is supported by its effectiveness in battling hypoglycemia in rat neonates (Nagaya & Widmaier, 1993). In contrast to their reaction to diet or stress, hormones can themselves influence dietary preferences. Hyperglycemia decreases circulating levels of dehydroepiandrosterone (DHEA) and dehydroepiandrosterone-sulfate (DHEA-S), which are the major androgens secreted by the adrenal gland, independently of serum insulin level (Yamauchi *et al.*, 1996). Even at low concentrations, DHEA has been shown to cause taste aversion in Zucker rats (*Rattus norvegicus*) and its antiobesity effect, which is probably not mediated by the taste aversion, has been linked to a reduction of abdominal obesity, and the facilitation of insulin effects on glucose tolerance (Wright *et al.*, 1994). Also, rats that are supplemented with CCT consume more fat than those that are not, culminating in dramatic weight gain among previously lean rats through the remodeling of their macronutrient selection (Bligh *et al.*, 1993).

Fat and protein may influence the psychosocial parameters of an animal's stress physiology. For example, rats receiving a low protein diet are more hyperactive and less emotional than those receiving moderate and high protein diets (De Vicente *et al.*, 1991). The secretion of DHEA-S declines with increasing age in free-ranging male chimpanzees (Fontenot *et al.*, 2001). That adrenocortical function is influenced by dietary protein, as well as ageing, is supported by one study where a low (5%) protein diet similar to that of wild chimpanzees increased both plasma and adrenal CCT levels in 6-month-old male albino rats, but decreased the levels of adrenal CCT in 9-, 12-, and 18-month-old rats (Bandyopadhyay & Poddar, 1998). Micronutrient malnutrition can additionally influence the secretion of reproduction-enhancing steroids. For example, blood testosterone (TEST), semen ejaculate volume, and total seminal zinc loss per ejaculate are all sensitive to short-term zinc depletion in young men (Hunt *et al.*, 1992).

Fundamentally, food intake determines steroid synthesis, because the basic building blocks of gonadal and adrenal steroids are dietary cholesterols and glycerides in the form of vegetable oils and animal fats. Consequently, in primates as in many other species, food restriction is often associated with an acute suppression of reproductive hormones (Lado-Abeal, 1999). In many animals, feeding represents a Zeitgeber, or time keeper, regulating endocrine secretions in a manner similar to that of day length. For example, in men, spikes in the secretion of CORT sometimes coincide with the lunch and dinner hours (Goldman *et al.*, 1985). In a comparison of castrated and sexually viable male adult pigeons (*Columba livia*), the influence of androgens on courtship behavior was reciprocal to that for weight promotion (Pietras & Wenzel, 1974). This suggests that the processes by which androgens increase body weight, or vice versa, are not necessarily identical to those needed for triggering appetitive sexual behavior. Instead, it is likely that seasonal factors independently regulate increases in weight and testosterone, particularly in seasonally breeding species, which often experience dramatic fluctuations in the availability of food.

Hormones and Climate

Other nonsocial variables influencing neuroendocrine functions include factors that are related to climate, such as altitude, barometric pressure, temperature, weather (i.e., wind and precipitation), photoperiodicity (day length), and season. In a comparison of adolescent males living near the Dead Sea and in Amman the amounts of serum insulin and growth hormones were higher in the males living below sea level (Khraisha, 1990). Contrastingly, serum glucose and CORT levels were lowest among the males living above sea level (Khraisha, 1990). In addition to altitude and associated barometric pressures, the ecophysiological factors of extreme heat and cold can significantly modulate hormone secretions. In contrast to hypothermia (cold stress), hyperthermia (heat stroke), which seriously threatens cell survival, has a tendency to increase the biologically active form of TEST, in vitro (Obminski, 1998). While TEST tends to increase with heat, corticosteroids can increase or decrease in cold conditions. For example, plasma CCT increases with exposure to cold in rats (Bligh-Tynan *et al.*, 1993). In humans, a decrease in body core temperature is often accompanied by an increase in CORT (Hennig *et al.*, 1993). Finally, in a study of people swimming in winter around the southern Baltic Sea, CORT dramatically increased and mild elevations of serum glucose were observed following the stress of cold water immersion (Hermanussen *et al.* 1995). The proportion of biologically active or free CORT to inactive CORT, which is bound to corticosteroid-binding globulins, is directly related to ambient temperature and blood acidity—but in a different manner. In people, biologically active CORT has also been found to rise in response to hyperthermia (Obminski & Stupnicki, 1996). These significant alterations of CORT, in response to heat and cold, hint at its role in innate homeostatic mechanisms that exist to prevent extreme metabolic and somatic stress.

That season impacts physiology is well supported by circannual variation in the levels of circulating steroid hormones and the compartmental morphology of lymphoid tissues. Owing to the temperature dependence of membrane homeoviscosity, both the thymus and spleen decrease in size during the winter (Leceta & Zapata, 1985). Season dramatically influences the temporal patterns of gonadotrophins, prolactin, androgens, and CCT secretion in the rat (Wong *et al.*, 1983) by modulating hypothalamic functions through the influence of light. In mammals, production of melatonin by the pineal gland fluctuates with day length, keeping time with circannual as well as circadian rhythms, as these change throughout the year (Goldman & Nelson, 1993). Night and darkness cause the levels of melatonin to increase and this, in turn, dampens the secretion of gonadal and adrenal hormones. In contrast, light detected by the suprachiasmatic nucleus of the hypothalamus through its connections with the retina inhibits the secretion of melatonin, driving peaks in gonadal and adrenal hormone secretion during the early hours of the day. Much evidence for the circannual fluctuation of hormones has been obtained from seasonally breeding species, where TEST is higher in males during the annual mating season (copperhead snakes [*Agkistrodon contortrix*]: Schuett, 1996; Syrian hamsters [*Mesocricetus auratus*]: Vaughan *et al.*, 1994; white-tailed deer [*Odocoileus virginianus*]: Snyder *et al.*, 1983).

Changes in endogenous steroid secretion are moderated by temperature because ambient temperature is related to weather patterns and photoperiodicity, which together indicate the season. As ultimately related to day length, season, temperature, and weather are particularly important endocrine modulating factors in avian species. Here, they represent major factors regulating reproductive behavior, due to their combined influence on both the hypothalamic–pituitary–adrenal and hypothalamic–pituitary–gonadal axes (Wingfield, 1994; Wingfield *et al.*, 1996; Romero, 1998; Romero *et al.*, 2000). As the weather warms and breeding season approaches, seasonally reproducing male birds typically exhibit increasing levels of TEST and declining levels of glucocorticoids. In arctic-breeding passerine birds, weather conditions associated with season explain a great deal of the individual variation in the levels of CCT (Romero, 2000). Being often released in response to the somatic stress caused by inclement weather, CCT contributes to the physiological and behavioral changes that enhance thermal adaptations. For example, male white-crowned sparrows (*Zonotrichia leucophrys*) show increased CCT with decreasing temperature (Wingfield *et al.*, 1996). Testosterone enhances sexually competitive and courtship behaviors of males in most vertebrate species. For example, although not absolutely essential for territorial aggression to occur, high plasma TEST supports the reproductive activities of king penguins (*Aptenodytes patagonicus*) during the breeding season (Burger & Millar, 1980). Conversely, outside of the breeding season, in winter, both territorial and nonterritorial adult male penguins have extremely reduced levels of TEST (Burger & Millar, 1980).

In addition to circadian cycles, a circannual rhythm for the corticosteroids typically examined in relation to psychosocial stressors has been observed in young men (Touitou *et al.*, 1983). Reproductive competition in primates also involves important biological rhythms. According to Ewing (1982), "In both nonhuman primates and humans, photoperiod is an important impeller of seasonal reproduction, and seasonal reproduction is driven by a complex interaction between the animal and the social environment." The seasonal boost in TEST is particularly marked in primate species that only breed during particular times of the year. For example, male squirrel monkeys exhibit seasonal patterns in TEST secretion that parallel with breeding (Winslow & Miczek, 1988). Also, a seasonal rhythm of TEST that is independent of age has been observed for the male lesser mouse lemur (*Microcebus murinus*) (Aujard & Perret, 1998). In male rhesus macaques (*Macaca mulatta*), cerebrospinal fluid and plasma TEST dramatically increase during the reproductive season (Mehlman, 1997). Furthermore, in wild male rhesus macaques, the amount of appetitive and consummatory sexual behavior, TEST production, and spermatogenesis peak during the mating season, and later precipitously fall to pre-mating levels around the time of parturition (Ewing, 1982).

Not all nonhuman primates show an annual rhythm in their reproductive hormone profiles. Male langur monkeys (*Semnopithecus entellus*), for example, lack an annual reproductive hormonal rhythm, which is perhaps due to the increased male–male competition that typically arises as a result of female defense, or harem, polygyny (Lohiya *et al.*, 1998). Another example can be found in stump-tailed macaques (*Macaca arctoides*), where 28 months of data on 13 adult males living in a large male–female group has failed to show any seasonality in either plasma TEST, testis size, or body weight (Nieuwenhuijsen, 1987). Still, the interactions between social experience, sexual, and seasonal variables influence hormone levels in male primates. Demonstrating a dual influence of sex and season, isolated male lesser mouse lemurs (*M. murinus*) free of competition for access to females are characterized by a unimodal peak in seasonal TEST (Perret, 1985). Compared to these isolated males, pair-housed males that show a bimodal peak, in conjunction with the sexual activity of their female partners (Perret, 1985). One study of adult male rhesus monkeys showed that TEST rises following an introduction to females, regardless of season; but this elevation is more manifest during the breeding season (Bernstein *et al.*, 1977). Adult male rhesus monkeys housed in all-male social groups with only visual contact with females continue to show modest annual TEST rhythms that cannot be distinguished from the pattern observed in male rhesus monkeys that are part of male–female communities (Gordon, 1978). This fact supports the combined influence of seasonal and sociosexual modulators of endocrine function.

Despite the wealth of information, from human and other animal studies, the relationship between temperature and hormones is not entirely clear. For example, in one study of men, no significant changes in the levels of TEST and CORT were observed to occur in response to repeated periods of exercise in a warm environment (Hoffman *et al.*, 1997). Also, in a study on Norwegian soldiers performing military training in a cold, arctic setting, CORT and TEST responding were found to be moderate and generally transient in nature (Hackney & Hodgdon, 1991). Nevertheless, it can generally be said that extreme weather poses an intense somatic challenge to the survival and breeding success of free-ranging primates.

Seasonality in Chimpanzees

The most comprehensive studies on feeding behavior in wild chimpanzees have been conducted in Gombe National Park, Tanzania. Chimpanzees living in the

Gombe Reserve of Tanzania (*Pan troglodytes schweinfurthii*) consume a seasonally variable diet that is 45–76% fruit, 12–45% leaves, 1–18% flowers, 1–11% seeds, and 0–5% animal protein (Goodall, 1986). This, wild chimpanzee diet is largely vegetarian, as well as low in fat and protein. Chimpanzees living in the Sonso region of Budongo Forest, Uganda, are thought to consume a diet that is broadly similar to that of Gombe chimpanzees, eating 64.5% of fruit and 19.7% of arboreal leaves (Newton-Fisher, 1997, 1999b). Over 250 food items (Wrangham, 1977) have been identified for chimpanzees. Astonishingly, the pharmacodynamics of the vast majority of these foods is unknown. As a result, it is not possible to exactly predict relationships between hormonal secretions and the foods that are consumed by chimpanzees. Consequently, this represents an exploratory investigation on the potential impact of diet on the hormonal excretions of male chimpanzees in Budongo Forest. Sexual activity is slightly seasonal for chimpanzees living in Gombe National Park, Tanzania, and is probably more so for chimpanzees living in Budongo Forest (Wallis & Reynolds, 1999). Therefore, this represents an ideal chimpanzee population in which to study the relationships between hormonal secretions and natural ecological variables such as climate.

Predictions

Thus, we predicted that the general finding whereby CORT and CCT increase with decreasing temperature will be borne out in the male chimpanzees of Budongo Forest. Rising TEST is typically associated with an abundance of food and the onset of reproduction for seasonal breeders. Although convincing data on the relationship between DHEA/S and ecological factors are limited, we also predict that both of the androgens examined, DHEA/S and TEST, will increase with increasing temperature.

METHODOLOGY

Fifteen juvenile, subadult, and adult male Budongo chimpanzees (*Pan troglodytes schweinfurthii*) were selected as focal animals in this study. Fecal steroids were employed as an integrated measure of unconjugated cortical and gonadal steroid secretory activity over a 24-h period that lends itself to the noninvasive study of wild animals. Over 100 fecal samples were opportunistically collected from focal animals, between the 0730 and 1330 h, over 17 weeks

(124 days) spanning the months of April through August 2000, which is considered a "wet season" (Newton-Fisher, 1997). Soon after collection, each fecal sample was preserved by baking, as described by Seraphin (2000). Using a previously validated extraction method (Whitten *et al.*, 1998a,b), DHEA-S, TEST, CORT, and CCT were extracted from each fecal sample and then measured by radioimmunoassay, as described by Seraphin (2000). In the Budongo male population, average CORT was 0.457 ng/g (SD = 0.28 ng/g), average CCT was 1.55 ng/g (SD = 0.98 ng/g), average TEST was 2.27 (SD = 1.63 ng/g), and average DHEA-S was 28.16 ng/g (SD = 23.61 ng/g). A general trend wherein all four hormones fluctuated together, over the study period, was observed. In this regard, only the relationship between CCT and CORT were not significantly correlated, at either the 0.01 or 0.05 level, over the course of this 17-week investigation.

To test the relationship between climatological factors and hormonal secretion, information on the daily minimum, average, and maximum temperature was collected, using a standard outdoor thermometer, for each of the 124 days. Spearman's rho (ρ_S) was used to test for correlations between hormone levels in each individual fecal sample and the minimum, average, and maximum forest temperature from the day prior to its collection. To test the impact of foods on the excretion of hormones, data on the collective dietary composition of the 15 subjects were collected. These records were later compiled on the basis of the number of observations in which subjects consumed the food items, per week. The weekly average was calculated for the population, for each fecal hormone. Spearman's rho was also used to test for correlations between the weekly average hormone levels and the food items consumed during each week.

RESULTS

Analysis of Diet Information

The male chimpanzees of Budongo were witnessed consuming 28 different food items over the 17-week span of this study. One of these was meat obtained by hunting a small rat. Figs comprised 34.5% of this diet. Among these, *Ficus sur* (*capensis*) was the most frequently consumed species. Next in preference was *Broussonetia papyrifera*, which comprised 10.6% of the male diet.

When a weekly average level ($n = 17$) was calculated for each hormone and compared with the foods consumed over the course of each week, only one significant correlation was found. Average weekly CCT excretion was positively correlated with *Ficus barteri* consumption by Budongo male chimpanzees ($\rho_S = 0.560$, $P = 0.01$). This was primarily the result of increases in CCT during 2 weeks of intensive use. There was a near significant correlation of CCT ($\rho_S = 0.477$, $P = 0.053$) and TEST ($\rho_S = 0.477$, $P = 0.053$) with the consumption of *Myrianthus holstii*.

Analysis of Temperature Data

Over the 17 weeks during which Budongo Forest temperatures were monitored, the minimum temperatures ranged from 15 to 18.5°C, the maximum temperatures ranged from 23 to 32°C, and the average temperatures ranged from 20.25 to 24.75°C. When each fecal hormone sample was compared with forest temperatures from the day prior to its collection, five significant correlations were observed. CCT ($\rho_S = 0.233$, $P = 0.02$), TEST ($\rho_S = 0.328$, $P = 0.001$), as well as DHEA-S ($\rho_S = 0.311$, $P = 0.002$) were positively correlated with average temperature. Also, the minimum temperature positively correlated with DHEA-S ($\rho_S = 0.279$, $P = 0.005$) and the maximum temperature positively correlated with corticosterone ($\rho_S = 0.235$, $P = 0.018$).

CONCLUSIONS

For the last couple of decades, primatologists have grappled with the potential impact of dietary quality on broad social organization and degree of sociality in primate communities. In particular, the dramatically different behavioral ecologies of bonobos (*Pan paniscus*) and common chimpanzees have been explained on the basis of bonobos' largely subsisting on rich terrestrial herbs that are densely distributed, thus decreasing food competition between group members (Wrangham, 1986). Our data suggest that, in addition to their nutritional quality and spatial distribution, the foods that are consumed by primates may affect their social behavior by modulating their endocrine secretions. A definite relationship between the consumption of products from one tree species (*Ficus barteri*) and CCT excretion was discovered through this exploratory investigation on the potential influence of dietary components on hormone secretion in male Budongo chimpanzees. CCT has the effect of increasing plasma glucose levels. Consequently, it is considered clinically effective for battling hypoglycemia (Nagaya & Widmaier, 1993). Figs have been described as fall-back foods for many chimpanzee communities (Wrangham, 1986). However, they may be considered as staples of the Budongo Forest chimpanzee diet (Newton-Fisher, 1997, 1999b). Perhaps the consumption of more figs is related to metabolic adaptations to food shortage, such as the mobilization of protein and glucocorticoid secretion, as suggested in this study. Conversely, the significant correlation between *F. barteri* consumption and CCT could have arisen as an artifact of the coincidental fruiting of this tree and the occurrence of some other environmental factor(s) shaping CCT secretion, such as temperature. Only a study on the pharmacodynamics of this plant could confirm its bioactivity with regard to CCT excretion in chimpanzees. It is interesting, however, that *F. barteri* consumption did not correlate with levels of CORT, the other glucocorticoid that we examined. In addition to curbing urinary zinc and chromium losses, carbohydrate loading dampens CORT responding in exercising men (Anderson *et al.*, 1991). In this light, one could speculate that the levels of CORT would not increase when less salty, but more sugary, foods are consumed by male chimpanzees. However, with regard to the different correlations observed for CORT and CCT, it is important to note that both endocrine measures have the potential to reflect metabolites of cortisol, rather than the hormones themselves.

Salt retention is also enhanced by CORT (Clore *et al.*, 1988). Dietary sodium must be a limited element in the diet of mainly frugivorous chimpanzees such as those in Budongo. Although insignificant, there was the suggestion that CORT rises in response to the consumption of meat. Although reports on the relationship between CORT secretion and sodium intake diverge, urinary CORT excretion is generally said to decrease with sodium restriction (Lewicka *et al.*, 1998). Thus, the subtle and insignificant increase in CORT could have occurred as a consequence of hunting itself and might not be perfectly correlated for this reason. Certainly, the many questions left unanswered by our investigation warrant the continued monitoring of Budongo chimpanzee diets over a more extended period of time.

The prediction that CORT and CCT would increase with decreasing temperature in the Budongo Forest was not satisfied by our results. That CCT levels instead correlate with maximum and average temperature came as a surprise, in light of the literature previously discussed. However, the prediction that TEST would increase with increasing temperature, as is typically associated with an abundance of food and the onset of breeding, was supported by its significant correlation with maximum temperature. That the adrenal androgen DHEA-S can increase with increasing average temperature supported our prediction that this will rise in synchrony with ambient temperatures. Despite the expected seasonality of copulations in Budongo, which were low between the months of May and August (Wallis & Reynolds, 1999), this study seems to have captured a slice of the temporal fluctuations in androgen secretion. However, the correlation coefficients were low and can only be said to explain a small percentage of the variance between these ecophysiological factors. Once again, these results beg for an expanded study, where samples are collected for a considerably longer period. For example, the brief duration of this study makes the data collected herein generally inadequate for testing whether the amplitude of DHEA-S experienced by chimpanzees is more circannual than circadian in nature, as is the case with elderly women. Others should address this question, in addition to that concerning the pharmacodynamics of foods consumed by chimpanzees, in the future. Nevertheless, it is fair to conclude that our investigation duly demonstrates the significance of interactions between ecophysiological factors, such as food and forest temperature, and the secretion of gonadal and adrenal hormones in wild common chimpanzees.

CHAPTER SEVEN

The Role of Diet in Self-Medication Among Chimpanzees in the Sonso and Kanyawara Communities, Uganda

Paula Pebsworth, Sabrina Krief, and Michael A. Huffman

INTRODUCTION

With mounting evidence, the idea that primates obtain medicinal benefits from plant ingestion (e.g., Wrangham & Nishida, 1983; Huffman & Seifu, 1989; Wrangham & Goodall, 1989; Huffman *et al.*, 1993; Wrangham, 1995; Huffman, 1997; Huffman & Caton, 2001) is gaining acceptance among primatologists. The medicinal component of a plant is found in its secondary

Paula Pebsworth • ppebsworth@mac.com **Michael A. Huffman** • Primate Research Institute, Kyoto University, Kanrin 41-2, Inuyama, Aichi 484-8506, Japan **Sabrina Krief** • Eco-anthropologie et Ethnobiologie—Chimie des Substances Naturelles, Muséum National d'Histoire Naturelle, 43, rue Buffon, 75231 Paris, France.

Primates of Western Uganda, edited by Nicholas E. Newton-Fisher, Hugh Notman, James D. Paterson, and Vernon Reynolds. Springer, New York, 2006.

compounds, which taken in large doses are often toxic to most animals. They also reduce the palatability or digestibility of the plant. Scientists have discussed food selection according to the presence of secondary compounds in their diet and how animals cope with them while in search of food (Glander, 1975, 1982; Hladik, 1977a,b; Oates *et al.*, 1977, 1980; McKey, 1978; Milton, 1979; Wrangham & Waterman, 1981a,b). But it was Janzen (1978) who first suggested that animals' ingestion of plants rich in secondary compounds might actually help them fight pathogens and parasites. Subsequently, it was proposed that these compounds may also improve the reproductive fitness of an individual (Hart, 1990; Holmes & Zohar, 1990), and lessen the many diseases caused by parasites (cf. Allison, 1982; Toft *et al.*, 1991). Why primates ingest these secondary compounds has sparked interest in the fields of ethology, pharmacology, and parasitology, to name a few, and opened the door to the field of zoopharmacognosy (Rodriguez & Wrangham, 1993) also commonly referred to as primate self-medication (Huffman, 2007).

The term *zoopharmacognosy* was coined after evidence appeared supporting the idea that self-medication among primates existed. The basic argument is that animals exploit plant secondary compounds or other nonnutritive substances for curative purposes. In the field of primatology, chimpanzees have provided more evidence of self-medicative behaviors than any other primate species. Two types of self-medication behavior have been described in detail. One involves ingestion of an item rare to the diet and/or of little nutritional value (e.g., leaf swallowing, bitter pith chewing). Use of these plants tends to be restricted to certain seasons on the basis of reports to date, in particular when parasite reinfection is greatest. The individual ingests the plant item when infected with parasites and/or is showing related signs of illness (e.g., Huffman & Seifu, 1989; Wrangham, 1995; Huffman *et al.*, 1996). In some cases it has been shown that subsequent to the ingestion of the plant, the individual recovers from symptoms associated with the illness and/or expels the parasites in question (e.g., Huffman & Seifu, 1989; Huffman *et al.*, 1993, 1996). The second type of self-medicative behavior includes the ingestion of plants that are more common to the diet, but are also used ethnomedicinally or have demonstrated biological activity, suggesting a medicinal component. Huffman and colleagues (Huffman, 1997; Huffman *et al.*, 1998) proposed the term *medicinal foods*, borrowing the concept of food as medicine in traditional human societies (e.g., Etkin & Ross, 1982).

The following descriptions all fall into the first category of self-medicative behaviors. Whole leaf swallowing was the first such behavior reported by Wrangham (1977) and then later described in detail by Wrangham and Nishida (1983). These two researchers found folded, undigested leaves of *Aspilia mossambicensis* (Oliv.), *Aspilia pluriseta* (O. Hoffm.), and *Aspilia rudis* (Oliv. & Hiern) in chimpanzee feces at Gombe and Mahale. It was noted that the chimpanzees did not masticate the leaves but instead carefully placed them in their mouths one at a time, folding them with their tongue and palate and then swallowed them whole. This type of consumption suggested that ingestion of the leaves incurred no nutritional benefit, and so a medicinal or curative function was suspected (Wrangham & Nishida, 1983). Later, a phytochemical hypothesis based on the reported presence of thiarubrine A in *Aspilia* spp. leaves was proposed (Rodriguez *et al.*, 1985). Evidence for this hypothesis based solely on the presence of thiarubrine A in *Aspilia* spp. leaves providing strong nematocidal activity were not replicated by others (Huffman *et al.*, 1996, 1997; Page *et al.*, 1997). Moreover, cross-site comparisons of chimpanzees (Wrangham & Goodall, 1989; Huffman & Wrangham, 1994) and an extensive multiple ape species comparison (Huffman, 1997) revealed that many different plant species were being used in leaf swallowing by great apes across Africa. Huffman and colleagues (Huffman & Wrangham, 1994; Huffman *et al.*, 1996; Huffman, 1997) first realized that the leaves of all different species consumed in this manner have one peculiar trait in common: they are rough, and the surfaces are covered with bristly trichomes. At Mahale, a consistent pattern for the expulsion of parasites (the live adult worms of *Oesophagostomum stephanostomum*) along with leaves swallowed whole, was recognized by Huffman in the 1993–1994 rainy season at Mahale (Huffman *et al.*, 1996, 1997), while Wrangham at Kibale reported the relationship between the expulsion of proglottids of *Bertiella studeri* with whole leaf swallowing during a period of high tapeworm infection in 1993 (Wrangham, 1995). In addition, at Kibale, the chimpanzees swallow the rough leaves of *Rubia cordifolia* without chewing them: experiments conducted on different stages of nematodes from the genus *Strongyloides* showed that the leaves' extract had no effect on their motility, supporting the hypothesis of a physical effect via leaf-swallowing (Messner & Wrangham, 1996). On the basis of the observations, it was hypothesized that the leaves were consumed to flush the intestinal tract of nematodes or tapeworms, keeping infections at manageable levels (Huffman *et al.*, 1996). Leaf swallowing typically occurs in the early morning prior to eating or otherwise on an empty stomach (Wrangham & Nishida, 1983; Huffman *et al.*, 1997). A detailed analysis of the rapid time (6 h compared to the typical 30–40 h passage time) in which these unchewed leaves were passed through the gastrointestinal tract prompted Huffman and Caton (2001) to propose that consumption of these types of leaves on an empty stomach stimulate rapid gut motility, which flushes out the parasites.

A second example of this first type of self-medicative behavior was seen while following habituated chimpanzees in the Mahale Mountains. Huffman and Seifu (1989) opportunistically came across an adult female whom they observed to deliberately seek out and ingest the young pith of a tree, *Vernonia amygdalina*, commonly known as bitter-leaf. The chimpanzee bent down several shoots and meticulously stripped away the leaves and outer bark, revealing the inner pith, which she chewed and sucked for approximately 20 min. Further, detailed behavioral observations showed that she was unable to keep up with the group, lacked appetite, and her urine was darker than normal and stools were loose. Twenty hours after consumption, they were able to verify that she no longer showed any of these signs of illness from the previous day (Huffman & Seifu, 1989). A few years later, these observations were further supported by subsequent, more detailed observations of another female chimpanzee at Mahale (Huffman *et al.*, 1993). At this time, longitudinal parasitological studies were underway (Huffman *et al.*, 1997) and it was shown that, after ingesting the bitter pith of *V. amygdalina*, a significant drop in the parasite load of *Oesophagostomum stephanostomum* had occurred. This was accompanied by recovery from the visible symptoms of ill health (lack of appetite, malaise, diarrhea) within 24 h. In vitro the plant has demonstrated medicinal value, with activity noted against the parasites responsible for malaria, schistosomiasis, amebic dysentery, and leishmaniasis (Toubiana & Gaudemer, 1967; Kupchan *et al.*, 1969; Asaka *et al.*, 1977; Gasquet *et al.*, 1985; Jisaka *et al.*, 1992, 1993; Ohigashi *et al.*, 1994).

Another putative self-medicative behavior seen in chimpanzees is the consumption of soils from termite mounds (Mahale, Gombe, Budongo) and from other specific places as natural holes and root masses of fallen trees (Kanyawara, Budongo). This behavior, called geophagy, was first shown in chimpanzees to provide low mineral intake compared to other chimpanzee food by Hladik and Gueguen (1974), and then suggested as a possible means of detoxifying secondary compounds present in the diet (Hladik, 1977a,b).

Mahaney *et al.* (1996a,b) later suggested that soil consumption might also be beneficial to individuals suffering from intestinal discomfort, for example associated with parasite infections. Mahaney *et al.* (1996a,b, 1997) found that the soils consumed by chimpanzees in the Mahale Mountains and the Kibale Forest contain a type of clay comparable to kaolinite, not unlike Kaopectate™, a popular stomach medicine. Kaolinite can allay gastrointestinal upset, adsorb toxins and bacteria (Aufreiter *et al.*, 2001), and form a protective coating along the gastrointestinal tract (Johns, 1990; Mahaney *et al.*, 1996a,b, 1997; Stambolic-Robb, 1997). In addition, Ketch *et al.* (2001) demonstrated that the soils that chimpanzees selected for consumption were qualitatively different from other soils present in the chimpanzees' habitat. Soils selected by chimpanzees contained a higher proportion of soil microorganisms that produce antimicrobial and/or antiparasitic properties. One organism found was a filamentous bacteria (actinomycetes), which accounts for 75% of all known medicinal antibiotic compounds (Kutzner, 1981; Ketch, 2001). The second type of behavior can be illustrated by the consumption of a food item with interesting pharmacological significance such as the bitter-tasting berries of *Phytolacca dodecandra* or leaves of *Trichilia rubescens.* The Kanyawara group of chimpanzees at Kibale frequently eats these berries that are known to contain at least four toxic triterpenoid saponins (lemmatoxin, lemmatoxin-C, oleanoglycotoxin-A, phytolacca-dodecandra glycoside) capable of controlling schistosomes (Kloos & McCullough, 1987; Abbiw, 1990). The berries also possess antiviral, antibacterial, antifertility, spermicidal, and embryotoxic properties (Kloos & McCullough, 1987). In addition, compounds with highly significant in vitro antimalarial activity, trichirubines A and B, have been isolated and identified from *Trichilia rubescens* leaves, following the observation of unusual feeding behavior of Kanyawara chimpanzees: usually only one individual of the party selects and eats a few leaves from a young tree, even when more trees are present and available for itself and other individuals of the party (Krief *et al.*, 2004).

These are just a few examples of the potential health benefits from consuming plants with known medicinal properties already underlining the importance of comparing habits from different communities to highlight similarities and specificities related to potential local behavioral traditions in plant use. The aim of this chapter is to elucidate the potential medicinal value of dietary items consumed by the chimpanzees at two Ugandan sites, the Budongo Forest Reserve and Kibale National Park, and to discuss their possible roles in health maintenance. Demonstrating why chimpanzees consume the foods they do is difficult, but a solely nutritional role for foraging should not be assumed a priori. One way to identify and better understand the possible therapeutic benefits from plant ingestion is to document the foods consumed and the state of health of individuals before and after they consumed them. We review hereafter ethnomedicinal uses of plant parts eaten, which might provide interesting information from the empirical wisdom of indigenous people. A compilation of such data over a 9-month period in Budongo and a 5-month period in Kibale attempts to shed light on how the diet of chimpanzees might help to keep pathogens and parasites at manageable levels. We also preliminarily explore the possibility that even though the flora composition overlaps at these two sites, each community may have unique medicinal cultures.

METHODOLOGY

Study Sites

The Budongo Forest is approximately 428 km^2 and is described as a moist semi-deciduous tropical forest. Vegetation types include secondary mixed forest, swamp forest, and wooded grassland that lies between latitudes 1°35′–1°55′ N and longitudes 31°18′– 31°42′ E in the Bunyoro District of Western Uganda (Eggeling, 1947; Synnott, 1985; Plumptre, 1996). The Sonso community study site was established in 1990 by Vernon Reynolds and the Budongo Forest Project staff (Reynolds, 1992). The forest is drained by the Sonso and Waisoke rivers, which flow into Lake Albert. This study was carried out by PP from February to October 1998. During this time the total rainfall was 1845 mm. The rainiest months were April–May and September–October. Altitudes range from 910 to 1100 m above sea level. Mean annual minimum temperatures range from 17 to 20°C to maximum temperatures of 27–29°C.

The Kibale National Park covers 766 km^2 located in the Kabarole district of Western Uganda, between 0°13′–0°41′ N and 30°19′–30°22′ E. The area lies between an elevation of 1300 and 1500 m, and the rainfall averages 1700 mm per year. Vegetation of this midaltitude moist forest also includes secondary forest, grassland, swamp, Eucalyptus and pine plantations, and elements of lowland tropical rainforest. Mean daily temperatures range between 14.9 and 20.2°C and rainy seasons occur from March to May and from September to November (Chapman & Chapman, 2004). The study conducted in the Kibale Chimpanzee Project by SK comprised a 3-month period in the dry season (December–February 2001) and two 1-month periods in the rainy season

(October 2001 and October 2003) of observation of the Kanyawara community. Facilities at the study site of Kanyawara were provided by Makerere University Biological Station.

Fecal and Urine Analysis

In the field, urinalysis can be an important noninvasive tool available to researchers when monitoring the health status of an individual (Kaur & Huffman, 2004; Kelly *et al.*, 2004).

Sonso

A total of 299 fecal samples were collected primarily from 14 known adult individuals and stored in three preservatives when possible. Immediately after defecation, the fecal sample was examined macroscopically for presence of whole leaves and proglottids; the state of the feces (firm, soft, or diarrheic) was documented. A representative sample free from soil was then collected and stored individually in 5.0-ml sterile Corning^{TM} vials. In camp, vials and feces were weighed and 1-g samples were fixed within 3 h of collection. The primary preservative was 10% neutral formalin; secondary preservatives were polyvinyl acetate and Proto-fix. The contents were mixed and stored in a cool dark room. The samples were later analyzed microscopically by Alpha Tec, Inc., and Dr. S. Gotoh at the Primate Research Institute, Kyoto University. Dr. Gotoh also measured parasitic load via the MGL (formalin ether sedimentation) and MacMaster techniques (expressed as eggs/g feces [EPG]). A few samples were examined on site by the MacMaster flotation method using zinc sulfate and direct examination. Owing to time constraints, expertise level, and field conditions, it was not possible to examine on site all samples collected.

Some parasites can be determined from microscopic examination; others, like *Oesophagostomum* eggs, are difficult to distinguish from hookworm eggs. Without expertise, the eggs need be cultured and examined at the larval stage, which is morphologically unique (Krepel, 1994). Twenty of the 299 samples were cultured using the Harada–Mori technique (Harada & Mori, 1955). Of these 20 samples, 10 were analyzed for the presence of *Oesophagostomum* by the laboratory of Prof. Ton Polderman and Coby Blotkamp from the Department of Parasitology, Leiden University, The Netherlands.

Urine was analyzed opportunistically to detect potential illness. This was done using urinalysis reagent strips (Roche Chemstrip 9©) that tested the following parameters: glucose, bilirubin, ketones, specific gravity, blood, pH, protein, nitrites, and leukocytes. The urine was pipetted off the surface of leaves; it was never collected when possibly contaminated with feces or soil. The analysis was performed while in the field. A total of 15 samples were collected from 7 females and 3 males.

In humans, highly elevated levels of leukocytes may signal a urinary tract infection, kidney infection, cystitis, or urethritis (e.g., Pfaller *et al.*, 1987; Pezzlo, 1988). Normal levels range from 0 to 10 leu/μl. A pathological condition is thought to occur when levels are greater than 20 leu/μl.

Kanyawara

A total of 252 fecal samples from 38 known chimpanzees were collected, consisting of 187 samples collected during the dry season and 65 during the rainy season; 127 samples came from 18 females, and 125 samples came from 20 males. Methods are detailed in Krief *et al.* (2005a,b,c). They differed from those used in Sonso as MacMaster flotation was performed on fresh material, using $MgSO_4$. Two grams of each sample, stored in 18 ml of 10% formalin, were also analyzed by direct examination and diphasic ether–formalin concentration to search for rare eggs. Protozoan cysts were searched for in 0.5 g of feces fixed in merthiolate-iodo-formalin (MIF staining). Then, according to stool consistency, parasite loads as counted by the MacMaster method and direct examination were corrected by multiplying the count by a coefficient of 2 if the dung was soft and pasty and 3 if it was diarrheic or liquid (Hercberg *et al.*, 1986). These counts were called the "corrected parasite load."

Fresh urine from chimpanzees in trees was collected either on the concave surface of a plastic bag or by pipetting the urine off the surface of leaves as described in Krief *et al.* (2005b). Urine samples, when not contaminated by feces or soil matter, were stored in a clean dry container and analyzed immediately on returning to the field station. The samples were tested with commercial dry reagent dipsticks (Multistix 10 SG Bayer©) for 10 parameters, including those listed for the Sonso community plus urobilinogen. The identity of the chimpanzee, date and hour of collection were noted. Urine obtained when the chimpanzee urinated from its night nest tends to be highly concentrated and is thus the most likely to be diagnostic of abnormality. Place and means of collection (e.g., leaf, pipeting, urine-stick use), amount collected, macroscopic aspects of urine such as color or turbidity, and presence of crystals were considered as potentially useful information. The analyses performed on 76 urine samples from 32 chimpanzees, including 45 samples (21 from males and 24 from females) collected during the dry season and 31 (13 from males and 18 from females) collected during the rainy season are described in Krief *et al.* (2005b).

Behavioral Observations and Plant Consumption

Sonso

The Sonso community comprised 52 individuals; of these, 14 adult chimpanzees were followed, 7 females and 7 males. Focal-animal and ad libitum behavioral observations were made. Chimpanzees were followed as long as possible from the time they left their night nests (352 h). The level of habituation at this time did not permit dawn-to-dusk follows. All behaviors were noted using a continual scan method. Health documentation included respiratory, digestive, reproductive, locomotive, and urine functions. Also documented were any signs of illness or injury, to include wounds, snare injuries, decreased appetite, sneezing, coughing, nasal or eye discharge, and level of activity (Huffman *et al.*, 1997). Fecal and urine samples were collected from known individuals whenever possible.

When documenting feeding activities, all plant and nonplant items consumed were noted and samples collected when possible; the location the feeding activity took place was also documented. A total of 33 plant items from 28 plant species were collected. Chimpanzees were also observed to feed on soil from termite mounds. Four samples were collected and sent for analysis to Mahaney and colleagues of the Geophagy Research Group at York University (see Tweheyo *et al.*, Chapter 8, this volume).

Kanyawara

The Kanyawara community of chimpanzees comprised 50 individuals, well habituated to the presence of the observers on the ground at a distance of 5–10 m. Chimpanzees have been monitored daily since 1987 by the Kibale Chimpanzee Project team directed by Richard Wrangham. In June 1999, 10 adult males, 2 adult females without offspring, and 14 mothers with 22 dependants (10 females and 10 males, 2 young infants of unidentified sex) were counted in the Kanyawara community.

Observations were conducted from dawn to dusk when possible. The focal subject was observed for a 10-min period to estimate activity budgets and diet.

The target was changed every 10 min whenever possible. During this study, 450 h of observations were collected in the dry season and 195 h during the wet season.

In addition, ad libitum observation allowed accurate recording of particular sequences related to possible self-medication. Attention was focused on the diet of the identified chimpanzees; all items ingested were recorded in detail. Veterinary work consisted of daily clinical observations (respiratory, digestive, reproductive, locomotive, and urinary functions), looking for clear signs of probable illness such as decreased appetite, long and frequent resting, sneezing, coughing, or intestinal disorder, as described by Huffman *et al.* (1997). Urinalysis and intestinal parasite evaluation were carried out. Whenever possible, feces and urine were collected from all known individuals.

RESULTS

Fecal Analysis

Sonso

During the study, 299 fecal samples were collected. Of these, 100 were analyzed by Alpha-Tec, Inc. These results are summarized in Table 1. They identified seven different species of helminthes: *Anoplocephala* (tentative identification) (12%), *Strongyloides* (65%), *Ternidens* (20%), *Necator* (49%), *Trichostrongylus* (2%), and two that could not be identified. One species of protozoa, *Troglodytella* (76%), was also found.

From the 44 samples analyzed by Gotoh, three parasite species were identified: *Strongyloides* (84%), *Oesophagostomum* (23%) and *Troglodytella* (93%). Among the 20 coprocultures prepared, 10 were analyzed by Polderman and Blotkamp. All contained *Oesophagostomum* third-stage larvae (L3) regardless of sex, age, or month the sample was collected. The samples identified as *Ternidens* by Alpha-Tec are most likely *Oesophagostomum*.

Kanyawara

During both dry and rainy seasons, the mean parasite count by direct examination was low (96% of the samples contained less than 1000 helminthes/g of feces), uniform and not significantly different (301 [$n = 187$] and 197 [n = 65], respectively, ns). These results are summarized in Table 1 and presented by Krief

Table 1. Summary of fecal and urinalysis for both communities

Parasite species	% Fecal samples infected				Abnormal urine value	
	Kanyawara chimpanzees		Sonso chimpanzees		Kanyawara chimpanzees	Sonso chimpanzees
Method	MacMaster ($MgSO_4$)	DE	DE		Reagent strips	Reagent strips
No. of samples	239	247	144	No. of samples	76	15
No. of chimpanzees	38	38	14	No. of chimpanzees	32	11
Trichuris sp.	1.7	0.4		B	6	4
Strongyloides fulleborni	15.5	8.9	69	L	14	2
Oesophagostomum sp.	55	30.7	7	N	2	1
Probstmayria sp.	0.8	10.5		pH	2	
Necator americanus			49	P	1	
Ternidens deminutus			31	K	1	
Unidentified nematode	26.3	64	41	B & L	14	3
Bertiella studeri	0	2.8	12	B & N	3	1
Troglodytella abrassarti	0	66	80	L & N	2	1
Small entodiniomorphs	0	23		L & P	2	
				N & K	1	
				U & B	1	
				B, L, & N	1	1
				U, B, & N	1	1
				L, N, P, & G	1	

DE = direct examination; B = blood: G = glucose; K = ketones; L = leukocytes; N = nitrites; P = proteins; U = urobilinogen. Kanyawara data from Krief *et al.* (2005b).

et al. (2005b). Nevertheless almost all of the samples and the individuals were parasitized regardless of the method used. On the other hand, *Oesophagostomum* sp. and *Strongyloides fulleborni* were commonly observed and *Trichuris trichiura* and *Bertiella studeri* were rarely found (Krief *et al.*, 2003) as detected previously by Ashford *et al.* (2000). Among protozoa, two species of entodiniomorph ciliates were detected by direct examination: *Troglodytella abrassarti*, the more common protozoa, and a "small entodiniomorph," likely the same one described previously in studies in Kibale (Ashford *et al.*, 2000), Gombe (File *et al.*, 1976), Mt. Assirik (Mc Grew *et al.*, 1989) and La Lope (Landsoud-Soukate *et al.*, 1995).

Urine Analysis

Sonso

From August to October 1998, 15 urine samples were analyzed using a urinalysis reagent strip. These results are summarized in Table 1. This was not the primary objective of the study, and only a limited number of samples were obtained. Samples could only be collected when the chimpanzees were feeding low in the canopy as the urine was pipetted off the surface of leaves. Many samples had to be discarded as the fruits being consumed turned the urine bright yellow, causing the blank to fail. Of the samples analyzed, 13% gave a negative result for all nine parameters; both individuals were males. Forty percent of the samples, all from females, tested positive for leukocytes. Of those individuals that tested positive, only one had an activity budget that seemed abnormal. The sample obtained from KG on August 18, 1998, contained in excess of 500 leu/μl of urine. This chimpanzee slept more than 3 h in a 6-h focal; 1.5 h were spent grooming her son and the remaining time was spent feeding and moving only a short distance to obtain ripe fruits. Thirteen percent of the samples tested positive for nitrites. Only one chimpanzee had a urine pH of 7, all others had values of 8 or 9. One chimpanzee, KY, tested positive on 19 August 1998 for 3 parameters: leukocytes, nitrites, and hemoglobin. We estimate that this chimpanzee was approximately 4 months pregnant.

Kanyawara

As described in Krief *et al.* (2005b) leukocytes and blood were found respectively in 45 and 34% of the samples ($n = 76$) and were often associated together.

Fifty-three percent of the samples (cycling females excepted) had an abnormal value for at least one parameter. Half of the female samples were positive for blood versus 15% of the male samples ($P = 0.002$) but samples from noncycling females were also more often positive than samples from males ($P = 0.02$). Leukocytes were significantly more frequent in cycling females than in noncycling females. All but two urine samples had an alkaline pH (>7) (Table 1).

Plant Consumption

Sonso

From February to October 1998, the Sonso community of chimpanzees consumed 48 plant items from 41 plant species. Throughout the study period, fruits were the dominant food source consumed, with 21 species eaten corresponding to 58.5% of total times in the feeding budget; 10 species of leaves were consumed at 20.7%, seeds (two species) were ingested at 6.9%, and flowers (five species) at 3.6% (Table 2). *Ficus sur* fruits were so preferred during this study period that chimpanzees consumed them every month and at all stages of ripeness. Even during the month of October, when the figs were unripe, 16 observations were made of fruit consumption.

Kanyawara

Kanyawara chimpanzees consumed 46 plant items from 35 plant species during the study. Fruits were the dominant food consumed, with 19 species eaten corresponding to 81% of total time in the feeding budget. Leaves (16 species) were ingested at 15% and stems (seven species) at 3.5%. *Ficus natalensis* fruits were the most common food item in Kanyawara during the study period (Table 2).

Medicinal Plants Consumed

During the course of this study, the chimpanzees from both sites consumed a total of 69 different plants, all but 9 identified to species level (Table 2). Of these, 24 species are unique to Budongo, and 11 to Kibale, with 34 species found at both sites. Surprisingly, of the species present at both sites, the chimpanzees from both communities consumed only five items (14.7%) in common during this time period; the fruits of *F. sur* (= *capensis*), the leaves of *Ficus exasperata*,

Table 2. Plants consumed by the Sonso and Kanyawara chimpanzees during the studies

Plant	Part ingested in Sonso	% of feeding time	Part ingested in Kanyawara	% of feeding time
Albizia grandibracteata			B[a]	
Acanthus pubescens	FL[a]	0.2	St[a]	2.1
Afromomum sp.	P	1.1	F, St	0.5
Alstonia boonei	B[a]	0.2		
Antiaris toxicaria	F	0.5	L[a]	<0.5
Balsamocitrus dawei	F	0.2		
Broussonetia papyrifera	L, FL	19.4		
Celtis africana			L	11.7
Celtis durandii	L[a]	0.3	F, L[a]	8.9
Celtis mildraedii	L	5.2		
Celtis wightii	L	1.3		
Celtis zenkeri	L	3.1		
Chaetacme aristata			F, L[a]	<0.5
Chrysophyllum albidum	F	0.5		
Chrysophyllum perpulchrum	F	0.2		
Chrysophyllum gorungosanum	F	0.2		
Cleistopholis patens	F	0.3		
Cordia abyssinica			F	12.6
Cordia africana			F	<0.5
Cordia millenii	F	0.3	F	1.5
Costus sp.	P[a]	0.2		
Crassocephalum bojeri			L[a]	<0.5
Cynometra alexandri	S, B[a]	7.3		
Despatsia dewevrei	F	1.1		
Dialium excelsum	F	0.2		
Ekebergia senegalensis	FL	0.2		
Epiphytes		0.2		
Eucalyptus sp.			B	<0.5
Ficus asperifolia			F, L, St	<0.5
Ficus barteri	F	0.3		
Ficus brachylepsis			F	13.1
Ficus cyathistipula			F, L	<0.5
Ficus dawei			F	11.2
Ficus exasperata	L[a], B	3.9	F, L[a]	2.2
Ficus mucuso	F	4.7		
Ficus natalensis			F[a]	18.4
Ficus ottonofoli			F	2.5
Ficus polita	F	0.2		
Ficus sansibarica	F	0.8		
Ficus saussureana	F	0.2		
Ficus stipulifera			F	1.7

Table 2. *(Continued)*

Plant	Part ingested in Sonso	% of feeding time	Part ingested in Kanyawara	% of feeding time
Ficus sur (= F. capensis)	F[a], B	33.8	F[a]	0.6
Ficus thoningii			F	
Ficus varifolia	L	0.5		
Illigera pentaphylla			F	0.6
Jasminum sp.			L	<0.5
Khaya anthotheca	B[a]	0.2		
Laciodiscus mildbraedii	L	0.6		
Lannea welwitschii	FL, F	0.2		
Lepistemon owariense			L	1.1
Marantocholoa sp.	P	0.3		
Markhamia platycalyx			B[a]	<0.5
Mildbraediodendron excelsum	F	3.1		
Milletia dura			L	<0.5
Milletia sp.	F	0.2		
Morus lacteal	F	0.3		
Myrianthus arboreus			F[a]	0.8
Myrianthus holstii	F	1.0		
Parnari excelsa			F	1.8
Pennisetum purpureum			St	<0.5
Phytolacca dodecandra			F[a]	<0.5
Piper umbellatum			St[a]	<0.5
Pseudospondias microcarpa	L	0.2		
Psychotria capensis			F	<0.5
Raphia farinifera	W	1.9		
Sterculia dawei	F	0.2		
Strombosia scheffleri			L[a]	<0.5
Termite mound soil		1.0		
Tree cabbage	L	1.1		
Trichilia rubescens			L[a]	<0.5
Triumfetta sp.			L	<0.5
Urera camaroonensis	S, FL	0.6		
Urera sp.			FL	3.6
Unidentified THV	P	0.3		
Unknown		2.6		

F = fruits; L = leaves; FL = flowers; P = pith; B = bark; W = wood; St = stems.
[a] Plants with ethnomedicinal properties. Unknown items were typically climbers that were too high to obtain a sample. In the case of THV, what remained after consumption was inadequate to positively identify.

the pith of *Afromomum* sp., the leaves of *Celtis durandii* and the fruit of *Cordia millenii*. Both communities also consumed *Acanthus pubescens*, but different parts were ingested. It was the flowers at Budongo and the stems at Kibale. When the dietary items were expanded to include Newton-Fisher's 1994–1995 data, only two additional items were found to be common, the leaves of *Trichilia rubescens* and the pith of *Pennisetum purpureum* (Newton-Fisher, 1999b). It is interesting that of the seven items shared, six of them have ethnomedicinal uses, four of which (*F. exasperata*, *Afromomum* sp., *T. rubescens*, and *A. pubescens*) demonstrate bioactive properties.

Table 3 presents ethnomedicinal uses and pharmacological properties of the plants that were consumed at both sites during the study. The following behavioral and health-related observations were made in association with ingestion of these plants.

Observations from Budongo

Alstonia boonei bark. There are many ethnomedicinal uses for *Alstonia boonei*, including for diarrhea, nausea, worms, and stomachache. In addition, it possesses antimalarial, antiprotozoal, and antimetazoal properties (Table 3).

On May 5, 1998, at approximately 1700 h, four adult males traveled to a large *A. boonei* tree. They chewed the outer bark, then began to strip approximately 1/4 in of the outer bark away and consumed the inner bark. This continued for approximately 10 min. One member of the group, KK, an adult male, was followed that day for approximately 8.5 h. The majority of his time was spent foraging (44%), followed closely by resting (39%). No fecal sample was obtained from KK on the day of the focal, but one was collected the next morning that contained an anoplocephalid cestode [from direct laboratory examination, genus and species unknown] and the nematode *Necator*. Fecal samples were also collected from the other males who consumed the bark with KK. Samples from AY (April 28, 1998) and DN (April 27, 1998) both showed the presence of *Strongyloides fulleborni*, *Ternidens deminutus* and *Troglodytella abrassarti*. *Oesophagostomum* cultures were performed on three of the four males during the study (DN, AY, and NJ), and all tested positive for presence of this parasite.

Ficus sur (= capensis) bark. Traditional medicine uses the bark of *Ficus sur* to treat bronchitis, dysentery, and stomach ache (Table 3).

Table 3. Medicinal plants consumed by the Sonso and Kanyawara chimpanzees

Family, Genus species	Ethnomedicinal uses for plant part ingested (source)	Pharmacological properties (source)	Part of plant consumed	Community that consumes
Acanthaceae, *Acanthus pubescens*	Dermatosis, sterility (1)	Antibiotic (3)	Flowers	Sonso
Acanthus pubescens	Abcess, skin disease (2)		Stems	Kanyawara
Apocynaceae, *Alstonia boonei*	Diarrhea and nausea (4, 5) Snakebites (4) Stomachache and malaria (5) Stomach worms (6) Worms (7, 8) Measles (9)	Antiprotozoal and Antimetazoal (7, 8)	Bark	Sonso
Asteraceae, *Crassocephalum bojeri*	Malaria, rhinitis, detoxicant (10) 3-day fever (11)	Antimalarial activity of aerial parts (12)	Leaves	Kanyawara
Commelinaceae, *Commelina* sp.	Child's fever (5) Medicinally (14) Tumor (15)	ND	Leaves	Sonso
Aneilema sp. (*A.aequinoctiale* for Kanyawara)	Rash (16)	ND	Leaves	Kanyawara Sonso
Euphorbiaceae, *Acalypha ornata*	Relief of postpartum pain (14)	ND	Leaves	Kanyawara
Gramineae, *Pennisetum purpureum*	Infammation of mammary glands (17) Anthelminthic and for amoebiasis (18)	ND	Stems, piths	Kanyawara and Sonso (Newton-Fisher, 1999a,b,c).
Leguminosae, *Albizia grandibracteata*	Antiparasitic (19) Swollen belly (20, 21)		Bark	Kanyawara and Sonso
Cynometra alexandri	Wounds (4) Acute backache (6)	ND	Bark	Sonso
Meliaceae *Khaya anthotheca*	Headaches (6) Parasites (22) Fever (23)	ND	Leaves	Sonso
Khaya anthotheca	Wounds (6)	ND	Bark	Sonso
Trichilia rubescens	Gonorrhea (24) Soporific, bruises, lumbago, dysentery, purgative (14)	Antimalarial (25)	Leaves	Kanyawara and Sonso (Newton-Fisher, 1999a,b,c)

(cont.)

Table 3. *(Continued)*

Family, Genus species	Ethnomedicinal uses for plant part ingested (source)	Pharmacological properties (source)	Part of plant consumed	Community that consumes
Moraceae *Ficus exasperata*	Diarrhea (6) Antiulcer remedy (26) Kidney complaints (4) Colic, cough (27)	Anthelminthic, Analgesic (14), Antinematodal and insecticidal (27, 28)	Leaves	Kanyawara and Sonso
Ficus mucuso	Analgesic, bronchitis, convulsions, otitis (29)	ND	Leaves and fruits	Kanyawara and Sonso
Ficus natalensis	Pains and venereal disease (30)	ND	Leaves and fruits	Kanyawara and Sonso
Ficus thonningii	Bronchitis and urinary tract infection (29, 31)	ND	Leaves and fruits	Kanyawara
Ficus sur (= *F.capensis*)	Bronchitis, dysentery, antidote (32)	ND	Bark	Kanyawara and Sonso
	Stomach disorders (5) Sterility (33) Laxative, abortifacient, aphrodisiac (34) Lactogenic, dermatosis (14)		Fruits	Kanyawara and Sonso
Olacaceae *Strombosia scheffleri*	Abdominal complaints (13)		Leaves	Kanyawara
Phytolaccaceae *Phytolacca dodecandra*	Bilharziosis (35)	Triterpene saponins: molluscicidal, antiviral, antibacterial, spermicidal, antifertilizing activities (36, 37, 38)	Fruits	Kanyawara
Piperaceae *Piper capense*	External parasitism (39)	ND	Stems	Kanyawara
Piper umbellatum	Tonic (40)	ND	Stems	Kanyawara
Ulmaceae *Celtis durandii*	Cough and stomachache (6)	ND	Leaves and bark	Kanyawara and Sonso
Celtis africana	Indigestion (30)	NS	Leaves	Kanyawara

Table 3. *(Continued)*

Family, Genus species	Ethnomedicinal uses for plant part ingested (source)	Pharmacological properties (source)	Part of plant consumed	Community that consumes
Chaetacme aristata	Back wounds and spinal weakness (6) Antituberculosis (41)	Bacteriostatic (42)	Leaves	Kanyawara
Urticaceae *Myrianthus arboreus*	Hypoglycemic, analgesic, bronchitis, help to give birth (24)		Stems	Kanyawara
Myrianthus arboreus	Dysentery (43) Toothaches, bronchitis (44)	Triterpene acids (43) Triterpenoid (44)	Leaves	Kanyawara
Myrianthus arboreus	Emetic, purgative (24)		Fruits	Kanyawara

ND = No data; NS = Not significant; Sources: 1. Baerts & Lehmann (1991); 2. Ndayiwayeko & Ntungwanayo (1978); 3. Krief (2005a); 4. Terashima *et al.* (1991); 5. Ichikawa (AFLORA) (1998); 6. Howard *et al.* (1991); 7. Thomas & Mbenkum (1987); 8. Davies & Richards (1991); 9. Falconer (1991); 10. Kokwaro (1976); 11. Nyakabwa & Gapusi (1990); 12. Weenen *et al.* (1990); 13. Terashima *et al.* (1992); 14. Watt & Breyer-Brandwijk (1962); 15. Hartwell (1967/1971); 16. Altschul (1973); 17. Kayonga & Habiyaremye (1987); 18. Sugiyama & Koman (1992); 19. Balagizi Karhagomba, & Ntumba Kayembe (1998); 20. Defour (1994); 21. Heine & König (1988); 22. Jeanrenaud (1991); 23. Uphof (1968); 24. Bouquet (1969); 25. Krief *et al.* (2004); 26. Akah *et al.* (1998); 27. Abbiw (1990); 28. Ohigashi *et al.* (1991); 29. Bouquet *et al.* (1971); 30. Kokwaro (1976); 31. Iwu (1993); 32. Ayensu (1978); 33. Ake-Assi (1992); 34. Bouquet & Debray (1974); 35. Mesfin & Obsa (1994); 36. Taniguchi *et al.* (1978); 37. Kloos & McCullough (1987); 38. Katende *et al.* (1995); 39. van Puyvelde *et al.* (1985); 40. Polygenis-Bigendako (1990); 41. Ake-Assi *et al.* (1981); 42. Krief *et al.* (in press); 43. Ojinnaka *et al.* (1980); 44. Ngounou *et al.* (1988).

Consumption of *F. sur* bark was observed on two occasions. The first observation was made May 4, 1998, at 1440 h. Several chimpanzees were found on the ground biting off bark from the buttresses, and the inner bark was stripped away. The bark was then chewed. After several minutes the resultant wadge was discarded. On this day the chimpanzees were also eating unripe fruits of *F. sur.* The second observation occurred May 11, 1998, at approximately 1500 h. An adolescent female, SH, was found feeding on the bark. It was consumed in the same manner as described above. Both observations were ad libitum, and no further behavioral or fecal data were collected.

Khaya anthotheca bark. Ethnomedicinally, the bark from *Khaya anthotheca* is used to treat parasites, aid in the healing of wounds, and for fever (Table 3).

On October 3, 1998, GS, a subadult male, was seen feeding on the bark of *K. anthotheca*. The observation was made at 0913 h. No record of this individual's parasite burden was collected at the time.

Observations from Kanyawara

Albizia grandibracteata bark. This is traditionally ingested as a medicine in Uganda and in the Democratic Republic of Congo (DRC) against intestinal parasites and bloat (Heine & König, 1988; Defour, 1994). Bioactive saponins have been extracted and isolated from leaves of this species (Krief *et al.*, 2005a,b,c) as well as from bark (Krief *et al.*, 2006).

On October 16–20, 2001, OK, a 6-year-old female, was observed to be suffering from intestinal disorder. The diagnosis was based on alternately dry, soft, and liquid stools. In addition, fecal analysis revealed a high load of parasitic infection (strongyle species and *Probstmayria gombensis*) (Krief, 2004). We observed OK eating *Chaetacme aristata* leaves on October 16, 2001, *Albizia grandibracteata* bark at 0942 h on October 20, 2001, and *Myrianthus arboreus* stems on October 22, 2001. In these three cases, she was the only chimpanzee of the party to consume the items. This was also the first recorded time since the observations began in Kanyawara in 1987 that a chimpanzee had been seen to consume the bark of *A. grandibracteata*. She ate it for 3 min while her mother and siblings were waiting for her. Feces collected October 22 had a normal consistency and the parasitic load was nil. *A. grandibracteata*, *C. aristata*, and *M. arboreus* are used in ethnomedicine (Table 3). This observation raises the possibility that OK's bark eating associated with ingestion of other specific items were responsible for reducing the high parasite load and alleviating digestive symptoms seen since October 16.

Phytolacca dodecandra fruit. This item is known for its antiparasite bioactivities as described in the Introduction and is considered to be toxic by the traditional healer of Kanyawara. Consumption was observed for three individuals (KK, LK, and NS) of a large party on October 29, 2003.

Trichilia rubescens leaves. Bioassays revealed a strong antimalarial activity of the leaf extract and led to the isolation of two new limonoids having an

IC_{50} on *Plasmodium falciparum* in culture roughly equivalent to chloroquine (Krief *et al.*, 2004). Kanyawara chimpanzees feed only occasionally and in short bouts (3 to 7 min for our five observations) on *Trichilia rubescens* leaves. For each observation, only one chimpanzee ate a few leaves (ca. 5/min) in each bout, whereas this individual was always included in a party of several individuals who were resting, feeding, or traveling. Four fecal samples collected from KK on February 2, 2001, a day when he was observed ingesting *T. rubescens* leaves, contained all helminthes larvae or eggs. Of the four samples collected in the three following days, only one was positive, with one *Trichuris trichiura* egg.

Piper capense stems. Stems of this species are used as an antiparasite in traditional medicine (Table 3). TU was observed consuming the stems of this item, which is an uncommon food in Kanyawara (February 16, 2001). A fecal sample was collected the same day and analyzed (MacMaster method), revealing seven *Oesophagostomum* eggs. A sample analyzed from February 20, 2001, was negative.

This item was also consumed by BB along with *Pennisetum purpureum* stems (February 15, 2001). Fecal samples collected prior to consumption contained *Oesophagostomum* sp. (February 14, 2001, MacMaster) and *Strongyloides fulleborni* (February 15, 2001, MacMaster).

In Both the Sonso and Kanyawara Communities

Celtis durandii leaves. These leaves are used ethnomedicinally for coughs, stomach ache, and edema from trypanosomiasis, back wounds, spinal weakness, and antituberculosis. Bacteriostatic properties are also present.

Sonso: On May 9, 1998, focal observations were made on KL, an adult female. Duration of the focal observation was approximately 3 h 30 min. The majority of the time was spent foraging (50%), followed by resting (26%) and grooming her infant (18%). Between 0926 h and 1001 h, KL was seen feeding on young *C. durandii* leaves. A fecal sample taken that day contained eggs of *S. fulleborni* and *T. abrassarti*.

Ficus exasperata leaves. These leaves are used in traditional medicine for diarrhea, as an ulcer remedy, for kidney complaints, colic, and cough. Antihelminthic, analgesic, and insecticidal properties are known.

Sonso: The mature leaves are usually not consumed, as the surface of the leaf is rough. One ad libitum observation was made of the mature leaves of *F. exasperata* being consumed. On August 13, 1998, ZT began to pick 4+ leaves at a time. He bit off the stems, rolled the leaves up, bit them in half, chewed, and then consumed the remaining half. This behavior lasted approximately 12 min. A fecal sample taken at the time contained *T. abrassarti* and *Oesophagostomum* sp.

Kanyawara: On February 15, 2001, a fight was observed between two adult males, YB and LB. YB bit LB's foot. The fifth toe was severely cut, hanging from his foot by only a strip of skin. Leaves and stems of *F. exasperata* were the only unusual food consumed in the following days.

Ficus sur (= capensis) immature fruits. Uses include a treatment for sterility, a laxative, abortifacient, or aphrodisiac, and have known lactogenic and antidermatosis properties.

Sonso: Consumption of immature *F. sur* fruit is common.

Kanyawara: On February 15, 2001, a 17-year-old male, KK, was weak and had a deep cough. Sneezing was frequent and analysis of his feces showed a large number of *P. gombensis* (1750 parasites/g) and was positive for strongyle eggs and larvae and *Trichuris* eggs. The activity budget of KK when compared to 13 other individuals from the same party shows that KK rested 77% of the time (compared to 33% for the rest of the group) and fed during only 16% of the time (compared to 48% for the other individuals). KK was the only individual from the party to feed on immature figs from *F. sur*, which are usually consumed only when mature. Urinalysis on February 15 revealed a low pH value.

Commelinaceae leaves. Chimpanzees at Budongo and Kanyawara are known to swallow leaves whole from the Commelinaceae family (C. Bakuneeta, V. Reynolds, personal communication, for Budongo; Messner & Wrangham, 1996, for Kibale). However, no personal observations were made during the study period. Unchewed leaves from the genus *Commelina* or *Aneilema* were found twice in the dung. The first instance was an ad libitum observation on May 16, 1998, of the adult male VN. In addition to the presence of adult worms on the surface of the leaf, microscopic examination detected *S. fulleborni* and *T. abrassarti*. The second time whole leaves were found was also an ad libitum observation on June 24, 1998, from an adult female, KL. No adult worms were detected; the fecal sample collected did contain *T. abrassarti*.

Soil Consumption

Sonso

A detailed report of the soil analysis from Budongo can be found in Chapter 8 of this volume. Ingestion of termite mound soil was recorded six times by five individuals during the study period. Of these six cases, two were recorded during focal-animal observations.

On August 22, 1998, the adult female KY, approximately 4 months pregnant, ingested soil at 1125 h. She spent the majority of this 10-h focal observation foraging (52%), followed by resting (28%) and moving (14%). Two fecal samples were obtained that day. The first specimen taken at 0655 h detected no parasites. The second taken at 1602 h contained *T. abrassarti*. On August 25, three days later, this same chimpanzee was seen eating soil at 0912 h (ad libitum observation). Neither behavioral data nor fecal analyses are available for this day.

On August 27, 1998, the adult female KW and her infant son KZ were observed sharing soil at 0856 h. After the mother began to consume the soil, the infant placed his hand near her mouth and requested soil from her. Initially she pushed some soil out from her mouth, which he took and placed in his mouth. When subsequent requests were made, she broke off a small piece from a larger piece and handed it to him. This interaction lasted approximately 9 min. An overwhelming amount of time from this 255-min focal was spent foraging (93%). The remaining time was spent moving (3%) and resting (4%).

Three fecal samples were collected from KW: one on August 19, one on August 25, and another on August 27, 1998, the day geophagy took place. All showed the presence of *T. abrassarti*. Two of the three samples contained *Oesophagostomum* sp.

Kanyawara

Geophagy was observed three times during the study period. In all cases, two individuals were eating soil. In each case, one of the two individuals had eaten *Myrianthus arboreus* before this. In two cases, fruits (ripe or immature) and young leaves on one occasion were ingested before this.

On December 22, 2000, ingestion by the old female LP, who was suffering from bloat and abdominal distension, was observed. Moreover, her hand was in pain, and could not be used for tree climbing. She built her nest early (1720 h) in spite of the feeding activity of the other individuals. The following day, she went out of her nest at 0750 h, which was late compared to her offspring, who had climbed down from the tree at 0700 h.

In the beginning of February 2001, this old female exhibited concomitant abnormal urinalysis (proteins and leukocytes), coccidiosis, and a high parasitic load. On February 5 at 1405 h and 1415 h, she was observed rummaging through fresh elephant dung, removing, crunching, and swallowing unidentified seeds from it. At 1540 h, she ate soil for 2 min. Five days later, at 0950 for 5 min, she ate several handfuls of fine fibrous material from an old fallen hollow trunk. A dung sample from this day was soft and contained high amounts of *T. abrassarti* (about 32,000/g) (Krief *et al.*, 2005b).

DISCUSSION

During the two study periods, the chimpanzees of both communities consumed 24 plant species used in traditional medicine. Of them, eight species possess known pharmacological properties (Table 3) that could have aided in the medicinal treatment for some of the symptoms or illnesses identified in the particular chimpanzees at the time of ingestion. They include *Acanthus pubescens* stems and flowers, *Alstonia boonei* bark, *Crassocephalum bojeri* leaves, *Albizia grandibracteata* bark, *Ficus exasperata* leaves, *Phytolacca dodecandra* fruit, *Trichilia rubescens* leaves, and *Chaetacme aristata* leaves.

The consumption of these eight plants was looked at in detail. Behavioral observations and fecal and urine analysis were used to gain insight into the individual's state of health at the time that the plant in question was ingested. Only the consumption of *A. grandibracteata* bark fits the first type of self-medicative behavior described by Huffman (1997). Although they possess bioactive properties, the remaining plants appear to fall into the second type of self-medicative behavior, which use "food as medicine" (see Huffman, 1997, 2003), with a special place to *T. rubescens* leaves because the low amount consumed and the high bioactivity do not really fit with a "food" category. These plants contain secondary compounds, which could play a role in health maintenance. All are food items rare to the diet. One curious aspect is that many times, only a few individuals in the groups ate these plants while others ignored or looked on. In addition, in the case of *Acanthus pubescens* flowers, a plant that possesses in vitro antibacterial properties, the chimpanzees were food grunting, so taste appeared to reinforce their consumption of this item. The difference between food and medicine is often difficult to detect even in humans. The concept of medicinal foods introduced by Etkin and Ross (1982) is supported by Johns (1990), who believes these nonnutritional components, once part of our diet, have now been replaced with herbal medicine and modern pharmaceuticals. Within traditional human societies worldwide, there is much overlap between food and medicinal items. The preliminary observations of our study further suggest this to be true for chimpanzees as well. Owing to secondary compounds present in some of these infrequently ingested plants, the medicinal value of these plants may exceed their nutritional value (Huffman, 2003). Observations recorded in this study that fit this second type of self-medicative behavior include *A. pubescens* stems and flowers, *A. boonei* bark, *P. dodecandra* berries, and the leaves of *C. bojeri*, *T. rubescens*, *Strombosia scheffleri*, and *C. aristata*. While we cannot quantitatively assess the underlying motivation for consumption, on the basis of our observations of health at the time of ingestion in some cases, it can be concluded that the individuals were possibly ill. Future research should search for pharmacological properties in the remaining 16 plants found in their diet that are known to be used ethnomedicinally.

Soil may have also provided medicinal benefits. Analyses of the soils consumed at Budongo and in Kibale strengthen the argument of consumption to alleviate gastrointestinal distress, suppress diarrhea, or possibly as a detoxification agent as the samples all contained clay similar to Kaopectate™. The first two observations of geophagy in Budongo occurred 3 days apart. There were times during the study when diarrhea was prevalent on the trails, and it was usually found in association with the consumption of unripe *F. sur* fruits. At this time the chimpanzees were eating the leaves, ripe fruits, and flowers of *B. papyrifera*, the ripe fruits of *F. sur*, and the seeds of *Urera camaroonesis*. The chimpanzee that consumed the soils was then 4 months pregnant. We do not believe chimpanzees experience nausea and gastrointestinal distress associated with pregnancy (Dr. Rick Lee, personnel communication), but the clay present in the soils could have decreased the levels of metabolic toxins such as steroidal metabolites associated with pregnancy (Johns & Duquette, 1991). Another possibility is that her parasite burden was higher than normal. In humans, pregnant and lactating females may be immunologically compromised and thus more susceptible to parasite infection (Kalema, 1995a).

Kalema felt this was a contributing factor in higher egg counts of fecal parasites in females than males in a Ugandan population of mountain gorillas. The next observation discussed was a mother and infant sharing soil. The mother had *Oesophagostomum* and a large number of *T. abrassarti* eggs present in two fecal samples surrounding the consumption of soil. *T. abrassarti* is believed to be a symbiont that aids in the digestion of cellulose. Fluctuation in numbers is likely to be related to corresponding changes in dietary fiber intake. The soils consumed by the Kanyawara chimpanzees during this study were not analyzed, but analysis from previous studies at Kibale (Mahaney *et al.*, 1997, in press) showed it to be similar in composition to those analyzed at Mahale and Budongo (Mahaney *et al.*, 1996; Tweheyo *et al.*, Chapter 8, this volume).

The differences in plants selected seem pronounced between these two communities with regards to species with potential phytochemical benefits for maintaining health. Although there are many food plants shared in common between the two sites, which are separated by only 200 km (Chapman & Chapman, 2004), they exploited different plants with medicinal properties. Twenty-three percent of the 117 food species corresponding to 35 items ingested by Kanyawara chimpanzees are used in traditional medicine (Krief *et al.*, 2005a). Even when they consumed the same plant, the frequency of consumption varied as well as the part eaten. An example is *A. pubescens.* At Budongo the chimpanzees ate the flowers, which contain important antihelminthic and antibiotic properties (Krief *et al.*, 2005a) (Table 2) but no other part of the plant was consumed. Even when the flowers were present, the chimpanzees at Kibale ate only the stems.

Variation in the consumption of barks used in traditional medicine is another difference that exists between the Sonso and Kanyawara chimpanzees. It is a nonseasonal item, so availability is not an issue. During this study, the Sonso community consumed bark from five species of trees. Three of the five species (*F. sur*, *F. exasperata*, and *Cynometra alexandri*) exist at Kibale, however they were not eaten by the Kanyawara chimpanzees during this study nor has Wrangham ever observed their consumption at Kibale (unpublished data). Alternatively, during the study at Kibale, the bark of *A. grandibracteata* and *Markhamia platycalyx* was consumed. Both of these species exist at Budongo, but they were not eaten during the 1998 study period, nor during Newton-Fisher's 1994–1995 study (Newton-Fisher, 1999a,b,c). The possible role these barks play in health maintenance is unknown. Future research in the area of bark consumption and its underlying motivation is needed. But for *A. grandibracteata*, the potential role in health maintenance has been evidenced by bioactive properties against helminthes. Furthermore, the parasite levels of the individual that ate the bark of this species abruptly dropped after its consumption.

Goodall (1986) felt that variation in dietary items between sites was primarily due to food availability. Our analysis found that between Kibale and Budongo, 58% of all dietary items consumed during our studies are present at both sites. However, of these items only 8.45% were commonly consumed at both sites. In this case, availability alone cannot explain the differences we found. Wrangham offered another theory for intersite variation in diet. He believes that important differences may exist in plant chemistry (Goodall, 1986). Goodall offered a third hypothesis, which suggested differences are due to different group feeding traditions (1986). We suggest that both tradition and difference in plant chemistry may play a role in the unique "medicinal culture" found at each site. Future studies conducted in parallel are warranted to confirm the presence of secondary compounds and the amounts of each compound detected across sites.

In addition, it is plausible that regional differences in health status brought about by variability in parasites, pathogens, and other causes of illness could shape the different "medicinal cultures" among sites. We suggest that future studies looking for intersite variation perform both the MacMaster and direct examination methods of parasite detection. Given the demonstrated importance of *Oesophagostomum* as potential motivation for self-medication at some sites (Huffman *et al.*, 1996; Dupain *et al.*, 2002), knowing its prevalence and intensity of infection is important. Repeated infections can cause significant complications such as secondary bacterial infections, diarrhea, severe abdominal pain, weight loss, and weakness, which can result in high mortality (Brack, 1987). Therefore, we also recommend Harada–Mori coprocultures be performed. The combined impact of multiple-species infections needs to be looked at in more detail. Besides parasitological methods, fecal samples may also be used to look for fecal antigens and antibodies using molecular techniques (PCR) to aid in identifying a variety of pathogens responsible for disease. In addition to such noninvasive methods, we emphasize the necessity of knowing the etiology of death. Sick chimpanzees have to be followed carefully to find the carcass if the disease is lethal. Results of a necropsy may enable us to associate the real cause of death with the behavioral, fecal, and urine analysis from when it was alive. Urine analysis also provided insights into potential health problems. The sample size for the Sonso community is small, but the Kanyawara study was able to draw several conclusions based on their results. We would also advise Budongo researchers to implement periodic pregnancy testing of cycling females, a practice that already exists at Kibale. During the Budongo study, two positive pregnancy tests were obtained. Once a positive pregnancy test was recorded, careful monitoring of these chimpanzees took place, providing insight into behavioral changes associated with pregnancy and allowing the field assistants the opportunity to view one of the births (Kiwede, 1999).

In conclusion, our results suggested that unique medicinal cultures might exist at each of the sites in question. These two communities exploited different plants, different parts of the same plant, with varied degrees of frequency. We suggest future research continue to analyze dietary items to look for secondary compounds useful in health maintenance. Obvious illness is difficult to observe, so further research must focus on the chimpanzees' ongoing efforts toward consuming items that help it reach and maintain health homeostasis. Potential health problems can be detected by behavioral observation, fecal and urine analyses. Future research of this type may also lead to the discovery of new medicinal compounds for human medicine and help to explore questions regarding hitherto little explored aspects of primate medicinal culture.

ACKNOWLEDGMENTS

Our sincere thanks go to Prof. Vernon Reynolds (founder of the Budongo Forest Project) and on-site directors of Sonso Jeremy Lindsell, Lucy Bereford-Stooke, Mark Attwater, Fred Babweteera and current acting director Dr. Nick Newton-Fisher. Staff of the Project provided invaluable assistance; in particular Joseph Kalamagi, Zephyr Kiwedi, and Gershom Muhumuza. Without the permission to live and work in the Budongo Forest Reserve, this study would have not been possible. SK wishes to give her deepest gratitude to Richard Wrangham (director of Kibale Chimpanzee Project), Claude-Marcel Hladik, and Thierry Sévenet for their invaluable contribution in the study in Kanyawara and in France. We also give thanks to the Uganda Council for Science and Technology, the Uganda Wildlife Authority, the President's Office, and the Forest Department. This research was supported by generous grants from the J. William Fulbright Foreign Scholarship Board and the United States Information Agency to PP, from the Leakey Foundation and in part from a Grant–In–Aid for Research

(C-16570193 to M.A. Huffman) from the Japan Society for the Promotion of Science, to MAH during the manuscript's preparation. Centre National de la Recherche Scientifique and Museum National d'Histoire Naturelle, France, to SK. Much thanks to Dr. Richard Williams of Alpha Tec, Inc., Dr. S. Gotoh, and the laboratory of Prof. Ton Ponderman for their generous gift of fecal analysis. PP wishes to thank Drs. Rick Lee and Mark Gold for their expertise in the area of veterinary medicine. PP also wishes to thank her family for their support. Facilities in Kanyawara were provided by Makerere University Biological Field Station and we are most thankful to John Kasenene and Gilbert Isabirye-Basuta for their collaboration in the fieldwork. For their invaluable assistance in the fieldwork, SK expresses her sincere thanks to Jean-Michel Krief, Kim Duffy, Katie Pieta, the late Donor Muhangyi, Francis Mugurusi, Christopher Katongole, Christopher Muruuli, Peter Tuhairwe, John Barwogeza, and Moses Musana. SK sincerely thanks the Institut de Chimie des Substances Naturelles and its teams, CNRS, Gif-sur-Yvette, which supports her PhD work in phytochemistry, and P. Grellier and P. Loiseau for their help and support for biological analyses. For their help in the laboratory for coprological analysis, SK is indebted to Jacques Guillot and Christian Bories.

CHAPTER EIGHT

Geophagy in Chimpanzees (*Pan troglodytes schweinfurthii*) of the Budongo Forest Reserve, Uganda

A Multidisciplinary Study

Mnason Tweheyo, Vernon Reynolds, Michael A. Huffman, Paula Pebsworth, Shunji Goto, William C. Mahaney, Michael W. Milner, Anthony Waddell, Randy Dirszowsky, and Ronald G. V. Hancock

Mnason Tweheyo • Department of Forest Biology and Ecosystems Management, Makerere University, Kampala, Uganda **Vernon Reynolds** • School of Anthropology, University of Oxford, 51 Banbury Road, Oxford OX2 6PE, United Kingdom **Michael A. Huffman, Paula Pebsworth, and Shunji Goto** • Primate Research Institute, Kyoto University, Kanrin 41-2, Inuyama, Aichi 484-8506, Japan **William C. Mahaney, Michael W. Milner, and Anthony Waddell** • Geomorphology and Pedology Laboratory, York University, 4700 Keele St., North York, Ontario, Canada M3J 1P3 **Randy Dirszowsky** • Geomorphology and Pedology Laboratory, York University, 4700 Keele St., North York, Ontario, Canada M3J 1P3, and Department of Chemistry and Chemical Engineering, University of Toronto, Toronto, Ontario, Canada K7K 7B4 **Ronald G. V. Hancock** • Department of Chemical Engineering and Applied Chemistry, University of Toronto, Toronto, Canada M5S 1A4.

Primates of Western Uganda, edited by Nicholas E. Newton-Fisher, Hugh Notman, James D. Paterson, and Vernon Reynolds. Springer, New York, 2006.

INTRODUCTION

Geophagy occurs widely among primate species (Krishnamani & Mahaney, 2000). While reported for chimpanzees in the wild since the 1960s (Hladik, 1977; Nishida & Uehara, 1983; Goodall, 1986), the geochemical and behavioral study of geophagy in relation to self-medication (Huffman, 1997) was not initiated until the mid-1990s, the first being that of Mahaney and Huffman. This work began in Tanzania with the analysis of termite mound soils, behavioral and parasitological data collected from the Mahale Mountains National Park (Mahaney *et al.*, 1996b; 1998; Aufreiter *et al.*, 2001; Ketch *et al.*, 2001). Further analyses have included termite soils eaten by chimpanzees in Gombe National Park, Tanzania, and exposed subsurface clays eaten by chimpanzees in the Kibale National Park, Uganda (Mahaney *et al.*, 1997, 1998; Aufreiter *et al.*, 2001). Geophagy has recently been noted to occur in a fourth East African population, the Sonso community in the Budongo Forest Reserve, Western Uganda. Early published studies from Budongo did not report any kind of soil eating by chimpanzees. However, more recently, Reynolds *et al.* (1998) referred to the eating of riverbank soil and other authors have noted sporadic termite mound soil eating by chimpanzees in this forest (e.g., D. Quiatt in Reynolds *et al.,* 1998:335; Newton-Fisher, 1999a,b). Termite mounds of the species *Cubitermes speciosus* are present in the Budongo forest (Newton-Fisher, 1999b).

At Gombe, chimpanzees consume *Macrotermes* with the aid of termite fishing tools inserted in a mound's ventilation ducts (Goodall, 1986). Reference is made to the consumption of mound soils of *Pseudacanthotermes spiniger* in Mahale, as being distinct from the consumption of termite mound soil there (Uehara, 1982). In the case of *Cubitermes* at Budongo, however, chimpanzees consume termites along with lumps of earth wrenched from termite mounds. While information exists on the consumption of termites, little consideration is given to the depth reached by termite species. Pomeroy (1976) cites *Pseudacanthotermes* as a builder of smaller mounds in Uganda. *Cubitermes humiverus* is also a builder of small mounds that are characteristically mushroom-shaped. This species' shallow activity in the soil, unlike the other mound builders, is likely to produce high organic contents in mound soils, a characteristic antithetic to geophagy. Furthermore, nowhere is there a detailed analysis of soils that provides information on the different structural components of these mounds. When considering the ingestion of termite mound soils, this information is important for increasing our understanding of their selection by chimpanzees.

A central theme has been to explain geophagic behavior from the perspective of the ingested soils' physical, chemical, and mineralogical properties. Theoretically, there should be a common adaptive property or properties of the soil being selected for by primates that helps explain why they spend considerable time searching for and ingesting soil, sometimes on nearly a daily basis (e.g., Goodall, 1986; Mahaney *et al.*, 1998; Wakibara *et al.*, 2001). Previous work has drawn attention to the high percentage of clay in every instance of geophagy studied among chimpanzees (Wrangham, 1977; Mahaney *et al.*, 1996a,b, 1998), gorillas (Mahaney *et al.*, 1990, 1995a; Mahaney, 1993), orangutans (Mahaney *et al.*, 1996a), and macaques (Mahaney *et al.*, 1993, 1995b; Wakibara *et al.*, 2001). In addition, the clay mineral components have a near-perfect crystallinity in almost every detailed analysis carried out on these samples by Mahaney and colleagues of the Geophagy Research Group at York University. All the soil samples they have analyzed to date, they have identified a pharmaceutical-grade clay mineral of low Si composition (Si:Al = 1:1) belonging to the kaolinite, halloysite, and metahalloysite group.

It has been suggested that the ingestion of small quantities of clay-rich earth may assist in nutrition, serve as a dietary supplement, or even have pharmaceutical properties beneficial to chimpanzees (Mahaney *et al.*, 1999; Aufreiter *et al.*, 2001; Ketch *et al.*, 2001; Mahaney & Krishnamani, 2003). Behavioral studies have yet to be fully incorporated into the research program, in part because of the rarity of occurrence of the behavior at some of these sites. The behavior is short in duration, hard to predict when it will occur, and thus difficult to sample completely. This paper reports the first attempt at Budongo to analyze the physicochemistry and mineralogy of soils eaten by chimpanzees and presents behavioral, dietary, and parasitological data in an attempt to assess the possible benefits of geophagy for chimpanzees at this site.

METHODOLOGY

The Study Site

Behavioral and chemical analyses presented here are from data and samples collected while pursuing other behavioral and ecological studies in the Budongo Forest Reserve of Western Uganda. The chimpanzees observed were members of the Sonso community, which has been investigated since 1990 under the direction of Vernon Reynolds. The Budongo Forest Reserve is a medium-altitude, moist, semideciduous forest, a mixture of tropical high forest with a large population of mahoganies, woodland, and savannah grassland (Eggeling, 1947; Reynolds & Reynolds, 1965; Howard, 1991). The mean annual rainfall is 1780–1900 mm, with a short dry season from mid-December to mid-February. The mean monthly minimum and maximum temperatures are 17–20°C and 27–29°C, respectively (Newton-Fisher, 1999b; Tweheyo, 2003).

Field Protocols

The detailed behavioral observations and soil samples we analyze here were collected during two research periods. Period I in 1998 covers the period between February 23 and October 14, 1998. During this period, PP conducted behavioral observations using ad libitum and focal-animal sampling. In this period, we recorded all social interactions, activity patterns, diet, and visible cues of health status. We observed seven adult males and seven adult females for a total of 352 h over 105 observation sessions. We used these data to evaluate possible relationships between health status and geophagy. We also used this focal data to analyze for possible changes in diet around the months in which we observed geophagy. Additional ad libitum records of geophagy made by field assistants and other researchers are included in the general analysis and discussion, but not in calculations of relative frequency or diet involving total hours of observation.

Period II covers the period between June 10, 2000, and August 24, 2001, and was conducted by MT and field assistant Monday Gideon Mbotella for a total of 286 observation days per person, for a total of 572 h. Three days a week we used scan sampling, and 2 days a week we used focal sampling. From the 54-member community, a total of 34 (16 males, 18 females) adult and juvenile members in the group were observed. Focal sampling was done from dawn to dusk on one specific chimpanzee per day. Scan sampling was used to record chimpanzee diet, behavioral activities, and habitat use. Among juveniles and adults, both sexes were equally considered. Over a period of 176 days, 2641 scans were recorded, 2107 of which involved feeding. Period II focused on the food sources and abundance and ecology of food trees fed on by adult chimpanzees in the Sonso community.

Observations of geophagy made by project field assistants during the course of their daily observations after the completion of MT's study in early October 2001 up to July 2002 are grouped into Period III for convenience. Nine additional cases of geophagy were observed by field assistants ad libitum in this period (between December 2000 and July 2002), and are included in some general analyses presented below. The forest is demarcated into compartments according to logging activities and the study area is demarcated into blocks by a system of N–S and E–W observation trails that intersect each other at 100 m intervals. The locations of observation and collection sites were noted on this grid system.

Samples of ingested and control soil samples were collected at the time geophagy was observed. All samples were collected in plastic bags and taken to the camp laboratory, where they were air dried at room temperature and subsequently mailed to Mahaney, York University, for analysis.

Laboratory Protocols

Part of the protocol for behavioral observations in Period I included the collection of fecal samples from focal individuals during observations and ad libitum from other community members when possible. One gram of feces was weighed, and stored in 5.0-ml Corning plastic tubes and fixed with 10% neutral formalin. SG performed the parasitological analysis at the Primate Research Institute using the McMasters flotation and formol–ether concentration techniques. Eggs/gram (EPG) fresh dung was calculated for each sample as the mean value derived from three trials and is used here only as a relative measure of infection level.

Soil samples were analyzed at York University for particle size following procedures established by Day (1965). Electrode and electrical conductivity following Bower and Wilcox (1965) determined the pH. Carbon and nitrogen were analyzed on a Leco apparatus. Elemental analysis was undertaken at the SLOWPOKE-2 reactor at the Royal Military College of Canada using a modified version of the instrumental neutron activation analysis (INAA) procedures outlined by Hancock (1984). In this preliminary investigation, the concentrations of both short-lived and long-lived isotope-producing elements were determined.

Data Analysis

We analyzed the data using Fisher's Exact Test and Kruskal–Wallis ANOVA by rank. Significance was set at $P < 0.05$, and all analyses were two-tailed. Data elaboration was carried out using the package Statistica (Statsoft Inc., 1998) and InStat GraphPad (Ver. 2.01).

RESULTS

Behavior

General Description

In total, 23 cases of geophagy by 17 individuals (6 females, 11 males) were observed at Budongo, of which detailed information was obtained for four cases analyzed in greater detail in this paper. In all instances, chimpanzees removed soil from termite mounds. Chimpanzees broke open the termite mound of *Cubitermes speciosus* from any height of the mound. Both active and inactive mounds were targeted for geophagy. In 60% of these cases, termites were ingested along with the soil by breaking a clump of soil with termites inside. In such cases it was difficult to determine whether chimpanzees were mainly after the soil, the termites, or both.

We observed 6 of the 15 cases of geophagy during Periods I and II by five individuals on four different days. We observed one case of active sharing by a focal adult female with her infant male. On two more occasions, another individual approached and fed, or attempted to feed, from the same mound after seeing the first individual feeding from it. Time taken to ingest the soil ranged from less than 1 min to 12 min in duration, depending on the number of pieces consumed (range 1–4 pieces). Further behavioral details of the four cases of geophagy observed under focal-animal sampling are shown in Table 1.

Relative Frequency and Temporal Distribution

On the basis of focal observations, we calculated the relative frequency of occurrence of geophagy per 100 h for research periods I and II (Table 2). We noted a higher frequency of occurrence in Period I than in II. The combined mean relative frequency of occurrence was 0.79 instances of feeding on soil per 100 h of observation.

All but one of the 23 cases of geophagy were observed before 1300 h, with a peak time of occurrence between 0900 and 1000 h. Interannual difference in the daily time of occurrence was negligible.

Table 1. Details of termite mound soil ingestion observed in Sonso group chimpanzees during study Periods I and II

No.	Individual, date	Description
1	Kewaya (KY, adult female), August 22, 1998	This adult female, approximately 4 months pregnant, feeds on soil from a termite mound in block 7E at 1125 h. She exhibited no signs of illness. The whole process of eating soil took less than a minute. The soil that was eaten was not mixed with leaves or any other vegetation. KY feeds on the soil of the upper part. (soil sample Bud 2)
2	Kewaya (KY, adult female), August 25, 1998	At 0911 h, KY removes a piece of soil and feeds on it while en route to another tree. The soil consumed was found in block 5B and had previously been knocked down. One soil sample (Bud 3) was collected from a portion discarded by the chimpanzee and a second sample (Bud 3a) from an intact, active mound nearby.
3	Kwera (KW, adult female) and Kwezi (KZ, infant male of KW), August 27, 1998	At approximately 0854 h, KW climbs down from a tree, leaving KZ above, and breaks off a piece of soil with her hand from a termite mound, located in block 5B. KW rejoins her infant KZ up in the tree and begins to consume the soil. KZ stares intently at his mother. KW breaks off a piece of soil and hands it to KZ. KW holds one piece with her hand and one with her foot. Once that mouthful of soil is consumed, the infant puts his hand on his mother's mouth. KW then pushes the soil forward between her lips and KZ removes it and puts it in his mouth. KW continues to feed on the soil while KZ moves away. KW consumes the piece in her hand and then begins feeding on the one held in her foot. KZ reapproaches and reaches for a third piece of soil. KW bites off a piece and hands it to him. At this point the mother drops the remaining soil and climbs down at 0905 h. The discarded soil is collected for analysis as sample Bud 4.
4	Tinka (TK, adult male) and Gashom (GM, subadult), April 28, 2001	Two males, TK and GM, were observed to ingest soil at around 1200 h. The termite mound was about 0.5 m tall and built between the buttresses of a large *Cynometra alexandri* "ironwood" tree about 15 m from the north line transect of block GD. There was no vegetation growing in the mound, indicating it had been occupied by termites until fairly recently. The soil consumed was a lighter brown color than the surrounding soil. They each removed a sizable piece of soil, using only their teeth. Chewing on the soil took about 2 min before swallowing. TK ate four pieces and GM ate three. The soil was eaten in a normal way, neither reluctantly nor with speed. (soil sample Bud 1)

Table 2. Relative frequency of geophagy observed across chimpanzee study sites in East Africa

Site	Frequency, per 100 h	Reference
Uganda		
Budongo, Period I	1.42	This study
Budongo, Period II	0.17	This study
Kibale	0.52	Mahaney *et al.*, 1997
Tanzania		
Mahale	4.07	Mahaney *et al.*, 1996a,b
Gombe	8.33	Wrangham, 1977

Budongo: Period I (February 1998–October 1998): five cases in 352 h of focal observation; Period II (June 2000–August 2001): one case in 572 h of observation; Kibale (not specified): four cases in 767 h of focal observation; Kibale (January 1995–July 1996, 68% wet months): five times in 824 h; Mahale (November–December 1991, both wet months): five cases in 123 h of focal observation; Gombe: extrapolated from figure of 1/12 h year-round as estimated by Wrangham, 1977.

Monthly Distribution, Interannual and Regional Variation

Geophagy was observed in August 1998 (four observations; 8 months' study, February–September), December 2000 (one observation), March 2001 (two observations), and August 2001 (two observations) during a 16-month study period (June 2000–September 2001), and again in 2002: January (two observations), March, April, May, and July (one observation in each). There was no consistent trend in the occurrence of geophagy for any particular month of the year. The intermonthly pattern of occurrence and relative frequency of occurrence of geophagy was not consistent. Furthermore, we found no significant difference in the number of months in which geophagy was observed between the three study periods (Fisher's Exact Test, two-tailed, Period I–II: 4/8–3/16 months, $P = 0.39$, NS; Period I–III (4/8–5/10, $P = 1.00$, NS; Period II–III (3/16–5/10, $P = 0.19$, NS). It appears to us that the stimuli inducing geophagy are dynamic, and suggests that geophagy at Budongo is not simply a habitual year-round behavior but a condition-specific reaction or craving brought on by changing external environmental factors that can affect the physiology of the chimpanzees.

Our data suggest that, compared to other East African study sites for which such data are available, the frequency of occurrence at Budongo is relatively low (Table 2). Kibale, another Ugandan site, also has a relatively low frequency of occurrence. Great variability exists between these Ugandan and Tanzanian sites. In Tanzania, Mahale and Gombe have much higher rates of occurrence. These two sites are highly seasonal in their annual rainfall patterns, with as much as half of the year classified as the dry season (<100 mm). Budongo and Kibale on the other hand have only 1–2 months a year with less than 50 mm of rainfall. Seasonality of rainfall affects food availability, which in turn is expected to affect dietary choice. If geophagy is influenced by diet, interregional differences in the seasonality of food availability may be responsible in part for this interregional variation in the relative frequency of occurrence of geophagy.

Diet and Health

Food Selection and Geophagy

Here we analyze changes in food item selection to test for possible group-level dietary shifts that may help explain the fluctuating pattern of geophagy observed in this study. We used focal observation data from Period I to analyze for possible differences in the amount of time spent feeding on food items before (July), during (August, the month we observed geophagy in this study period), and after (September). We found no significant differences in the amount of time spent feeding on three major food items: seeds (Kruskal–Wallis H (2, $n = 29$) = 2.38, $P = 0.31$), fruits (Kruskal–Wallis $H(2, n = 29) = 3.18$, $P = 0.20$) and leaves (Kruskal–Wallis $H(2, n = 29) = 1.08$, $P = 0.58$).

During Period II, we collected a total of 2107 scan samples involving feeding behavior. In total, 72% of the scans represented feeding on fruits, 15.1% on young leaves and 7.4% on flowers. Feeding on other items such as bark represented the remaining 5.5%, pith, seeds, wood, soil and insects. We conducted a preliminary analysis of the monthly dietary change over the course of Period II using this scan sample feeding data to evaluate the effects of changes in the amount of fruit, leaf, or flower consumption by month as possible stimuli for geophagy. The monthly average was 69.92% (SE = 17) for fruits, showing the strong preference for fruits and their high availability year round. The monthly average was 14.08% (SE = 9) for leaves and 8.0% (SE = 11) for flowers.

As noted above, we observed geophagy in the months of December, March, and August. Months in Period II were classified as being high or low months of consumption of each of the three food items, based on whether they fell beneath or above the mean monthly average rate of consumption for each item. On the basis of this ranking, no significant relationship was found for the occurrence of geophagy and the relative amount of time spent feeding on any of these items (fruit, flowers: $P = 1.00$; leaves: $P = 0.52$; Fisher's Exact Test, two-tailed test).

The above results from Periods I and II must be interpreted with caution, however, because the level of analysis is at group level. The limited amount of data for the individuals observed eating soil preclude us from conducting any further detailed analyses during either period. Finer-grained analysis at the individual level is needed to properly address this question any further.

Health Status and Geophagy

We conducted a preliminary evaluation of parasite infection in individuals observed during Period I to evaluate the effects of intestinal parasite infection as a possible stimulus for geophagy (Table 3). In Period I all cases occurred in August, a wet month (rainfall >100 mm). According to these parasite profiles, we verified all individuals to be infected by at least two nematodes (*Oesophagostomum* sp. and *Strongyloides fulleborni*) and a protozoan *Troglodytella abrassarti*. Fecal samples were not available for the two adult males, TK and

Table 3. Parasitological profiles of individuals observed eating soils and control individuals sampled around the same period

		Identified parasite species		
Subject	Date[a]	*Troglodytella abrassarti*	*Oesophagostomum* sp.	*Strongyloides fuelleborni*
Observed eating soil				
KW	19	+ + +		
	25	+ + +	+ (3)	
	27	+ + +	+ (4)	
KY	19	+	+ (1)	+ (3)
	22	+ + +		
	22	−		
Controls				
ZA	29	+++	+ + (39)	
MG	27	+++	+ (2)	
TK	19	+++	+ (5)	+ (5)
VN	17	+	+ (5)	
ZT	13	+++	+ (2)	

Profiles based on modified MGL methodology: + = 1–9 eggs per preparation; ++ = 10–99 eggs/protozoa per preparation; + + + = 100 + eggs/protozoa per preparation; (EPG count) per preparation (18 × 18 mm); − = Negative.

[a] All dates from August 1998.

GM, but they did not appear to be overtly ill. No sign of physical illness, such as coughing or diarrhea, were noted in any of the individuals observed, although one female (Kewaya) was pregnant at the time. Compared to parasite levels in individuals observed during the same period, but for which geophagy was not observed, no marked difference in infection levels or species number were noted.

Physicochemistry

Characteristics of Ingested and Control Soil Samples

In Period I, four soil samples (Bud 2, 3, 3a, and 4) from termite mounds ingested by chimpanzees were collected (Table 4). In Period II, one sample (Bud 1) and a control (Bud 5) were collected. The control sample was collected 10 m from the termite mound to avoid any contamination from the mound itself. Soils surrounding the termite mound were observed to be quite uniform and the materials collected representative of uneaten soils. The control sample was collected at a depth of 15 cm, just sufficient to avoid the topsoil organic matter covering the forest floor. Samples were collected from a piece of soil discarded by the chimpanzee or from the same mound. In one instance (Bud 3a), the chimpanzee had chosen soil from a previously knocked down, inactive mound. Soil from a nearby active mound was also sampled, so the species of termite could be determined.

Table 4. Particle-size distributions in the Budongo termite mound soil and a ground soil control sample

Sample	% Sand (2000–63 μm)	% Silt (63–2 μm)	% Clay (<2 μm)
Ingested			
Bud 1	47.8	40.1	12.0
Bud 2	41.7	23.3	35.0
Bud 3	25.3	18.7	56.0
Bud 3a[a]			
Bud 4	27.9	34.9	37.2
Mean	35.9	29.3	35.1
SD	10.8	9.9	18.0
67% Range[b]	24.9–46.5	19.3–39.2	17.0–53.1
Control			
Bud 5	50.5	35.7	13.8

[a] Insufficient sample material available for analysis.
[b] ±1 standard deviation.

Soil Structure

Observations on aggregates of soil preserved in the laboratory sample show three or four distinct subsets (Table 4). A dark brown color is prominent on the concave sides of cavities that appear to be feces of round flat forms, not unlike cow dung in form. Similar dark-colored material is present on the convex sides of the above shells that appear to be openwork adobe material, including some quartz and magnetite grains up to 2 mm in size. The binder appears to be the darker silty clay variety of soil. Within the shells is a light-brown clayey silt. The shells appear to be remnant structures from the termite nest. Tube-shaped openings provide access to the chamber that is of the order of centimeters in size. In some samples, large quartz grains as well as some structures of darker soil stand above the level of the concave surface. Occasionally, a layering in the finer-grained concave walls is apparent, and in some cases a concentric structure reminiscent of drop-forms is apparent. Occasionally termite body parts appear in the matrix of the soil, including both mandibles and head of *Cubitermes.*

Particle Size

The particle size distributions shown in Table 4 indicate vastly different proportions of sand (25–48%) and clay (12–56%) among the ingested samples. Buds 2–4 range in texture from clay loam to clay. The Bud 3 sample is essentially a claystone. The control samples contain more sand and less clay, and may be classified as sandy loam, with little clay-size material. Although the limited number of samples available precludes formal statistical testing, the textural differences between ingested and uneaten materials are considered significant. For example, according to the sample standard deviation (ingested samples), there is a less than 16.5% chance that clay content would be less than 17.0% or sand content greater than 46.5% (Table 4). Silt content varies among samples, but differences between the ingested and control soils do not appear to be significant.

Mineralogy of the Sand and Silt Fraction

The sand and silt includes a mixture of strongly cemented soil and angular quartz in the coarse fraction. Medium sands are composed of angular quartz, representing basement gneiss mineralogy. Fine sands and silts include an assortment of round worn mineral grains, including rutile, zircon monazite, and Ti–Fe oxides. These minerals come from tillite beds (ancient glacial materials) near the head of the watershed south of the limit of the forest. At the site, this fine-grained worn material represents alluvium deposited within locally derived residual grit.

Mineralogy of the Clay Fraction

The <2-μm fraction of the samples analyzed in the ingested group has a clay mineral component that is exclusively kaolinite, halloysite, and metahalloysite. These clay minerals all belong to the 1:1 (Si:Al = 1:1) group and in the present case exhibit excellent crystallinity. Kaolinite is the most abundant, followed by metahalloysite and halloysite. The primary mineralogy of the ingested samples includes small amounts of quartz, mica, and plagioclase feldspar. While these minerals could supply small quantities of Si, Al, O, and a range of metal cations, there is no known nutritional/dietary/pharmaceutical significance to their presence in the sample suite. Within the control group (Bud5), only one sample had sufficient mass to warrant clay and primary mineral analysis. The trace from Bud 5 showed moderate amounts of metahalloysite but no kaolinite or halloysite, along with limited quartz and virtually no feldspar or mica within the primary minerals.

Soil Chemistry

Colors of the dried samples shown in Table 5 range from a reddish brown hue (5 YR) for the ingested samples to a lighter 10YR color for the control (Bud 5). The colors indicate advanced liberation of Fe and, in some cases, incorporation of organic matter in small quantities. Indeed, the Bud 1 clay slurry in the laboratory showed the presence of white-colored microbes, presumably bacteria, after dispersion and particle size analysis. The pH of the ingested samples ranges from alkaline (Bud 4) to slightly (Bud 1 and 3) and moderately (Bud 2) acidic. The control sample is slightly acidic, with a pH of 6.1 recorded.

The total salt content as indicated by electrical conductivity (Table 5) is low in Bud 1, 3, and 4 and somewhat higher in Bud 2. In general the conductivity is close to the control sample, as expected in well-drained and leached tropical soils.

Table 5. Selected physical and chemical characteristics of soils in the Budongo sequence

Sample	Dry color	pH (1:5)	Electrical conductivity (μS/cm)	C (%)	*n* (%)
Ingested					
Bud 1	7.5YR 5/6	6.20	164	3.7	0.40
Bud 2	7.5YR 4/3	5.55	734	[a]	[a]
Bud 3	7.5YR 3/4	6.12	367	[a]	[a]
Bud 3a	10.0YR 5/1	[a]	[a]	[a]	[a]
Bud 4	10.0YR 5/3	7.58	420	[a]	[a]
Control					
Bud 5	10YR 5/4	6.12	153	2.9	0.27

[a] Insufficient sample material available for analysis.

The elemental chemistry of ingested and uneaten soils is shown in Table 6. Consistent with the clay mineralogy, the most abundant measured element is Al, comprising 5.8–7.5% by weight of each sample. Iron is also abundant (1.7–5.7%), particularly in the ingested soils. While Mg comprises about 1.0–1.25% by weight of the soils examined, other major elements (Ca, K, Na, and Ti) are relatively rare. Trace elements detected using the INAA procedures include As, Br, and Cr. Iodine was usually below detection limits.

Here too, statistical testing is limited by the small sample sizes available; however, once again considering the standard deviation of the ingested materials,

Table 6. Concentration of chemical elements in the Budongo sequence

Sample	Al (%)	Ca (%)	I (ppm)	Mg (%)	Mn (ppm)	Na (%)	Ti (ppm)	K (%)	As (ppm)	Br (ppm)	Fe (%)	Cr (ppm)
Ingested												
Bud 1	7.54	<0.20	<7.7	–	1376	0.06	8998	0.27	3.24	15.0	5.70	102.0
Bud 2	6.37	0.34	<6.3	1.24	1281	0.07	8560	0.32	1.71	14.1	4.20	86.0
Bud 3	6.91	0.30	<4.6	1.23	878	0.04	6716	0.19	2.43	11.4	4.64	91.1
Bud 3a	6.01	0.39	<6.9	–	1181	0.05	6980	0.17	2.67	17.9	4.39	79.0
Bud 4	6.87	0.38	9.1	1.06	798	0.04	7014	0.21	2.85	11.6	4.65	89.7
Mean	6.74	0.33	–	1.18	1103	0.05	7654	0.23	2.58	14.0	4.72	89.6
SD	0.58	0.05	–	0.10	253	0.01	1054	0.06	0.57	2.68	0.58	8.39
Plus[a]	5.58	0.23	–	0.98	597	0.04	5564	0.11	1.44	8.70	3.56	77.8
Minus[b]	7.90	0.42	–	1.38	1609	0.07	9744	0.35	3.72	19.4	5.88	106.3
Control												
Bud 5	5.81	0.36	<5.6	1.00	1098	0.05	7550	0.89	1.00	8.17	1.67	52.9

[a] ±2 standard deviations.

significant differences between ingested and uneaten materials appear to occur with at least 5 of the 11 elements measured. Table 6 gives the two standard deviation ranges (95% confidence interval) for ingested sample materials assuming elemental concentrations are normally distributed. In the control sample As, Br, Cr, and Fe fall below this range while K falls well above. Al and Mg also appear to be less abundant in the uneaten control soil. The most significant differences between the two soils occur in the case of Fe and K. While the former is on average 2.8 times more abundant in the ingested materials, the latter appears to be depleted by approximately 75%. Given the range of variability in Ca, Mn, Na, and Ti among samples, there is no evidence of any differences between the ingested and uneaten materials on the basis of the concentration of these elements.

DISCUSSION

This is the first detailed report of geophagy in chimpanzees of the Budongo forest. While it is clear that chimpanzees are selecting termite mound clay, they also appear to be selecting termites themselves in many cases. The size of our data set is admittedly small, which prevents an in-depth analysis of the possible ecological or health-related factors responsible for geophagy in this population. Nonetheless, we were able to add new insights into geophagy in primates in general, provide new details from this site, and further confirm trends in the chemical and mineralogical contributions of the soils selected by chimpanzees for consumption across East Africa.

Anecdotal evidence suggests that on some occasions around the time geophagy was observed, individuals in the group were suffering from gastrointestinal upset (i.e., diarrhea), suffering from influenza-like symptom (i.e., coughing), or were feeding excessively on unripe fruits. Some of these symptoms might have been partially relieved by the ingestion of clay. From our analyses to date, we have established that, like other previous reports of geophagy in primates, a major self-medicative value of this behavior is likely to be its ability to soothe the stomach via the physical absorption of stomach acids and plant or pathogen-related toxins in the gut. Future studies of geophagy will require greater real-time correlation at the individual level between diet and geophagy to more adequately address the immediate stimuli for and effects of geophagy.

The geomorphic "flat" on which the termite mound occurs most likely is an alluvial landform—either a floodplain or terrace. The abundance of monazite in the heavy mineral fraction of the Bud 1 soil implies probability of a Ce anomaly in the light rare-earth elements. Chromium, which may be an important microelement in nutrition, identified by INAA is derived from Cr–Fe oxide in the heavy mineral suite. The geological source is not apparent in basement gneiss. It may be derived from the tillite in the headwaters of the drainage or from unmapped units in the basement complex.

The ingested soil is high in percent clay relative to controls. The clay mineral composition of the ingested material includes kaolinite, halloysite, and metahalloysite in varying proportions but with kaolinite making up more than 50% of the material in every case. The abundance of kaolinite specifically distinguishes ingested materials from uneaten controls and this appears to be a common phenomenon at sites throughout Africa where chimpanzees are attracted to ancient land surfaces in their quest for earth materials for ingestion (Mahaney *et al.*, 1998; Mahaney, 1999; Mahaney & Krishnamani, 2003). It may be no coincidence that older soils also contain better-developed clay mineral crystals, since refined (pharmaceutical-grade) crystallinity is characteristic of over-the-counter remedies for gastrointestinal upset, such as Pepto-Bismol™ and Kaopectate™. Chimpanzees may consume clay, and especially kaolinite-based soils, to offset gastric upsets and diarrhea and not to negate the positive effects of this for seed dispersal (e.g., Plumptre *et al.*, 1994).

Differences in the chemistry of the ingested and uneaten soils largely correspond to changes in clay content, and support its possible role in stimulating geophagic behavior. Modest increases in Al as well as trace elements, which may occur as adsorbed cations (As, Br, and Cr), are consistent with the observed increase in overall clay content. The differences in Fe and K, however, reflect changes in clay mineralogy, and specifically a shift to more advanced weathering products such as kaolinite or iron oxides (which may occur in trace amounts). While clay content may provide an ultimate (i.e., medicinal) explanation for soil ingestion, it should be noted that differences in color (e.g., reddish hues due to Fe) and potentially odor or taste, as well as site context (i.e., termite mound centennials), may assist chimpanzees in identifying suitable soils for ingestion.

It is noteworthy that subjects were observed attempting to exploit more indurate soils at the base of the termite mounds as well as at the top, since color, odor, and taste rather than texture would distinguish this material from the uneaten control soils. The olfactory response to this material may be the clayey soil, characterized as having an unctuous odor. Similarly, maillot has a distinctive smell and is prominent in the soil in association with the remains of termites within the cell walls of in situ soil crumbs as well as on grains retrieved in sieve analysis. The fungus *Penicillium* is prominent on mounds of *Odontotermes* and *Pseudacanthotermes* (Ketch, 1998; Ketch *et al.*, 2001).

It is uncertain from the limited number of elements analyzed whether the ingested soils might help counter nutritional or other dietary deficiencies. While the carbon and nitrogen analyses are a minimum, the trend reported here indicates that the ingested material is higher in carbon (possibly because *Cubitermes* is humiverous, and both building material food and feces tend to be richer in carbon), which means the bacteria, mold, and fungi counts are higher as well, a factor we have not seen at other geophagy sites. This may mean the organisms have found a microbe that is beneficial to them, possibly one that fights off disease, and chimpanzees may benefit from this too. Further research is required to determine if there is a microbial substance in the ingested material that is of pharmaceutical importance to the chimpanzees (e.g., Ketch, 1998; Ketch *et al.*, 2001). Consideration should be given as to the effect of the tree species (*Celtis durandii* syn. *C. gomphophylla* Bak.) found in association with this eaten *Cubitermes* mound soil as it differs from species documented elsewhere (e.g., *Cynometra alexandri* Wright (Ironwood); Newton-Fisher, 1999b). The contiguous root system around the respective mounds may also impart a biogeochemical character to the soil that prompts its consumption.

Future work at Budongo on aspects of self-medication and disease are strongly encouraged, and are expected to provide a body of information invaluable for comparison with other long-term great ape study sites across Africa where similar data is now being collected. Beyond the direct value of such studies in better understanding the ecological and disease-related impacts on behavior, these studies are expected to add essential information for the conservation of great apes and their habitats across Africa.

ACKNOWLEDGMENTS

NORAD funded Mnason Tweheyo's PhD research work in Budongo. Michael A. Huffman's work in Budongo was funded by a grant from the Leakey Foundation, and the J. William Fulbright Scholarship Foundation funded Paula A. Pebsworth. The physicochemical and mineralogical analyses were funded by Quaternary Surveys, Toronto. NORAD, National Geographic Society, the Uganda National Council for Science and Technology, and the Uganda Forest Department provided invaluable support for all research at Budongo. We are grateful to Professor Kaare Lye of the Agricultural University of Norway, Dr. John Kaboggoza of Makerere University, and Monday Gideon Mbotella and Kiwede T. Zephyr for their direct input and assistance. We extend our thanks and appreciation to all the research assistants and support staff of the Budongo Forest Project. Dr. Tim Myles, University of Toronto, provided assistance in the identification of *Cubitermes* remains in the soil samples. Last but not least, we thank Massimo Bardi for his valuable advice in the preparation of the paper and advice on data analysis.

CHAPTER NINE

Nutritional Aspects of the Diet of Wild Gorillas

How Do Bwindi Gorillas Compare?

Jessica M. Rothman, Alice N. Pell, J. Bosco Nkurunungi, and Ellen S. Dierenfeld

INTRODUCTION

The foraging behaviors and diet selection of animals in their natural habitats have been the focus of ecological field studies for decades. Diet affects the lifetime fitness of primates and reproductive performance, an essential consideration for highly endangered species (Altmann, 1991, 1998).

The abundance of tropical arboreal folivores is positively correlated with the protein content of leaves and negatively correlated with fiber (Waterman *et al.*, 1988; Oates *et al.*, 1990; Ganzhorn, 1992; Chapman *et al.*, 2002; Chapman *et al.*, 2004). Mountain gorillas in Rwanda prefer foods that are high in protein and their movement patterns are related to the quality and availability of foods (Vedder, 1984). There is evidence that some primates in tropical forests

Jessica M. Rothman and Alice N. Pell • Department of Animal Science, Cornell University, Ithaca, NY 14853, USA **J. Bosco Nkurunungi** • Department of Zoology, Makerere University, Kampala, Uganda **Ellen S. Dierenfeld** • Department of Animal Science, Cornell University, Ithaca, NY 14853 and St. Louis Zoo, 1 Government Drive, St. Louis, MO 63110, USA.

Primates of Western Uganda, edited by Nicholas E. Newton-Fisher, Hugh Notman, James D. Paterson, and Vernon Reynolds. Springer, New York, 2006.

have difficulty obtaining required amounts of certain minerals such as sodium (Rode *et al.*, 2003). In western forest clearings, gorillas groups congregate with other animals to seek out water plants that are high in minerals (Magliocca and Gautier-Hion, 2002). Understanding of nutrient requirements and which plants are preferentially consumed can be used to establish priorities for habitat conservation.

Approximately 320 gorillas live in Bwindi Impenetrable National Park (BINP), a 360-km^2 forest "fragment" in southwestern Uganda (McNeilage *et al.*, 1998, 2001). Bwindi gorillas are currently classified as mountain gorillas (*Gorilla beringei*) (Groves, 2001). They are genetically similar to populations of mountain gorillas in the Virunga Volcanoes area in Rwanda (Garner & Ryder, 1996; Jensen-Seaman & Kidd, 2001), although some have questioned their taxonomic classification at the subspecies level (Sarmiento *et al.*, 1996).

Here, we review the nutritional ecology of gorillas through a discussion of diet, food choice, and digestive anatomy. We compare the nutritional content of foods eaten by gorillas living in different habitats to those consumed by the Bwindi gorillas and discuss the nutritional factors that may influence food selection. The effects of the physical environment on the nutritive value of gorilla foods and some of the analytical challenges that arise on cross-study comparisons are also considered.

DIET AND FORAGING STRATEGY OF GORILLAS

The composition of the diet of wild gorillas has been studied at several sites across central Africa. An initial understanding of gorilla feeding ecology was obtained from the mountain gorillas (*Gorilla beringei beringei*) in the Virunga region on the borders of Rwanda, Democratic Republic of Congo, and Uganda, an area characterized by afro-alpine vegetation with few or no fruiting trees. These gorillas are folivores whose diet consists primarily of a few species of herbaceous leaves, vines, stems, and shoots (Schaller, 1963; Fossey & Harcourt, 1977; Vedder, 1984; Watts, 1984, 1990, 1996; McNeilage, 1995, 2001). Thistle, wild celery, and *Galium* spp. are primary components of the diets of these gorillas (Watts, 1984). Diet remains relatively constant throughout the year, except for the seasonal consumption of bamboo shoots. Groups of mountain gorillas at lower elevations in the Virungas eat different, more varied foods and more fruit than those at higher altitudes (McNeilage, 2001).

In the Kahuzi-Biega National Park of the Democratic Republic of Congo, eastern lowland gorillas (*Gorilla beringei graueri*) eat more fruit than the Virunga population, reflecting the greater distribution and availability of fruit in the Kahuzi-Biega habitat (Yamagiwa *et al.* 1994, 2003). When fruits are unavailable, these gorillas supplement their diet with herbaceous vegetation and the bark of trees and woody lianas (Casimir, 1975; Yamagiwa *et al.*, 1994).

Overall, the western gorilla (*Gorilla gorilla gorilla*) diets are more diverse than those of the mountain gorillas; as many as 230 different plant parts of 180 species are consumed at some locales (e.g. Williamson *et al.*, 1990; Tutin & Fernandez, 1993; Nishihara, 1995; reviewed in Doran & McNeilage, 1998; Doran *et al.*, 2002; Rogers *et al.*, 2004). If fruits are available, they are consumed regularly. Sweet fruits are preferred, but astringent fruit is eaten when sugary fruits are scarce (Tutin & Fernandez, 1985; Rogers *et al.*, 1990; Williamson *et al.*, 1990; Nishihara, 1995; Remis, 1997; Remis *et al.*, 2001; Goldsmith, 1999; reviewed in Doran & McNeilage, 1998, Doran *et al.*, 2002, Rogers *et al.*, 2004). Along with fruit, western gorillas also eat large quantities of terrestrial vegetation throughout the year (Doran & McNeilage, 1998; Rogers *et al.*, 2004). In some areas, gorillas forage in swampy open forest clearings regularly to consume the readily available ripe fruits (Doran-Sheehy *et al.*, 2004), and/or mineral-rich water plants (Kuroda *et al.*, 1996; Magliocca & Gautier-Hion, 2002).

The altitude of the range of the Bwindi gorillas, between 1160 and 2600 m above sea level is similar to that of the gorillas living in the lower ranges of the Virungas and the higher ranges of the eastern lowland populations. Bwindi gorillas eat more diverse diets than do the gorillas in the Virungas. Between August 2002 and 2003, a single Bwindi group (Kyagurilo) consumed at least 160 parts (leaves, stems, fruits, etc.) of 105 different plant species (J. M. Rothman, unpublished data), and, in earlier years, the same group consumed at least 133 parts of 96 species, some of which differed from those eaten in the later study (Stanford & Nkurunungi, 2003). Gorilla groups at different sites within Bwindi consumed a more varied diet than the Kyagurilo group (Ganas *et al.*, 2004).

The Bwindi gorillas consume more fruit than the Virunga gorillas, but less fruit than the eastern and western lowland gorillas (Goldsmith, 2003;

Robbins & McNeilage, 2003), an observation based on the presence of seeds in fecal samples. Greater fruit consumption is likely a result of fruit availability (Goldsmith, 2003; Robbins & McNeilage, 2003; Stanford & Nkurunungi, 2003; Nkurunungi, 2005). Bwindi gorillas depend heavily on terrestrial herbaceous vegetation (Goldsmith, 2003) and consume it daily, even when fruit is in abundance.

In addition to herbaceous vegetation and fruits, Bwindi gorillas also consume mature and young true leaves, woody lianas, bark, epiphytes, moss, and a fungus. Bwindi gorillas spend some of their foraging time eating dry, decaying wood and bark (Goldsmith, 2003; Stanford & Nkurunungi, 2003), which has little obvious nutritional value (J. M. Rothman *et al.*, in press).

Insectivory and geophagy have been recorded at most study sites (Insectivory: Tutin & Fernandez, 1983, 1992; Harcourt & Harcourt, 1984; Carroll, 1986; Watts, 1989; Nishihara & Kuroda, 1991; Yamagiwa *et al.*, 1991; Kuroda, 1992; Nishihara, 1992; Kuroda *et al.*, 1996; Remis, 1997; Deblauwe *et al.*, 2003; Geophagy: Schaller, 1963; Fossey & Harcourt 1977; Watts, 1984, 1989; Mahaney *et al.*, 1990, 1995a,b; Williamson *et al.*, 1990). Insect consumption has been observed rarely, among both Bwindi and Virunga gorillas (Fossey & Harcourt, 1977; Harcourt & Harcourt, 1984; Watts, 1984, 1989; Stanford & Nkurunungi, 2003; Ganas & Robbins, 2004). In Bwindi, groups at lower elevations in Bwindi seasonally ate ants more frequently than did a group living at a higher elevation (Ganas & Robbins, 2004). Western gorillas eat insects more frequently than do mountain gorillas. In Cameroon, 74% of 34 fecal samples contained insect remains (Deblauwe *et al.*, 2003). Geophagy may be a means to adsorb plant toxins in the diet or relieve stomach discomfort (Krishnamani & Mahaney, 2000). Occasionally, gorillas consumed rocks and soil in Bwindi, either deliberately or accidentally (Stanford & Nkurunungi, 2003; J. M. Rothman, unpublished data).

Coprophagy by wild gorillas is rare. In the Virungas, only 25 instances of coprophagy occurred over thousands of hours of observation (Harcourt & Stewart, 1978). Similarly, the Bwindi gorillas were observed reingesting their own feces only twice during a year of regular observations and a juvenile ate the feces of another individual once (J. M. Rothman, unpublished data). Because this behavior is so uncommon, its nutritional significance is limited. Proposed reasons for gorilla coprophagy include boredom or the craving for warm food on cold days (Harcourt & Stewart, 1978). The rare instances of coprophagy in Bwindi occurred when the gorillas were eating fruit, and the ingested feces contained large amounts of seeds. It is possible that the gorillas were reingesting the feces to gain nutrients from the seeds (Krief *et al.*, 2004).

DIGESTIVE ANATOMY AND PHYSIOLOGY

Gorillas are the largest extant nonhuman primate. Male gorillas weigh between 139 and 278 kg, and mature females weigh between 72 and 98 kg (National Research Council, 2003). Mountain gorillas are larger in size and mass compared to lowland gorillas. Captive gorillas weigh more than wild gorillas, but the weight of the latter tends to be estimated (Leigh, 1994).

The anatomy of the gorilla provides a large capacity for microbial fermentation and subsequent energy gain through the absorption of volatile fatty acids and microbial protein (Stevens & Hume, 1995). Gorillas have a large, pouched colon, which, based on a male specimen, is about 200 cm long and has a maximum width of approximately 30 cm in the lower ascending colon (Steiner, 1954). While gorillas are frequently considered to be ceco-colonic fermenters, their cecum is relatively small: cecal volume is only about 14% that of the colon, typical of animals that are folivorous and frugivorous (Chivers & Hladik, 1980).

On a captive zoo diet of cultivated fruits, commercially prepared primate biscuits, mixed greens, and hay, captive lowland gorillas had a mean retention time of 50 h, with a range of 16.5–136 h measured with a chromium marker (Remis, 2000). Particulate material is retained longer than liquid material (Remis & Dierenfeld, 2004). When gorillas were fed a diet of 30% neutral detergent fiber (NDF) and 7% acid detergent fiber (ADF), NDF digestibility was 70% but, when the diet contained 30% NDF and 19% ADF, the apparent digestibility of the diet decreased to 45% (Remis & Dierenfeld, 2004). Energy derived from microbial fermentation in the form of short-chain fatty acids may provide between 30 and 60% of the maintenance energy requirements of gorillas (Popovich *et al.*, 1997). Most fiber digestion in the gorilla probably occurs in the colon, but there is no research on sites of fermentation in the gorilla. Although gorillas preferentially consume nutritious, readily digestible fruits and herbaceous leaves, they can subsist on fibrous fruit and vegetation when succulent fruits are not in season or available in a habitat. Mobilization and repletion of body reserves provide nutritional safety nets to many species of animals faced with recurring feast or famine conditions. The extent to which gorillas lose weight during lean times and regain it during times of abundance has not been established.

CROSS-SITE COMPARISONS

Environmental Effects on Nutritional Composition

The nutritional composition of plants is affected greatly by external factors in the environment. Soil, water, carbon dioxide, soil nutrients, stress, disease, predation, and weather all influence plant metabolism and the production of resistant structures (Van Soest, 1994). For example, plants that grow at higher temperatures accumulate structural cell wall more quickly than those that grow at lower temperatures. In cooler eastern afro-alpine gorilla habitat, leaves contained more protein and less fiber, were more digestible and richer in phosphorous than foliage from western rain forest trees (Waterman *et al.*, 1983).

Considerable between- and within-plant variation in chemical composition has been observed in Kibale forest trees (Chapman *et al.*, 2003). Canopy leaves eaten by lemurs in Madagascar contained more protein than leaves from undisturbed areas in the same forest (Ganzhorn *et al.*, 1995). The nutrients in the leaves eaten by colobus monkeys changed seasonally, which influenced the monkeys' food choices (Baranga, 1983). Therefore, a chemical analysis provides only a snapshot of the nutrient content of a particular plant at a single point in time. Careful sampling protocols are needed to ensure that the samples collected reflect what the target animal actually consumed, and that seasonal and spatial variation in food composition are accounted for (Chapman *et al.*, 2003).

Standardizing Analytical Techniques

Standardization of the techniques used to measure the nutrient content of primate foods and nonfoods in habitats is necessary to ensure that cross-site comparisons are valid (Chapman *et al.*, 2003). Freeze-drying, a very effective means of preserving plants, frequently is not an option at field stations, but drying at temperatures less than 45°C in low light minimizes the effects of temperature and oxidation (Palmer *et al.*, 2000). Physical processing, including the type of mill and size of the screen used, also is an important consideration because particle size affects many chemical analyses (Mullin & Wolynetz, 1995).

Choosing an appropriate analytical technique is essential. While a comprehensive analysis of appropriate techniques for measuring the nutrient content of primate foodstuffs is beyond the scope of this review, there are a few problems with commonly used analytical techniques that merit consideration.

Condensed Tannins

Many different methods are used to estimate the condensed tannin content in animal diets (Waterman & Mole, 1994), each of which has its strengths and limitations. The butanol–HCl assay to measure condensed tannins has been the most commonly used for the tannin analysis of primate foods (Porter *et al.*, 1986). In this spectrophotometric assay based on oxidative depolymerization of the condensed tannin and measurement of red anthocyanidins, the tannin usually is extracted with acetone or methanol. The quantity of tannin is estimated using a standard curve (absorbance vs. tannin concentration) of a known compound, usually quebracho. However, reliance on external standards may pose analytical problems because of the great variation in the structure and biological activity of tannins. When external standards are used, it is implicitly assumed that all tannins have the same absorbance per unit mass as the standard. In some plants with high levels of biologically active tannins such as *Desmodium ovalifolium*, the estimates of tannin content may be off by a factor of 5 if a quebracho standard is used instead of relying on an internal standard (Giner-Chavez *et al.*, 1997). The case for extracting tannins from each plant species to develop internal standard curves has been forcefully made by Giner-Chavez *et al.* (1997) and Schofield *et al.* (2001). Controlling for color interference in samples is also necessary because many fruits and flowers contain red pigments that interfere with condensed tannin measurements (Watterson & Butler, 1983).

In Vitro Dry Matter Digestibility

Choo *et al.*'s (1981) in vitro digestibility method using a fungal cellulase is often used to estimate the digestibility of primate foods. Although fungal cellulase assays permit easy comparisons of foods and minimize interlaboratory variation, fibrolytic enzymes of fungi differ from those in bacteria, resulting in different estimates of digestibility. The various fungal and bacterial enzymes in an animal's gut act synergistically to digest a mixed diet (Béguin & Aubert, 1994). In vitro assays that use digestive fluids from domesticated ruminants have the advantage of using live microbial cultures to simulate digestion, but microbial populations and their activities are affected by the physiological status and diet of individual animals (Palmer *et al.*, 1976; Pehrson & Faber, 1994). It has not been established that the species, and activities, of ruminal microbes are similar to those from the ceco-colon, which is likely the primary fermentation site in the gorillas.

Digestibility also can be estimated on the basis of plant lignin content. Lignin is largely indigestible by both mammals and microbes, so high lignin foods are poorly digested (van Soest, 1994). In vitro methods may provide a means of ranking feed ingredients by digestibility, but the actual data may differ from in vivo observations. Animals eat mixed diets and the associative effects of one ingredient often affects the digestibility of another. Therefore, the application of these digestibility numbers to estimate intake or other parameters of interest should be performed cautiously.

Lipids

In the few studies with measurements of the fat content of primate diets, the ether extract method has usually been used (Horwitz, 2000). While this method is appropriate for the measurement of triglycerides, plants have non-fat components that are extracted by ether, such as wax, cutin, galactose, chlorophyll, glycerol, and other compounds that cannot be saponified and that are frequently indigestible (Palmquist & Jenkins, 2003). In a study by Palmquist and Jenkins (1980), forage leaves contained 5.3% fat as determined by ether extract, but 57% of the ether extract comprised nonnutritive substances. In the gorilla literature, clarification of how much of the ether extract consists of fatty acids is important to resolution of the ongoing debate about whether gorillas choose fatty fruits (Calvert, 1985) or avoid them (Rogers *et al.*, 1990). The fatty acid protocol suggested by Sukhija and Palmquist (1988) (protocol described in Palmquist & Jenkins, 2003) circumvents this problem.

NUTRITIONAL COMPOSITION OF FOODS EATEN BY GORILLAS

The nutritional composition of foods eaten by gorillas in different habitats has been studied in some detail across study sites. Remis (2003) summarized the roles of habitat, body size, and preference on gorilla nutrition, with an emphasis on captive and wild western gorillas. In this section, we provide general comments on the nutritional characteristics of leaves, fruits, and stems eaten by gorillas and review the nutritional composition of foods and aspects of food selection in different habitats across Africa. The mean values for nutrients at each site for leaves, fruit, and stem are compared in Table 1.

Nutritional Characteristics of Gorilla Foods

Across gorilla habitats, leaves contained more crude protein (16–26% on a dry matter [DM] basis) than stems (3–17% DM) and fruits (6–11% DM) (Casimir, 1975; Goodall, 1977; Waterman *et al.*, 1983; Calvert, 1985; Rogers *et al.*, 1990; Plumptre, 1995; Popovich *et al.*, 1997; Remis *et al.*, 2001; Rothman *et al.*, in press) (Table 1). This is not surprising, since leaves contain appreciable amounts of rubisco, the most common protein in the world that is involved in photosynthesis.

Fruits are high in energy because they usually contain large amounts of easily digested nonstructural, water-soluble carbohydrates in the form of sugars. Fruits eaten by gorillas across sites varied in their sugar content (<70% DM), but fruits contained more sugars than leaves (<12% DM), stems (<31% DM), and bark (<10% DM) across sites (Calvert, 1985; Rogers *et al.*, 1990; Remis *et al.*, 2001; Rothman *et al.*, in press).

All populations of gorillas select some stems in their diet and eat the whole stem or inner core (pith) after removing the bark. Usually, whole stems are fibrous, with a succulent inner core. Young bark may is protein-rich and contains soluble carbohydrates, but, as the bark ages, it becomes less digestible because of the synthesis of complex structural carbohydrates.

Western Gorillas

The available data on the chemical constituents of western lowland gorilla foods highlight the nutritional variability in foods eaten by gorillas living in lowland rain forests (Rogers *et al.*, 1990; Popovich *et al.*, 1997; Remis *et al.*, 2001) and logged/disturbed forests (Calvert, 1985) in West Africa.

In coastal Cameroon, gorillas forage for foods that are higher in protein, more digestible, and lower in lignin compared to foods not chosen. Shoots

Table 1. Mean nutrient content of foods eaten by gorillas across sites

Subspecies	*Gorilla g. gorilla*				*Gorilla b. graueri*		*Gorilla b. beringei*		
Site	Campo, Cameroon[1]	Lopé, Gabon[2]	Bai Hakou, CAR[3]	Bai Hakou, CAR[4]	Kahuzi, DRC[5]	Tshibinda, DRC[6]	Karisoke, Rwanda[7]	Karisoke, Rwanda[8]	**Bwindi, Uganda**[9]
No. of species in diet/No. of parts in diet[†]	50/69	134/144	>200/>300	138/239[10]	121/194[11]	>104/>160	75/38[12]	NA**	**>133/>96**[13]
No. of parts analyzed[†]	36	95	31	68	14	39	33	18	**125**
Leaves									
DM digestibility	45.0	—	—	—	—	66.4	56.5[b]	56.7[b]	—
Protein	16.6[a,b]	17.8[a]	17.3[a]	18.9[a]	20.4[a,b]	25.3[b]	15.5[a]	25.6[b]	**16.9**[a]
Fiber:									
NDF	46.1[a]	—	64.2[b]	64.0[b]	—	—	—	—	**40.2**[a]
ADF	42.6[a,b]	30.1[b,c]	47.7[a]	47.5[a]	—	—	35.5[b]	—	**25.5**[c]
Lignin	19.4[a]	—	—	25.2[a]	—	—	—	—	**9.1**[b]
Condensed tannin[‡]	50%	68%	—	13%***	—	—	14.3%	—	**38%**
Fruits									
DM digestibility	28.4[a]	—	—	—	—	70.7[b]	—	—	—
Protein	6.2[a,b]	6.11[a]	5.7[a,b]	8.4[b,c]	7.5[a,b,c]	13.1[b,c]	—	—	**9.8**[c]
Fiber:									
NDF	64.6[a,b]	—	78.7[a]	59.3[b]	—	—	—	—	**42.9**[c]
ADF	44.8[a,b]	23.9[c]	65.4[a]	45.3[b]	—	—	—	—	**28.0**[c]
Lignin	26.9[a]	—	—	24.9[a]	—	—	—	—	**13.3**[b]
Condensed tannin[‡]	38%	93%	—	41%***	—	—	—	—	**38%**

Stems									
DM Digestibility	41.1[a]	—	—	—	—	—	41.1[a]	—	—
Protein	7.2[a]	5.1[a,b]	3.4[b]	16.9[c]	—	—	6.2[a,b]	—	**6.4[a,b]**
Fiber:									
NDF	54.4[a]	—	80.4[b]	67.4[b]	—	—	—	—	**53.3[a]**
ADF	41.6[a,b]	44.6[a,b]	54.5[a]	49.2[a,b]	—	—	49.3[a,b]	—	**34.3[b]**
Lignin	9.4[a]	—	—	26.0[b]	—	—	—	—	**5.5[a]**
Condensed tannin[‡]	10%	100%	—	0%***	—	—	9%	—	**25%**

* Mean nutrient and dry matter (DM) digestibility values are reported on a DM basis (% DM of the diet) and were recalculated to permit uniform comparisons (Casimir, 1975; Goodall, 1977; Waterman *et al.*, 1983; Calvert, 1985; Rogers *et al.*, 1990; Plumptre, 1995a,b; Popovich *et al.*, 1997; Remis *et al.*, 2001; Rothman *et al.*, in review).

** Study based on food intake of large mammalian herbivores, based on foods identified from gorilla dung.

*** Based on Remis *et al.*, 2001, results of radial diffusion assay.

† As reported by the study unless otherwise noted.

‡ Percent of species analyzed that contain condensed tannins.

[a,b,c] Denotes statistical significance, using Student Neuman-Keul's multiple range and Mann–Whitney U test at $P < 0.05$.

[1] Calvert, 1985; [2] Rogers *et al.*, 1990; [3] Popovich *et al.*, 1997; [4] Remis *et al.*, 2001; [5] Casimir, 1975; [6] Goodall, 1977; [7] Waterman *et al.*, 1983; [8] Plumptre, 1995; [9] Rothman *et al*., in review ; [10] Remis, 1997; [11] Yamagiwa *et al.*, 1994; [12] Watts, 1984; [13] Stanford & Nkurunungi, 2003.

and stems eaten frequently, were lower in lignin (shoots: 11.3%; stems: 11.3%) and higher in moisture (shoots: 89.0%; stems: 86.6%) compared to the rest of the diet, and both rarely contain phenolic compounds (Calvert, 1985). Gorillas preferentially ate mature *Aframomum* fruits that did not contain tannins over immature fruits with tannins (Calvert, 1985). The gorillas also eat the pith of wild and cultivated banana trees, which were higher in protein and gross energy than the discarded outer stem/bark.

In Lopé, Gabon, Rogers *et al.* (1990) determined the nutritional composition of foods eaten by gorillas in lowland rainforest. Gorillas in Lopé typically ate sugary fruits; protein-rich leaves; and succulent, fibrous inner stems. Leaves eaten were lower in fiber and higher in protein than those that were not eaten. Like the gorillas in Cameroon, the Lopé gorillas ate large quantities of pith, which were high in water-soluble carbohydrates (<20.3%) (Rogers *et al.*, 1990). Bark, which contains up to 17% crude protein, is a source of protein for Lopé gorillas, particularly when young leaves are not available (Rogers *et al.*, 1990). Lopé gorillas ate some foods with high levels of ADF (~50%), especially when succulent, sweet fruits were not available, but the most fibrous parts of the plants were discarded. The fruits that Lopé gorillas ate do not have significantly higher sugar content than those eaten less frequently. Unripe fruits that were avoided did not have significantly lower sugar content, but were tougher and more astringent, which may have contributed to their unpalatability. The parts of eaten fruits that were discarded, such as the outer skin of some fruits whose flesh was eaten, contained less sugar and had higher levels of fiber and secondary compounds (Rogers *et al.*, 1990).

Foods eaten by gorillas in the lowland rainforests of Dzangha-Sangha Reserve, Central African Republic, were collected when preferred fruits were not available and analyzed for their nutrient content by Remis *et al.* (2001). The most frequently eaten fruits in the diet of Bai Hakou gorillas were higher in sugar and lower in protien than other eaten fruits. Although the most succulent, sweet fruits were not available during seasons of fruit scarcity, the Bai Hakou gorillas tolerated fruits that contained relatively high amounts of fiber and tannins (Remis *et al.*, 2001).

Tannins are present in the diets of the western gorilla population, but, because of the difficulties in interpreting date from assays standardized with quebracho, whether they affect food selection, whether they are present in large quantities, or whether they have high levels of biological activity is not known.

Eastern Gorillas

Casimir (1975) completed a study of the feeding ecology of eastern lowland gorillas living in the Mt. Kahuzi region in the Democratic Republic of Congo in 1971–1972. The crude protein content of nine frequently eaten leaves was 19.7%. According to the micromineral (Na, K, Ca, Mg) analyses of eight foods, the sodium requirement (generally 0.25% of the diet) was not met, so gorillas may seek mineral supplements and/or consume mineral rich soil (Casimir, 1975).

Goodall (1977) analyzed foods eaten by the eastern lowland gorillas. Bamboo shoots eaten seasonally contained 47% crude protein. The percentage of crude fiber in the samples was relatively low (range: 7.3–43.1%), with the highest fiber found in bark (Goodall, 1977).

In terms of diet composition and variability in feeding habits of study groups, and individuals within those groups, the Virunga gorillas are well-studied compared to other populations (Vedder, 1984; Watts, 1984; McNeilage, 2001). Nutritional analyses of frequently consumed gorilla leaves and stems were completed by Waterman *et al.* (1983): leaves contained more protein (leaves: 15.5%, stems: 6.2%) and were more digestible (leaves: 56.5%, stems: 41.1%) (digestibility estimated using fungal cellulases) than stems. Digestibility was negatively correlated with ADF. Minerals in the mountain gorilla diet (P, K, Ca, Mg) were more concentrated in leaves than in stems (Waterman *et al.*, 1983). The protein content, digestibility, and the ratio of protein to digestion inhibitors were higher in preferentially consumed foods than in foods that were not favored (Watts 1984, 1990). Most of the plants consumed were low in phenolics, and few plants contained condensed tannins (Waterman *et al.*, 1983). Plumptre (1995), who analyzed foods eaten by large mammalian herbivores and related these data to their consumption and proportion in the diet, found that gorillas selected the most digestible foods in the habitat.

Bwindi Gorillas

When compared to gorillas at other sites, the Bwindi gorillas ate leaves that have similar levels of protein (16.9%) and lignin (9.1%) (Table 1). Bwindi gorillas consumed fruits that were higher in crude protein than the fruits eaten by the western gorillas. In the Bwindi gorilla diet, fruits (15.3%) and pith (8.4%)

have the most water-soluble carbohydrates compared to other plant parts, and the most frequently eaten fruit (Robbins & McNeilage, 2003) is the sweetest (*Myrianthus holstii*: ~50% sugar). Bwindi fruits were significantly less fibrous than those eaten by western gorillas in Campo and at Bai Hakou (Table 1). Both the leaves and fruits eaten by Bwindi gorillas have less lignin than those eaten by other groups across sites and are more digestible: cooler temperatures in Bwindi may contribute to these differences. Thirty-eight percent of the leaves consumed in Bwindi contained condensed tannins compared to 14% of the leaves eaten by gorillas in the Virungas. The occurrence of condensed tannins was not as widespread in the Bwindi gorilla diet as in the diets of western gorillas (Table 1).

Bwindi gorillas frequently consumed dry, decaying wood stumps, an interesting phenomenon that is difficult to explain. During 1 year of observations, a group of gorillas ate wood from up to 12 species of trees (J. M. Rothman, unpublished data). Often individuals carried the wood for several meters, and one juvenile carried wood for 100 m before eating it. The wood is high in lignin (~40%) and low in protein (~3%) and water-soluble carbohydrates (<1%) (Rothman *et al.*, in press). It is possible that wood is eaten for nonnutritional reasons, possibly serving a prophylactic medicinal purpose (Cousins & Huffman, 2002). The wood could also be used to aid digestion by altering passage rate or filling the gut. Nevertheless, why Bwindi gorillas eat wood is not obvious and is of interest for future research.

Some groups of Bwindi gorillas range outside of the park and raid crops in nearby villages. They eat the pith of banana stems (*Musa* spp.) and the midrib of banana leaves (J. M. Rothman, unpublished observation). Both of these food items are high in water; banana pith is 96% water and the midrib of the leaves is 90% water (Rothman *et al.*, in press). Otherwise they were lower in protein compared to the rest of the diet (pith: 8.6%, midrib: 4.2%) and similar in NDF and ADF content (pith: NDF = 50.0%; ADF = 30.9%), but lower in lignin (pith: 2.0%; midrib: 2.6%). The pith of *Musa* sp. has 5.0% water-soluble carbohydrates (Rothman *et al.*, in press). The gorillas may be raiding crops to eat the pith for its water content, but minerals and other compounds were not analyzed.

Nkurunungi (2005) examined the habitat use patterns of the Bwindi mountain gorillas and assessed the availability of food plants through study of one group. According to the analysis of feeding sites using an indirect method and calculations of Ivlev's electivity index, Bwindi gorillas are selective feeders that consume some herbs, shrubs, and fruits at a higher frequency than their availability in the environment would predict. Nkurunungi (2005) identified the foods that the gorillas preferred, on the basis of frequency of consumption, and the foods selected, the basis of the availability of these plants in their range. These estimates were not based on intake, and so the results are biased toward food items that were eaten in small quantities on a regular basis.

To better understand food selection, we sought to determine whether the preferred nonfruit foods that were frequently consumed were higher in protein, fiber (NDF, ADF, and lignin), and/or water-soluble carbohydrates than other foods. We used the chemical analyses of gorilla plants (Rothman *et al.*, in review) and compared the mean values of nutrient content between preferred foods and other foods using the Mann–Whitney test. The 10 most frequently eaten nonfruit foods (preferred foods) were significantly higher in crude protein (21.8%) compared to the other nonfruit foods (14.1%) in the diet ($P < 0.05$) (Table 2). There was no difference between the frequently eaten nonfruit foods compared to other nonfruit foods with respect to fiber (NDF, ADF, and lignin). Water-soluble carbohydrates in the preferred foods compared to others were similar.

Selected nonfruit foods (those that were eaten at a higher frequency compared to their availability in the environment) and selected fruits were similar in their contents of protein, NDF, ADF, lignin, and water-soluble carbohydrates (Table 2). Additional studies are underway in our nutritional ecology laboratory (Cornell University) to provide more complete understanding of the nutritional aspects of food preference and selection with respect to estimates of food intake.

Table 2. Nutritional aspects of preference and selection of foods eaten by Bwindi gorillas

	N	Protein	NDF	ADF	Lignin	WSC
Nonfruit foods						
Frequently consumed	10	**21.8**	41.3	28.2	7.4	3.4
Other diet items[a]	69	**14.1**	44.4	29.0	9.2	3.9
Selected	12	16.6	44.3	30.1	9.0	3.6
Not selected[b]	16	14.1	42.6	27.2	9.2	4.8
Fruit foods						
Selected	5	8.3	46.3	26.8	10.1	26.4
Not Selected	5	10.0	44.9	28.6	14.1	15.3

Bold values are significantly different as determined by the Mann–Whitney test ($P < 0.05$), nutrient values are from Rothman *et al.* (in press); all data is presented as percent dry matter.

[a] Dry wood diet items omitted.

[b] Only items sampled by Nkurunungi (2005) were used in the analysis.

NUTRITION OF GORILLAS: HOW DO BWINDI GORILLAS COMPARE?

Bwindi gorillas feed heavily on terrestrial herbaceous plant material, which constitutes the bulk of their diet, but they also eat ripe fruit when it is available. Aside from the common, most frequently eaten foods in their diet, the Bwindi gorillas eat a range of tree barks, leaves, orchids, and the bark and stems of uncommon herbs that they come across on their daily paths.

The preferred foods eaten almost daily by the Bwindi gorillas are leaves, bark, and stems from lianas and shrubs that grow mainly in the large open forest gaps that are common throughout the rugged terrain of the forest. These have more protein than other items in their diet, but similar amounts of fiber. Because of their digestive strategy and large body size, the gorillas probably have the ability to use the lower quality items in their diet to a larger extent than other primates. Fiber estimates in Bwindi gorilla foods were generally lower compared to foods at other sites (Table 1). All gorillas, including those in Bwindi, appear to tolerate foods that contain condensed tannins. The function of dry wood and bark in the diet of Bwindi gorillas is puzzling and studies are ongoing to try to understand its role in the gorilla diet (J. M. Rothman, in preparation).

NUTRITIONAL ECOLOGY: FUTURE DIRECTIONS FOR GORILLA CONSERVATION

All subspecies of gorillas are endangered, with population estimates for *Gorilla gorilla* of ~90,000), and for *Gorilla beringei* of ~3700. For some subspecies, the numbers are even more daunting, *Gorilla beringei beringei*: ~700; *Gorilla gorilla diehli*: ~150. The management of gorilla habitats is more likely to be successful if we know which plants are most important to protect and how many gorillas can be sustained in a specific area (McNeilage, 1995). Baseline studies should address nutrient intake and selection along with assessments of food distribution so that models of carrying capacity and habitat use can be developed to establish management guidelines and to determine which areas are essential for gorillas' survival.

In the Virungas, food distribution and density are important factors for mountain gorillas with respect to habitat selection (Vedder, 1984; Watts, 1991, 1998a). Whether the nutrient content of food has an impact on their density and range has not yet been determined, but these gorillas favor parts of the habitat where they find the most nutritious plants (Vedder, 1984; Watts, 1991) and they revisit areas that have a greater density of high quality food species (Watts, 1998b).

Future studies of nutritional ecology should emphasize the relationship between nutrient requirements and nutrient supply across different habitats with consideration of seasonal and spatial variation. Presently, the entire habitat of the Bwindi gorillas is well protected, but recommendations from nutrition studies could be used to develop habitat management and protection strategies. Although logging and overharvesting are not currently problems for Bwindi gorillas, the same cannot be said of other gorilla populations. Changes in habitat structure and food availability may have serious implications for diet adequacy, gorilla health, and, ultimately, their continued existence. By understanding the nutritional demands imposed by gorillas on their habitat, we will be better able to ensure their continued survival.

ACKNOWLEDGMENTS

We thank the Uganda Wildlife Authority and Uganda Council of Science and Technology for permission to conduct research in Bwindi Impenetrable National Park, and the Institute of Tropical Forest Conservation for logistical support in the field. The Department of Animal Science at Cornell University, Robert G. Engel Family Foundation, Mario Einaudi Foundation, Institute of African Development and the Cornell University Graduate School provide financial support for J. Rothman's research. We thank the editors of this volume and two anonymous reviewers for comments on this chapter.

CHAPTER TEN

The Parasites of the Gorillas in Bwindi Impenetrable National Park, Uganda

Jessica M. Rothman, Dwight D. Bowman, Gladys Kalema-Zikusoka, and John Bosco Nkurunungi

INTRODUCTION

Detecting disease threats to endangered species and their ecosystems plays a crucial role in the survival of a population (McCallum & Dobson, 1995). As human pressure increases around and within habitats that contain endangered species, so does the potential for disease transmission. Communities and wildlife managers must act proactively to discourage and prevent zoonotic disease transmission between humans and endangered wildlife.

Jessica M. Rothman • Department of Animal Science, Cornell University, Ithaca, NY 14853, USA **Dwight D. Bowman** • Department of Veterinary Microbiology & Immunology, College of Veterinary Medicine, Cornell University, Ithaca, NY 14853, USA **Gladys Kalema-Zikusoka** • Conservation Through Public Health, P.O. Box 4483, Kampala, Uganda **John Bosco Nkurunungi** • Department of Zoology, Makerere University, Kampala, Uganda.

Primates of Western Uganda, edited by Nicholas E. Newton-Fisher, Hugh Notman, James D. Paterson, and Vernon Reynolds. Springer, New York, 2006.

Highly endangered mountain gorillas (*Gorilla beringei*) are found in only two isolated populations: the Virunga Volcanoes located on the borders of Uganda, Rwanda, and the Democratic Republic of Congo, and Bwindi Impenetrable National Park (BINP) in southwestern Uganda. Both areas are surrounded by some of the most densely populated and intensively cultivated areas in Africa. Approximately 700 mountain gorillas remain; 320 live in BINP (McNeilage *et al.*, 2001). While mountain gorillas in the Virunga Volcanoes have been extensively studied for the past 35 years (Plumptre & Williamson, 2001; Stewart *et al.*, 2001), those in BINP have not been a focus of the scientific community until recently.

Approximately 20% of the mountain gorilla population in BINP is habituated to humans, for either tourism or research purposes. While strict health guidelines and preventative regulations for park staff, tourists, and researchers have been recommended and/or implemented specifically for mountain gorillas (Homsy, 1999; Mudakikwa *et al.*, 2001; J. M. Rothman, personal observation), they are sometimes not practiced and all recommendations have not been yet been implemented (Woodford *et al.*, 2002; J. Rothman, personal observation). A medical survey of the local human population revealed that there was a high prevalence of disease symptoms compatible with infectious diseases that could be potentially transferable to the gorillas, and, when local people enter the park, more than half do not bury their feces (Guerrera *et al.*, 2003). Bwindi gorillas were victims of *Sarcoptes scabies*, which was likely the result of contact with infected humans (Graczyk *et al.*, 2001a; Kalema-Zikusoka *et al.*, 2002). As a result of habituation for tourism and research, gorillas are less fearful of ranging near park boundaries and outside the park into surrounding villages. This familiarity with people is likely to bring the gorillas into shared habitats more often, with an increased chance of shared disease. In addition, because of civil unrest and political instability in the region, wildlife populations are at higher risk of disease transmission from military presence in parks and human traffic (Mudakikwa *et al.*, 1998; Hamilton *et al.*, 2000; Dudley *et al.*, 2002). As habitats are destroyed, the home ranges of different groups are likely to overlap, and may be intensively overused by gorillas, which may increase their frequency of contact with pathogens.

To mitigate the transmission of parasites from humans to gorillas, park-established rules must be strictly enforced and reviewed in a timely manner (Homsey, 1999). Communities surrounding the park should be recipients of conservation education and they should be monitored and treated for infectious diseases. Occupational health programs should be implemented so that park staff, researchers (both foreign and local) are properly vaccinated and periodically screened for infectious diseases (Nutter & Whittier, 2000). Often overlooked, military and security personnel should be included in conservation plans (Ngabirano, 2005). In addition, local farmers should be educated on practices that would reduce overlapping ranges of wildlife and livestock, and take preventative measures against livestock disease. Crop raiding by gorillas should be discouraged. A population and habitat viability assessment report considered disease to be one of the most devastating threats to the gorilla population (Werikhe *et al.*, 1997), and therefore conservation plans should seriously address, implement, and stringently enforce their policies.

A critical need of disease control management is screening, which is helpful in recognizing and understanding the history of outbreaks (Woodroffe, 1999). The collection and analysis of fecal samples is a less invasive method of understanding aspects of the health of the gorilla population (Ashford *et al.*, 1996) compared to other diagnostic techniques that involve physical contact and/or the administration of anesthesia. Fecal samples can provide information about some parasites, bacterial populations, genetics, nutrition, and stress levels. Regular monitoring and observations of habituated groups by trained park staff already visiting the groups for other purposes can assist in identifying sick animals with clinical symptoms. Opportunistic collection of wild gorilla fecal samples provides important baseline data to compare with habituated groups, and allows for monitoring of rarely encountered gorillas. Protocols for the standardized collection and analysis of fecal samples have been developed by the Mountain Gorilla Veterinary Project (Cooper *et al.*, 1996; Cranfield *et al.*, 2002;). In addition, trained pathologists should necropsy dead animals (Lowenstine, 1990). The Uganda Wildlife Authority (UWA) has done an excellent job of providing veterinary care to the mountain gorillas, and, through their veterinary program, they employ an on-site veterinarian for BINP. The Mountain Gorilla Veterinary Program (MGVP) assists in monitoring the health of the mountain gorillas along with UWA (Cranfield *et al.*, 2002). MGVP's primary objective is to provide emergency care to injured and seriously sick gorillas, and monitor the health of mountain gorilla groups, but they have also contributed greatly to disease surveillance and monitoring, identifying parasites present in the Bwindi gorilla population and carrying out necropsies (Cranfield *et al.*, 2002). Conservation Through Public Health (CTPH) is a newly developed NGO that promotes the health of the local people, their animals, and wildlife. Several other NGOs work together with UWA to monitor the health of the gorillas.

This system-based review focuses on the parasites of the mountain gorillas in Bwindi Impenetrable National Park (BINP). These parasites have been identified at least once in different groups of Bwindi gorillas, thus it is expected that they may be encountered again. Anyone involved in the management of the gorillas, especially those monitoring the gorillas on a regular basis, will find information on the history, morphology, method of transmission, life cycle, and clinical signs of each parasite recorded in the Bwindi population. It is hoped that this review will be a useful reference for those who are interested in learning more about the parasites of Bwindi gorillas.

SURVEY OF PARASITES

The Stomach

Hyostrongylus kigeziensis *[Nematoda, Order: Strongylida, Family: Trichostrongylidea]*

Hyostrongylus kigeziensis was described from specimens collected from a necropsied Bwindi mountain gorilla (Durrette-Desset *et al.*, 1992), and was later described by Sleeman *et al.* (2000) in the Virunga population. Other hosts of this genus of nematodes include the Ethiopian Leporidae, Suidae, and a few Bovidae.

Living worms are often bright red in color. Females are about 1 cm long, males slightly shorter. The male has a distinct copulatory bursa and short stout spicules. *Hyostrongylus* can be differentiated from the *Paralibyostrongylus kalinae*, because the latter has large medial cuticular ridges at the level of the vulva.

At necropsy, the worms may be found on the mucosa of the stomach or in small, ulcerated areas on the mucosal surface. The eggs in the feces are typical strongylid eggs that are about 70–75 μm in length (Figure 1).

The mountain gorilla is the only known host of this species of *Hyostrongylus.* The life cycle of *Hyostrongylus kigeziensis* has not been described but it is expected to be similar to the life cycle of a swine parasite, *H. rubidus.* The life cycle is direct, and from eggs in the feces, first-stage larvae hatch. These feed on soil, grow, and molt to become nonfeeding third-stage larvae that are about

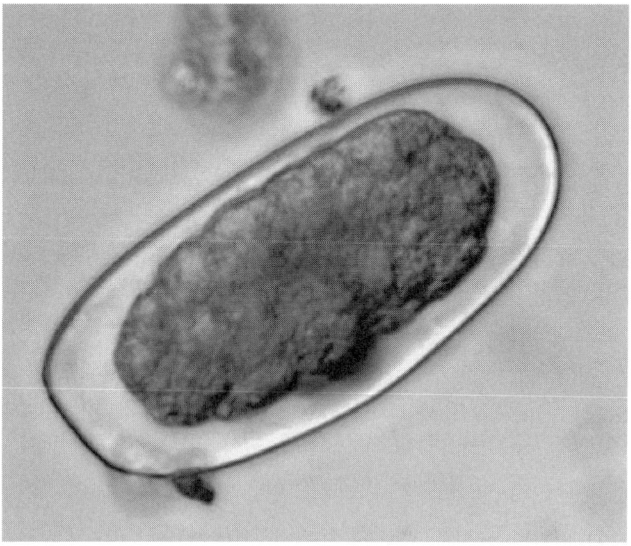

Figure 1. Typical strongylid egg found in the feces of the Bwindi gorilla population.

0.7 mm long and retain the second-stage larval cutical as a protective sheath. Nonfeeding third-stage larvae produced by the second molt are about 0.7 mm long and retain the second-stage larval cuticle as a protective sheath (Alicata, 1935). The gorillas are infected by the ingestion of larval contaminated foodstuffs. After ingestion, all development occurs in the stomach mucosa. Eggs appear in the feces about 3 weeks after infection. It is expected that the adult worms might induce a chronic catarrhal gastritis, leading to the formation of ulcers, as occurs in infected swine (Anderson, 2000).

Paralibyostrongylus kalinae *[Nematoda, Order: Strongylida, Family: Trichostrongylidae]*

This species was named after conservation biologist Jan Kalina, who with her husband Tom Butynski, facilitated the creation of the Bwindi Impenetrable National Park, and discovered this as a new species along with *Hyostrongylus kigeziensis* in the Bwindi population (Durette-Desset *et al.*, 1992).

The living adults are often bright red. There is a large dorsal esophageal tooth present. Females are about 1 cm long, males slightly shorter. The male has a distinct copulatory bursa, and short stout spicules. *Paralibyostrongylus kalina*, unlike *Hyostrongylus kigeziensis*, has large medial ridges at the level of the vulva. The worms are found in the stomach at necropsy. The eggs are typical strongylid eggs (Figure 1), and are very similar to those of *Hyostrongylus*.

On the basis of the life cycle of the related species *P. hebrenicutus*, the host is probably infected through the skin, and the larvae reach the stomach within 2 days. If larvae are ingested through contaminated food or soil, they reach the stomach within 1 day. Larvae localize within the stomach mucosa. Eggs appear in the feces about a month after infection (Cassone *et al.*, 1992).

Small Intestine

Trichostrongylus *sp. [Nematoda, Order: Strongylida, Family: Trichostrongloidea]*

Gorillas are probably not the normal host of these parasites. However, without more careful descriptions, it is not possible to identify the actual host, which is likely a ruminant or a lagomorph.

Trichostrongylus sp. are small, brownish worms that are difficult to discern at necropsy without special care to examine the mucosal scrapings of the small intestine. The eggs in feces are typical of other strongylid nematodes, and may be confused with other parasitic nematodes common in the mountain gorilla. To identify eggs as being those of *Trichostrongylus*, it is necessary to culture the eggs and to examine hatched, developed third-stage larvae. The sheath extending beyond the tip of the tail is shorter than that of most other trichostrongylid parasites (Durette-Desset, 1974).

The life cycle of *Trichostrongylus* in gorillas is not known, but many experimental studies have been conducted in other animals, particularly ruminants and lagomorphs (Audebert *et al.*, 2003; see summaries in Levine, 1980; Anderson, 2000). The host becomes infected by ingesting the infective third-stage larvae. The eggs passed in the feces hatch and develop to first-stage larvae in the feces, where they feed on microorganisms. They develop to the second stage in a few days and then to the third-stage, ensheathed, infective larva. When larvae are consumed, they develop in the small intestine, and adults inhabit the anterior part of the small intestine.

There have been no clinical signs of disease associated with *Trichostrongylus* sp. reported from gorillas; however, the disease manifestations associated with trichostrongylosis is diarrhea that may sometimes be quite severe. In humans infected with *Trichostrongylus*, slight abdominal discomfort is occasionally reported, but usually patients are asymptomatic (Boreham *et al.*, 1995).

Ascaris lumbricoides *[Nematoda, Order: Ascaridida, Family: Ascarididae]*

Ascaris lumbricoides is one of the most common parasites of people around the world and lives in an estimated 1.4 billion people (Crompton, 1999). In 1995, *A. lumbricoides* eggs were found in four fecal samples of Bwindi mountain gorillas (Kalema, 1995). In 1990, a study of the feces of people living around Bwindi forest revealed high loads of *Ascaris*, but there was no evidence of *Ascaris* in gorilla populations (Ashford *et al.*, 1990).

A. lumbricoides is a large cream-colored worm that will reach lengths of up to 30 cm or greater. The worms have three large lips on the anterior end. The tail is relatively short and comes to a rather abruptly pointed tip. The male tail curls ventrally and has no bursa.

A. lumbricoides worms would be difficult to confuse with any of the other parasites that might be found in the small intestine of gorillas at necropsy. There are no reports of necropsied gorillas infected with *Ascaris* sp.

A. lumbricoides eggs are ovoid and about 60 μm long. The eggs have a thick shell and a rough outer coat typically stained by bile to a golden brown color. When passed in feces, the eggs usually contain a single undivided cell. Humans and other primates are the hosts of *A. lumbricoides*, and become infected by ingesting infected soil or foodstuffs. The eggs persist in the soil for long periods and can remain infectious in soil for years. It takes about 2 weeks under optimal conditions for eggs to become infectious through the embryonation of the larvae to the infective stage. After ingestion of an infective egg, larvae hatch. The hatched larvae migrate into the intestinal wall and travel through the portal circulation to the liver, heart, and, through the pulmonary vessels, to their destination, the interalveolar tissues of the lung. The larvae then make their way up the respiratory escalator and are then swallowed. The remainder of the development takes place in the lumen of the small intestine. The prepatent period is about 2 months, and the worms live about 1 year. The females produce prodigious amounts of eggs, having been shown to produce about 200,000 eggs per day (Khera, 2000).

There are no clinical signs reported in mountain gorillas. In humans, light infections usually produce little in the way of signs or symptoms. Heavy infections can cause intestinal obstruction and related signs. It has been shown in people that *A. lumbricoides* infections do have some effect on the nutrition of their human hosts (see review by Crompton & Nesheim, 2002).

Strongyloides fulleborni *[Nematoda, Order: Rhabditida, Family: Strongyloididae]*

Strongyloides fulleborni eggs have been reported in the feces of mountain gorillas in both Bwindi and the Virungas (Ashford *et al.*, 1990, 1996; Kalema, 1995a,b; Mudakikwa *et al.*, 1998; Nkurunungi, 1999; Sleeman *et al.*, 2000).

The parasitic parthenogenetic female is about 3.6–4.6 mm long. Like other *Strongyloides* sp., the female is slender and has an esophagus that is very long: one-fourth to one-third of the total body length. The ovary of *S. fulleborni*, like other species that produce eggs rather than larvae in the feces, spirals around the intestine. The eggs are clear, contain a larva, and measure 50–60 μm by 25–35 μm. It is easy to miss the transparent eggs in fecal samples. If the feces sit for any length of time at room temperature, the larvae are likely to hatch. The larvae of *Strongyloides* can be identified by their possession of a short rhabditiform esophagus with a distinct corpus, isthmus, and bulbus, and a genital primordium that is quite large, being longer than the body is wide (Premvati, 1958).

These worms will be overlooked at necropsy unless techniques are used that are designed specifically for their collection. It is best to use a small portion of fresh bowel (proximal small intestine) and suspend it with weights in a graduated cylinder of saline at 37°C overnight. The small worms will migrate out of the tissue and drop to the bottom of the cylinder. The next morning, the intestine can be removed from the cylinder, and the sediment examined for the presence of the small *Strongyloides.*

S. fulleborni is much more common in nonhuman primates than in humans even where the ranges overlap; it is believed that typically nonhuman primates are the major host of this parasite. The most common mode of infection is penetration of the skin by infective third-stage larvae. In people of the Democratic Republic of Congo, 26 of 76 infants less than 200 days of age were infected with this parasite (Brown & Girardeau, 1977). Examination of milk from nursing mothers revealed three *Strongyloides* larvae in one 2-ml sample of milk from a nursing mother 2 weeks postpartum (Brown & Girardeau, 1977). It is expected that *Strongyloides fulleborni* would also be transmitted in nonhuman primate milk.

The infective larvae that develop in soil are 575–640 μm long and penetrate the skin. When larvae penetrate the skin, they enter the bloodstream and are carried to the lungs. From the lungs, the larvae make their way to the intestinal tract by entering the respiratory openings, being coughed up, and swallowed. It takes 9 days before eggs are passed in the feces. Larvae that penetrate the skin may also enter into muscle and other parts of the body where they are capable of persisting as larvae for extended periods. These larvae form the reservoir of larvae that cause transmammary transmission, which accounts for infection in newborn animals (Premvati, 1958).

The clinical sign of most significance in humans is the "swollen belly" syndrome that has been reported in neonate humans in New Guinea (Ashford *et al.*, 1992). It appears that the infection is well tolerated by adult humans. In the case of primates, there appear to be no reports dealing with the presentation of disease in infected animals.

Anoplocephala gorillae *[Cestoda, Order: Cyclophyllidea, Family: Anoplocephalidae]*

Nybelin (1924) described *Anoplocephala gorillae* from a mountain gorilla inhabiting Mount Sabinio, Kivu Volcano (Virunga Region). Ashford *et al.* (1990) found that Bwindi gorillas had an 85% prevalence of these cestodes in the populations sampled. This finding was confirmed through fecal examination of Bwindi gorilla groups in later surveys (Kalema, 1995a,b; Nkurunungi, 1999). Gorillas in the Virunga region were infected as well, and that prevalence between age–sex classes differed, with infants containing the lowest prevalence (57%) compared with adults (77–100%) (Sleeman *et al.*, 2000). In a study in 1992–1993, Nkurunungi (1999) found that all Bwindi gorillas sampled were infected with *A. gorillae*, with the exception of an infant.

Anoplocephala gorillae are relatively large cestodes with widely segmented thick bodies having a scolex with large muscular suckers and no rostellum. The tapeworms that have been described are about 10 cm long and about 1.5 cm wide (Nybelin, 1924).

At necropsy, large cream-colored worms will be found in the small intestine. Fossey (1983) recovered these worms from the large intestine in one necropsied

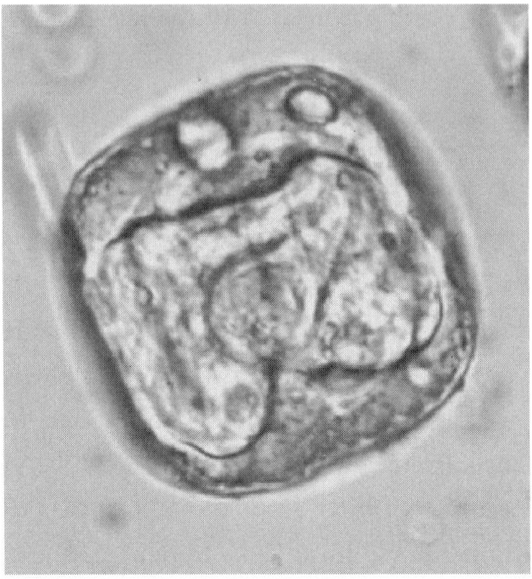

Figure 2. Anoplocephalid egg found in the feces of the Bwindi gorilla population.

Virunga mountain gorilla, and from the small intestine in another. Six of eight Virunga mountain gorillas were found to harbor this parasite (Fossey, 1983).

In sedimentations or direct smears, the eggs of anoplocephalid tapeworms appear rather dark, with a clear central area containing the hexacanth embryo (Figure 2). The shape tends to be that of an irregular spheroid. In sugar flotations, the eggs appear clearer, but the hexacanth embryo will be more evident.

The gorilla is probably infected by the accidental ingestion of infected orabatid mites present in vegetation or soil. In *A. perfoliata*—a related parasite that uses the horse as a final host—soil-dwelling oribatid mites are the intermediate hosts. The mite ingests the egg passed in the feces, and a larval tapeworm matures and grows within the body cavity of the mite. The horse is infected by the accidental ingestion of the infected oribatid mite. All development takes place within the lumen of the horse's intestinal tract (Denegri *et al.*, 1998). It is expected that the life cycle of *A. gorillae* would be similar.

Mudakikwa *et al.* (2001) attributed small-intestinal cestodiasis to *Anoplocephala* in six of eight mountain gorillas. Most Anoplocephalidae are relatively harmless. *A. perfoliata* is associated with ulceration and inflammation of the intestinal wall (Beroza *et al.*, 1983). There are no clinical symptoms described, and on the basis of related species, no significant pathology would be expected.

Giardia lamblia *[Protozoan, Order: Diplomonadida, Family: Hexamitidae]*

Giardia lamblia is a very common parasite and the cause for a reemerging infectious disease (Thompson, 2000). Hastings *et al.* (1992) first described *Giardia* in the feces of necropsied Virunga gorillas. Nizeyi *et al.* (1999) found *Giardia* in two fecal samples, with an overall prevalence in the population of 2% in nonhabituated Bwindi gorillas. The prevalence of *Giardia* in cattle living outside park boundaries in Bwindi can reach 35%, and may be a significant means of transmission to the gorillas (Nizeyi *et al.*, 2002a). People living in communities around the park could also be a source of *Giardia*, and although infection levels of humans living near gorilla habitat were generally low (5%), they should still be considered a potential source of transmission (Graczyk *et al.*, 2002a).

Giardia exists as both a trophozoite and a cyst. The trophozoite of *Giardia* is a 12–15-μm bilaterally symmetrical pear-shaped flagellate. There are two nuclei with central karyosomes, two axonemes, two blepharoplasts, two parabasal bodies, and four pairs of flagella. An ovoid sucking disc occupies approximately three-fourths of the ventral surface. The cyst is ellipsoidal, 9–12-μm, and contains two to four nuclei, with several of the structures of the trophozoite. This flagellate inhabits the small intestine (duodenum and upper jejunum), and at times the bile duct and gall bladder (Meyer, 1994).

At necropsy, fecal samples can be assessed for the presence of trophozoites and cysts.

Trophozoites may be detected in direct smears of diarrheal feces. Cysts may be found through fecal flotation in zinc sulfate.

Humans and other primates, domesticated pets, and livestock are all hosts of *Giardia*. It is probable that mountain gorillas are infected by ingesting contaminated food or water, or through accidental contact with infected feces of man and/or other gorillas. *Giardia* has a direct life cycle (Meyer, 1994). After the infective cyst is ingested, *Giardia* excysts and trophozoites attach to the epithelial cells of the small intestine by the use of its sucking disc. Trophozoites probably feed by absorbing nutrients through their surface and multiply by longitudinal binary fission. Encystation occurs as trophozoites move into the colon. Recently formed cysts have two nuclei, but the trophozoite will divide so that the mature cyst contains two trophozoites. Cysts are the infective stage passed in the feces. The prepatent period ranges from 6 to 15 days, and infections can last months to years. *Giardia lamblia* is the only flagellated parasitic protozoa found in the small intestine, all others are found in the cecum and colon.

There are no published reports of clinical signs of giardiasis in wild or captive gorillas. Clinical signs in infected humans include diarrhea, malabsorption syndrome, nausea, flatulence, and weight loss. Nizeyi *et al.* (1999) report that the stool of gorillas harboring *Giardia* was normal (i.e., containing no blood or mucus).

Encephalitozoon intestinalis *[Protozoan, Order: Pleostphoridida, Family: Encephalitozoon]*

This intracelluar, protozoan parasite recently rectassified as fungi (Thomorat *et al.*, 2004), was found at 3% prevalence in fecal samples of the human-habituated groups of gorillas in BINP (Graczyk *et al.*, 2002b). It was also found in the human population living around BINP at a prevalence of 3% (Graczyk *et al.*, 2002b). *Encephalitozoon intestinalis* infects a wide range of mammals.

Microsporidia are obligate, intracellular, single-celled, spore-producing parasites that are about 1.2×2.0 μm in size (Wasson & Peper, 2000; Weiss, 2001). Diagnosis is through the identification of spores in infected fecal material, biopsy of infected area, and sometimes through the identification of spores in urine, bile, or nasal fluids. PCR and FISH can be used to detect the preserved spores in the biopsied tissues and fluids.

E. intestinalis is transmitted through the ingestion of spores and infects the enterocytes, macrophages, fibroblasts, and endothelial cells. The life cycle is direct. Ingested spores extrude a polar filament that injects parasite DNA into a host cell. Within the host cell, the sporoplasm divides to yield meronts. These mature and yield infective spores. The host cell then bursts and the infective spores are released (Wasson & Peper, 2000). Gorillas infected with *E. intestinalis* had stool containing blood and mucus (Graczyk *et al.*, 2002b). In humans, *E. intestinalis* is most prevalent in immunocompromised hosts and causes ulcerative enteritis and cystitis, colitis, and hepatitis.

Cryptosporidium parvum *[Protozoan, Order: Eucoccidiorida, Family: Cryptosporidiidae]*

Cryptosporidium is a minute coccidian parasite with a large distribution. There are more than 20 species described from fish, birds, mammals, and reptiles.

Nizeyi *et al.* (1999) were the first to describe the presence of *Cryptosporidium* oocysts in the feces of wild gorillas. Using immunofluorescence staining, 11 of 100 Bwindi gorilla fecal samples were positive for cysts (Nizeyi *et al.*, 1999). Molecular analysis revealed that these samples contained *Cryptosporidium parvum*, which can be transmitted zoonotically between mammals and humans (Graczyk *et al.*, 2001b). The species *C. hominis* is restricted to humans, and it is not known if it occurs in gorillas. Captive populations of lowland gorillas are also known to rarely harbor *Cryptosporidium* (Gomez *et al.*, 1996). Cattle and people living outside park boundaries in Bwindi may be a reservoir of *Cryptosporidium* (Nizeyi *et al.*, 2002a,b).

The oocysts when passed in the feces contain four sporozoites and a residual body. The wall of oocysts are red with acid-fast staining. The oocysts reported from gorillas are 4.3–5.6 μm in diameter (Nizeyi *et al.*, 1999). Various fluorescent antibody test kits have been developed for the detection of this parasite in human feces, and there are also ELISA kits that can be used to detect free antigen in the fecal matter of infected hosts. More recently, PCR techniques have been used to distinguish species of oocysts present in fecal matter using extracted DNA or RNA.

In histological sections at necropsy, the life-cycle stages of *C. parvum* are found just under the surface of enterocytes. These very small stages can be observed with routine histopathology of the intestinal mucosa, although the probability of sectioning the appropriate area within the intestine is small.

Cryptosporidium is transmitted through direct fecal–oral transmission of the oocyst stage. Excystation occurs in the upper gastrointestinal tract, and the developmental stages are found in the brush border of the mucosal epithelium of the intestine. Sporozoites develop into trophozoites, which undergo schizogony to produce Type I schizonts (containing eight merozoites each) and then Type II schizonts (containing four merozoites each). The Type II schizonts initiate a gametogenous cycle, which, by forming micro and macro gametocytes, form mature gametes. Fusion of the gametes occurs, and oocysts are produced.

Nyzeyi *et al.* (1999) report that 3 of 11 mountain gorillas found to be infected with *Cryptosporidium* had blood and mucus in their stool; however, neither sign is typically associated with cryptosporidiosis. In humans, acute, watery diarrhea is often a clinical sign of cryptosporidiasis and, in some cases, nausea, vomiting, abdominal cramps, and fever. In infants and young children, dehydration from diarrhea and vomiting can have a fatal outcome. Fortunately, there have been no reports of infant gorillas dying of symptoms associated with cryptosporidiosis.

Large Intestine

Murshidia devians *[Nematoda, Order: Strongylida, Family: Strongylidae]*

Campana-Rouget (1959) described *Murshidia devians* from a lowland gorilla in the Republic of Congo and suggested that the parasite was an accidental parasite of the gorilla owing to its close affinity with parasites of elephants and its recovery from ectopic sites. Hastings *et al.* (1992) report finding *M. devians* in the large intestine of a mountain gorilla at necropsy. Ashford *et al.* (1996) identified the worms during the necropsy of a mountain gorilla in the Bwindi forest.

Members of the genus have a large buccal capsule with an external leaf crown that has about 80 elements and no teeth at the base of the buccal capsule (Campana-Rouget, 1959). The worms are about 20 mm long. The vulva of the female is located just anterior to the anus. The bursa of the male has a well-developed dorsal lobe (Lane, 1914).

The adults are typically found free within the lumen of the large intestine (Hastings *et al.*, 1992). Fossey (1983) reported finding the worms in the small intestine of a necropsied older female mountain gorilla. Campana-Rouget (1959) found worms in ectopic sites. Thus, it is possible that worms may be found in sites other than the intestine.

Eggs of the typical strongylid type may be detected in the feces (Figure 1); the eggs are 60 by 30 μm. When third-stage larvae are cultured in feces, they are found to have a long sheath that extends beyond the tip of the tail (Bhat & Manikam, 1998).

There is no information on the life cycle of any *Murshidia* sp. Like other members of the subfamily Cyathostominae, *Murshidia* probably has a direct life cycle. Eggs passed in the feces hatch on the ground to release the first-stage larvae, which feeds on bacteria. The first-stage larvae molt to second-stage larvae that also feed on bacteria. The second stage molts to the third-stage larvae that are ensheathed. The development in the final host has not been studied. The finding of the adults of *Murshidia* in ectopic sites in the gorilla would indicate that there might be some form of extraintestinal development associated with the development of these worms in their normal hosts.

Oesophagostomum stephanostomum *[Nematoda, Order: Strongylida, Family: Chabertiidae]*

This species was first described from gorillas in 1904 (Stossich, 1904). The worms were preserved in the zoology museum of Cambridge University, with the designation "from large intestines of gorilla," without any indication as to where the host originated. Rousselot and Pellissier (1952) reported on the presence of *Oesophagostomum stephanostomum* in western lowland gorillas from the Republic of Congo. Specimens of this genus were later found at necropsy of Virunga gorillas (Hastings *et al.*, 1992), in Bwindi gorillas (Ashford *et al.*, 1996) and in larval cultures (Sleeman *et al.*, 2000). Surveys found eggs identified as *Oesophagostomum* in the feces of gorillas from the Bwindi population (Kalema, 1995a,b; Ashford *et al.*, 1990, 1996).

Adult females are 18–30 mm in length, and males 18–24 mm. The worm has a relatively small buccal cavity compared to other members of strongyloidea and a marked transverse cervical groove at the excretory pore. The stoma has both external and internal leaf crowns, with the external crown having 30–38 petals. There are two to three times as many petals on the internal leaf crown. The base of the buccal capsule contains six rather large chitinous plates. The male has a prominent bursa and two rather long spicules. The vulva of the female is near the anus and the vagina connects to the kidney shaped ovejector (Stossich, 1904).

Adults live in the lumen of the large intestine. Larvae form nodules in the intestinal wall. These nodules are small raised areas about 1 mm in diameter in the small and large intestine. The lesions can grow into small abscesses, and eventually the entire intestine may be inflamed and edematous. Nodules may reach 4–5 mm in diameter, and contain eosinophils and leukocytes. There can be a significant number of nodules at necropsy; 40 nodules were found in one of nine gorillas in the Congo (Rousselot & Pellissier, 1952). Eggs are 60–80 μm by 40–55 μm and are typical strongylid eggs (Figure 1). When cultured to the infective stage, the third-stage larva is about 1 mm long, has 16–24 triangular intestinal cells, and a sheath that extends 150 μm beyond the tip of the larva.

The hosts of *O. stephanostomum* are gorillas, chimpanzees, and humans (reported from Uganda, Senegal, and, surprisingly, Brazil) (Chabaud & Lariviere, 1958). There is no information on the life cycle of *O. stephanostomum*. The life cycle is probably similar to related species such as *O. bifurcum*, another species found in Old World primates. After larvae are ingested, it takes approximately 3 months until eggs appear in the feces (Eberhard *et al.*, 2001). The molt from third to fourth stage occurs sometime before 19–22 days after infection. Some young adult worms were still present in nodules within the bowel wall after almost a year from the time of infection. Animals can shed eggs for almost a year.

Disease was first associated with gorillas captured for zoological collections in Gabon and the Republic of Congo (Rousselot & Pellissier, 1952). The gorillas died within a few months of capture following ill health. Fecal examinations revealed the eggs of strongylid nematodes, and necropsies later revealed that the gorillas were infected with both hookworms and had nodular disease due to oesophagostomins. The clinical signs manifested as anorexia and lack of grooming, with an unkempt appearance. There was mucoid diarrhea similar to amebic dysentery in humans. The gorillas remained lying or sitting, holding their heads with both hands in an attitude of desperation. In free-ranging mountain gorillas, nodules of oesophagostomiasis and a fatty degeneration of the liver were first described from gorillas captured for European zoos (van den Berghe *et al.*, 1964), and the authors attributed two of the captured gorillas' death to *O. stephanostomum*. More recently, nodules have been observed at necropsy by the Mountain Gorilla Veterinary Center (cited in Sleeman *et al.*, 2000; Mudakikwa *et al.*, 2001), and Hastings *et al.* (1988) speculate that feces containing blood and mucus may indicate infection caused by *Oesophagostomum*.

Probstmayria gorillae *[Nematoda, Order: Ascardida, Family: Atractidae]*

In 1955, Kreis described *Probstmayria gorillae* based on specimens collected from a gorilla that died in a zoo in Basel, Switzerland. Specimens belonging to the genus *Probstmayria* have been described from mountain gorillas in both the Virunga region (Fossey, 1983; Sleeman *et al.*, 2000) and in groups of gorillas in Bwindi Impenetrable National Park (Ashford *et al.*, 1990; Rothman *et al.*, 2002). Ashford *et al.* (1990) found that gorillas had a 100% prevalence of *Probstmayria*, while Rothman and others (2002) found a 13% prevalence in a single group sampled over 7 weeks. Specimens collected from the feces of lowland gorillas in Gabon have been described as two species distinct from *Probstmayria gorillae*; these two species were named *Probstmayria goodallae* and *Probstmayria gabonensis* (van Waerebeke *et al.*, 1988).

Probstmayria gorillae are small worms with an esophagus that has a large valved bulb at the base and a vestibule between the stoma and the beginning of the muscular esophagus proper. The female gives birth to large larvae that are almost one-third her total length.

At necropsy, larvae and adults are found in the lumen of the cecum and colon. The adults are not large (females are only about 2 mm long). Fecal examination will reveal both adults and larvae. The birth of highly precocious larvae with no requirement for an intermediate host means that very large populations can develop within an infected animal.

The hosts of *Probstmayria* sp. include gorillas, chimpanzees, pigs, horses, and tapirs. The life cycle is direct; the host ingests the third-stage larvae, usually through infected foodstuffs or by fecal–oral contamination. Transmission is probably assisted by the fact that larvae can remain alive in feces for 4 or 5 days. This species has not been found to be pathogenic.

Trichuris trichiura *[Nematoda, Order: Enoplida, Family: Trichuridae]*

Ashford *et al.* (1990) reported that the game guards at Bwindi Impenetrable National Park were infected with *Trichuris trichiura* and that it did not appear to be infecting the gorillas at that time. However, a later report (Kalema, 1995a,b) found that the gorillas in the same area where Ashford *et al.* (1990) conducted their study were infected. In Nkurunungi's study (1999), *Trichuris* was found in the Bwindi population. Nkurunungi claimed that the eggs found were morphologically dissimilar to *T. trichuria*. Sleeman *et al.* (2000) found 1 of 74 fecal samples from the Virunga population of mountain gorillas positive for *Trichuris* eggs. They were of the opinion that it was the same as *T. trichiura* found in man, and were concerned that it was a case of transfer from humans to habituated gorillas (Sleeman *et al.*, 2000).

Trichuris adult parasites live with their anterior ends threaded through the mucosa of the large bowel. The worms are about 2–3 cm long and have a very thin anterior end that encloses the stichosome esophagus characteristic of this group of worms.

At necropsy, the worms will be found as small cream-colored worms attached to the wall of the cecum. Uncommonly, worms in humans are found in the wall of the appendix and in the colon, probably when present in large numbers. The exact location of the worms in gorillas at necropsy has not been described.

The eggs of *T. trichiura* are characteristic and would be difficult to confuse with other eggs passed in the feces of gorillas. The eggs contain a single cell when passed in the feces and are brown and lemon shaped, with polar plugs; the eggs are about 50 μm long.

The hosts of *T. trichiura* are humans and other primates. *T. trichiura* is a common parasite of people, and there is a good chance that it could move into the gorilla population if the gorillas were to become infected and the soils were of an appropriate type to support the transmission of this parasite.

Transmission is by the ingestion of eggs containing an infective-stage larva. Eggs can persist in the soil for many years. Eggs passed in the feces enter the soil environment, where they embryonate and become infectious. Infection is obtained by the ingestion of an egg either in soil or soil-contaminated foodstuffs. Once ingested, the eggs hatch within the large intestine, and the larvae penetrate the wall, where they develop. The adult worms will lay eggs 2–3 months after eggs are ingested. Worms probably live somewhere between 1 and 4 years. Clinical signs have not been reported in gorillas. Clinical symptoms in people include anemia, growth stunting, and, in massive infections, prolapse of the rectum.

Skin

Pthirus gorillae *[Arthropod, Order: Anoplura, Family: Phthiriidae]*

This sucking louse was first identified from specimens of mountain gorillas collected during a game hunting trip in 1927 to the Democratic Republic of Congo (Ewing, 1927). Kim and Emerson (1968) further described and illustrated the morphology of these sucking lice found by Ewing, and from captive eastern lowland gorillas collected from the Democratic Republic of Congo. Dian Fossey found this louse during necropsies of mountain gorillas in the Virunga region (Fossey, 1983). A female gorilla in BINP necropsied shortly after her death was found to have several *Pthirus gorillae* in her stomach; she was probably engaged in social or allo-grooming shortly before her death (Ashford *et al.*, 1996).

This species resembles the human pubic louse (*Pthirus pubis*), but unlike *P. pubis*, it has large eyes that are placed on large lateral protuberances. It has not more than 20 small abdominal setae. *Pthirus gorillae* is about 2.20 mm long, and it is a short and broad sucking louse with sprawling legs (Kim & Emerson, 1968).

The life cycle of *P. pubis* is direct, and in humans is usually transmitted through sexual contact. It would be expected that gorillas would transmit *Pthirus gorillae* to each other by social grooming, shared bedding, and through sexual contact. *P. pubis* spends its entire life on the widely spaced hairs of its host, and the cycle from egg to egg takes about a month. *P. pubis* causes pruritus and papular dermititis.

Sarcoptes scabiei ***[Arthropod, Order: Acariformes, Family: Sarcoptidae]***

The genus *Sarcoptes* causes a condition described as scabies in humans and non-human primates, and sarcoptic mange in domestic and wild animals. *Sarcoptes scabiei* was first described in mountain gorillas in Bwindi Impenetrable National Park in 1996 (Kalema-Zikusoka *et al.*, 2002) and then again in a later outbreak (Graczyk *et al.*, 2001a). To date, this condition has not been observed in the Virunga mountain gorillas. This parasite was later reported in the chimpanzees (*Pan troglodytes schweinfurthii*) of Gombe National Park (Pusey, 1998).

A tourist-habituated group of four gorillas developed varying degrees of pruritis, alopecia, and white scaly skin that appeared to be age- and size-related. The male infant gorilla had approximately 75% alopecia and was emaciated and making frequent crying sounds. The male juvenile gorilla had approximately 90% hair thinning, with alopecia and skin thickening. The adult female gorilla and mother of the infant had alopecia, and was scratching frequently. The adult male silverback gorilla only showed signs of pruritis, with no clear visual signs of alopecia. In the immobilized juvenile male, skin scrapes and biopsies revealed mites typical of *S. scabiei* (Kalema-Zikusoka *et al.*, 2002). The infant died and necropsy revealed severe emaciation, with a body weight of 2.8 kg at the age of 8 months. The skin was thick and scaly, with crusts and alopecia on the neck, back, legs, chest, and arms. The only part of the body with hair was the head. The skin showed numerous mites and eggs. The skin biopsies revealed severe cutaneous acariasis associated with marked acanthosis, spongiosis, ballooning degeneration, and hyperkeratosis, with few signs of inflammation. The primary cause of death was scabies, with secondary pneumonia, dehydration, and inanition (Kalema-Zikusoka *et al.*, 2002). The rest of the group recovered with one dose of intramuscular Ivermectin, and could not be immobilized for a further dose.

The affected group occasionally ranged in people's gardens and ate banana plants. Three months prior to showing clinical signs, the group spent several days in gardens where people had rough skin due to scabies. The *S. scabiei* mite is ubiquitous, normally causing mild disease in the normal host, but can settle in temporary hosts and cause severe disease especially if the host is closely related (Ibrahim & Abu-Samra, 1985) and naive to the mites. In the case of the mountain gorillas, the most likely source of the mites would be people. However, the possibility of the source of scabies coming from other animals sharing the Bwindi forest habitat cannot be ruled out. Stress may have contributed to the severity of scabies. This group was being visited by tourists. Cheetahs in Masai Mara, Kenya, under greater pressure from tourist visitation developed sarcoptic mange that was not observed in other cheetahs (Mwanzia, 1995). Four years later another scabies outbreak occurred in another group of gorillas, which was being habituated for tourism (Graczyk *et al.*, 2001a), causing morbidity, but no mortality. The group recovered with Ivermectin treatment. The Nkuringo group also spent time outside the park in people's gardens and public pathways. Further research revealed that the prevalence of scabies in the community surrounding this gorilla group is high, where 45.7% admitted to having had the disease and three (37.5%) of the eight people who agreed to have skin scrapes and were also showing clinical signs were positive for scabies (Nziza, 2003). In addition, the scabies mites in the skin scrapes from these people were found to be genetically similar to the mites found in the mountain gorillas (Ndizihiwe, unpublished data; Nziza, 2003). These findings suggest that the mountain gorilla scabies probably came from the surrounding local community.

S. scabiei mites are circular in shape and the females (330–600 μm by 250–400 μm) are slightly larger than the males (200–240 by 150–200 μm). Mellanby (1944) first described the life cycle of *S. scabiei* in man. *Sarcoptes* are burrowing mites, and the female mites lay 40–50 eggs in the tunnel they form in the epidermis. In 3–5 days, these eggs hatch into six-legged larvae; some of which develop into nymphs within the tunnels and eventually become sexually active males and females. The time from egg production to development of adult mites is an estimated 17 days. The adult female is then fertilized by the male, and extends the tunnel laying more eggs. These fertilized females are responsible for the main spread of infection together with the wandering larvae, nymphs, and fertilized young. The scabies mites are susceptible to dryness and cannot survive for more than a few days off their host. The burrowing mites suck lymph and feed on epidermal cells, causing marked irritation, leading to further itching and scratching. The inflammation causes exudation from the skin that coagulates to form crusts, excessive keratinization and proliferation of connective tissue, and resultant thickening and wrinkling of the skin. Pruritis appears after 2–6 weeks (Mellanby, 1944).

SUMMARY

Parasite surveys in Bwindi Impenetrable National Park have revealed that at least 16 parasites infect Bwindi gorillas, 10 helminths, 1 cestode, 3 protozoa, and 2 arthropods. Of these, there are 14 endoparasites and 2 ectoparasites. A summary comparing the parasites found in wild populations of the three subspecies of gorillas can be found in Table 1. Parasites found in captive gorillas were not included (not a complete list, but see Lane, 1923; Sandosham, 1950;

Table 1. The parasites of free-ranging gorillas

	Study population			
	Gorilla beringei[a] (Bwindi)	*Gorilla beringei beringei*[b] (Virunga)	*Gorilla beringei graueri*[c]	*Gorilla gorilla gorilla*[d]
Helminths				
Cestodes				
Anoplocephala gorillae	X	X	X	
Tapeworms (unidentified)				X
Nematodes				
Ascaris sp.				X
Ascaris lumbricoides	X			
Capillaria hepatica		X		
Chitwoodspirura wehri		X[e]		X
Gongylonema sp.				X[e]
Hyostrongylus sp.		X		
Hyostrongylus kigeziensis	X			
Impalaia sp.		X		
Loa loa sp.				X
Mansonella lopeensis				X
Mansonella perstans				X[e]
Murshidia devians	X			
Oesophagostomum sp.	X	X	X	
Oesophagostomum stephanostomum	X			
Paralibyostrongylus kalinae	X			
Probstymaria gabonensis				X
Probstymaria goodallae				X
Probstymaria gorillae	X	X		
Probstymaria sp.	X	X		
Strongyles	X	X	X	
Strongyloides fulleborni	X	X		X
Strongyloides sp.				X
Trichostrongylus sp.	X	X		
Trichuris trichuria	X	X		

(cont.)

Table 1. *(Continued)*

	Study population			
	Gorilla beringei[a] (Bwindi)	*Gorilla beringei beringei*[b] (Virunga)	*Gorilla beringei graueri*[c]	*Gorilla gorilla gorilla*[d]
Protozoa				
Balantidium coli			X	X
Chilomastix mesnili		X		
Cryptosporidium	X			
Encephalitozoon intestinalis	X			
Endolimax nana		X		
Entamoeba coli		X		
Entamoeba hartmanni		X		X
Entamoeba histolytica		X		X
Enteromonas hominis			X	
Giardia lamblia	X	X	X	
Iodamoeba buetscheli		X	X	X
Pentatrichomonas hominis				X[e]
Arthropods				
Pthirus gorillae	X	X		
Psoroptes sp.		X[e]		
Sarcoptes scabies	X			
Other				
Various ciliates (commensal)	X	X	X	X

[a] Ashford *et al.*, 1990, 1996; Durette-Desset *et al.*, 1992; Kalema, 1995a,b; Nizeyi *et al.*, 1999; Nkurunungi, 1999; Graczyk *et al.*, 2001a,b; Kalema-Zikusoka *et al.*, 2002; Rothman *et al.*, 2002.
[b] Nyeblin, 1924; Fossey, 1983; Redmond, 1983; Hastings *et al.*, 1992; Graczyk *et al.*, 1999; Mudakikwa *et al.*, 1999; Sleeman *et al.*, 2000.
[c] Eilenberger, 1998.
[d] Chabaud & Rousselot, 1956; Garin *et al.*, 1982; Goussard *et al.*, 1983; van Waerebeke *et al.*, 1988; Imai *et al.*, 1991; Bain *et al.*, 1995; Landsoud-Soukate *et al.*, 1995; Lilly *et al.*, 2002.
[e] Author was unable to make a definitive conclusion as to the diagnosis of the parasite.

Amberson & Schwarz, 1952; Fain, 1957; Rousselot & Pellissier, 1952; Chabaud & Rousselot, 1956; Yamashita, 1963; Noda & Yamada, 1964; van de Berghe *et al.*, 1964; Mortelmans *et al.*, 1970; Flynn, 1973; Paciepnik, 1976, Graber & Gervey, 1981; Grady *et al.*, 1982; Taere & Loomis, 1982; Rehmann *et al.*, 2003) since the animals were not under natural conditions. In captive environments, animals are limited to a confined space and are sometimes under increased stress, thereby facilitating the transmission of natural parasites and introduced parasites from other hosts that may be housed nearby. However, the importance of captive studies should not be overlooked; information from zoo and captive animals provides a resource base for which parasites are capable of infecting wild gorillas. We encourage the regular monitoring of all populations of *Gorilla*.

CHAPTER ELEVEN

Preliminary GIS Analysis of Range Use by Sympatric Mountain Gorillas and Chimpanzees in Bwindi Impenetrable National Park, Uganda

J. Bosco Nkurunungi and Craig B. Stanford

INTRODUCTION

A central premise of evolutionary theory is that ecological competitors act as important agents of natural selection, molding species into their current phenotype. When two closely related, morphologically similar species occur in the same habitat, it is reasonable to infer that they are different in some critical

J. Bosco Nkurunungi • Department of Zoology, Makerere University, Kampala, Uganda **Craig B. Stanford** • Departments of Anthropology and Biological Sciences, Jane Goodall Research Center, University of Southern California, Los Angeles, CA 90089-0032, USA.

Primates of Western Uganda, edited by Nicholas E. Newton-Fisher, Hugh Notman, James D. Paterson, and Vernon Reynolds. Springer, New York, 2006.

ecological features that have allowed coexistence. It remains for field researchers to identify these ecological differences.

A clear understanding of primate ranging patterns is essential to understanding feeding and foraging behavior and consequently the behavioral ecology of the species. Ranging patterns may be influenced by climate, territoriality, distribution of water resources, location of sleeping sites, distance traveled the previous day, intergroup encounters, competition for food and mates, predation pressure, diet, and other ecological constraints. Spatial and temporal distribution and abundance of food resources are perhaps the most important environmental determinant of primate movements (Milton & May, 1976; Clutton-Brock & Harvey, 1977; Wrangham, 1980; Isbell, 1983; Boinski, 1987). Diet is the most crucial factor suggested to explain the differences in ranging patterns of animals (Mitani & Rodman, 1979) bearing in mind that an animal's energetic needs determine its home range size (Mace & Harvey, 1983).

The degree of frugivory or folivory has been shown to influence primate ranging patterns (Milton & May, 1976; Clutton-Brock & Harvey, 1977; Doran & McNeilage, 1998). Seasonal fluctuations in resource abundance often cause primates to change their feeding behavior and ecology, with frugivores tending to have larger home ranges and longer day ranges for their size than do folivores (van Schaik, 1983; Chapman, 1988; Janson & Goldsmith, 1995).

Previous studies of gorilla ranging behavior indicate that home range areas of Virunga gorillas (Fossey, 1974; Fossey & Harcourt, 1977; Watts, 1984; Vedder, 1984; Yamagiwa, 1987; McNeilage, 1995) are small compared to eastern lowland gorillas (Casimir, 1975; Goodall, 1977; Yamagiwa *et al.*, 1994) and western lowland gorillas (Bai-Hokou: Remis, 1994; Goldsmith, 1999; Lopé: Tutin, 1996). In addition, foraging effort varies more over time for eastern lowland gorillas and western lowland gorillas than with mountain gorillas (Watts, 1996). The distances of the foraging path taken by an animal from dusk to dawn (day range length [DRL]) are longer for eastern lowland gorillas (Goodall, 1977; Yamagiwa & Mwanza, 1994) than for western lowland gorillas (Remis, 1994, Goldsmith, 1996; Lopé: Tutin, 1996). This distance has been shown to increase when gorillas used seasonal fruit crops than when they ate mostly terrestrial herbaceous vegetation (THV) or bamboo (Watts, 1996).

Studies of chimpanzee ranging behavior have tended to be less systematic, because of the logistical difficulties imposed by the chimpanzee social system. While gorillas travel in cohesive groups, chimpanzees travel in temporary subgroups, or parties, in which membership is highly variable (Goodall, 1986). Because of this fission–fusion system, there is no one direction or distance of group travel. In practice, daily range estimates are based on individual path lengths. Male chimpanzees tend to travel farther (Goodall, 1986) and faster (Wrangham 2000) than females, presumably because of the high cost of female locomotion when carrying offspring, and the tendency for males to travel to home range perimeters to patrol territorial boundaries.

In this chapter, we present preliminary results of a long-term study of range use by sympatric mountain gorillas (*Gorilla gorilla beringei*) and chimpanzees (*Pan troglodytes schweinfurthii*) in Bwindi Impenetrable National park, Uganda. The primary objectives of the study were to (1) map and measure the home range area for one group of gorillas and one community of chimpanzees in Bwindi and (2) measure mean DRL for one group of gorillas. We describe our use of GIS/GPS technology in collecting, analyzing, and presenting ranging data.

METHODOLOGY

The study area covers approximately 25 km^2 of Afromontane forest in the Ruhija section of Bwindi Impenetrable National Park in southwestern Uganda (from approximately 0°53′–1°08′ S and 29°35′–29°50′ E). The study site is of rugged, wet terrain, with an elevational range from 2000 to 2300 m. The forest is quite heterogeneous, with at least 163 tree species recorded (Butynski, 1984). Bwindi possesses a complex floristic composition that includes eight botanical communities, among which *Parinari*-dominated forest, *Chrysophyllum*-dominated upland forest, *Newtonia*-dominated forest, swamp, and a small bamboo zone are the most widely distributed (Howard, 1991; Bitariho, 1999). Unlike the Virunga Volcanoes, where more than 50% of gorilla habitat is bamboo forest (Bitariho, 1999), the bamboo zone in Bwindi covers no more than 1–2% of gorilla habitat, located primarily in the highest elevations of the park (outside the study site) between 2400 and 2600 m (Bitariho, 1999). Annual rainfall averages 1100–2400 mm (Butynski, 1984), and the climate is characterized by two dry seasons lasting from about May to July and from late December to February.

Gorillas in Bwindi Impenetrable National Park number approximately 300, or about 1/km^2 (McNeilage *et al.*, 1998). Mitochondrial DNA studies have shown them to be virtually indistinguishable from their sister population in the

Virunga Volcanoes (Garner & Ryder, 1996; Jensen-Seaman & Kidd, 2001) with which they occupied continuous forest until 400–500 years ago (Hamilton *et al.*, 1986; Stanford, 2001). Little was known about the Bwindi chimpanzee population before the present study was begun. Research on Bwindi chimpanzee–gorilla sympatry began in late 1996 and is ongoing. Until 1999, the research was carried out at two sites: Nkuringo in the southwestern corner of the park, and Ruhija, in the eastern section. Political instability forced closure of the Nkuringo site and the suspension of data collection in Ruhija in early 1999. In January 2000, the project started again in Ruhija. The data presented in this paper were collected in Ruhija, primarily from January to December 2000.

The study population was the Ruhija chimpanzee community and one gorilla group, the habituated Kyagurilo "research" group. One of us (JBN) plus field assistants have studied the gorilla group since 1997. The animals are habituated and individually identified; both direct observational and indirect data were collected. During the study period, the group consisted of 13 animals (1 silverback male, 1 blackback male, 5 adult females, 6 immatures) and was monitored daily by research staff of the Institute of Tropical Forest Conservation. The Ruhija chimpanzee community is partially habituated; many of its members tolerate approach by observers to within 25 m when feeding in trees, but animals cannot be followed or watched at close range on the ground. This community consists of at least 25 individuals, including at least 5 adult males whose identities and dominance ranks are known. CBS and three field assistants collected data on this community.

Using GIS Technology for Range Use Analysis

A recently emerged key technology that can be used to study habitat use patterns in primates is geographic information systems (GIS) technology. Using handheld global positioning system (GPS) units, which interface with orbiting satellites, researchers can precisely map the locations of animals, nests, feeding sites, and other habitat features. Although this technology has been available for more than a decade, its use has been limited in primate field studies by the lack of digitally mapped field sites onto which GPS coordinates could be plotted. In Uganda, conservation-oriented GIS research has been carried out since the 1990s, and many of the conservation areas in the country have been mapped from Landsat images.

A contour map of Bwindi Impenetrable National Park in southwestern Uganda, converted to a digitized image, is used for ecological monitoring purposes by research staff at the Institute for Tropical Forest Conservation, a nongovernmental conservation organization. This map is analyzed using Arcview GIS software application, and contains features such as elevation contours, stream drainage, vegetation type, trails, and political boundaries. We gather chimpanzee and gorilla ranging data by obtaining coordinates with handheld GPS units wherever known animals are observed, and wherever nests are found within the known home range. These coordinates are downloaded from the units onto the Arcview image of the study site, where they can be sorted and analyzed by date, location, and a number of other variables.

GPS readings were obtained with handheld Garmin 2+ units, often facilitated by using 2-m remote antennas suspended overhead. One GPS recording for gorillas and one for chimpanzees was chosen per day for home range mapping and analysis. Readings were normally taken from gorilla nest sites, and from either fresh chimpanzee nests or feeding sites. Poor reception under thick vegetation or on cloudy and rainy days sometimes made it impossible to obtain GPS readings.

A major problem in studying chimpanzee ranging patterns is the uncertainty whether nests in border areas of the known range were made by the Ruhija or a neighboring community. For this reason there is some bias in the chimpanzee range sampling toward samples taken within the estimated core area of the range. A total of 264 GPS points in 2000 for gorillas and 483 for chimpanzees were recorded for use in range analysis.

To map home range area, GPS readings were downloaded into Arcview software application (version 3.1) and analyzed at both the GIS laboratory at Makerere Institute of Environment and Natural Resources and the Jane Goodall Research Center at the University of Southern California. The GPS points were overlaid onto Bwindi Impenetrable National Park database of CARE-DTC map 1997 (adapted from Cahusac, 1958). The total home range during 2000 was mapped to show the area size and location of nest and contact sites and superimposed over the vegetation map of Bwindi Impenetrable National Park. The map was also divided into two blocks, Kagyembagyembe (K) and Nyaruchundura (N), on the basis of vegetation characteristics of the gorillas' home range.

Gorilla home range was computed in two ways. First, the peripheral GPS points for the entire period of study were connected to yield the total home range area. In addition, the peripheral GPS points for each year were connected to show inter-annual variation in home range area. This is the minimum convex polygon (MCP) method and was used for home range estimates for mountain gorillas in the Virungas (McNeilage, 1995). The MCP method does not take into account the empty spaces that the animals will not have visited, thus overestimating the home range size. In addition, it gives no indication of how the range is utilized, and is subject to bias by both small sample sizes and extreme outlying locations (Hooge *et al.*, 2000).

Using a second method, the map was divided into 1-km^2 quadrats. The number of quadrats with GPS points was then counted and used to estimate home range. The grid method has been used for mountain gorillas in the Virungas (Vedder, 1984; Watts, 1997), although without GPS data. It is the most suitable method for animals with irregular-shaped home ranges or home ranges with several areas of concentrated use, which was the case with gorillas. In this study, both methods were used to avoid bias.

RESULTS

Figure 1 shows the distribution of chimpanzee and gorilla nesting sites from GPS points collected during 2000. The gorilla group used various vegetation types within their home range. They avoided the large swamp, dividing their range into two forest blocks. The frequency of quadrat use varied significantly among core, regularly, and frequently used areas (Kruskal–Wallis test $H(2, n = 34) = 27.9$, $P < 0.001$). Overall, quadrat use ranged from 0.2 to 12.7%, with lower percent frequencies towards the periphery of the home range. Only 7 km^2 of the total home range was used more intensively (core area), accounting for 20.6% of the total home range area. Ten square kilometers of the remaining area (29.4%) was regularly used and most of the remaining area (50% of the total home range) was less frequently used. Range use by both species varied widely by month, and range use appears to diverge (note that chimpanzee ranging data in July 2000 were not available). Chimpanzee and gorilla ranges showed extensive overlap at a broad scale, but in only 2 months (March and October) did nest sites overlap extensively. In other months, gorilla nests were tightly clustered in distribution, while chimpanzee nests were found in a more scattered pattern that fell entirely or almost entirely outside the range of gorilla nests. Chimpanzee and gorilla range use was similar, but overlapped little, despite their tendency to feed on the same fruit resources (Stanford &

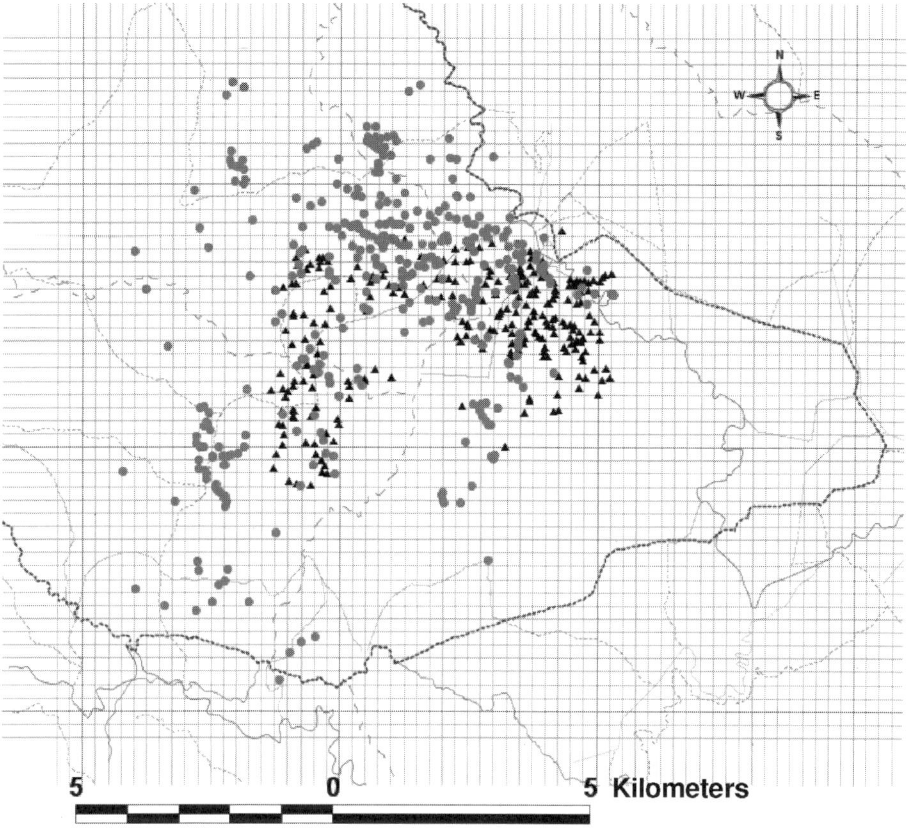

Figure 1. Range use during 2000 by chimpanzees and gorillas in the Ruhija study area. Black triangles indicate GPS locations for gorilla nest sites; grey circles indicate the locations for chimpanzee nests.

Nkurunungi, 2003). Monthly presence of chimpanzee activity was, however, generally to the east or south of gorilla activity (Figure 2A and B).

The Ruhija chimpanzee community ranges over roughly the same area as the Kyagurilo gorilla group, although DRL of the gorilla group (~800 m) is shorter than that of most chimpanzee parties (>1.0 km; Stanford & Nkurunungi, 2003). Chimpanzee ranging patterns as recorded during 2000 should be regarded as the minimum known range area for the Ruhija community. Because of the fission–fusion community structure, individuals and small parties frequently go undetected and may travel well outside the known range of the community. In addition, some nests recorded within the known community range could have been made by members of neighboring communities rather than the study community. However, the likelihood of serious

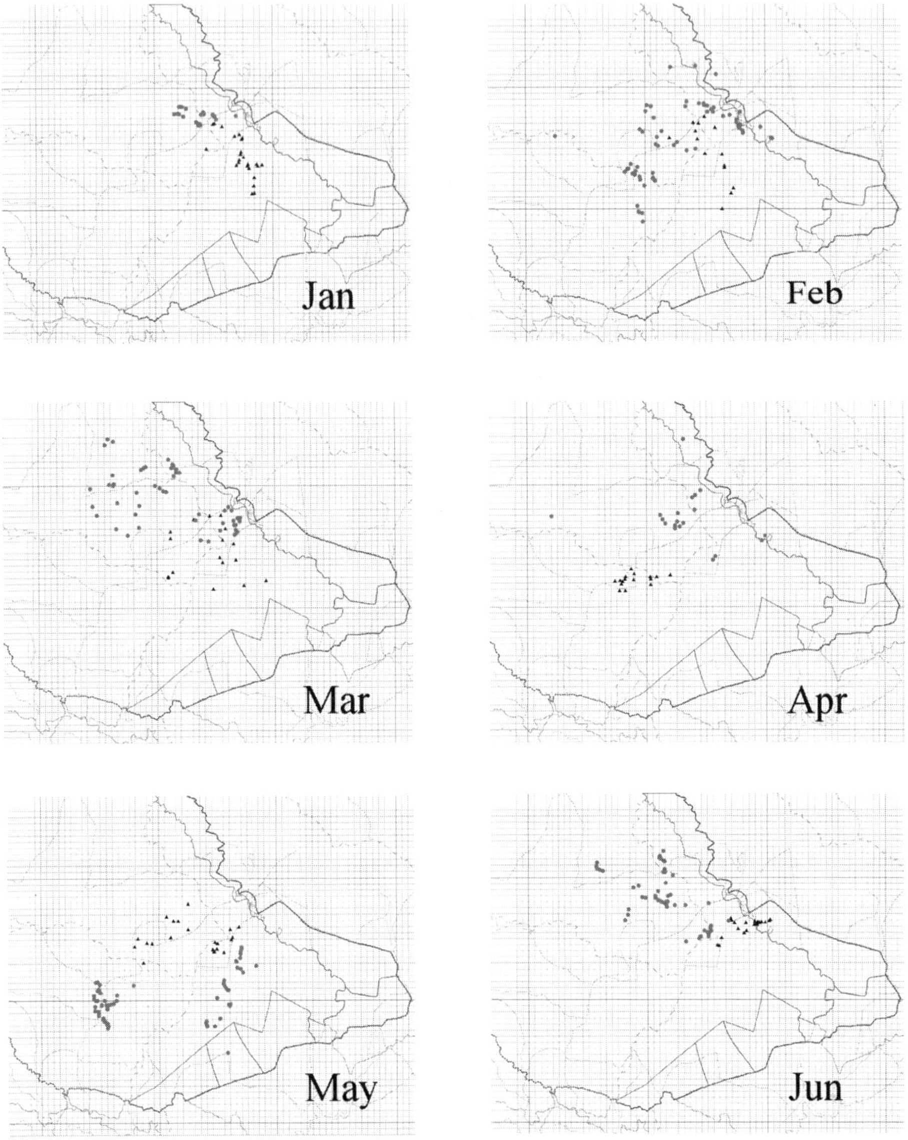

(A)

Figure 2. (A) Monthly range use by chimpanzees and gorillas in the Ruhija study area, based on GPS data (January–June 2000). (B) Monthly range use by chimpanzees and gorillas in the Ruhija study area, based on GPS data. (July–December 2000). Black triangles indicate GPS locations for gorilla nest sites; grey circles indicate the locations for chimpanzee nests.

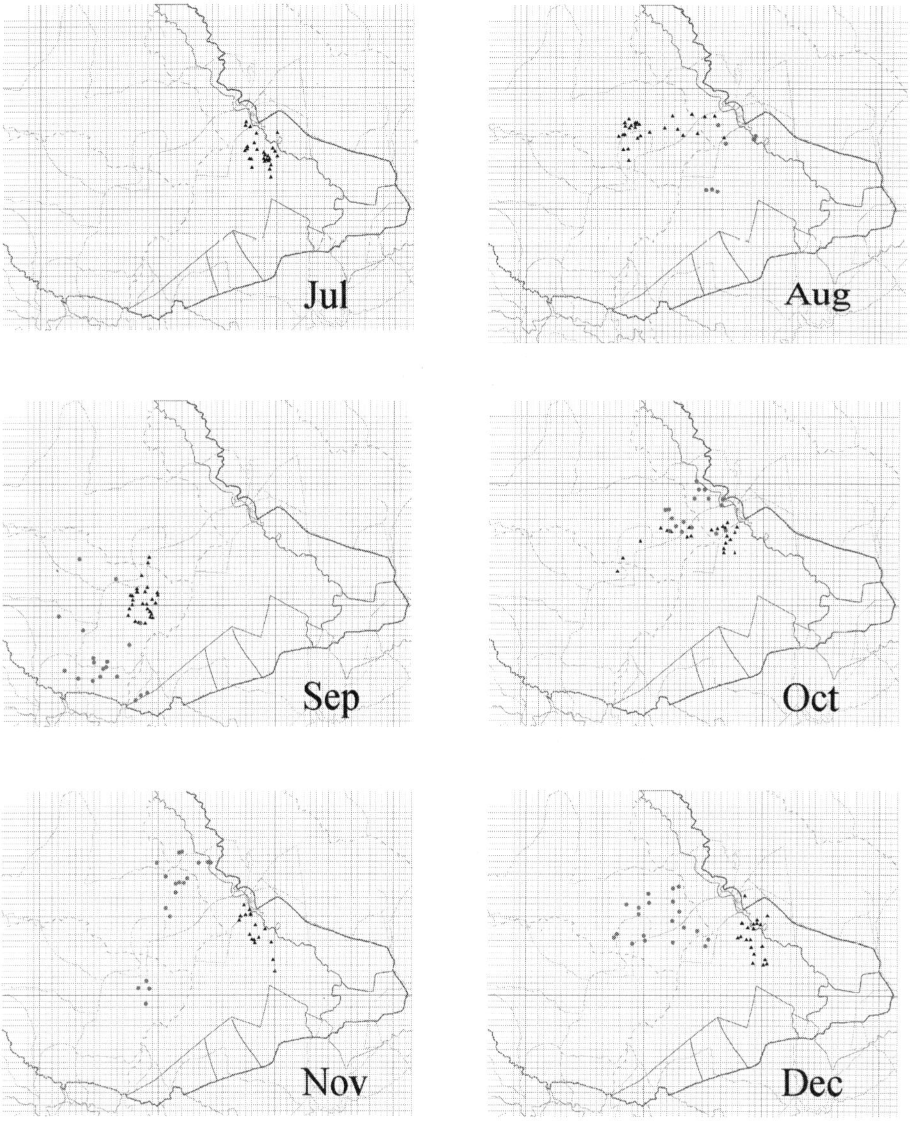

Figure 2. *(Continued)*

Location		Nyarucundura (17 km^2)			Other (4 km^2)	Kagyembagyembe (13 km^2)			
9883000		0.6	1.7	2	1.1	3.3	2.9	0.2	
9882000		0.9	2.2	2.9	3.3	5.5	12.7	7.2	
9881000		1.5	5.5	1.3	1.7	2.9	8.7	7	
9880000		5.2	3.7	1.3	1.1	1.3	1.5	2.6	
9879000		3.1	4.6	0.6			0.2		
9878000		0.6	0.4						
9876000									
Coordinates		803000	804000	805000	806000	807000	808000	809000	810000

(dark shading)	Over 5% use (core area)—7 km^2
(medium shading)	2–4.9% use (regularly used)—7 km^2
(light shading)	Less than 2% use (less frequently used)—7 km^2

Figure 3. Differential use of 1-km^2 quadrats over a 2-year period (n = 544 GPS nest/feeding sites). The values shown are percent frequencies of nest/feeding sites for each quadrat.

error in this regard is low owing to the highly territorial nature of chimpanzee communities.

Figure 3 shows the distribution of gorilla nest/feeding sites during 1997–2000. The gorillas were recorded 264 times in 2000, and in 37 quadrats total (= 37 km^2 from 1997 to 2000). By connecting the peripheral points and assuming that all complete quadrats within the boundary were 1 km^2 each and all the incomplete quadrats were halved, the total area was 26 km^2, close to that determined from Arcview. Thus from quadrat estimates, the home range was between 26 and 37 km^2, which is similar to range estimates obtained using Arcview computations (25.8 and 34.7 km^2). Using the same assumptions for Ruhija chimpanzees, we estimate a minimum home range size during 2000 of 17 km^2.

Both ape species nested on the ground as well as in trees. Unlike gorillas in the nearby Virungas Volcanoes, Bwindi gorillas sometimes nest in trees. When gorillas nested in trees, they nearly always used the same species, *Alcornea floribunda* (Euphorbiaceae). *A. floribunda* is a common understory tree species in Bwindi. These nests were easily identified by their large size and by the presence of gorilla dung in and around them. Nkurunungi (2003) found that approximately 22% of all gorilla nests during 1997–2000 were made in trees.

Chimpanzees nested in trees most (but not all) nights during 2000, but their use of nest trees was more varied. In the northern sector of Bwindi, outside the study area, approximately 6% of all chimpanzee nests were recorded on the ground. These appeared to be night nests rather than day nests, according to the style of construction (Maughn & Stanford, in press). Within the study area, ground nests were found only 10 times in 2000. Whether ground nesting is an adaptation to local conditions or a cultural tradition as seen in many other chimpanzee behaviors, is unknown.

DISCUSSION

The total home range size measured in this study was larger relative that for groups in the Virungas (Fossey, 1974; Fossey & Harcourt, 1977; Vedder, 1984; McNeilage, 1995). However, it appeared to be larger than for *G. g. gorilla* (Jones & Sabater-Pi, 1971; Remis, 1997; Tutin, 1996) and comparable to *G. b. graueri* at Kahuzi-Biega (Casmir and Butenandt, 1973; Yamagiwa *et al.* 1994). Even within Bwindi, there were variations in home range size. Achoka (1993), using a different method and a limited number of nest sites, estimated only slightly smaller home range sizes for two other groups in Bwindi.

These results are not surprising given that one of the factors that influence home range size is the availability and distribution of food resources. In Bwindi, fruit trees and food resources are patchily distributed compared to the more uniform habitat in the Virungas, where preferred foods are abundant and widely distributed (Watts, 1984). This supports the observation that frugivores tend to have larger home ranges than folivores (Milton & May, 1976). Virunga gorillas are strict folivores, presumably because they inhabit an area lacking in fruit tree species compared to Bwindi.

There are likely costs involved for gorillas traveling long distances in search of fruit. First, gorillas are large-bodied animals, and using a large range would be energetically costly. They may therefore eat nonherbaceous foods, which are more readily available and less temporarily distributed, as an adjustment to fruit scarcity. Second, a larger home range could influence the rates of encounter with other groups, perhaps leading to intergroup male agonism. For example,

Watts (1998b) argued that close proximity to other social units can strongly affect short-term movements in mountain gorillas and a group moved farther on days of and after interactions with other groups or lone males than on other days. Although gorilla encounters may lead to female transfer, they often result in fierce agonistic behaviors. Evidence of fights has been observed between groups in this area. There is also a risk of male infanticide from other groups when groups interact.

Finally, longer DRL may lead to competition with other gorillas or other animals for food. There are other groups of wild gorillas in the study site, which are the most likely intraspecific competitors. Chimpanzees are also possible food competitors, which show considerable dietary overlap with gorillas (Stanford & Nkurunungi, 2003). Our research has shown substantial dietary overlap between the two ape species (Stanford & Nkurunungi, 2003), and at least one bout of aggressive interspecific food competition.

Watts (1998b) and Yamagiwa *et al.* (1996) point out that even where gorilla groups and chimpanzee communities share the same area of forest, the two species exploit resources differently. Gorilla groups tend to use small parts of their home range each month, covering the entire home range only over the course of an annual cycle. Chimpanzees, on the other hand, forage widely for fruit on a daily basis, covering large portions of their home range in a shorter time period. When important chimpanzee foods are scarce, the community disperses into small subgroups, with larger foraging parties forming mainly when ripe fruit is abundant (Goodall, 1986). These divergent foraging strategies may also allow the two species to avoid feeding competition for fruit when sympatric. It has been hypothesized that since there is a greater seasonal abundance of fruit in Bwindi than in the Virungas (Butynski, 1984), Bwindi gorillas should be more frugivorous, and should travel further each day, than those in the Virungas. Since fruit resources tend to be more widely and ephemerally distributed, gorillas in Bwindi are expected to respond to fruit in their diet by increasing their home range size and daily ranging behavior. Consequently, gorillas at Bwindi are expected to have bigger home ranges, longer day ranges than their Virunga relatives. Achoka (1993) and Sarmiento *et al.* (1996) suggested that gorillas in Bwindi exhibited larger home ranges and traveled farther than their counterpart subspecies in the neighboring Virungas. Bwindi gorillas, however, have DRLs only slightly longer (800 m; Goldsmith *et al.* 1999; Nkurunungi, 2003) than those in the Virungas, despite a seasonal pattern of foraging for ripe fruit. It is not clear why Bwindi gorillas do not travel longer distances each day than they do. Goldsmith and Moles (2003) found no correlation between terrain and travel distance, but their study was of gorillas that traveled outside the forest for lengthy periods and reflected abnormal, human-influenced foraging patterns and DRLs.

A future goal of the study made possible by the use of GIS is to test whether the range use pattern of one ape species influenced that of the other. This should be possible once a larger database of GPS readings over multiple seasons and annual cycles is available. For example, range use of areas within the overall home range by one ape might be nonrandomly correlated, either positively or negatively, with range use of the same area by the other species. A positive correlation in use of an area would suggest that one species was drawn to the same food resource and was attracted to the area by the presence of the other, perhaps through detection of food calls. A negative correlation would imply avoidance, either mutual or by one species of the other. Range use patterns may detect patterns of the relationship between the two species that dietary studies do not, provided a fine-grained level of detail were available.

ACKNOWLEDGMENTS

The authors gratefully acknowledge the support of the Uganda Wildlife Authority, the Ugandan National Council for Science and Technology, and the Institute of Tropical Conservation and its staff. In particular, we thank Dr. Gilbert Isabirye-Basuta, Dr. Hannington Oryema-Origa, Dr. Eric Edroma, Caleb Mgambaneza, Mitchell A. Keiver, Dr. Richard Malenky, Chief Warden Keith Masana, Evarist Mbonigaba, Dr. Alastair McNeilage, Chief Warden Christopher Oryema, Dr. Martha Robbins, Dr. Nancy Thompson-Handler, and Gervase Tumwebaze. For help in analyzing the data, we thank Terrelita Price and Tatiana White. We express our gratitude for the funding support of the Fulbright Foundation (CIES), the LSB Leakey Foundation, the National Geographic Society, the Wenner-Gren Foundation for Anthropological Research, the Max Planck Institute for Evolutionary Anthropology, and the Faculty Development Fund of the University of Southern California.

SECTION THREE

Behavior and Physiology

Figure iv. Chimpanzees *Pan troglodytes.* Above: alpha male grooming a key alliance partner; below: adult female playing with her first-born infant (photos, N. E. Newton-Fisher.

CHAPTER TWELVE

Comparison of Sex Differences in Gregariousness in Fission–Fusion Species

Reducing Bias by Standardizing for Party Size

Melissa Emery Thompson and Richard W. Wrangham

INTRODUCTION

Fission–fusion grouping means that individuals leave and join others according to their own decisions. It occurs in all chimpanzee populations as well as many other animals, including primates, cetaceans, carnivores, and ungulates. In chimpanzees, temporary associations ("parties") may last from a few minutes to several days. The individuals who form parties together all come from the same social network ("community" or "unit-group").

Melissa Emery Thompson and Richard W. Wrangham • Department of Anthropology, Harvard University, 11 Divinity Avenue, Cambridge, MA 02138, USA.

Primates of Western Uganda, edited by Nicholas E. Newton-Fisher, Hugh Notman, James D. Paterson, and Vernon Reynolds. Springer, New York, 2006.

The fission–fusion system is associated with variation in parameters such as the average party size, the average duration of time for a party to be stable, and the amount of time spent alone. The amount of variance in such parameters is also of potential interest. Using such variables, studies have examined how grouping tendencies in chimpanzees vary across seasons, individuals, sexes, ages, communities or populations (Symington, 1990; White, 1992; Chapman *et al.*, 1994a; Sakura, 1994; Doran *et al.*, 2002). Such comparisons are full of potential, but confidence in their conclusions is reduced by various methodological problems. Here we discuss methods for standardizing the comparison of grouping tendencies among chimpanzees.

First, while it is clear that parties must be defined in the same or comparable ways, this is often hard to arrange. For example, in relatively open forest, a definition like "all visible chimpanzees" is appropriate, whereas in thicker forest, observers may be forced to use a distance criterion such as "all individuals seen or thought to be within 50 meters." As a result, different observation conditions can lead observers to use different definitions. However, different forest habitats may also be expected to affect chimpanzee grouping variables such as party spread. This problem will probably never be completely solved, but efforts have been made to ensure standardization (Chapman *et al.*, 1993). In this chapter, we assume that parties have been satisfactorily defined in comparable ways.

Second, to reveal differences in grouping tendency, studies need to take account of differences in demographic factors such as the sex ratio or community size. For example, Boesch (1996b) concluded that variation in party size is partly a consequence of variation in community size. Boesch's approach implies that the appropriate dependent variable is sometimes the residual of party size on community size. Similar approaches could be taken with other independent variables, such as community sex ratio. The Ngogo research team attempted to remove this bias by using a randomization technique to calculate indices of gregariousness for each age–sex class (Pepper *et al.*, 1999; Mitani *et al.*, 2002). However, their results may have been biased by introducing the assumption that all individuals of an age–sex class behave in an equivalent manner.

Third, parties may be formed either because individuals have a general tendency to be gregarious or because they have an affinity for a specific other individual (or individuals). Both tendencies lead to grouping, which means that what appear to be similar results may have different causes. Thus, similar average party sizes between sexes or communities could therefore in theory represent different underlying tendencies. For example, Pepper *et al.* (1999)

concluded that chimpanzee males at Ngogo are more gregarious than females, but that females have higher specific affinities for each other.

The fourth problem is that it is impossible to select parties for observation at random. This is because individual chimpanzees move unpredictably through large areas of thick forest, which means that researchers do not have the luxury of scheduling their observations of individuals within the community according to a predetermined design. Instead, they must accept the opportunities that chimpanzees offer. Unfortunately, some parties are more easily located or observed than others (e.g., larger parties are often found more easily, because they are noisier); some individuals are better habituated than others (e.g., males are usually better habituated than females); and some individuals may be more gregarious than others (e.g., males more than mothers). In addition, some observers may be more concerned to obtain data on one class of individuals than another (e.g., males compared to females). For all these reasons, parties that contribute to monthly or yearly data sets are invariably a nonrandom representation of the community. Attempts to evaluate chimpanzee grouping behavior by simply calculating average party sizes observed are therefore misleading and provide little useful information to understand an inherently dynamic process. However, the extent of the bias and the variation in the bias between studies has not been described.

In order to make meaningful comparisons of grouping tendencies, therefore, we need to understand what the biases are, and how to control them. In this chapter we focus on controlling for party size and community composition.

In many studies, observers obtain data by recording from parties located by search. But this method inadvertently biases toward large parties (because they are more easily found and pursued than small parties). If party composition varies systematically with party size, therefore, grouping parameters will be biased also.

Such bias is well known from several sites. It was analyzed for Kanyawara chimpanzees by Wrangham (2000), who showed that mothers tended to be found in smaller parties than males or nulliparous females. Accordingly, apparent differences in party size or sex differences in gregariousness between sites could be a result entirely of observer differences in the degree of bias toward collecting data on large parties.

This means that comparisons between sites can be improved by controlling for party size. Accordingly, we here compare grouping tendencies between two communities, Kanyawara (Kibale) and Sonso (Budongo), by relating party composition to party size. The communities have similar size and composition, and live in habitats of somewhat similar overall structure (lightly disturbed high forest).

METHODOLOGY

Study Sites

This chapter compares party compositions in two communities of chimpanzees in Western Uganda: the Sonso community of Budongo Forest Reserve and the Kanyawara community of Kibale National Park. Individuals of both communities are well habituated, having been subject to continuous research for more than 10 years. Kanyawara is predominantly moist evergreen forest, with small segments of swamp, grassland, and colonizing forest (Chapman *et al.*, 1994a,b). Although the Budongo Forest Reserve is predominantly a moist, semideciduous forest, the home range of Sonso chimpanzees is a mosaic of forest types owing to a history of managed logging (Reynolds, 1992; Newton-Fisher, 1997). At Kanyawara, major chimpanzee food species include figs *Ficus sansibarica* and *Ficus natalensis*, as well as the drupe fruits *Mimusops bagshawei* and *Uvariopsis congensis*, which can comprise up to 75% of the diet during their short fruiting seasons; Kanyawara chimpanzees also spend an average of 20% (ranging up to 50%) of their monthly feeding time consuming piths. Sonso chimpanzees favor a variety of figs, particularly the abundant *Ficus sur* (*capensis*) and *Ficus mucuso*, as well as the fruits of *Celtis durandii* and *Broussonetia papyrifera*. Piths are rarely eaten at Sonso, while a variety of leaves (e.g., *B. papyrifera*, *Celtis mildbraedii*), flowers (*B. papyrifera*), and seeds (*Cynometra alexandri*) make up more than 20% of the diet (Newton-Fisher, 1999b).

Party composition data from the Sonso community were collected from September 2001 to November 2002; the data set consists of 4021 15-min scan samples (~1005 h). During this study period, the Sonso community comprised 51–58 individuals, including 12–13 males (8 adult, 4–5 subadult), 15–18 adult females, and 1–3 subadult females.

Data from the Kanyawara community were collected from January to December 2000; this data set consists of 8586 fifteen-minute scan samples (~2150 h). Community composition during this period was comparable to that of the Sonso community. The Kanyawara study community contained approximately 50 individuals, including 13 males (11 adult, 2 subadult), 15 adult females, and 2 subadult females.

Observation Methods

The same observation method was used to monitor party composition in each community. At 15-min intervals, observers recorded the number and identity of all individuals present in the party during that period. In both cases, observations were conducted primarily by senior field assistants with the assistance of students and other researchers. For the purposes of this analysis, we consider only nondependent individuals in our calculation of party size, as juveniles and infants can be easily missed or confused in counts.

Party composition analyses were performed by calculating the average number of individuals of each age–sex class present in scans of party size X. This was repeated for all party sizes. To control for the community's composition, numbers of chimpanzees were converted to the percentage of available members of the age–sex class that they represent. For instance, such data would state that an average of 20% of the community's adult females were present in a party size of 10, while at a party size of 20, 55% of the community's adult females were present.

Our analysis began by comparing three methods of quantifying party composition. First, all scan observations were used. Second, to control for any tendency by different observers to stay with certain types of parties longer, we recalculated these results using only the first observation of each party. In this case, we used the strict definition that scans after fissions and fusions were not included if individuals from the initial party were present. Third, to control for any bias in observation times, we performed the same analyses using only one observation per day, that occurring closest to 9:00 AM.

This comparison was instructive in that there was very little difference between plots produced using any of these three methods, and thus little evidence that minor biases affect this method of party analysis. Figure 1 gives an example of such a comparison, here illustrating the number of Kanyawara mothers present at each party size using the three different methods. As a compromise between maximizing data and minimizing potential bias, the results presented in this chapter use the first observation of each party ($n = 507$, Sonso; $n = 499$, Kanyawara).

In a separate analysis, we examined the composition of parties containing at least one fully swollen female. Since party size and the number of swollen females should be expected to vary over the year with food availability (Wrangham, 1977; Wrangham *et al.*, 1992, 1996; Chapman *et al.*, 1994a; Wallis, 1995, 1997; Doran, 1997; Isabirye-Basuta, 1998; Matsumoto-Oda, 1999a,b; but

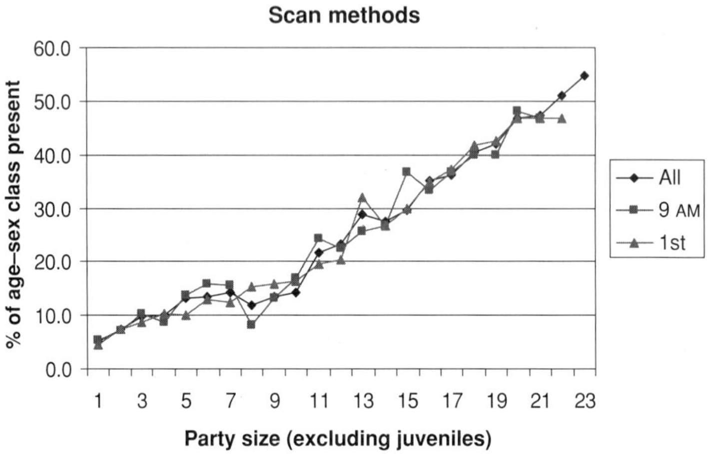

Figure 1. Three methods for evaluating party composition. Example data from Kanyawara adult females.

see Newton-Fisher *et al.*, 2000; Hashimoto *et al.*, 2001, 2003; Anderson *et al.*, 2002), we calculated average numbers for each month of adult and subadult males and nonestrous females in parties containing 0, 1, 2, etc., swollen females. Kanyawara data on estrous parties were collected from January 1999 to April 2001, and Sonso data from September 2001 to November 2002.

Definitions

"Party size" is the number of nondependent individuals considered to be traveling as a group, associating together, or feeding in the same patch of food. Subadult males are those whose testes have descended and who sometimes travel independently from their mothers. Males are adults when their testes have grown to adult size; their faces typically darken at this time as well. Subadult females in this analysis are those who have begun sexual cycling, while adult females are those who have given birth.(Goodall, 1986). A female is considered "fully swollen" when genital tissues are tumescent enough to form a tight slit and there is swelling in both the vaginal and circum-anal tissues (Dahl *et al.*, 1991). "Peripheral" or rare females (7 of 15 at Kanyawara; 6–8 of 15–17 at Sonso) are those who are observed in fewer than 15% of party scans; at Kanyawara these females have been shown to be spatially peripheral in their ranges, maintaining core areas to the north or the south of central females.

Statistics

Linear regression analyses were performed to calculate slopes of best-fit lines through the average party composition data points. The difference between regression slopes was tested using t statistics. To test if regression lines were significantly separated, an adjustment was made to a common slope and the difference between mean y values calculated using t statistics.

RESULTS

Sex Differences in Gregariousness

Both communities showed a significant difference in the gregariousness of males and females. A greater percentage of community adult males than adult females were present in all but the smallest parties and this effect was exaggerated as party sizes grew larger (Figure 2). The sex difference was more pronounced at Kanyawara, with a mean difference of 31% versus 16% at Sonso. This is consistent with previous reports of greater male general gregariousness in wild chimpanzees (Pusey, 1980; Wrangham *et al.*, 1992; Doran, 1997; Pepper *et al.*, 1999; Wrangham, 2000; but not Boesch, 1996b; Boesch & Boesch-Achermann, 2000).

Intercommunity Differences

We compared the two communities for differences in the representation of each age–sex class at various party sizes. These regressions revealed some interesting population differences, while showing, in some cases, striking similarities in patterns of party composition.

Adult males at Kanyawara were more gregarious than those at Sonso (Figure 3). Slopes of these regressions were virtually identical; that is, males were not joining parties at a significantly faster rate in Kanyawara. However, there was a small (9%) but significant difference in the mean y for the two communities, so that at any given party size more of the community's males were present at Kanyawara. At a party size of 11, which could contain all of the community's adult males in either case, an average of 63.6% (7) of Kanyawara's adult males were present, while only 43.8% (3.5) of Sonso's males were present.

When all males (adults and subadults) were considered, a significant difference was still detected (Kanyawara: $y = 4.38x + 3.30$; Sonso: $y = 4.20x + 0.60$;

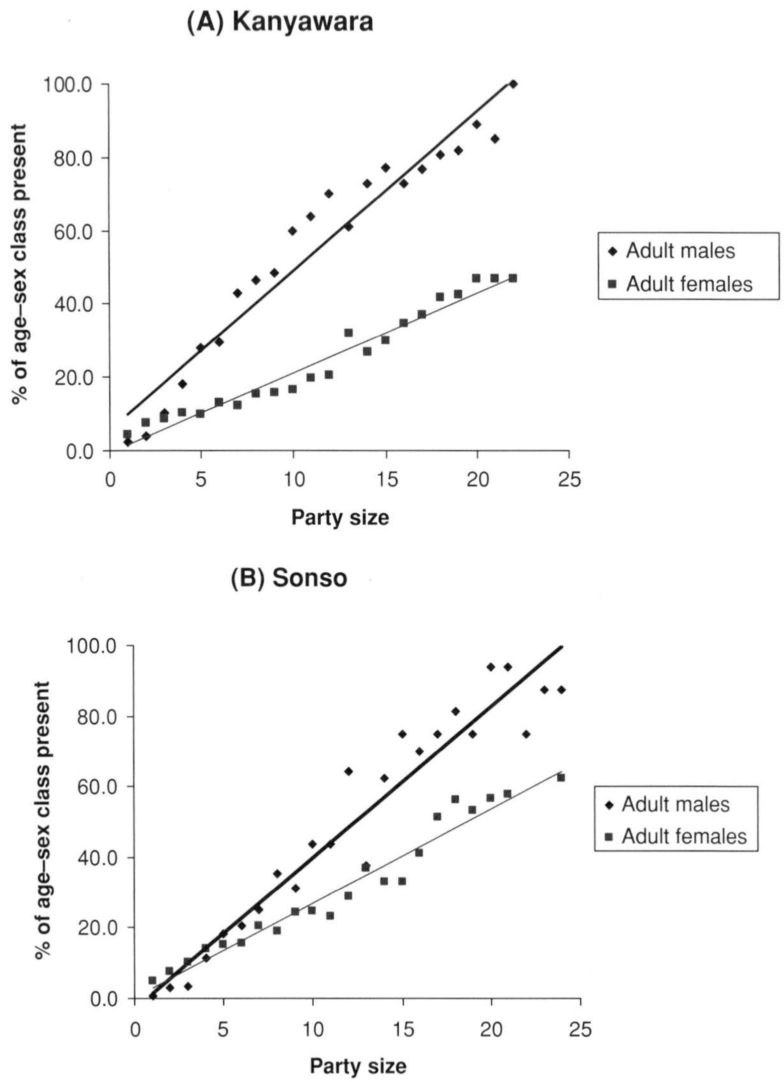

Figure 2. Regressions: sex difference in gregariousness. (A) *Kanyawara chimpanzees.* Males: $y = 4.35x + 5.48$; females: $y = 2.17x - 0.51$; differences between slope $= -2.17$, $t = -8.33$, df $= 40$, $P < 0.0001$; adjusted differences between mean $y = 30.97$, $t = 11.46$, df $= 41$, $P < 0.0001$. (B) *Sonso chimpanzees.* Males: $y = 4.26x - 2.60$; females: $y = 2.66x + 0.50$; differences between slope $= -1.60$, $t = -5.44$, df $= 42$, $P < 0.0001$; adjusted differences between mean $y = 16.04$, $t = 6.31$, df $= 43$, $P < 0.0001$.

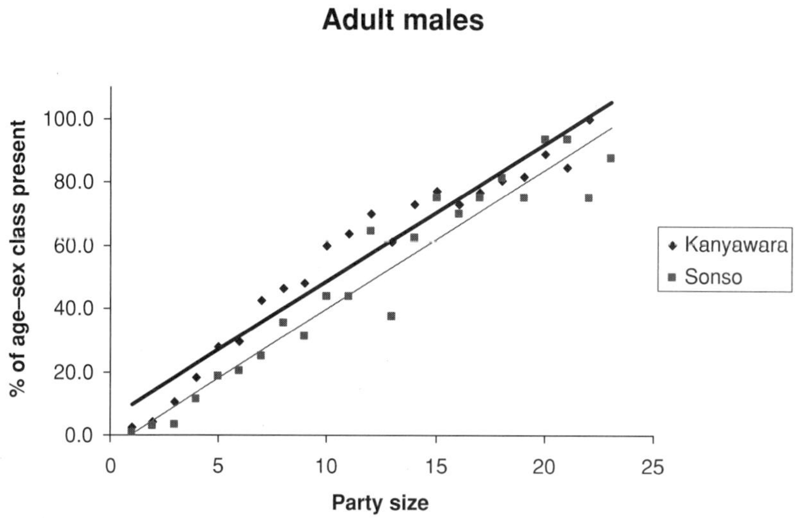

Figure 3. Regressions: intercommunity difference in gregariousness of adult males. Kanyawara: $y = 4.35x + 5.48$; Sonso: $y = 4.26x - 2.60$; differences between slope $= -0.09$, $t = -0.25$, df $= 42$, $P = 0.81$; adjusted differences between mean $y = 9.11$, $t = 3.95$, df $= 43$, $P = 0.0003$.

differences between slope $= -0.18$, $t = -0.68$, df $= 40$, $P = 0.50$; adjusted differences between mean $y = 4.80$, $t = 2.82$, df $= 41$, $P = 0.007$). However, we do not consider subadult males separately here because of a strong potential for individualistic effects given the small size of this age–sex class. Population differences observed for subadults may result from the age of the particular subadults considered and the status of their mothers. For example, the two subadult males in the Kanyawara data for this period still associated frequently with their rarely seen peripheral mothers; the five Sonso males who were subadults during this period were more diverse, including three who were near the transition to adulthood—two without living mothers and one who was the son of the dominant female in the community.

We found the reverse result for adult females. A significantly higher percentage of the parous females at Sonso were present in parties of a given size, and a significant difference in slopes revealed that this effect was greater as party size increased (Figure 4). Further investigation of this phenomenon revealed that the intercommunity difference in female gregariousness was primarily the result of peripheral female behavior. While an analysis of the percentage of central females in parties produced virtually identical regression lines (Figure 5), peripheral females showed much greater gregariousness at Sonso (Figure 6).

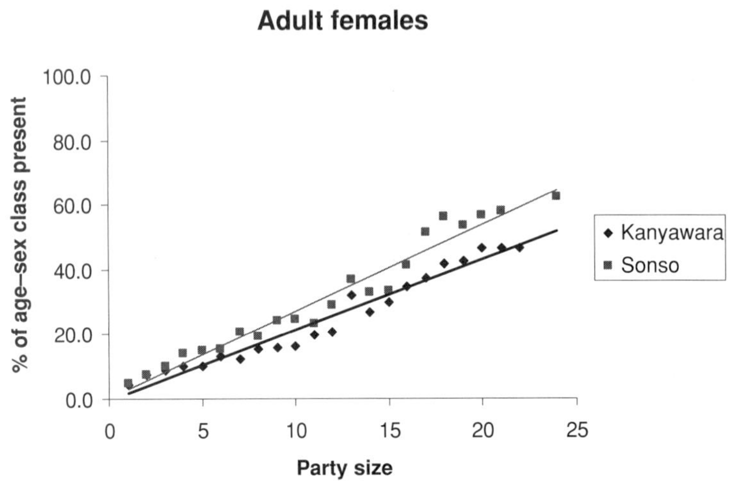

Figure 4. Regressions: intercommunity difference in gregariousness of parous females. Kanyawara: $y = 2.17x - 0.51$; Sonso: $y = 2.66x + 0.50$; differences between slope = 0.49; $t = 3.07$; df = 40, $P = 0.004$; adjusted differences between mean $y = -6.67$, $t = -5.92$, df = 41, $P < 0.0001$.

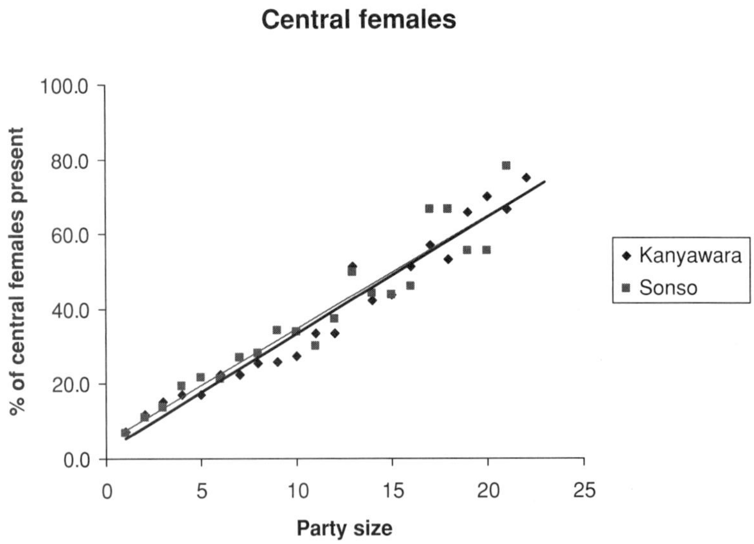

Figure 5. Regressions: intercommunity difference in gregariousness of central parous females. Kanyawara: $y = 3.12x + 2.09$; Sonso: $y = 3.17x + 3.21$; differences between slope = 0.05; $t = 0.22$, df = 40, $P = 0.83$; adjusted differences between mean $y = -1.72$, $t = -1.11$, df = 41, $P = 0.27$.

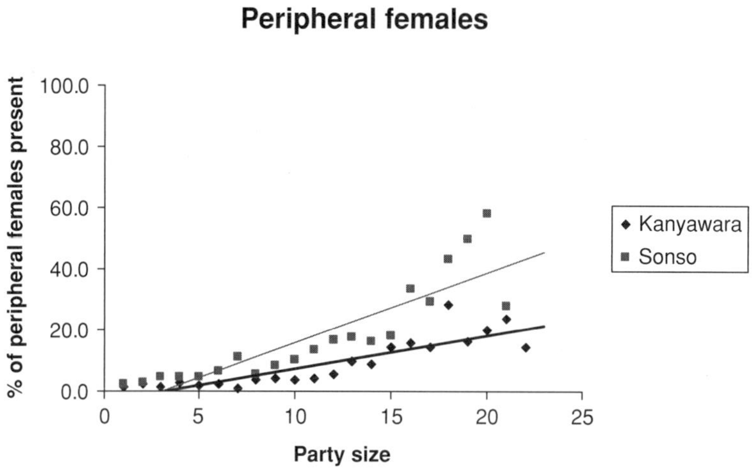

Figure 6. Regressions: intercommunity difference in gregariousness of peripheral parous females. Kanyawara: $y = 1.09x - 3.44$; Sonso: $y = 2.03x - 4.66$; differences between slope $= 0.94$; $t = 2.91$, df $= 40$, $P = 0.006$; Adjusted differences between mean $y = -9.59$; $t = -4.26$, df $= 41$, $P = 0.0001$.

Significant differences were detected in both slope and intercepts for peripheral females, who showed the greatest difference in larger parties. A party size of 20, which could be comprised without any peripheral females in either community, contained an average of nearly 60% (4.2) of the 6–8 peripheral mothers at Sonso but only 20% (1.4) of the 7 peripheral mothers at Kanyawara.

As with males, intercommunity differences in subadult female gregariousness are difficult to describe because of large potential for individualistic differences. In this case, we found interesting evidence of such individual effects when comparing the current Kanyawara data set to a previous analysis of party composition at Kanyawara (Wrangham, 2000). In the examination of 1995–1996 data, using the same method, two nulliparous females seemed to behave more like adult males than like adult females (Figure 7A). This is consistent with the hypothesis that adolescent females suffer fewer costs of grouping than do mothers owing to lack of dependent offspring and decreased travel time (Wrangham, 2000). However, our current examination of 2000 data produced the opposite result: the two nulliparous females in this analysis were more similar in grouping behavior to adult females (Figure 7B). This difference is probably due to the subadults considered. In the 1995–1996 study, the subadult females were both immigrants, while the 2000 subadults were the daughters of central mothers. The Sonso subadults also showed gregariousness more similar to adult females

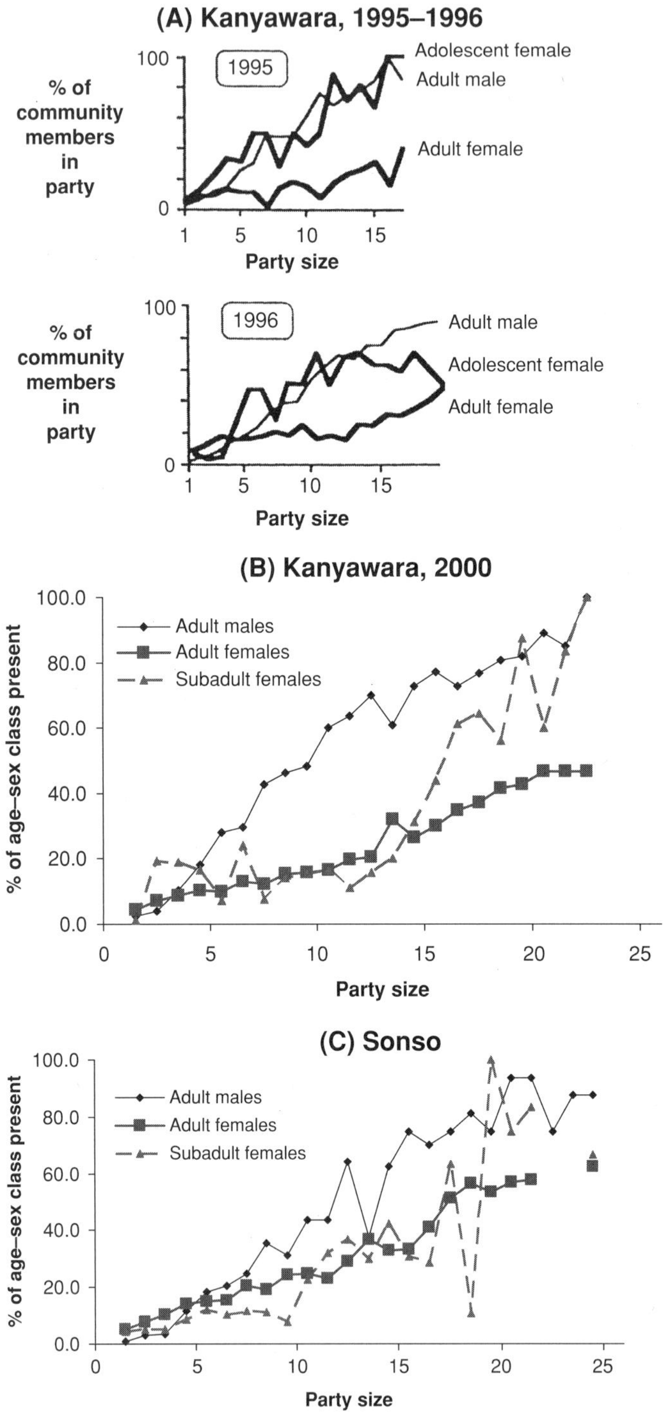

Figure 7. Subadult female grouping behavior in comparison with adult females and adult males: (A) Kanyawara 1995–1996 (Chart from Wrangham, 2000), (B) Kanyawara 2000, (C) Sonso 2001—2002.

(Figure 7C). Only one of the three Sonso females had a mother in the community, while another immigrated 4 years previous, and another with unclear immigration status had been in the community for at least 9 years. Although based on a few individuals, these current data do not support the hypothesis that adolescent females, free from the constraints of offspring, are more gregarious than mothers; subadult females' degree of gregariousness may vary according to the presence of their mother or their length of tenure in the community.

"Estrous" Parties

Next, we examined the grouping behavior of the two communities as the number of fully swollen females increased. There was no discernable difference in the affiliation of nonswollen females with these parties. Kanyawara males, however, associated in greater numbers (average of 2.25 more) with swollen females than did Sonso males, and their numbers rose more rapidly as the number of

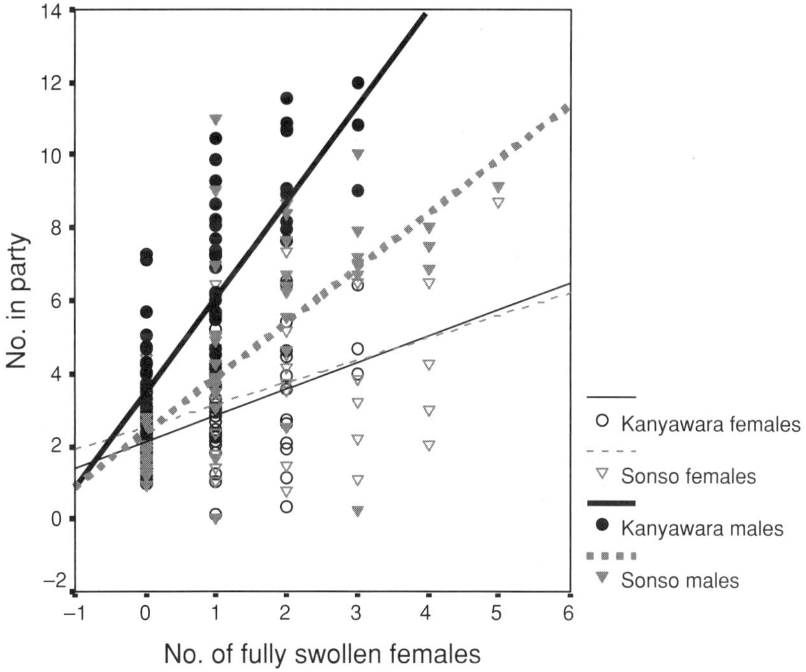

Figure 8. Regressions: intercommunity difference in party affiliation with swollen females. Kanyawara males: $y = 2.63x + 3.51$; Sonso males: $y = 1.51x + 2.45$; differences between slope $= -1.12$, $t = -3.15$, df $= 111$, $P = 0.002$; adjusted differences between mean $y = 2.25$, $t = 5.50$, df $= 112$, $P < 0.0001$. Kanyawara females: $y = 0.73x + 2.10$. Sonso females: $y = 0.61x + 2.52$; difference NS.

swollen females increased (Figure 8). Newton-Fisher (1999a) recorded similar averages for Sonso, with the number of males increasing for parties of one and two swollen females but not increasing significantly with the addition of more swollen females.

Is the Sonso–Kanyawara Difference Due To Behavior of Males or Females?

Our analyses of party composition in these two communities of chimpanzees show differences in the gregariousness of both males and females. In this type of analysis, it is possible, however, that a difference in the social behavior of one sex would make it appear as if the other was behaving differently; that is, if males are more gregarious in one community, then a party of a given size would necessarily contain fewer females. However, these communities show evidence of differences in the behavior of both sexes.

Clearly, peripheral females at Sonso appear less socially isolated than peripheral females at Kanyawara. On the other hand, the analysis of estrous parties illustrated that there was a difference in male gregarious that was demonstrably separate from the effect of female gregariousness. In addition, when we examined the number of bisexual parties containing *x* number of males, regardless of party size, we found that the majority of Kanyawara parties have more males (Figure 9). In particular, parties with seven or more males were more frequent at Kanyawara (38% of all parties, compared with 20% at Sonso).

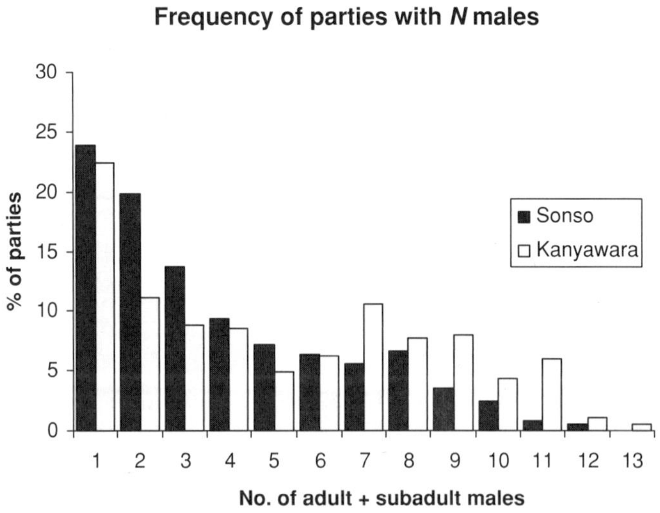

Figure 9. Observation frequency of parties with a given number of males.

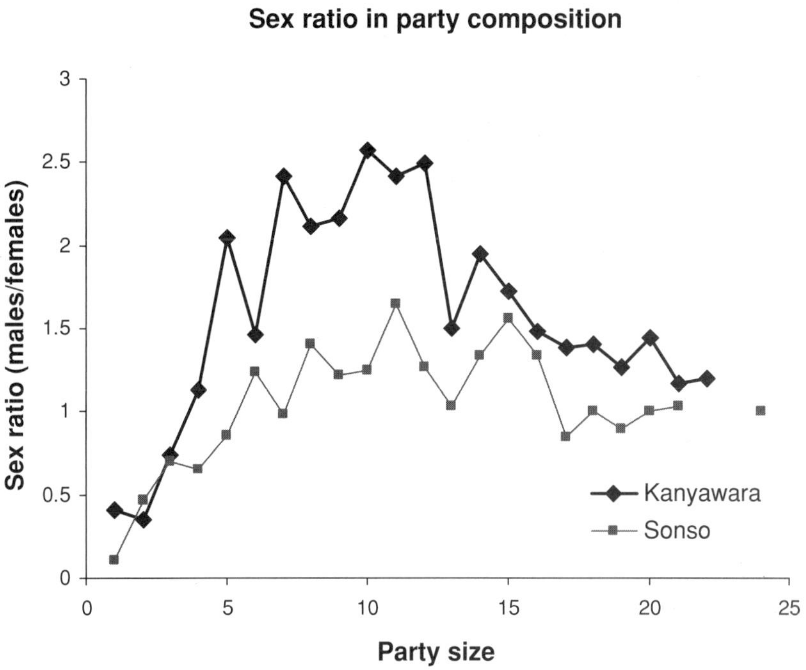

Figure 10. Sex ratio of parties. Males = adult + subadult. Ratios expected by community composition: Kanyawara = 0.76; Sonso = 0.68.

When we examine the cumulative effect of male and female differences, the contrast is pronounced. Figure 10 depicts the male-to-female sex ratio at each party size. Across party sizes, this sex ratio averaged 1.6 at Kanyawara and 1.0 at Sonso. All parties of greater than three are predominately male at Kanyawara. At Sonso, males tend to outnumber females in parties of 6–16 and, even then, not to the dramatic degree that they do at Kanyawara.

DISCUSSION

One way to look at data presented in this chapter is to point out the striking similarities in grouping patterns between the two sites. For instance, regression lines for percent central mothers present at each party size were virtually identical between the two communities. While parties of the same size contained a greater percentage of the community's males in Kanyawara, the regression slopes for the two communities were very similar.

On the other hand, the differences are clear and consistent. Kanyawara's central and peripheral females show a much greater contrast in their social strategies than do females at Sonso (see also Emery Thompson *et al.*, Chapter 13, this volume, for analysis of reproductive consequences). Patterns of male behavior indicate greater baseline gregariousness of males at Kanyawara and suggest a greater affinity for swollen females.

Several of our analyses illustrate the importance of incorporating party size into an analysis of grouping behavior. As in the case of understanding changes in male association with swollen females, the slopes of our regressions demonstrated meaningful changes in party composition as parties grew in size. Snapshot evaluations of mean party size would capture only a fraction of this intercommunity contrast.

There is not sufficient space here to test hypotheses to explain these intercommunity differences in gregariousness, but we can make some preliminary suggestions. There are likely to be many important interrelated factors.

Differences in gregariousness may simply be the accumulation of individualistic effects. We find this unlikely, particularly as both communities are well habituated and have been under observation for similar periods of time. These differences could represent variation in the costs and benefits of grouping for male and female chimpanzees at the two sites:

(i) Female cost of grouping at Sonso could be lessened by reduced scramble competition for food (Janson & Goldsmith, 1995; Wrangham, 2000; Williams *et al.*, 2002a), for instance if feeding trees or patches were larger and/or fruit abundance less seasonal, as proposed by Newton-Fisher *et al.* (2000). Increased use of nonfruit items could decrease cost of grouping, but this is unlikely the source of the difference in this case as Kanyawara chimpanzees could do this by utilizing THV. Furthermore, Wrangham *et al.* (1996) report that THV had little effect on feeding competition, as fruit and fig production continued to determine party size.

(ii) Female cost of grouping could be increased as an effect of male coercive behavior (Wrangham, 2002). Severe and prolonged male beatings of females, even with the use of large branches as weapons, can take place at Kanyawara even outside of the context of sexual coercion (Linden, 2002; Muller, 2002). Sonso females experience relatively little severe aggression from males, those few cases typically occurring in the context of failed consort attempts (personal observation, MET; Newton-Fisher, in press).

(iii) Male benefits of grouping could be increased because of (a) more frequent hunting and/or (b) increased intercommunity encounters. Larger parties should be more successful at or more likely to engage in these endeavors, and party membership may promote bonds that are beneficial for these activities (Stanford *et al.*, 1994; Mitani & Watts, 1999; Boesch & Boesch-Achermann, 2000; Wilson *et al.*, 2001, 2002; Watts & Mitani, 2002). Because they typically occur at the borders of the range, intercommunity encounters are so rarely observed that it would be difficult to judge this assertion. However, Newton-Fisher (1999a) suggests alternatively that male alliances may be better fostered in small parties with less interference from competitors.

(iv) Greater reproductive rates could alter benefits of grouping for males, particularly in parties with cycling females, as suggested by operational sex ratio theory (Mitani *et al.*, 1996a). That is, if fecund females are less common, males may be more likely to associate with females or with other competing males to increase their chances of encountering a fecund female.

(v) Differing distribution of fruiting trees (Newton-Fisher *et al.*, 2000) and/or size of home range might affect gregariousness in a chance manner. That is, when major food species are concentrated in a smaller area, individuals may be more likely to encounter one another. Newton-Fisher (2003) estimates the home range for Sonso chimpanzees to be half the size of those of Kanyawara, despite comparable community size. However, this could be expected to have the same effect on gregariousness of both sexes.

This comparison, as with many others, illustrates the further need to document the diversity of behavior in chimpanzees. It is increasingly unproductive to make overly restrictive generalizations about the behavior of chimpanzees as a species. Behavioral diversity in chimpanzees, as in humans, is likely to extend well beyond easily observable differences in material culture and foraging strategies, to more subtle differences in everyday social interactions.

Cost of grouping arguments typically are posed in the context of sex differences in party membership within a single community. These examinations have provided intriguing hypotheses (Wrangham, 2000; Williams *et al.*, 2002), but these data are limited in their hypothesis-testing capacity. We propose that intercommunity comparisons of gregariousness would be useful in testing these hypotheses across a range of ecological and sociosexual circumstances. These data collection and analysis methods provide a simple means to compare study communities; while controlling for varying composition of communities, they provide a dynamic view of party composition changes with increasing party size.

We encourage other chimpanzee researchers to assist us in documenting this variation in male and female gregariousness.

ACKNOWLEDGMENTS

Permission for research in Uganda was granted by Makerere University, the Ugandan National Council for Science and Technology, Uganda Forest Department, and the Uganda Wildlife Authority. Our gratitude to V. Reynolds and F. Babweteera for access to the Budongo chimpanzees and valuable project assistance. G. Isabirye-Basuta and J. Kasenene provided important support at Kanyawara. Data at Kanyawara were collected by J. Basigara, J. Barwogeza, C. Katongole, K. Clement, K. Deo, F. Mugurusi, D. Muhangyi, C. Muruuli, and P. Tuhairwe, with help from A. Arcadi, C. Chapman, K. Duffy, C. Hooven, M. Muller, J. Obua, S. Mugume, and M. Wilson. Thank you to Z. Kiwede for data collection at Sonso. RWW and the Kibale Chimpanzee Project were supported by the Leakey Foundation, the National Geographic Society (5626-96), the National Science Foundation (NSF SBR-9120960, BCS-9807448), and the Getty Foundation. MET was supported by Harvard University, the Leakey Foundation, and the Wenner-Gren Foundation for Anthropological Research.

CHAPTER THIRTEEN

Urinary Estrone Conjugates and Reproductive Parameters in Kibale (Kanyawara) and Budongo (Sonso) Chimpanzees

Melissa Emery Thompson, Richard W. Wrangham, and Vernon Reynolds

INTRODUCTION

Most of the information about reproduction in chimpanzees has come from studies of captive animals. However, long-term field studies have yielded very different statistics on reproductive parameters, which indicate more study of

Melissa Emery Thompson and Richard W. Wrangham • Department of Anthropology, Harvard University, 11 Divinity Avenue, Cambridge, MA 02138, USA **Vernon Reynolds** • Emeritus, School of Anthropology, University of Oxford, 51 Banbury Road, Oxford, United Kingdom.

Primates of Western Uganda, edited by Nicholas E. Newton-Fisher, Hugh Notman, James D. Paterson, and Vernon Reynolds. Springer, New York, 2006.

Table 1. Interbirth intervals in four wild chimpanzee communities

Study community	IBI average (mos.)	Range	Reference
Mahale (M Group) (*P. t. schweinfurthii*)	69.1 (*n* = 33)	52–87	Nishida *et al.*, 2003
Tai (*P. t. verus*)	69.1 (*n* = 33)	48–120	Boesch and Boesch-Achermann, 2000
Gombe (Kasakela) (*P. t. schweinfurthii*)	66.0 (*n* = 21)	48–78	Goodall, 1986
Bossou (*P. t. verus*)	61.2 (*n* = 15)	36–132	Sugiyama, 1994

All intervals were calculated as mean completed intervals when the first infant survived a minimum of 3 years. For each community, only the most complete published estimate is reported.

animals in their natural environment is needed to really understand reproduction in this species. As with humans, chimpanzees have already revealed considerable population variation in behavior, as well as in life history statistics. Examination of this variation can help us learn about the ecological and social factors that may influence the biology of chimpanzees.

In captivity, females have interbirth intervals that average approximately 4 years (Coe *et al.*, 1979; Courtenay, 1987; Orbell, in Tutin, 1994; Bloomsmith, in Wallis, 1997). Figures from wild studies are considerably longer (Table 1), suggesting that reproduction in the wild is likely hindered by relatively high disease burden or harsher energetic conditions (Tutin, 1994; Knott, 2001). The differences observed in average birth intervals between communities defy any simple explanation, for example, by subspecies. However, these average figures may also be deceptively similar, failing to capture the great variability within some communities. When only completed intervals are considered, short birth intervals tend to be overrepresented, as by their nature they are observed more frequently. For instance, interval estimates at Bossou have lengthened 10 months with the addition of seven new births (Sugiyama, 1989, 1994). Survival analyses, incorporating both complete and incomplete intervals, would be more instructive for examining intercommunity variation in that they more accurately incorporate intracommunity variation.

Within-community variation in reproductive parameters, including birth intervals, is considerable and should be an important component for our understanding of the pressures on reproduction in this species. At Gombe, high-ranking females had higher infant survival, faster-maturing daughters, and higher birth rates (Pusey *et al.*, 1997). Gombe researchers suggest this may be due to increased access to areas of high fruit productivity; females with core areas in certain "neighborhoods" had higher reproductive success (Williams *et al.*, 2002a). Peripheral females, who may have ranged in community border areas or associated with more than one community, did either very well or very poorly. In general, these rare or "peripheral" female chimpanzees are worthy of closer examination to document variation within females.

Thus far, all the information on endocrine physiology in female chimpanzees has come from captive studies. These data indicate patterns of ovarian function that are very similar to those in human females. Chimpanzee ovarian cycles are longer than a human's 28 days, averaging approximately 35 days in the wild and in captivity (Tutin, 1980; Nadler *et al.*, 1985; Wallis, 1997; Matsumoto-Oda, 1999a; Mahale: 31.5 days—Hasegawa & Hiraiwa-Hasegawa, 1983; captive: 36.7 days—Wallis, 1985; Graham, 1970). However, the overall pattern of ovarian hormone activity is very similar. After the follicular peak in estradiol production, chimpanzees, like humans, show an elevation of estradiol in the luteal phase (Graham, 1978; McArthur *et al.*, 1981; Nadler *et al.*, 1985; Dahl *et al.*, 1991). Such an elevation is common to the apes, but reduced or absent in Old World monkeys (Hess & Resko, 1973; Weick *et al.*, 1973; Kling & Westfahl, 1978; Nadler *et al.*, 1985; Czekala *et al.*, 1988). While the pattern of progesterone is similar across primates, average and maximal levels in the chimpanzee are much closer to human females than those of orangutans, gorillas, or other Old World primates (Nadler *et al.*, 1979, 1984, 1985). In addition, endometrial histology (Graham, 1973) and several specific features of the pattern of gonadotropin secretion over the cycle are common to humans and chimpanzees (Nadler *et al.*, 1985; Ross *et al.*, 1970).

Human reproductive ecologists have documented considerable variation in ovarian function both within and between populations (Ellison, 1993; Ellison *et al.*, 1993b). Factors such as age, health, and energetic condition can affect the production of ovarian hormones. Compatible studies of salivary hormones reveal differences in age-specific ovarian function between populations, with American women producing higher estradiol than women in nonindustrialized settings (Ellison *et al.*, 1989; Panter-Brick *et al.*, 1993). While genetic effects have not been ruled out, it is likely that much of this difference is the cumulative result of chronic and periodic energy shortages, intensive workloads, and high disease burdens experienced by women in developing countries. There may also be developmental effects, setting the range or sensitivity of ovarian function in accordance with energetic conditions during maturation (Apter & Vihko,

1985; Ellison *et al.*, 1993a; Ellison, 1996; Vitzthum, 1997; Lipson, 2001). Interbirth intervals also vary in a similar manner among human populations (reviewed in Wood, 1994), although these comparisons are difficult owing to variation in lactation and contraceptive practices. Postpartum infecundity in urban women can be a year or more shorter than in rural populations (Wood, 1994); nutritionally stressed women undergo longer periods of postpartum infecundity than do well-nourished women (Huffman *et al.*, 1987; Valeggia & Ellison, 2001).

Variation in ovarian steroid levels has a significant influence on reproduction, as has been demonstrated in humans and in nonhuman primates. Estrogens have a variety of effects during the menstrual cycle that are immediately relevant to fertilization and implantation: stimulating proliferation of cells in the uterine lining, increasing penetrability of cervical mucus, promoting progesterone receptor formation, and fueling growth of the oocyte. Follicular estradiol levels are correlated with ovarian follicle size (Eissa *et al.*, 1986) and increase the fertilizability of the ovum (Yoshimura & Wallach, 1987). At blastocyst development, higher estradiol levels are associated with greater probability of implantation in *in vitro* fertilization studies .(Akman *et al.*, 2002). Studies of humans have convincingly demonstrated correlations between follicular estrogens and conception success (Lipson & Ellison, 1996; Li *et al.*, 2001). Similarly, estrogen levels have been tied to increased fecundity in captive (Markham, 1995) and wild orangutans (Knott, 1997a,c; 1999) and baboons (Wasser, 1996).

This chapter presents new data on interbirth intervals and urinary estrone conjugates (E_1C) in wild chimpanzees from Kibale National Park and Budongo Forest Reserve. We examine whether interbirth interval differs between Kanyawara and Sonso communities of chimpanzees, test the hypothesis that interbirth intervals are associated with ovarian function by comparing E_1C levels between communities, and assess whether central females have shorter birth intervals and higher E_1C levels than peripheral females.

There are some suggestions that differing energetic conditions in these two locations might affect reproduction. First, Kanyawara has been observed to be seasonally stressed, thus Wrangham and colleagues (1996) hypothesized that high-quality THV may buffer them during periods of low fruit availability. While Sonso chimpanzees rarely utilize THV, phenology and feeding data indicate that the Sonso chimpanzees maintain a diverse diet throughout the year without severe seasonal deficits in ripe fruit availability (Newton-Fisher, 1999b; Plumptre, Chapter 20, this volume). Second, home range estimates indicate higher population densities at Sonso (Chapman & Wrangham, 1993; Newton-Fisher, 2003); this suggests greater or more uniform habitat quality, including high fruit density, in the Sonso home range (Newton-Fisher, 2003). Gombe reproductive statistics indicate that there can be enough disparity even in local conditions within a community to differentially affect the reproduction of individual females (Williams *et al.*, 2002a).

METHODOLOGY

Study Populations

The Sonso community in the Budongo Forest Reserve has been the subject of continuous research by the Budongo Forest Project since 1990. Budongo Forest Reserve is predominantly a moist, semideciduous forest, although the Sonso region has been the target of selective logging activities and thus comprises a mosaic of forest types (Reynolds, 1992). Budongo undergoes one dry season from approximately December to February (Newton-Fisher, 1999a,b,c; Wallis, 2002). During the period of urine collections (October 2001 to December 2002), the community size ranged from 51 to 58, including 8 adult males, 16–18 parous females, and 3–4 adolescent females. Data presented here concern approximately 740 urine samples collected from 22 adult and adolescent females.

The Kanyawara chimpanzees in Kibale National Park have been studied by the Kibale Chimpanzee Project since 1987. The home range of these chimpanzees comprises approximately 60% moist, evergreen forest, with small areas of swamp, grassland, and colonizing forest (Chapman *et al.*, 1994a,b). Kibale undergoes two short dry seasons, from approximately June to August, and December to February (Ghiglieri, 1984). Over the study period from November 1997 to November 2002, the Kanyawara community size comprised 40–56 individuals, including 9–11 adult males, 11–17 adult females, and 3–6 adolescent females. Data presented here include more than 1100 urine samples from 22 adult and adolescent females.

In each community, there was a clear subsample of females who were encountered far less frequently than other individuals. These females, categorized as "peripheral," were observed on fewer than 15% of 15-min scan samples taken regularly at both sites. Wilson (2001) demonstrated that, in addition to being socially peripheral, these rare females at Kanyawara were also spatially peripheral. Females who were observed less frequently maintained home ranges distinctly to the north or the south of the largely overlapping ranges of central females, and they spent at least twice as much time in the community's border areas. At Gombe, too, range analyses demonstrated that those females who associated infrequently with males were also spatially peripheral (Williams *et al.*, 2002b). Home range analyses have not been performed for Sonso females. Eight of the mothers sampled at Kanyawara and seven at Sonso were considered peripheral. Central and peripheral categories were used as a gauge of intracommunity variability.

Birth Intervals

Owing to the length of study at these field sites, it is now possible to calculate interbirth intervals on the basis of a set of birth dates known to the nearest month. To include the considerable variability in incomplete birth intervals, we calculated median intervals using Kaplan–Meier survival analysis. Intervals were not considered if the first offspring died before the mother resumed cycling. These results include 10 complete and 15 incomplete intervals for Sonso females, and 11 complete and 13 incomplete intervals for Kanyawara females.

Endocrine Methods

Noninvasive urine sampling was attempted daily by researchers and field staffs of Budongo Forest Project and Kibale Chimpanzee Project. The staff at Kibale has been collecting urine samples from females since November 1997 (Muller & Wrangham, 2004); MET initiated Sonso collections in September 2001. Urine was collected on plastic sheets, in a catch-pole extended into the stream, or by pipetting directly from undergrowth (Knott, 1997b; Robbins & Czekala, 1997). Care was taken to avoid or discard contaminated samples, although other researchers (Knott, 1999; Muller & Wrangham, 2004) have demonstrated that small amounts of plant matter, soil, and feces do not meaningfully alter counts of steroid conjugates or creatinine from urine. Samples were frozen as soon as possible (0–14 h) in field laboratories and transported on ice to the United States. Laboratory studies have confirmed stability of steroids and steroid conjugates at room temperature for at least 24 h (Grant and Beastall, 1983; Kesner *et al.*, 1995).

Estrogens are excreted in the urine in metabolized form and can be measured without extraction by assay for these estrone conjugates (E_1C). MET performed enzyme immunoassays (EIA) for E_1C in the Primate Reproductive Ecology laboratory at Harvard University's Department of Anthropology. Laboratory protocols followed established methods (Czekala *et al.*, 1986; Shideler *et al.*, 1990), using R522 estrone glucuronide antibody and conjugate (HRP) from the Department of Population Health and Reproduction at UC Davis (CJ Munro). After incubation, thawed samples, controls, and standards (6.25–200 pg/40 µl, Sigma Co.) were added in duplicate aliquots of 0.04 ml to antibody-coated plates. After a short incubation, 0.05 ml of E_1C-HRP was added to each well and the plates incubated overnight. Samples and controls were diluted with phosphate buffer in amounts suitable to place them within the standard range. After addition of substrate solution, optical densities were read with 405- and 630-nm filters, and concentrations estimated after fitting to standard curve run in duplicate (AssayZap, Biosoft, Cambridge, UK). All results were standardized to creatinine (Cr) levels to control for concentration (Taussky, 1954).

Because of the opportunistic nature of subject availability in the wild, it is not feasible to limit time of day for sample collection. However, more than 70% of samples were collected before 12:00 PM at either site, and a circadian effect was not apparent in E_1C excretion from chimpanzee urine samples (excluding pregnant females: $r^2 = 0.0004$, $n = 1603$, $P = 0.45$; including: $r^2 = 0.001$, $n = 1804$, $P = 0.12$). Samples from pregnant females, who exhibit extremely high and variable E_1C levels, did show a significant decline over the day ($r^2 = 0.069$, $n = 201$, $P = 0.0002$), but sampling time distributions were not different between sites; this effect may be an artifact of low sampling in the late afternoon. Some samples were discarded because of low Cr levels (<0.07), which appear to overcorrect and produce artificially elevated steroid levels.

Data Treatment

Subject reproductive records were consulted, and each sample was assigned to the subject's reproductive state. Samples from cycling females were further classified as to their timing in relation to maximum sexual swelling. Five categories of females are considered: adolescent, lactating, pregnant, cycling with maximal swelling, and cycling with no swelling. In these analyses, "adolescent" refers to peripubertal females: at least 8 years old but not yet having displayed their first maximal swellings. "Lactating" refers here to those females still undergoing postpartum amenorrhea, from parturition until the month before maximal swellings reappear. Onset of pregnancy was detectable by sharp increase in E_1C levels (Shimizu *et al.*, 2003a), and was confirmed by hCG test strips (AimStick, Craig Medical, Vista, CA). Females were considered to have maximal swellings when both the vaginal and circumanal portions showed tumescence and tissues were turgid enough that they pressed together in a tight vaginal slit (Dahl *et al.*, 1991). Samples from maximally swollen females were assigned to the month in which the swelling began. To examine a baseline cycling level, the category of "flat," or nonswollen, females excluded any samples from the 10 days following detumescence, as such samples might include luteal estrogen peaks (Nadler *et al.*, 1985; Dahl *et al.*, 1991). Budongo data include samples from 2 adolescents, 16 lactating females, 8 pregnant females, 9 swollen females, and 8 nonswollen females. Kanyawara data includes samples from 6 adolescents, 14 lactating females, 7 pregnant females, 12 swollen females, and 12 nonswollen females.

In handling the data, we wanted to minimize bias from frequently sampled individuals while also avoiding any possible seasonal effects on E_1C levels or sampling success. Repeated samples from an individual in the same day were averaged together to give a single figure for that day. Then, a monthly average level was calculated for each female in a reproductive state. For analyses comparing communities and reproductive states, the median E_1C level was calculated for each month across all females in the reproductive state being considered. For analyses comparing central and peripheral females, median E_1C levels were calculated for each female across all months she was sampled in the reproductive state being considered. Statistical tests were performed using these medians so that *N*s for these analyses refer to months or to females. With the exception of EIA validation data, nonparameteric statistics were used. Two-tailed probabilities were reported for comparisons of Sonso and Kanyawara; one-tailed probabilities (labeled P_1) were used to test the hypothesis that central females have better reproductive statistics than peripheral females.

RESULTS

Assay Validation

EIA monitoring of E_1C levels has been utilized for captive chimpanzees (Shimizu *et al.*, 1999, 2003a,b) and a variety of other primate species (Czekala *et al.*, 1986; Shideler *et al.*, 1990; Heistermann *et al.*, 1995, 1996; Fujita *et al.*, 2001). Test procedures to verify the validity of the assay illustrated the efficacy of this method for quantifying ovarian function in this species. Accuracy was tested by recovery of known E_1C amounts added to urine samples in duplicate; mean ± standard deviation recovery was 94.5% ± 12.7 (regression of observed to expected doses: $r^2 = 0.989$, $n = 5$, $P = 0.0005$). Interassay coefficients of variation for high and low controls were 9.8 and 12.8%, respectively. Intraassay CVs of four controls assayed in six replicates were 3.1–5.6%. Parallelism was confirmed by serial dilution of a urine sample, which produced a curve with slope not significantly different from that of the standard curve ($t = 0.894$, df $= 8$, $P = 0.40$).

Interbirth Intervals

Median length of completed birth intervals was very similar between these two communities: 63.0 months at Sonso ($n = 10$), 68.0 at Kanyawara ($n = 11$). However, these figures disguise much of the interesting variation in this reproductive parameter. Birth intervals at Sonso were clustered around the median (range: 57.6–83.7 months, Table 2), while Kanyawara intervals were significantly more variable, spread across a range of more than 7 years (27.5–98.4 months; $F = 6.76$, $n_K = 11$, $n_S = 10$, $P = 0.004$). In addition, there were two very long incomplete intervals in the Kanyawara data set; one of these mothers died before giving birth again, so this important data would not be included in a standard birth interval analysis. These examples illustrate why a survival analysis is a more useful tool for examining birth intervals.

When Kaplan–Meier survival analyses were conducted, figures for these two communities diverged. Figure 1 illustrates results of survival analyses contrasting these two communities of chimpanzees. Known birth intervals at Sonso were considerably shorter (median 65.0 ± 3.7 months, $n = 25$) than at Kanyawara (79.1 ± 15.2 months, $n = 24$). Log rank tests approached significance (log rank $= 3.11$, df $= 1$, $P = 0.08$), but because of variability within the data sets could not reject the hypothesis that all birth intervals were from the same population.

We also considered how interbirth intervals varied with community status (Figures 2 and 3). Kanyawara peripheral females (median 93.5 ± 40.0 months, $n = 8$) had much longer intervals than central females (median 69.7 ± 7.5 months, $n = 16$) and exhibited great variability. At Sonso, on the other hand, peripheral females (59.4 ± 1.5 months, $n = 8$) actually had slightly shorter birth intervals than central females (65.0 ± 5.1 months, $n = 17$).

In sum, the birth interval data lead us to some interesting conclusions. The small range of Sonso birth intervals contrasts with a wide range for Kanyawara.

Table 2. Individual interbirth intervals used in Kaplan–Meier analyses

Sonso					Kanyawara				
Mother	Female status	Incomplete IBI (months)	Completed IBI (months)	Mother dead	Mother	Female status	Incomplete IBI (months)	Completed IBI (months)	Mother dead
KG	Central	4			BL	Central	3		
KU	Central	5			PU	Peripheral	15		
ML	Central	7			BL	Peripheral		28	
NB	Central	11			KL	Central	31		Yes
MK	Peripheral	13			OU	Central	34		
ZM	Central	14			OU	Central		38	
RH	Peripheral	17			TG	Central	38		
CL	Central	21			OU	Central		39	
KW	Central	26			UM	Peripheral		40	
ZA	Peripheral	30			UM	Peripheral	46		
HT	Peripheral	35			NL	Central	47		
KL	Central	41			AL	Central	51		
JN	Central	51			EK	Peripheral	54		
RD	Peripheral	52		Yes	TG	Central		63	
RH	Peripheral		58		AR	Central	66		
KU	Central		58		LP	Central		68	
BN	Peripheral		59		BL	Central		70	
KG	Central		60		MU	Peripheral	72		
KY	Central	61			FG	Central		79	
KU	Central		61		LP	Central	86		
ZM	Central		65		AR	Central		93	
KL	Central		68		PU	Peripheral		94	
ZA	Peripheral		73		KL	Central		98	
KW	Central		82		JO	Peripheral	100		Yes
NB	Central		84						

Intervals are calculated as of January 2004.

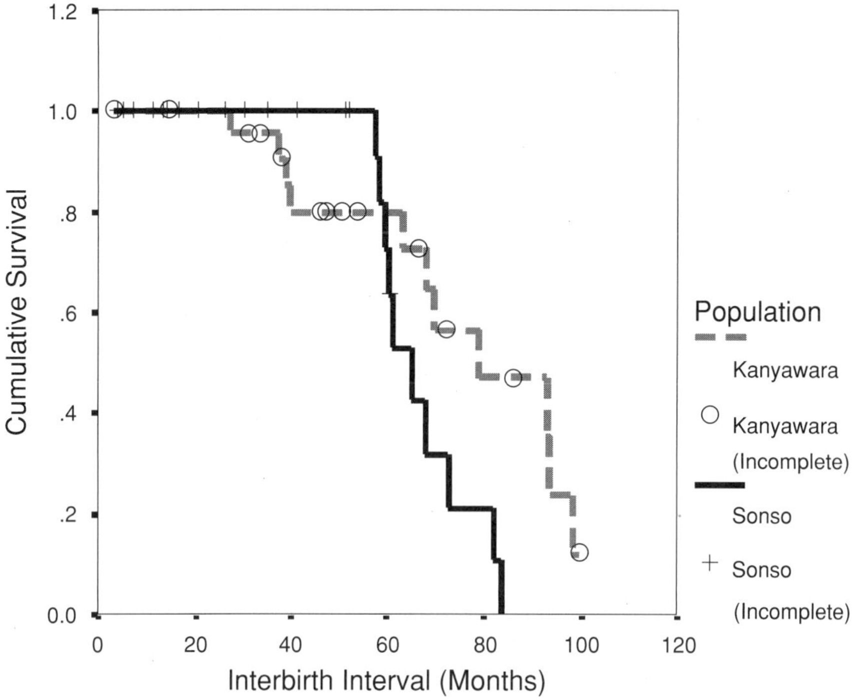

Figure 1. Survival plot: known interbirth intervals at Sonso and Kanyawara.

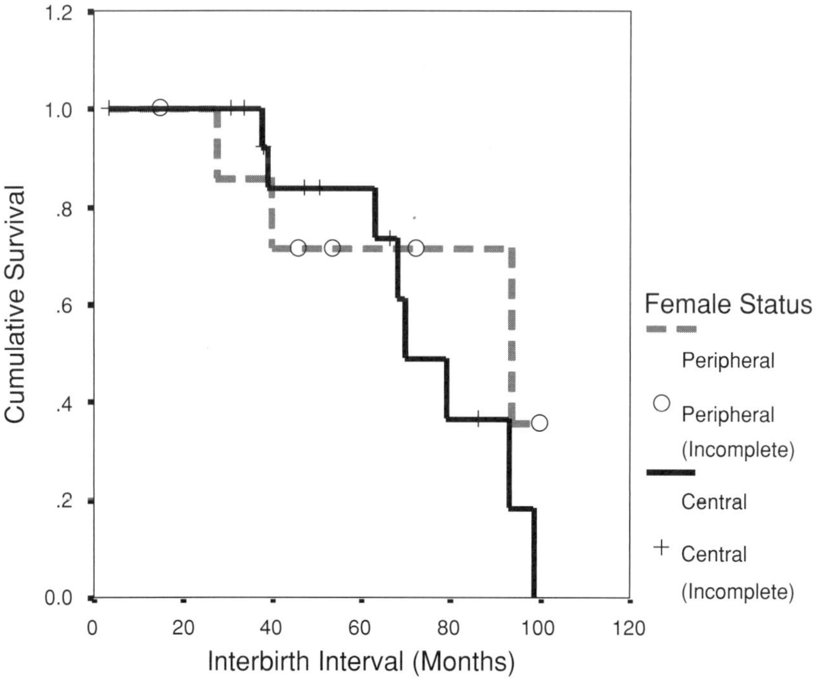

Figure 2. Survival plot: known interbirth intervals at Kanyawara by community status.

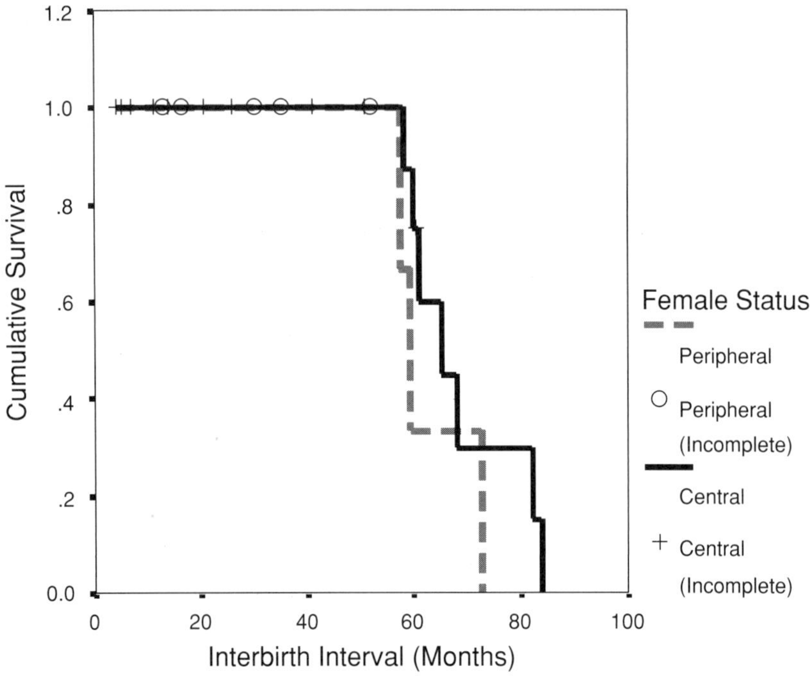

Figure 3. Survival plot: known interbirth intervals at Sonso by community status.

Because of several long complete and incomplete Kanyawara intervals, it appears that Sonso females, as a group, reproduce at a faster rate than Kanyawara females. Central females at Kanyawara appear to experience better, or at least more consistent, reproductive conditions than peripheral females. Many birth intervals for Kanyawara's peripheral females were extremely long, but one (27.5 months) may be the shortest known wild birth interval for chimpanzees. This accords with the Williams *et al.* (2002a) finding that peripheral females at Gombe either had very good or very poor reproductive statistics. By contrast, Sonso's peripheral females may even be doing better than their central counterparts.

Estrone Conjugates

E_1C and Reproductive State

To further validate the endocrine method, we examined E_1C values by reproductive state, with both communities combined. Reproductive state was a significant source of variation in E_1C levels (Kruskal-Wallis ANOVA, $X^2 = 156.02$, df $= 4$, $P < 0.001$), with relative values as should be expected.

Adolescents exhibited the lowest levels (median ± standard deviation = 6862 ± 4149 pg/mg Cr), significantly lower than the closest category of lactating females (Mann–Whitney *U*, $z = -4.43$, $n_{AD} = 51$, $n_{LA} = 71$, $P < 0.001$). Lactating females (8661 ± 3350 pg/mg Cr), in turn, had significantly lower levels than cycling females without swellings (11,363 ± 10,932 pg/mg Cr; $z = -2.96$, $n_{LA} = 71$, $n_{FL} = 45$, $P = 0.003$). Females excreted more than twice as much E_1C during the swelling phase (31,930 ± 23,815 pg/mg Cr), when they were undergoing the late follicular estradiol peak, than they did when nonswollen ($z = -6.07$, $n_{FL} = 45$, $n_{SW} = 41$, $P < 0.001$). E_1C levels in pregnant females were nearly 10 times those of any other category (289,493 ± 226,311 pg/mg Cr; vs. Swollen: $z = -6.80$, $n_{SW} = 41$, $n_{PG} = 32$, $P < 0.001$). There was marked variation within the reproductive categories, although lactating females exhibited considerably less variation than other categories (CV: adolescents, 56%; lactating, 36%; flat, 76%; swollen, 67%; pregnant, 65%).

Do Estrone Conjugates Vary Between Sonso and Kanyawara?

No significant differences in estrone conjugate levels were detected between communities for these reproductive categories (Mann–Whitney *U*, $0.24 < P < 0.77$; Figure 4). In fact, median values for each category were remarkably similar considering the variation within each site (Table 3). Shorter birth intervals at Sonso did not seem to be associated with higher ovarian function in general. However, the most apparent difference between sites was in the high variability of birth intervals at Kanyawara. So, we looked further to find out whether this variability was apparent in the E_1C data.

Do Central Females Have Higher Estrone Conjugate Levels Than Peripheral Females?

We examined whether female status was a systematic source of variation in E_1C levels by looking at individual medians across the months in which they were sampled. Individual central females at Kibale had higher E_1C levels than peripheral females in each category (Wilcoxon Paired: $z = -2.02$, $n_1 = n_2 = 5$, $P_1 = 0.02$; Figure 5); at Sonso, peripheral females actually had higher levels in three of the five categories ($z = -1.21$, $n_1 = n_2 = 5$, $P_1 = 0.11$; Figure 6). Within reproductive categories, central cycling females at Kanyawara had significantly higher E_1C levels than peripheral cycling females (Nonswollen: $z = -2.03$,

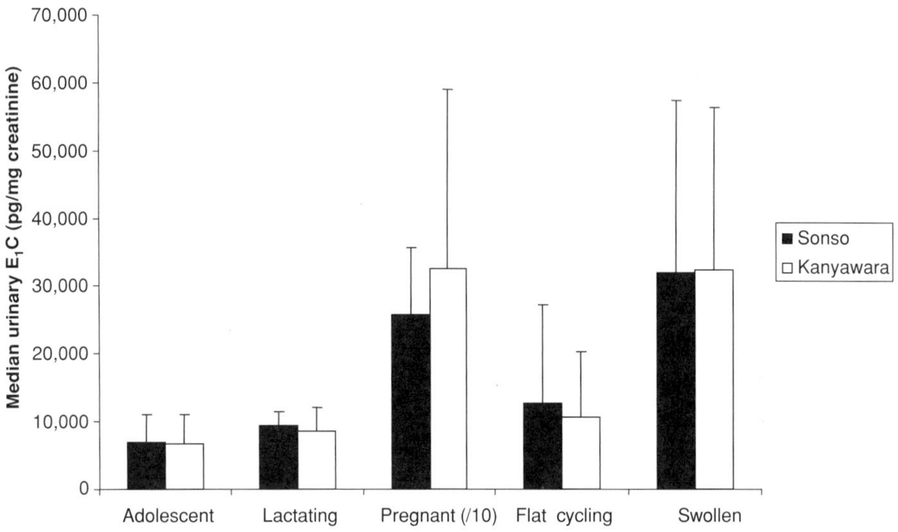

Figure 4. Estrone conjugate levels compared between Sonso and Kanyawara communities. Median ± standard deviation. For this graph, pregnant values were divided by 10 in order to fit scale.

$n_C = 7$, $n_P = 5$, $P_1 = 0.02$; swollen: $z = -1.868$, $n_C = 8$, $n_P = 4$, $P_1 = 0.04$). Peripheral lactating females at Sonso had significantly higher levels than central lactating females ($z = -2.81$, $n_C = 9$, $n_P = 7$, $P_1 = 0.002$).

DISCUSSION

Our data indicate a difference in interbirth intervals between the Sonso and Kanyawara communities when incomplete intervals are considered. In particular, the lengthy birth intervals of several peripheral females and the wide variability of completed intervals at Kanyawara suggest less consistency in reproductive conditions. Were mortality conditions to be equal, this amount of variation would lead to a meaningful difference in birth rates over time both within and between communities. Mortality rates are not yet available for Sonso, although cross-community comparisons demonstrate sizeable differences between sites (Hill *et al.*, 2001). An expanded comparison of reproductive parameters across study communities would further expand our understanding of population dynamics and life history in this species.

We hypothesized that such a difference in interbirth interval should be reflected in the ovarian hormone levels, as proximate reproductive indicators.

Table 3. Median ± standard deviation E_1C levels by community and community status

	Kanyawara	Sonso	Kanyawara		Sonso	
	Central and peripheral (months/samples)		Central (females/samples)	Peripheral (females/samples)	Central (females/samples)	Peripheral (females/samples)
Adolescents	6839 ± 4140 (38/166)	6862 ± 4275 (13/62)	6572 ± 884 (5/62)	6545 (1/6)	6707 (1/42)	6686 (1/19)
Lactating	8540 ± 3575 (57/650)	9346 ± 2172 (14/460)	8672 ± 1250 (8/598)	8071 ± 1984 (6/52)	8015 ± 1516 (9/402)	11189 ± 6905 (7/58)
Flat cycling	10554 ± 9742 (35/123)	12736 ± 14463 (10/32)	12929 ± 3632 (7/106)	7865 ± 2042 (5/17)	9444 ± 6992 (3/19)	13861 ± 40982 (5/13)
Swollen parous	32399 ± 23882 (34/133)	31930 ± 25382 (7/45)	39308 ± 21512 (8/127)	28035 ± 11406 (4/8)	26489 ± 6312 (5/27)	24009 ± 19672 (4/18)
Pregnant	325831 ± 264653 (20/94)	257926 ± 98031 (12/142)	413110 ± 306780 (6/87)	174852 (1/7)	219510 ± 213960 (5/120)	297870 ± 163250 (3/22)

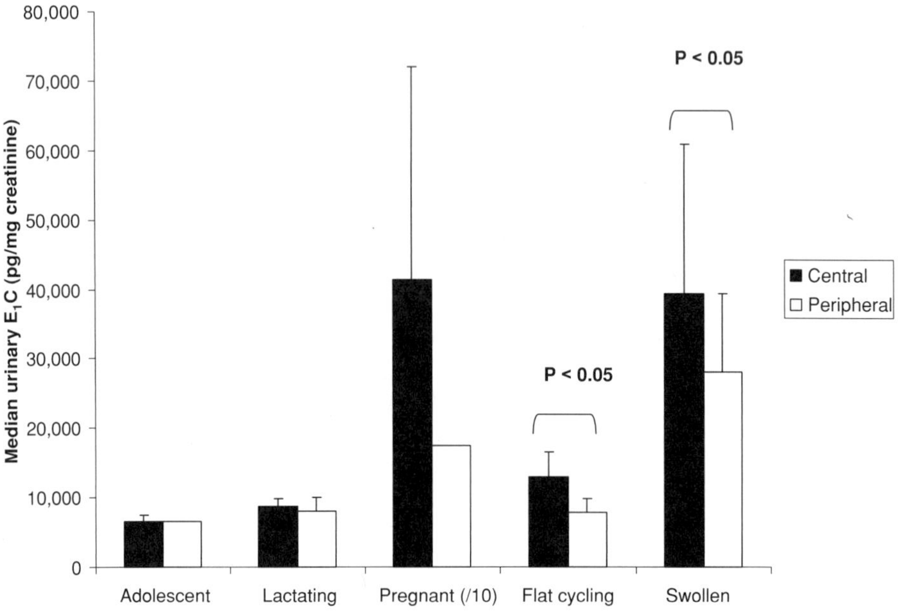

Figure 5. Estrone conjugates by female status at Kanyawara. Median ± standard deviation. Pregnant values divided by 10 to fit scale.

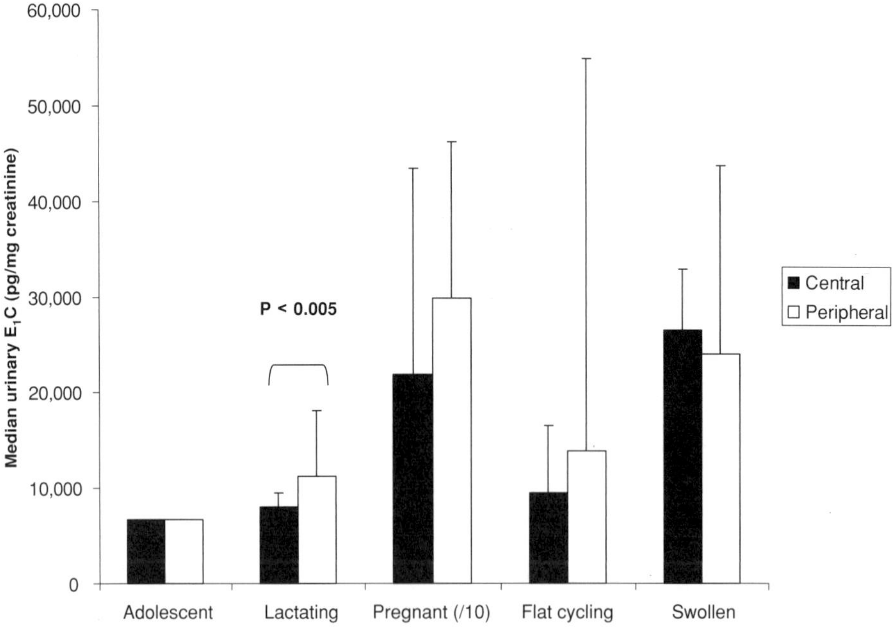

Figure 6. Estrone conjugates by female status at Sonso. Median ± standard deviation. Pregnant values divided by 10 to fit scale.

We might then expect E_1C levels to be higher on average at Sonso than at Kanyawara. Considering the community as a whole, these data do not support that hypothesis. While variability in E_1C levels was very high within communities, median levels were quite similar between communities. However, it should be noted that while peripheral and central females are more similarly represented in the birth interval data, central females are overrepresented in the hormonal data as urine samples from peripheral females were quite rare.

Indeed, when hormone levels are analyzed by females' community status, the data do indicate a correspondence of urinary E_1C levels and birth intervals. Peripheral females in the Kanyawara community had consistently lower estrone conjugate levels than central females in that community and their peripheral counterparts at Sonso. Peripheral females at Sonso, by contrast, exhibited ovarian function at or above the level of more common females. Data on party composition in these communities (Emery Thompson & Wrangham, Chapter 12, this volume) provides a complementary result, illustrating a disparity in gregariousness of the central and peripheral females at Kanyawara that is less pronounced at Sonso.

The fact that both reproductive parameters and ovarian hormones follow a similar pattern of variation, with a large disparity within the Kanyawara community, leads to some important inferences. Population or individual differences in ovarian function are unlikely to be due to genetic effects. Rather, these data suggest the influence of local conditions on reproductive hormones. This lends further support to the work of researchers at Gombe, who have drawn attention to the reproductive consequences of female competition for core areas (Pusey *et al.*, 1997; Williams, 1999; Williams *et al.*, 2002b). Females with quality home ranges at Gombe showed strong site fidelity and a degree of territorial behavior that has been informally linked to feeding competition and the heterogeneity of fruit availability in the habitat. The correlation of energetic conditions to ovarian function in humans (Ellison *et al.*, 1993b) and orangutans (Knott, 1999) suggests that in chimpanzees, too, variation in diet may have a significant impact on reproduction. Further study of seasonal variation in diet can be used to test this hypothesis. Our data do not yet allow us to presume whether peripheral females may have exhibited lowered ovarian function as a result of continued exposure to poor energetic conditions (Vitzthum, 2001) or whether individuals experience long-lasting effects on the reproductive axis as a result of poor conditions during development (Lipson, 2001).

An alternative hypothesis is that peripheral females experience reproductive consequences not from lowered access to fruit but as the result of heightened stress from living in border areas. Females ranging near community borders experience increased threat of attack or infanticide by neighboring males (Pusey, 1980; Goodall, 1986; Watts & Mitani, 2000, 2001). Repeated coalitionary attacks have been demonstrated to have significant reproductive consequences, including long interbirth intervals and increased spontaneous abortion, in yellow baboons (Wasser & Starling, 1988).

In addition to this interesting contrast within the community at Kanyawara, the data point to a notable intercommunity difference in the relative status of peripheral females (see also Emery Thompson & Wrangham, Chapter 12, this volume). Unlike Kanyawara or Gombe, rare females at Sonso do not seem to experience deleterious effects on reproduction. One possible explanation is that the Sonso habitat has higher fruit productivity or more heterogeneity in fruit production over the home range, as suggested by Newton-Fisher (2003). His recently published home range estimate for the Sonso community (approximately 7 km^2: Newton-Fisher, 2003) is actually lower than individual female core area estimates at Kanyawara (Wilson, 2001), although different sampling methods were used and the borders of the Sonso range are not yet firmly defined. This would leave little room for Sonso females to establish distinct core areas or at least would lessen the potential for habitat quality differences between core areas.

Chimpanzee females are already relatively difficult to locate and follow. Thus, documenting the wide variation among females has been and will be a complicated process. However, we now have several lines of recent data that indicate further complexity in the status and relationships of females, including formal dominance hierarchies (Pusey *et al.*, 1997; Wittig & Boesch, 2003) and differential reproductive success by rank or range location (Williams, 1999; Williams *et al.*, 2002b) in some communities. Whether females are marginalized in terms of territory or social status, the sum of preliminary evidence suggests that being a peripheral female can carry important correlates in terms of social behavior, physiology, and reproductive success. Furthermore, the data indicate that the degree of disparity among females may vary markedly among communities.

These are introductory results in a continuing hormonal study of wild female chimpanzees, which we hope will provide valuable details about this species' reproduction. The data presented here indicate that the EIA procedure for quantifying estrone conjugates is a reliable method for tracking ovarian steroid levels in wild chimpanzees. They also reveal a great deal of variability worthy of further examination. The major goal in further research will be to understand the impact of a variety of factors in producing variation in ovarian function. Further examinations will include comparison to a third population in Gombe National Park and will examine in detail the patterns of steroid secretion typical of wild females throughout various life history stages.

ACKNOWLEDGMENTS

The Ugandan Wildlife Authority, Uganda National Council for Science and Technology, Makerere University, and the Ugandan Forest Department provided local support of our research in Uganda. We thank the field staffs of the Budongo Forest Project (Z. Kiwede, G. Muhumuza, J. Kakura, J. Karimagi, M. Gideon, and K. Andama) and Kibale Chimpanzee Project (J. Basigara, J. Barwogeza, C. Katongole, K. Clement, K. Deo, F. Mugurusi, D. Muhangyi, C. Muruuli and P. Tuhairwe) for their hard work in collecting urine samples and behavioral data on a daily basis; a number of students, including Juliet Craig, Jean Ryoo, Sonya Kahlenberg, Katherine Pieta, Lucy Bates, and Ross Wrangham, assisted in urine collections. Martin Muller was instrumental in training the KCP field staff for long-term urine collection. Peter Ellison, Cheryl Knott, Susan Lipson, and Claudia Valeggia provided laboratory facilities, equipment, and training; Dipti Kandlikar, Jean Ryoo, Katherine Burmon, and Eleanore Humphries provided valuable assistance in the laboratory. Budongo Forest Project has been funded by the Norwegian Agency for Development Cooperation, the Department for International Development, and the National Geographic Society. Kanyawara demographic data were collected thanks to funds from the Leakey Foundation, the National Geographic Society (5626-96), the National Science Foundation (NSF SBR-9120960, BCS-9807448), and the Getty Foundation; urine collections were also assisted by the National Science Foundation (SBR-9729123 and SBR-9807448). MET's field and laboratory work is supported by Harvard University, the Leakey Foundation, and the Wenner-Gren Foundation for Anthropological Research.

CHAPTER FOURTEEN

Frequent Copulations by Females and High Promiscuity in Chimpanzees in the Kalinzu Forest, Uganda

Chie Hashimoto and Takeshi Furuichi

INTRODUCTION

The Kalinzu Forest Reserve is contiguous to the Maramagambo Forest in the Queen Elizabeth National Park on the west and the Kashoha-Kitome Forest Reserve on the northeast (Figure 1). These forests comprise the largest forest block in Uganda, which is one of the major chimpanzee habitats in Uganda (Howard, 1991; Plumptre *et al.*, 2003). We made a pilot study on chimpanzees in the Kalinzu Forest in 1992, and found that a considerable number of chimpanzees inhabited this forest (Hashimoto, 1995). The northern, eastern, and southern boundaries of the Kalinzu Forest is dotted with highly populated villages,

Chie Hashimoto • Primate Research Institute, Kyoto University, Inuyama, Aichi 484-8506, Japan
Takeshi Furuichi • Faculty of International Studies, Meiji-Gakuin University, Totsuka, Yokohama, Kanagawa 244-8539, Japan.

Primates of Western Uganda, edited by Nicholas E. Newton-Fisher, Hugh Notman, James D. Paterson, and Vernon Reynolds. Springer, New York, 2006.

Figure 1. Location of the Kalinzu Forest Reserve.

agricultural fields, and tea plantation. The only probable reason the Kalinzu Forest has been able to sustain the chimpanzee population is because people living in this area have no habit of hunting for chimpanzees and other primates.

After a 5-year break, we started an ecological survey of the chimpanzees and other primates in the Kalinzu Forest in 1997. From the distribution of nests and vocalization, we concluded that at least three groups of chimpanzees inhabited the Kalinzu Forest. We followed one group that was ranging around the Nkombe sawmill (the M group), and they became habituated over the years. We identified most male members of the group by 2001, and started intensive behavioral studies based on a focal animal sampling method.

When we started direct observation of the sexual behavior of female chimpanzees, we found that females showed very positive behaviors leading to copulation and performed very frequent copulations compared to bonobos

(Furuichi & Hashimoto, 2002). The copulation rate of estrous females in this forest also seemed to be considerably higher than those reported for chimpanzees in other study sites, although only a little data were available for comparison (Tutin, 1979; Hasegawa & Hiraiwa-Hasegawa, 1990). By 2003, we confirmed that all but one observed estrous female performed frequent and promiscuous copulation in a periovulatory period (POP; Goodall, 1986).

Female chimpanzees show maximal swelling of the perineum for approximately 12 days of each menstrual cycle (Tutin & McGinnis, 1981; Hasegawa & Hiraiwa-Hasegawa, 1983; Goodall, 1986; Wallis, 1997). Most chimpanzee copulation occurs during this period (Hasegawa & Hiraiwa-Hasegawa, 1983; Goodall, 1986; Takahata *et al.*, 1996; Nishida, 1997). A study in Gombe reported that a female during the maximal swelling period copulated 0.52 times per hour per male in association (Tutin, 1979), and another study in Mahale reported that she copulated 0.71 times per hour (Hasegawa & Hiraiwa-Hasegawa, 1990); both of these seemed to be fairly lower than the copulation rate in the Kalinzu Forest.

Other studies have reported some cases of much greater frequency of copulation. Goodall (1986) noted that Flo in Gombe copulated 50 times in 1 day. Hasegawa (1992) observed that Wakiruhia in Mahale copulated a maximum of 3.7 times per hour. Takahata *et al.* (1996) reported that a female in Mahale copulated 2.4 times per hour. However, these cases were reported as unusual, extraordinary cases.

Tutin (1979) and Hasegawa and Hiraiwa-Hasegawa (1983) reported that promiscuous copulation tended to occur frequently during the early term of the maximal swelling period, but high-ranking males tended to monopolize copulation during the POP. However, female chimpanzees of our study group performed highly promiscuous copulation during the POP. Goodall (1986) and Wrangham (2002) suggested that frequent and promiscuous mating might occur when dominance relationships among male members were unclear. Our observation in the Kalinzu Forest might be showing a case of chimpanzees' reproductive interactions when dominance relationships among males were unclear for as long as several years, probably due to a large number of male members.

METHODOLOGY

We observed the M group of wild chimpanzees living in the Kalinzu Forest Reserve, which is located in western Uganda (30°07′ E, 0°17′ S) (Figure 1;

Table 1. Number of individuals observed in each study period

	2001	2002	2003
Adult male	17	19	18
Adolescent male	1	2	2
Adult female	5	13	11
Adolescent female	3	4	0
Immature individuals	5	9	8
Total	31	47	39

Howard, 1991; Hashimoto *et al.*, 2001). The M group had been habituated since 1997 and we were able to follow chimpanzees at distances of 10–20 m without disturbing them much. We identified all adult male members and a part of female and immature members (Table 1). For individuals other than those whom we had been observing from their birth, we estimated the age class from their body size according to the standard given by Goodall (1986).

Observations were conducted over 11 months: 4 months in July 2001–October 2001, 4 months in August–November 2002, and 3 months in August–October 2003. The study period for each year included both dry season (July and August) and rainy season (September, October, and November) (Hashimoto *et al.*, 1999).

We recorded the state of perineum for adult females when they were observed. A female was considered to be in a maximal swelling period, or in estrus, when her perineum appeared turgid and lustrous and lacked fine wrinkles on its surface (Furuichi, 1987). We observed the estrous females using the focal animal sampling method, and recorded copulations performed by the focal animal. We also identified the chimpanzees that were observed with the focal animal in the same party.

RESULTS

We observed four estrous cycles of three females in 2001, seven estrous cycles of six females in 2002, and two estrous cycles of two females in 2003 (Table 2). Total observation time of the focal animal sampling of estrous females was 229.8 h. All estrous females were fully mature; three females (Ha, Bk, and Kn) had dependent offspring, and two females (Gi and Um) were nulliparous.

Table 2. Observation time and number of copulations for each female

(1) 2001

Name	*Ha*									*Te*			Female*		Average
Date (mo/day)	8/18	8/20	8/21	8/22	9/15	9/18	9/19	9/20	9/24	9/11	9/12	9/13	9/5	9/6	
Obs. time (min)	22	258	198	91	298	103	178	280	181	181	94	343	237	105	
No. of copulation observed	1	6	5	2	1	6	13	13	7	16	7	39	8	7	
Frequency (/hour)	2.7	1.4	1.5	1.3	0.2	3.5	4.4	2.8	2.3	5.3	4.5	6.8	2.0	4.0	
Mean freq (/hour)	1.5				2.3					6.0			2.6		3.1

(2) 2002

Name	*Bk*					*Ha*						*Ga*					
Date (mo/day)	8/27	8/28	8/29	8/30	8/31	9/9	9/10	9/12	9/16	9/18	9/20	10/3	10/4	10/8	11/6	11/8	11/9
Obs. time (min)	401	95	50	330	213	298	69	324	93	78	62	61	251	50	71	132	391
No. of copulation observed	4	0	0	5	1	11	11	14	2.0	1	1	1	14	0	6	10	30
Frequency (/hour)	0.6	0.0	0.0	0.9	0.3	2.2	9.6	2.6	1.3	0.8	1.0	1.0	3.3	0.0	5.1	4.5	4.6
Mean freq (/hour)	0.6					2.6						2.5			4.6		

Name	Female*	Female*			*Ht*					Average
Date (mo/day)	10/24	11/2	11/5	11/6	11/16	11/18	11/19	11/20	11/21	
Obs. time (min)	32	225	109	59	124	325	109	433	281	
No. of copulation observed	3	2	6	2	4	26	9	20	15	
Frequency (/hour)	5.6	0.5	3.3	2.0	1.9	4.8	5.0	2.8	3.2	
Mean freq (/hour)	5.6	1.5			3.5					2.2

(3) 2003

Name	*Um*									*Kn*								Average
Date (mo/day)	10/8	10/10	10/11	10/13	10/14	10/15	10/16	10/17	10/18	10/20	10/21	10/22	10/23	10/24	10/25	10/26	10/27	
Obs. time (min)	161	26	72	58	133	388	457	505	446	613	621	434	612	617	426	514	470	
No. of copulation observed	2	2	2	1	10	28	31	28	20	24	32	32	22	32	31	29	19	
Frequency (/hour)	0.7	4.6	1.7	1.0	4.5	4.3	4.1	3.3	2.7	2.3	3.1	4.4	2.2	3.1	4.4	3.4	2.4	
Mean freq (/hour)	3.3									3.1								3.2

* Unidentified female.

According to the date of delivery, Te, Ha, and Gi seemed to have conceived during the observed estrous period in 2001, 2002, and 2003, respectively.

We observed 674 copulations in total and 2.9 copulations per hour on average (Table 2). All observed females except Bk copulated more than 1 time per hour. In one extreme case, Te copulated 39 times in 343 min (6.8 times per hour). She sometimes copulated twice with different males within one minute. Females copulated with most of the males observed to be in the same party. For example, Ha copulated with 9 of 11 males who were observed in the same tree within 52 min.

The high copulation rate seemed to occur in the last several days before the maximal swelling period was finished. In 2003, we observed two estrous females (Um and Kn) for continuous several days until we confirmed the end of maximal swelling period. High copulation rate was observed for 5 days for Um and 8 days for Kn just before the end of the maximal swelling period (Figure 2).

Adult males did not interact aggressively, except when young adult males approached the estrous females and were chased away by older ones. When adult males finished copulating, they rested quietly for a while until ready to

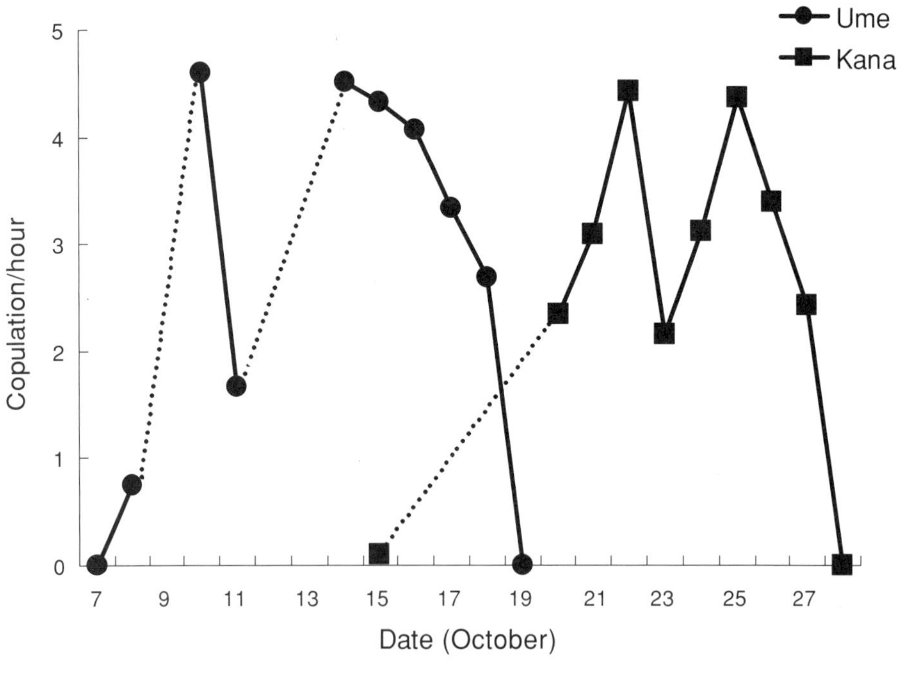

Figure 2. Change in the copulation rate of two estrous females (Um and Kn). Broken lines show the days when no observation was made on these females.

copulate again. When most males in the party had copulated and were resting, estrous females also rested for an hour or two, after which another series of copulations ensued. Figure 3 shows an example of such cycles of copulations for Kn, whom we followed from the morning to the evening for continuous 8 days.

There seemed to be no relationship between dominance rank and copulation rate of males. Because the dominance rank among males was unclear except that DO was the alpha male, we showed the males other than DO in the order of estimated age (Figure 4). DO did not copulate so frequently as compared to other males, and some young adult males copulated more frequently than DO did.

Only an adult female Bk copulated infrequently during one observed maximal swelling period. We observed her copulating only eight times, and five of which were with DO, the alpha male of the M group. During estrus, Bk always followed DO at a short distance. Other males could not easily approach Bk because of her proximity to DO, although more than five other males were observed in the party. During this period, copulation interference by DO against other males was observed three times.

DISCUSSION

Previous studies on chimpanzee mating have revealed three types of mating pattern: opportunistic mating, possessiveness by high-ranking males, and consortship (Tutin, 1979; Hasagawa & Hiraiwa-Hasegawa, 1983). In this study, we observed a high frequency of promiscuous and opportunistic mating. Although there have been many reports and discussion on frequency of copulation of female chimpanzees, only a few reports gave a copulation rate per female per hour (Hasegawa & Hiraiwa-Hasegawa, 1990; Takahata *et al.*, 1996). The present study reported the high copulation rate in the last several days of maximal swelling period, called the periovulatory period or POP (Goodall, 1986), based on the observation by focal animal sampling of estrous females.

A large number of adult males in a group may contribute to promiscuous copulation at a high frequency (Goodall, 1986; Furuichi & Hashimoto, 2002). As there is a physiological limit on males' ability to ejaculate, there may be a limit of copulation rate for males. It would seem likely that total number of copulations, and the rate, should increase with the number of males. The M group contained 17–19 adult males during the study periods. Such a large number of males may potentially allow frequent copulation.

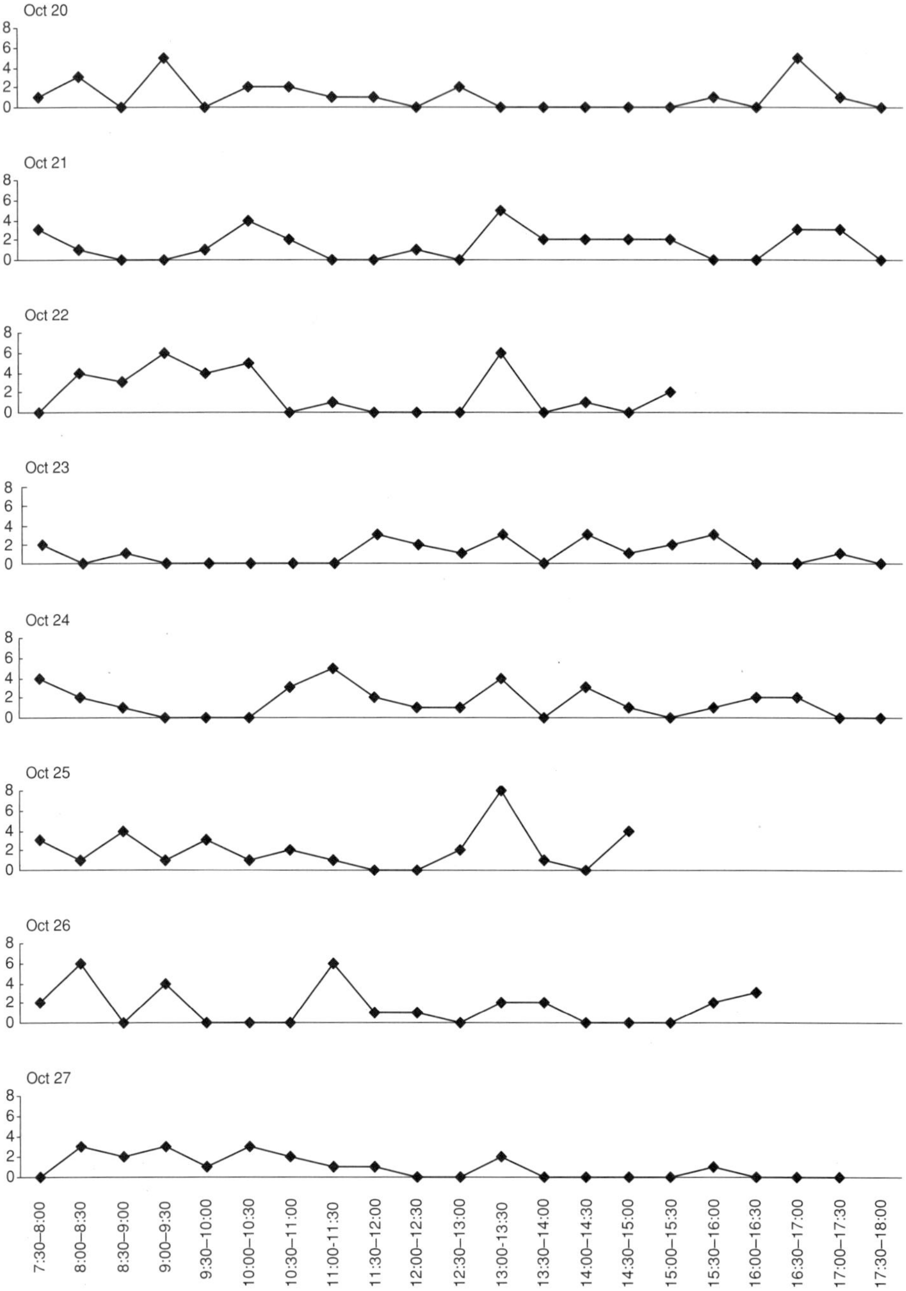

Figure 3. Change in the number of copulations from the morning to the evening. Each point shows the number of copulations observed for each 30 min.

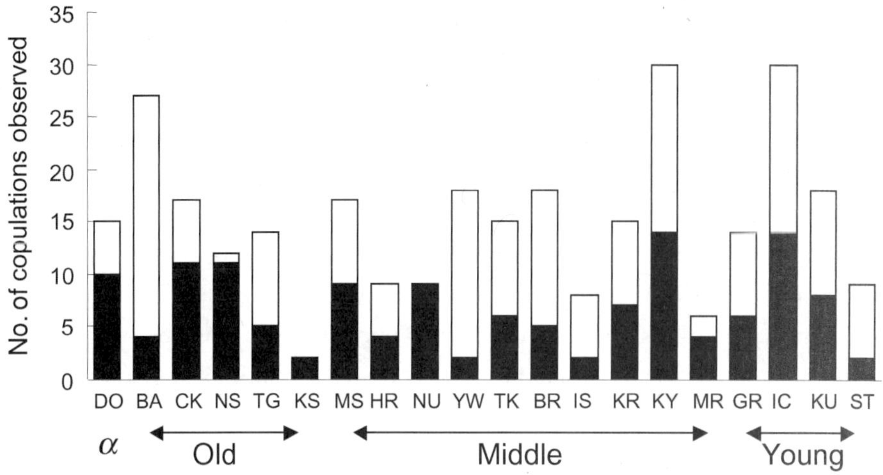

Figure 4. Number of copulations observed for each male. "Old," "middle," and "young" mean the estimated age classes for adult males. DO, the alpha male, was in the middle adult age class. White bars represent data from 2001; Black bars, data from 2002.

Moreover, a larger number of males may lead to unclear rank order among males, which may contribute to promiscuous mating due to the greater cost for possessive behavior. Previous studies suggested that promiscuous mating tends to occur when rank order among males is unclear (Goodall, 1986; Wrangham, 2002). Goodall (1986) reported that Flo copulated 50 times in 1 day when the male rank order was not clear. The frequent copulation by females in this study may be explained not only by the large number of males but also the unclear rank order among them. During the study periods, rank order among males was not clear in the M group. The pant–grunt behavior, a greeting of a subordinate individual to a dominant individual, was observed only infrequently. We could suppose that DO was the alpha male during the study period, according to a few cases of agonistic interactions and possessive behaviors toward an estrous female. However, even DO was observed to receive pant–grunts only a few times in a day.

In 2002, DO showed possessiveness toward Bk during a maximal swelling period, interrupting her copulation with other males. DO was also observed to stay with estrous females (Ga, Ht) for approximately an hour respectively, during which time the females copulated only with him. After DO left them, however, many males approached the females and copulated freely. On other occasions, DO joined the promiscuous mating as one of the males, or he himself, was absent from the party that included the estrous female. Only a few cases of aggressive interactions among males were observed when estrous females were copulating with many males at a high frequency. A comparison of number of copulations across males suggested that neither alpha status nor age class of males affected the number of copulation.

Watts (1998c) reported, however, that copulation tended to be monopolized by two high-ranking males of the Ngogo group in Kibale, which included adult males than our study group did. Whether or not the large number of male members leads to unclear dominance relationships may depend on the characteristics of the alpha male and social relationships among other males.

In the study group, frequent copulation by females was observed only during the POP. At other times during the maximal swelling period, females were not observed in the mixed party, or rarely copulated even when they were found in the mixed party. It is yet unclear why the copulation rate before the POP was so low in the study group.

Ha copulated frequently in both 2001 and 2002. If she maintained her observed copulation rate (2.3 times per hour), Ha would have copulated more than 1000 times from August 2001 to September 2002 when she conceived (assuming 10 hours of activity per day and copulation during 4 days of each 32-day menstrual cycle). Such a large number of copulation before a conception was also pointed out elsewhere (Wrangham, 2002). Although many hypotheses have been proposed, including the best-male hypothesis, the many-male hypothesis, the social-passport hypothesis, the paternity-confusion hypothesis, and the quick-conception hypothesis (see Boesch & Boesch-Achermann, 2000; Furuichi & Hashimoto, 2002; Wrangham, 2002), why chimpanzee females copulate so many times for one birth remains to be studied further and clarified.

ACKNOWLEDGMENTS

The Uganda National Council for Science and Technology, the Uganda Forest Department, and the Uganda Wildlife Authority gave us permission and valuable assistance for this study in the Kalinzu Forest. The staff of the Nkombe Forest Station, and Nkombe Sawmill Co. Ltd. helped in our study. We also thank our Ugandan assistants for their help in the field. Dr. H. Takemoto kindly gave us information on Bk. Dr. S. Uehara and members of the Primate

Research Institute, Kyoto University, and Dr. T. Nishida and members of the Laboratory of Human Evolution, Kyoto University, made valuable comments. Dr. T. Hasegawa and Dr. M. Nakamura also gave us valuable comments on the draft. We are very grateful to them all. This study was supported by a grant from the Monbusho International Science Research Program awarded to T. Furuichi (No. 12575017) and T. Nishida (No. 12375003) and a Monbusho Grant-in-Aid for COE Research (10CE2005) awarded to O. Takenaka.

CHAPTER FIFTEEN

Effects of Injury on the Locomotion of Free-Living Chimpanzees in the Budongo Forest Reserve, Uganda

Julie Munn

INTRODUCTION

Locomotion is a fundamental component of a chimpanzee's survival in the wild. Moving around in the canopy and on the ground is a skill much needed in the everyday life of chimpanzees in order to gain access to food and social partners, to assist young and to flee from potential danger. That chimpanzees in Uganda and elsewhere have survived disabling injury to one or more limbs suggests they are able to cope with locomotor deficiencies. In Uganda, at least 22% of chimpanzees (*Pan troglodytes schweinfurthii*) from five study sites have injuries that are known or suspected to have resulted from "snares" (Munn, 2003).

Julie Munn • Australian National University, Canberra, ACT 0200, Australia.

Primates of Western Uganda, edited by Nicholas E. Newton-Fisher, Hugh Notman, James D. Paterson, and Vernon Reynolds. Springer, New York, 2006.

Hunting of wildlife in all Ugandan forest reserves is illegal (Howard, 1991), but is nonetheless a common practice. The snaring of chimpanzees in the majority of cases has been assumed as "accidental" as chimpanzees are not typically target species for hunters in Uganda, in part because primate meat is not a traditional food item (Johnson, 1996a). Methods for hunting all bush meat include the setting of both snares and traps; wire nooses and spring snares are the most frequently used devices for catching terrestrial forest quadrupeds in the Budongo Forest Reserve. These snaring methods involve placing a length of wire or cord over a small hand built pit, which is then camouflaged with forest litter. The loop is either attached to a bent over sapling, or is secured by upright branches (Waller, 1995). Animals that subsequently tread on a snared pit invariably pull and, as a result, the snare tightens and embeds deeper into the victim's flesh. Steel jaw traps (or man-traps) are another trapping method less commonly used. The victim places weight on a metal plate that springs two parallel jaws together with great force, firmly trapping the limb. The force of the steel jaws is great enough to sever the limb completely (Waller, 1995). Throughout this text, the term "snare" is used throughout to describe all trapping methods including spring snares, wire nooses and steel jaw traps, except where otherwise stated.

Injuries are not unique to chimpanzees. Nonhuman primates, throughout their ranges, are exposed to the dangers of injury to limbs whether by accident (e.g., falling from a tree), disease, predation, through an aggressive interaction, or from being caught in a hunter's snare. Limb defects are also caused by congenital deformities. These injuries, regardless of their cause, will have impacts on various aspects of an individual's behavior.

Locomotion

Chimpanzees in the wild spend approximately 14% (Taï National Park, Côte d'Ivore) and 18% (Gombe National Park, Tanzania) of their time in locomotor activities (Doran, 1993). Chimpanzees are adapted to travel both on the forest floor and in the canopy, and the mode and pattern of locomotion differs depending on the substrate utilized. In Taï National Park, arboreal activities were found to account for 16%, and terrestrial activities for 84%, of all locomotor activities (Doran, 1993).

Very little has been documented about the effects of injuries on the types of locomotion used by chimpanzees. Ghiglieri (1988) and Goodall (1986) both describe cases of chimpanzees with deformed limbs having difficulty keeping up with traveling groups. In Gombe National Park, Tanzania, six individuals with limb paralysis (almost certainly from poliomyelitis) were described as adapting well to their injuries (Goodall, 1968b). A study on bonobos (*Pan paniscus*) at Wamba, Democratic Republic of Congo, reported changes in locomotion techniques for individuals that had physical abnormalities (Kano, 1984). Reports from Kibale National Park, Uganda, describe an injured female with a stump as a right hand as an "unbelievable daredevil" (Ghiglieri, 1988:147). Despite her injury she seemed both an excellent climber and an acrobatic one, although there is no mention of her behavior prior to injury. Finally, in a study focusing on the effects of snare injury on the locomotion of the Sonso community, it was found that individuals with injuries to the forelimbs used modified, but very successful adaptations for arboreal locomotion. Those individuals with injuries to the hind limbs were more awkward and slow (Quiatt, 1996).

Here, I examine and descried how locomotor and behavior patterns differ between injured and noninjured chimpanzees. Aspects reported here are types of locomotion used, the time individuals spent moving both in the trees and on the ground, and the ability to carry dependants while traveling.

METHODOLOGY

Study Site and Subjects

The Budongo Forest Reserve, the study site for this research, lies in the west of Uganda between the latitudes of 1°37′ and 2°00′ N, and between the longitudes 31°22′ and 31°46′ E (Plumptre, 1996). The ground is undulating with an average altitude of about 1100 m (3600 ft) (Eggeling, 1947). The Budongo Forest Reserve covers an area of 793 km^2 of both moist, semideciduous forest and grassland; 428 km^2 of the area is forested (Howard, 1991). The ecology of this forest is detailed in an extensive study by W. J. Eggeling (1947). Five species of diurnal primates are found in Budongo, including the eastern subspecies of common chimpanzees, *Pan troglodytes schweinfurthii*.

The subjects of my study are habituated chimpanzees of the Sonso community, in the Budongo Forest Reserve, Uganda. Habituation began in 1991 and all members of the Sonso community were identified and named by July 1995 (Newton-Fisher, 1997). The community contained 49 individuals during the study period, of which 10 had serious deformities. One of these was possibly congenital, and the other nine were as a result of entrapment in snares (Waller, 1995). Adult females with dependants from the Sonso community were the focus of my study because at the onset of the study they were the age–sex class with the most injuries (since that time two males, plus a very young female, have been wounded seriously from snares). Only one age–sex class was chosen to study, as this greatly reduces natural variation in the data caused by behavioral differences seen between individuals. Five adult females with dependants were observed, and their injuries are outlined below:

Kalema had an injury resulting from a wire snare on her right hand; there are reports that she bore the remains of a wire snare on her wrist for some time after the injury (Waller, 1995). Her hand was paralyzed and hooked from the wrist, with the fingers bent so far forward they almost touched the inner part of the lower forearm. Her fingers were curled right over, and came together at a point; they appeared to have no strength or mobility.

Kewaya's right hand was injured from a wire snare (Waller, 1995). The hand was extremely deformed being both twisted and hooked. The wrist was so twisted that the fingers of her right hand pointed out to the side. The fingers came together at a point (similar to Kalema). Her hand was paralyzed in this position so her fingers lacked any function.

Kigere was missing the entire foot of her right leg. The injury was likely to have been caused by a wire snare that became gangrenous, or by a larger metal trap (Waller, 1995).

Both of Zana's hands were injured as a result of snares (Waller, 1995; Waller & Reynolds, 2001). Zana's left hand resembled the injury of Kewaya's right hand. The wrist was twisted and the fingers deformed, pointing out sideways and very floppy, seeming to lack any voluntary movement. Zana's right hand had only one finger, the forefinger, and she had the remaining stump of her thumb. The missing digits, or perhaps the snare itself, have caused a great weakening in the wrist (Waller, 1995).

Banura's left foot was considerably enlarged (appeared swollen), not capable of much movement, and had only two digits. One digit was like a small finger or thumb, and seemed to have some ability to grip. The other digit was larger and folded over toward the centre of the sole of the foot.

Banura was the only individual in the Sonso community thought to have a congenital deformity (Waller, 1995; Waller & Reynolds, 2001). She is included here because she has a physical disability, and this shows signs of affecting her

Table 1. Level of habituation of study individuals (injured individuals in bold)

Level of habituation			
1—High (relaxed)	2	3	4—Low (shy)
Nambi NB	Kwera KW	**Kigere KG**	**Zana ZA**
Zimba ZM	**Kalema KL**	Ruda RD	Ruhara RH
	Kewaya KY	**Banura BN**	Kutu KU
	Janie JN		

behavior. The main difference between her disability and the snare wounds is that she would have been born with her deformity. She is designated as "injured" in this chapter.

Habituation levels affected the amount of time I was able to observe individuals moving (and resting) on the ground. In this study the level of habituation for the 12 females did differ, with some individuals being relaxed in the company of researchers, and others shy. Table 1 shows my judgment of the level of habituation for the study females to both myself, and my field assistant, over the study period.

Definition of Terms

A dependant refers to the youngest offspring of the mother. Dependant is divided into four age categories: Infant 0, Infant 1, Infant 2, and Juvenile 1. Infant 0 refers to the youngest infants, less than 12 months of age. Infant 1 refers to young chimpanzees between 1 and 2 years, Infant 2 those between 2 and 4 years, and Juvenile 1 those 4–6 years old. No dependant changed age category during the study period, so any effects owing to the change in age of the dependant during the 8 months of this study are not determined in these results.

Moving and traveling (here treated synonymously)—the transit from one feeding or resting location to another, whether to a neighboring branch, an adjacent tree, or a site some distance away.

For some analyses, severity of injury has been included. Zana, with injuries to both hands, and Kigere, who was missing a whole foot, were regarded as having the most severe injuries. Kewaya and Kalema had only one hand injured, and Banura had a swollen foot: these three are seen as having less severe injuries. Zana and Kigere both showed difficulties in balancing when moving and feeding, which also led to their being included in the category of most severe injuries.

Data Collection

Data were collected during both wet and dry seasons from October 1999 to June 2000, with a brief intermission between 18th December 1999 and 4th January 2000. Data were collected for 3 months in the dry season (219 observation periods) and 5 months in the wet season (232 observation periods).

Chimpanzees were found each day either by calls heard early in the morning, by visiting the place that chimpanzees were last seen on the evening before, or by visiting known fruiting trees (a large system of trails surrounds the field station at Sonso and covers a large proportion of the community's home range).

Chimpanzees were observed for half-hour observation periods, using both focal animal sampling and scan sampling. Focal animal sampling (Altmann, 1974) was used to determine the amount of time a mother spent carrying her dependant. If a mother was recorded as not carrying her dependant, the dependant was traveling without assistance from its mother or from any other chimpanzee in the group. Focal animal sampling was also used to determine the amount of time individuals spent traveling. Total time spent in arboreal space (regardless of behavior) was calculated using scan sampling (Altmann, 1974), with data collected at 5-min intervals during the observation period. At each 5-min interval it was recorded if a focal subject was on the ground, or in the trees. Ad libitum sampling was used to describe the different methods of locomotion used by injured chimpanzees at Sonso. Number of observation periods, number of 5-min scans and total time observed are shown in Table 2.

Observation periods were rotated among focal subjects. Sampling rules were made to determine the order in which subjects were observed. The sampling rules and the rotation of focal subjects aimed to enable observations to be spread equally between all individuals, and for all times of the day, over the study period. To reduce dependence in the data, no individual was observed more frequently than once every 90 min, that is, a full hour was allowed to pass before the end of one observation period and the beginning of the next.

Data Analysis

Linear models were applied to each set of data (Davidian & Giltinan, 1995; McCullagh & Nelder, 1989). Continuous data collected during focal animal sampling were analyzed using mixed linear models. All data were analyzed as a proportion (e.g., time chimpanzee spent moving/total time chimpanzee

Table 2. Number of observation periods, scan samples, and total time each focal subject was observed during the study period

Focal subject (dependant)	Number of 30-min observation periods	Number of 5-min scans	Total time observed (min)
Injured females			
Banura (Infant 2)	33	231	990
Kewaya (Infant 0)	44	308	1320
Kigere (Infant 1)	36	252	1080
Kalema (Juvenile 1)	41	287	1230
Zana (Juvenile 1)	39	273	1170
Total	*193*	*1351*	*5790*
Noninjured females			
Janie (Infant 0)	38	266	1140
Kutu (Infant 1)	31	217	930
Kwera (Juvenile 1)	45	315	1350
Nambi (Juvenile 1)	38	266	1140
Ruda (Infant 2)	34	238	1020
Ruhara (Infant 2)	29	203	870
Zimba (Infant 2)	43	301	1290
Total	*258*	*1806*	*7740*
Combined total	451	3157	13530

Age category of focal subject's youngest dependant is included in brackets.

observed). Proportions are not usually normally distributed, so the proportions were transformed to the logit scale to normalize the data. Discrete data, collected by means of scan sampling, were analyzed using a binomial Generalized Linear Model. The statistics used in both cases were the Change in Deviance Statistic (F test) or the Wald Statistic, which are characterized by a chi-square distribution. Factors used in the models include injury status of the mother, severity of her injury (see Definition of Terms, above), habituation level of the mother, age of her dependant, and the season during which the observation was taken. It was appropriate in the data analysis to determine the effect different seasons had on the results; however it must be noted that data were not collected for a full 12-month period.

Significance levels of $P < 0.05$ were used throughout, except where simultaneous tests were carried out on the same data set (e.g., time spent moving and time moving in arboreal space). In these cases the Bonferroni correction method was used to change the significance level (Howell, 1997) using the formula: α / m, where $\alpha =$ normal significance level (0.05), and $m =$ the number of simultaneous tests. The resulting significance level is 0.025. Only significant results are reported here.

RESULTS

Methods of Locomotion Used by Injured Chimpanzees of the Sonso Community

Results demonstrate that injured chimpanzees of the Sonso community used modified forms of the locomotor patterns typical of noninjured chimpanzees and showed a tendency to avoid some locomotor patterns altogether. The use of modified forms of locomotion affected the balance needed for arboreal travel in some individuals, as well as the speed at which individuals traveled, especially when attempting to move quickly on the ground. In general, injured females were more awkward in their locomotion when compared to able-bodied chimpanzees. A description of the locomotor patterns of individual chimpanzees with injuries follows.

Banura

Banura walked on the ground using quadrupedalism, placing weight on her deformed foot as she moved. She usually placed weight on the tips of the deformed digits, rather than on the sole of her left foot. She has also been observed walking tripedally, holding her left (deformed) foot off the ground. This always occurred when she was trying to move quickly. Banura's hips tilted in order to compensate for both the awkwardness of her step and the different lengths of her legs. When climbing and scrambling in arboreal space, Banura could not use her injured foot to cling or grasp. This foot was instead used for balance, by placing it alongside a branch or holding it outwards at varying angles. When moving in arboreal space along wide horizontal branches she was able to use her foot as she did when moving on the ground.

Kalema

Kalema used both knuckle walking and tripedalism when traveling on the ground. She used a modified version of knuckle walking with her right (injured) hand; it was placed with the back of the wrist touching the ground, rather than the back of the middle phalanges. While in the trees she used tripedal locomotion when moving along a wide horizontal branch. When climbing and scrambling, Kalema used her injured hand, which had no ability to clasp, for balance by holding it out at varying angles or by placing it against the trunk of a tree or branch. She also used the crook of her right elbow, or her right hand hooked over a branch when climbing down, for balance, or to help pull her up. She was unable to use brachiation, as she could not grasp with her right hand.

Zana

When terrestrial, Zana walked quadrupedally. She used the back of her left wrist as the point of contact with the ground in a modified form of knuckle walking, with the fingers splayed out to the side. Zana moved very slowly and deliberately in the trees. Many times she was seen to hesitate when crossing from the extremity of one tree to the next (bridging and tree swaying); on one occasion she lost balance and hesitated five times before making it across on the sixth attempt. She was never seen to be involved in any suspensory behavior. She was observed walking tripedally along wide horizontal branches, holding her left hand up. When climbing and scrambling in arboreal space she used the back of her left hand against branches to help with balance. She also used her forearms and elbows to assist by leaning against them or using them to pull herself up.

Kewaya

When terrestrial, Kewaya used quadrupedalism. She used a modified version of knuckle walking on the right (injured) side, using the point just above the right wrist as the contact point with the ground, with her fingers splayed out to the side. When vertical climbing, Kewaya used the back of her injured hand, placed against the trunk, to help her balance. When climbing and scrambling she held her injured hand out at an angle to help with her balance. She also used her injured hand as a hook to support herself. She was unable to brachiate, as she could not grasp with her right hand.

Kigere

When moving on the ground, Kigere used knuckle walking, placing even weight on both her injured and noninjured leg. As a consequence her hips tilted considerably as she walked, compensating for the shortness of her injured leg. When observed moving quickly on the ground she was tripedal, with her right leg bent up toward her chest. When moving in trees she used her stump during quadrupedalism (as when terrestrial), or for balance (by lightly touching against branches), or held it out of the way. Twice, when arboreal, Kigere was observed to lose her balance.

Time Spent Moving

On average, adult females in this study spent 10.5% of their time in locomotor activities. Two injured individuals, Zana and Kigere, spent approximately 4% of their time in locomotion whereas all other individuals spent 8% or more (up to 15%) of their time in locomotion (Figure 1).

A linear mixed model revealed that both severity of injury (Wald statistic = 18.81, df = 2, $P < 0.001$) and season (Wald statistic = 16.90, df = 1, $P < 0.001$) significantly affected the time spent moving. The most severely injured individuals spent less time moving than either less injured or noninjured individuals. Both injured and noninjured chimpanzees moved less in the wet season, and more in the dry season, possibly because they were forced to travel extensively to locate food sources. All seasonal differences are based on the 8 months studied, not an entire year.

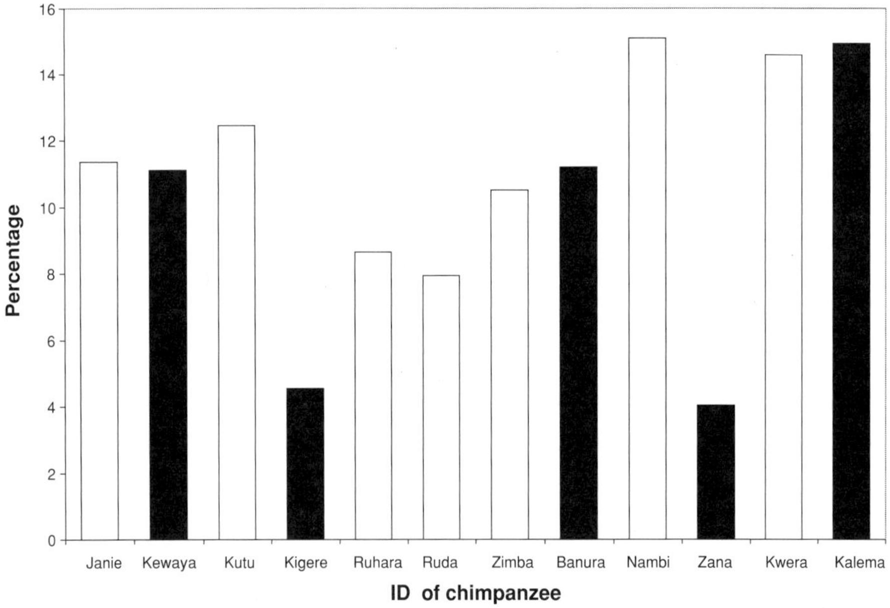

Figure 1. The percentage of total observation time that each focal subject spent in locomotion. Females are ordered from left to right along the base of the graph, with the female with the youngest dependant on the left through to the female with the oldest dependant on the right. Injured females are indicated by the black bars.

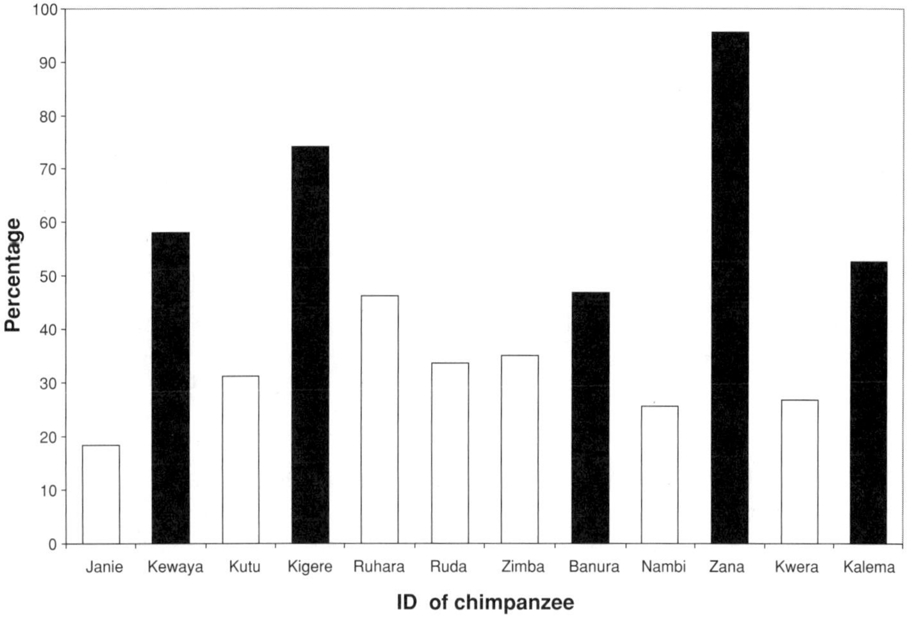

Figure 2. The time spent in arboreal locomotion as a percentage of the total time spent in locomotion. Females are ordered from left to right along the base of the graph, with the female with the youngest dependant on the left through to the female with the oldest dependant on the right. Injured females are indicated by the black bars.

The injured chimpanzees also spent more time moving in arboreal space, as a percentage of the total time spent moving, than did noninjured chimpanzees (Figure 2). A linear mixed model revealed that both severity of injury (Wald statistic = 10.58, df = 2, $P = 0.005$) and season (Wald statistic = 17.31, df = 1, $P < 0.001$) significantly affected the time spent in arboreal locomotion. The greater the severity of injury the larger the proportion of traveling time spent in arboreal locomotion (Figure 3). All individuals spent more time moving in the trees in the wet season than in the dry (Figure 3). A high number of *Cynometra* seeds were eaten off the forest floor in the dry season. *Cynometra* pods burst in the heat of the dry season, scattering hundreds of seeds to the ground. The time spent searching for *Cynometra* seeds was recorded as feeding, not moving, although the presence of these seeds meant that chimpanzees spent more time on the ground. Chimpanzees also tended to travel further distances to find food in the dry season.

These results show that when moving, injured chimpanzees move proportionally more in trees and proportionally less on the ground than noninjured

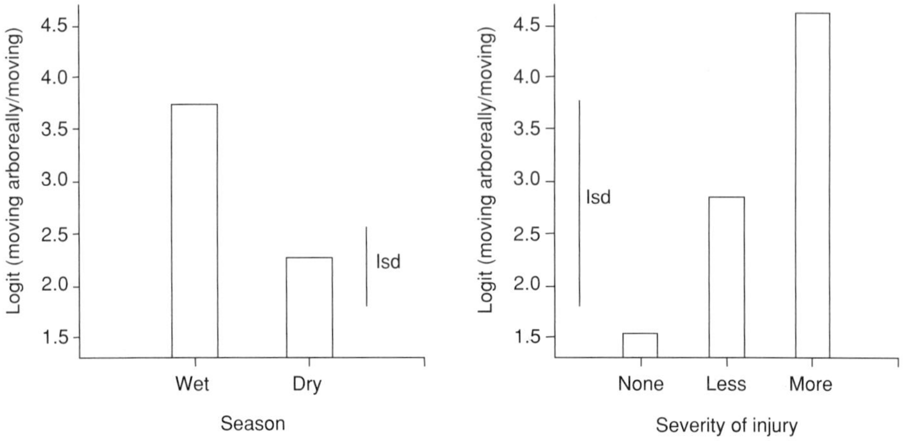

Figure 3. Time spent moving arboreal as a proportion of the total time spent moving. This graph explains the results found in the linear mixed model. The measure of error used in this figure is the least significant difference (lsd), which refers to the minimum distance needed to have a significant difference. Black shading indicates terrestriality.

chimpanzees. Injured chimpanzees, however, spent more time in arboreal space than noninjured chimpanzees, although there was some overlap (Zimba, Kutu, Ruhara). Kigere and Zana once again show the most distinct result, spending almost all of their time in the trees, and the least time on the ground (Figure 4).

A binomial GLM revealed that both the severity of injury ($F = 26.80$, $\mathrm{df} = 2, 422$, $P < 0.001$) and the degree of habituation ($F = 8.92$, $\mathrm{df} = 3, 422$, $P < 0.001$) significantly affected the proportion of time spent arboreal or terrestrial (Figure 5). The more severely injured individuals, as well as the least habituated individuals, spent the greatest proportion of time arboreal. This means that severely injured chimpanzees with low habituation are spending the most time arboreal, and noninjured chimpanzees with a higher level of habituation are spending the least amount of time in arboreal space. Injured chimpanzees are both high and low on the habituation scale (Table 1).

Mother Carrying Dependant during Locomotion

Injury may also affect a mother's ability to carry her offspring. Figure 6 shows a general decrease in the time a mother carries her dependant as the age of

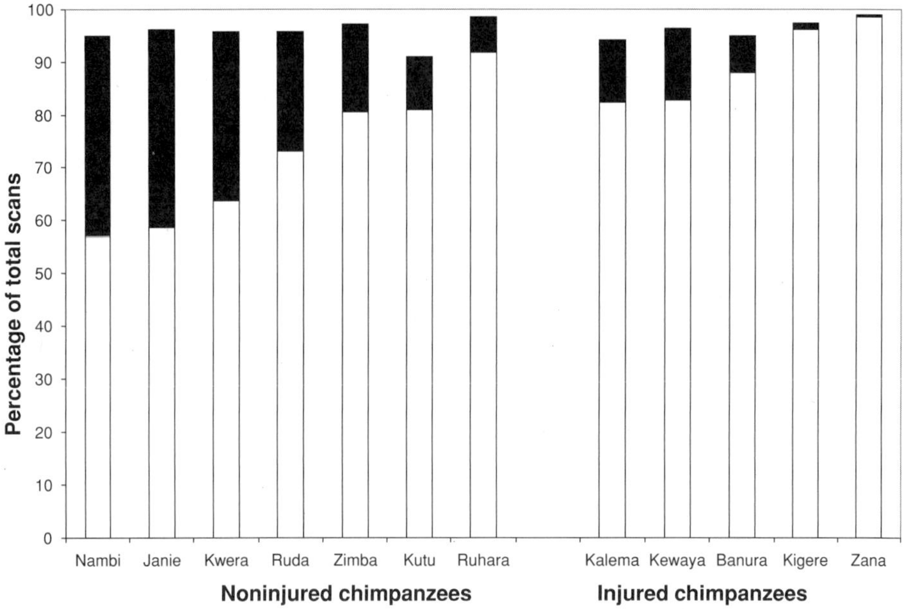

Figure 4. The percentage of 5-min scans that each individual was arboreal or terrestrial. The remainder of time not included on the graph represents time out of sight. The noninjured females are grouped on the left of the graph and injured females grouped on the right.

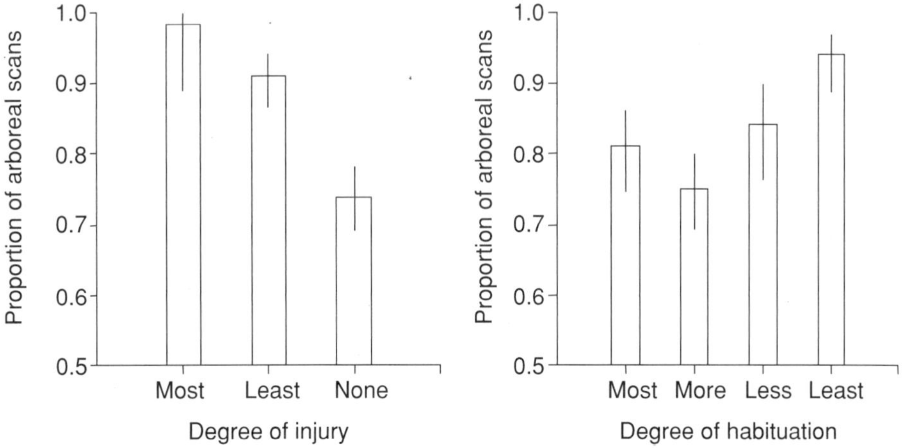

Figure 5. The proportion of scans spent arboreal, showing the effects of both severity of injury and habituation. This graph explains the results found in the binomial GLM. Error bars used in this figure refer to 95% confidence intervals.

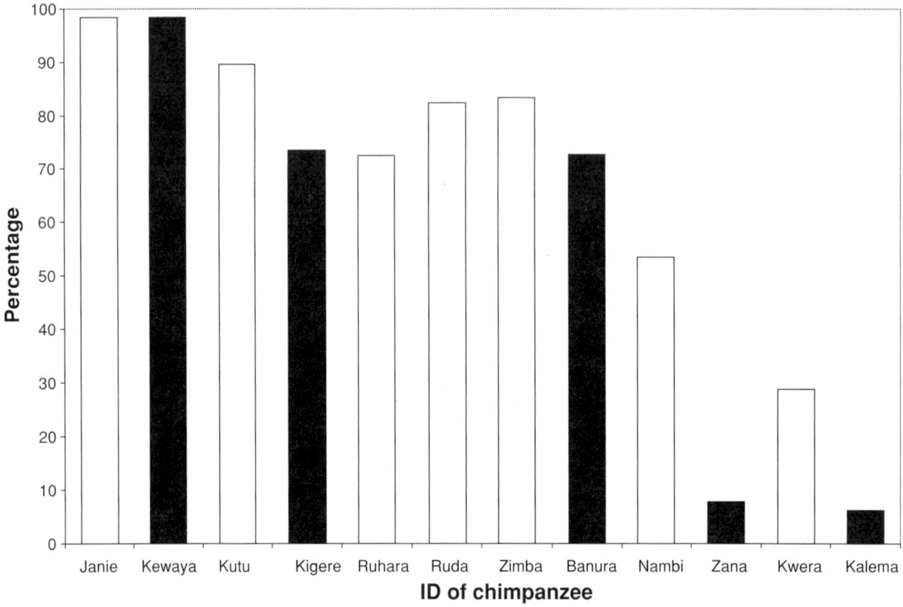

Figure 6. The time each mother spent carrying her youngest dependant as a percentage of the total time spent in locomotion. The mothers are ordered by age of dependant, with the youngest dependant on the left hand side of the graph and the oldest on the right. Injured females are indicated by the black bars.

the dependant increased. Two mothers (Janie and Kewaya) carried their infants (Infant 0) for almost 100% of their time during locomotion; these two mothers have the youngest offspring in the community. The mothers with Infant 1 and Infant 2, shown in the centre of the graph (Kutu through to Banura), show a small difference between injured and noninjured mothers, with injured mothers carrying their offspring for less time than noninjured mothers (with the exception of Ruhara). The four mothers with juvenile offspring (on the right of the graph) show a marked difference between the amounts of time that mothers spent carrying them, the injured mothers carrying their dependants for less time than the noninjured mothers. It appears that the effect of injury on the time a mother carries her dependant is more evident as the offspring gets older.

The linear mixed model revealed that both the age of the dependant (Wald Statistic = 40.43, df = 2, $P < 0.001$) and the injury status of the mother (Wald Statistic = 4.37, df = 1, $P < 0.05$) significantly affected the time a mother spent moving without carrying her dependant. The time the mother

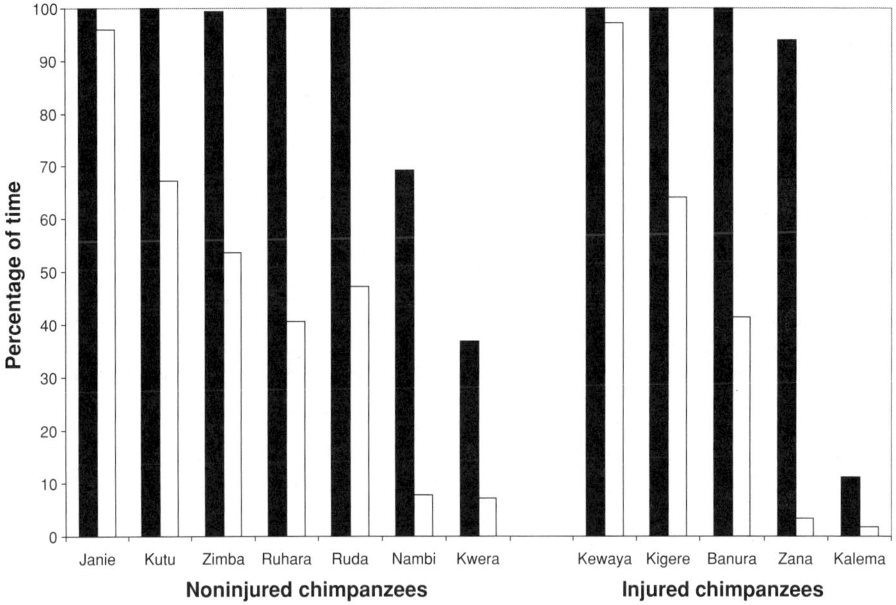

Figure 7. The percentage of time that each mother carried her youngest dependant in terrestrial and arboreal space. The noninjured females are grouped on the left of the graph and injured females on the right. Black shading indicates terrestriality.

spent moving without carrying her dependant increased significantly with the age of the offspring. Injured mothers spent significantly more time moving without carrying their dependant than noninjured mothers. Infant 0 group was found to be significantly different from other age classes (Wald Statistic = 66.90, df = 1, $P < 0.001$) with respect to the time they were carried by the mother.

Injured mothers may have more difficulty carrying their dependants, depending on whether they are traveling on the ground or in the trees. Figure 7 shows the percentage of time a mother carried her dependant, out of the total time she was moving, in both arboreal and terrestrial space. As the age of the dependant increased, the amount of time the mother spent carrying it decreased. The age at which an infant began traveling alone was different in arboreal and terrestrial space. Mothers continued to carry infants and juveniles in terrestrial space for far longer than in arboreal space; even Infant 0 moved on its own in the trees. This may be because long distances are most commonly traveled on the ground. There appears to be no difference between injured and noninjured mothers.

DISCUSSION

Methods of Locomotion Used by Injured Chimpanzees of the Sonso Community

Injured adult female chimpanzees of the Sonso community were more cumbersome in their locomotion, used many modified forms of locomotion, and employed novel methods to assist in locomotion in comparison to their non-injured counterparts. The character of each injury determined the type of accommodation adopted, and the number of injuries also affected the level of locomotor disability. A casual observer is likely to conclude that chimpanzees who survive their injuries, and any subsequent infection, can learn to cope adequately with their disability. That injured chimpanzees can move adequately is also reported in earlier studies (Ghiglieri, 1988; Goodall, 1968; Kano, 1984). Only after watching the injured females for the length of this study period did I witness events such as Kigere losing balance and Zana struggling with bridging and tree swaying. These two individuals are the most severely injured, and seem to be at the greatest risk of losing balance and falling, which is potentially fatal. This result could have serious repercussions for the individual, her dependant offspring, and her community. For example, a population modeling simulation of chimpanzees in Uganda determined that in a population of 100 individuals the annual removal, by human-caused mortality, of one single female and her dependant offspring could have "a real detrimental effect on the growth potential of chimpanzee populations" (Edroma *et al.*, 1997:71).

Time Spent Moving and Preference for Arboreal Space

Adult female chimpanzees at Sonso spent, on average, less recorded time in locomotion than has been recorded at other field sites (Doran, 1993). It is, however, difficult to compare these data, as the records from other field sites include both males and females, and females are known to travel less than males (Wrangham & Smuts, 1980).

The two females with the most severe injuries (Zana and Kigere) spent significantly less time moving than all other females in this study. They were often seen feeding by themselves, either remaining in the feeding tree after other group members moved off, or foraging alone. They may have been avoiding long-distance travel on the ground because of their more cumbersome locomotion.

All injured chimpanzees spent a significantly larger proportion of their traveling time in arboreal locomotion. This is likely a function of the general tendency for these individuals to be more frequently observed in trees. Injured chimpanzees may spend more time in arboreal space as they may move between feeding and resting sites in the trees, rather than come to the ground (although this was rarely observed for long distances), or they may spend more time feeding (mostly an arboreal activity). It may also be because they feel more vulnerable on the ground. The threat from predators and the avoidance of snares (which are all set at ground level) are two reasons why the ground could be perceived as more dangerous for injured individuals.

Feeding

The majority of food items in the diet of the Sonso chimpanzees are found in arboreal space (Stokes, 1999), so an increase in time spent feeding may be the reason why injured chimpanzees spent more time in the trees. No significant difference was found when proportion of time spent feeding was compared for injured and noninjured females (Munn, 2003). Also in results relating to feeding efficiency from Stokes' (1999) study of the Sonso community, feeding efficiency was reduced for only some injured individuals, and not for all food types. Therefore it is unlikely that injured chimpanzees spent more time in arboreal space because of time spent feeding.

That injured females did not spend more time feeding than noninjured females suggests they had a similar ability to process foods. However, they may have had physical difficulty in gaining access to the best foods when in the presence of a large group. Some foods, such as *Ficus sur*, a common fig species used by the chimpanzees at Sonso, were difficult for injured females to feed on since the figs are suspended from a branch in large clumps. Able-bodied chimpanzees commonly hang from the large clump by one hand, and one or two feet, while using the other hand to process the figs. An injured female may require a seated position to access these figs, and may be more confident doing this when other chimpanzees are not attempting to feed off the same clump. For other food types, it may be that injured females have one or two fewer limbs to help with balance in arboreal space, and so it is easier to access food when in the presence of fewer competitors. It is possible that injured chimpanzees spent more time in arboreal space to reduce feeding competition because of the physical difficulties associated with a limb injury; in this case, a lack of balance. Reducing feeding competition by foraging alone has been discussed elsewhere as a strategy used by female chimpanzees to increase food intake (Wrangham & Smuts, 1980). Injured females may behave in this way even more than noninjured females to further reduce their competition for food.

Threat from Predators

Whether arboreal space is "safer" than terrestrial is difficult to determine. Predators of chimpanzees include leopards (Hiraiwa-Hasegawa *et al.*, 1986; Boesch, 1991; Poppenwimer, 1999–2000), lions (Tsukahara, 1993), and humans. In many primate habitats, predation risk from terrestrial felids decreases as height above the ground increases (Boinski *et al.*, 2000), and one of the most common reactions of chimpanzees to feline predators is to climb up into a tree (Hiraiwa-Hasegawa *et al.*, 1986; Tutin & Fernandez, 1991; Tsukahara, 1993). Since raptors are not reported to be major predators of chimpanzees (Treves, 1999), these reports suggest that chimpanzees are safer from predators in the trees than on the ground.

No attacks by large carnivores on chimpanzees have been recorded for the Sonso community, and a recent survey of large mammals across the Budongo Forest found no evidence of large carnivore dung (Plumptre *et al.*, 2001), suggesting that these species are now rare. Lions do occur in the far northeast of the Budongo Forest at Kanyo-Pabidi, the site of one of the two tourist facilities within the reserve. On one known occasion in 2003, a lion was sighted within the Sonso campsite (Z. Kiwede, personal communication), and a lion has been heard from the Sonso campsite in 1992, although it was an extremely dry season and it was thought that the lions might have been looking for water (A. Plumptre, personal communication). Historically leopards were found in the Budongo Forest (Reynolds, 1965), and it is possible that some still exist, although they have not been sighted near Sonso for many years.

Although there are apparently very few feline predators in the area surrounding the Sonso community, it is possible that chimpanzees still respond to the potential risk of an attack by a predator. Dunbar (1988) commented that mortality due to predation can sometimes be low or negligible in primates, and he therefore proposes that it has little effect on social systems and group behavior. He argued that it is not the actual mortality rate that is important; rather it is the potential risk of encountering a predator that may be important in affecting the behavior of the prey species (Dunbar, 1988). It is possible, then, that injured chimpanzees in this study are responding to a potential risk of predation. Supporting this possibility, it has been found that male chimpanzees nest lower than female chimpanzees at Budongo. One explanation given for this behavior was that males strategically locate their nests so as to protect the females and young from ground predators (Brownlow *et al.*, 2001).

Injured chimpanzees at Sonso may be particularly vulnerable to predation owing to their more cumbersome locomotion and their difficulty moving quickly. Injured female chimpanzees at Sonso have been found to spend more time, than noninjured females, in small groups, and alone with their dependants (Munn, 2003); this may make them even more acutely sensitive to any risk of predation. It has been suggested elsewhere that female chimpanzees might be more vulnerable to attack by predators, as their habits are more solitary (Tsukahara, 1993).

Avoidance of Snares

Another factor that could be responsible for injured chimpanzees spending less time on the ground, and proportionally less time moving on the ground, is that they are avoiding snares. Injured chimpanzees may relate spending time on the ground with the painful experience of becoming trapped, and possibly avoid ground travel where possible for this reason. People with their wire snares and traps can be seen as yet another type of ground predator.

Mother Carrying Dependant during Locomotion

Injured mothers may find it more difficult to carry a dependant owing to their modified and sometimes more cumbersome locomotion.

The two youngest dependants (Infant 0) were carried for almost 100% of their mothers' traveling time, regardless of the mother's injury. Younger infants are more dependent on their mother and usually do not travel on their own until about 2 years of age (Hiraiwa-Hasegawa, 1990). Young infants also stay in close proximity to their mother; the two youngest infants in the Sonso community stayed within 1 m of their mother for more than 85% of their time (Munn, 2003). The results in this chapter show that differences in carrying are only apparent when infants become older.

Injured mothers may carry older dependants less because of the extra weight. As offspring weight increases, the effort imposed on the mother to carry must also increase. Furthermore, older dependants move around on the mother during locomotion—changing from ventral carrying to dorsal, reaching out to grab at passing-by objects, or by jumping on and off as the mother moves, creating a potentially unbalanced weight.

Results here show that as dependants get older and heavier, and able to travel independently, injured mothers are more likely to travel without them. One of the juvenile dependants with an injured mother (Zana) was observed being pushed off his mother while she was traveling along thin branches at the outer edge of a fig tree. It is possible that the mother was simply weaning the juvenile, but in context with the difficulties of arboreal travel it is also possible that the mother was avoiding carrying him. Once she reached the wider and more stable branches lower down in the fig tree, she carried the juvenile.

Injured mothers did not show any preference for carrying their dependants in arboreal or terrestrial space. All mothers carried their dependants until an older age when on the ground compared to when they were arboreal. This may be because there are fewer dangers in arboreal space (as described above), or may be because long-distance travel is more common on the ground and mothers are more likely to carry dependants over long distances. It may also be because dependants find it easier, or more fascinating, to clamber freely in trees, than on the ground, or because mothers find it easier to carry dependants when on the ground.

Snare Injuries vs. Congenital Deformity

Of the five injured chimpanzees in this study, four have injuries from snares and one, Banura, has a presumed congenital deformity. Banura's disability was equivalent to that of the snare-injured chimpanzees; she had no use of the digits of her foot, or the actual foot itself. This affected her movement and her balance, and made her locomotion more cumbersome, as was the case for the other handicapped chimpanzees. Socially, her disability may have been different. Since she was born with her injury, other community members might have accepted her to a greater or lesser degree, than mothers who received their injuries later in life. This was not the case. All handicapped chimpanzees, Banura included, were well accepted into the broader community. A lack of tolerance by other community members was argued not to be the causal factor of the differences seen between injured and noninjured chimpanzees' behavior (Munn, 2003).

Conservation Implications of Snaring

There are many threats to the survival of chimpanzees in Africa. Some, such as the bush meat trade in west and central Africa, result in the death of countless chimpanzees. In contrast, snaring has resulted in only a few known deaths in Uganda (Munn & Kalema, 1999–2000; Wrangham & Mugume, 2000). The low incidence of deaths is probably because the snaring of chimpanzees has mostly been accidental. All hunting of vertebrate species living within Uganda's forest reserves is against the law (Howard, 1991), and it is considered culturally unacceptable to eat primate meat for almost all Ugandans (Johnson, 1996). Snaring may, unfortunately, be on the increase in Ugandan forests. Firstly, this is because of the expanding human population. Uganda has an annual rate of population growth (without regard for migration) of 3% (Population Reference Bureau—World Population Data Sheet, 2002). The population in Uganda in mid-2002 was 24.7 million. This is expected to almost double to 48.0 million by 2025, and then by 2050 an expected 84.1 million people will be living in Uganda (Population Reference Bureau—World Population Data Sheet, 2002). Population pressure creates an increasing demand for all forest resources, including bush meat, and the accidental snaring of chimpanzees is likely to increase as a result. Secondly, there has been an influx of new ethnic groups into Uganda. The instability in the Democratic Republic of Congo (DRC) has led to hundreds of refugees crossing the Ugandan border in the hope of finding security. Nearly one-third of immigrants from the DRC living in a village near the Budongo Forest Reserve are known to eat primates (Johnson, 1996), although it is unclear if this includes chimpanzees. Until recently, no known deliberate hunting of chimpanzees for trade or meat in Uganda had been recorded. A recent unpublished report tells the tale of two bush meat traders who were arrested carrying chimpanzee carcasses in western Uganda (Moeller, 2000). It is possible that the bush meat traders were Congolese, and so hunting for themselves. Alternatively, they could have been hunting to supply a new market that may be opening in Uganda that reflects the traditions and habits of refugees, the eating of primate meat being one of them.

Deaths and injuries to chimpanzees will certainly increase as the human population increases, and as more immigrants, with no cultural taboo on eating primates, arrive in Uganda. Chimpanzees with existing injuries face the possibility of receiving further injuries, placing huge pressure on their ability to survive. Whether because of their more cumbersome locomotion, the potential threat from ground predators, or a reduction in feeding competition, injured female chimpanzees from the Sonso community are showing a preference for arboreal space. This leads to injured females spending more time alone or in small groups (Munn, 2003). Consequently, a reduction in social opportunities, less protection from other group members, and fewer opportunities to gain group knowledge, such as feeding sites, may occur. Injured mothers also carry their offspring less than a noninjured mother, particularly as the offspring gets older—this may have long-term behavioral implications for the offspring.

Finally, since snare wounds affect the behavior, survival (Boesch & Boesch-Achermann, 2000; Munn & Kalema, 1999–2000; Wrangham & Mugume, 2000), and the general welfare of chimpanzees, efforts must be made in order to reduce, or eliminate, the occurrence of snaring in Ugandan forests.

ACKNOWLEDGMENTS

I wish to thank the Norwegian Agency for Development, which provides core funding for the Budongo Forest Project, the Forest Department for permission to work in the Budongo Forest Reserve, and the National Council for Science and Technology for permission to study in Uganda. I wish to thank Prof. Vernon Reynolds and the Budongo Forest Project for allowing my study at Sonso. I wish to thank the Australian National University for funding to complete fieldwork. For assistance in the data collection I would like to thank Mr. Joseph Karamagi. I wish to thank Prof. Colin Groves, Dr. Robert Attenborough, Dr. Janette Wallis, Ms. Debby Cox, Dr. Duane Quiatt, Mr. Jeff Donne, and the Graduate Statistical Consulting Unit, Australian National University, for assistance with aspects of this chapter.

CHAPTER SIXTEEN

Effect of Snare Injuries on the Fig-Feeding Behavior of Chimpanzees of the Budongo Forest, Uganda

Behavioral Adaptations and Long-Term Implications

Emma J. Stokes and Richard W. Byrne

INTRODUCTION

Permanent injuries to chimpanzees (*Pan troglodytes*) as a result of snares have been recorded across their geographic range (Quiatt *et al.*, 2002). In August 1997, over 20% of the Sonso community of chimpanzees in the Budongo Forest, Uganda, showed some form of upper or lower limb injury. These injuries are extreme in the sense that they involve either the loss of a hand or foot or else partial or complete paralysis of at least one limb (Waller & Reynolds, 2001). The ability of an injured individual to overcome their injuries is paramount to

Emma J. Stokes • Wildlife Conservation Society, Nouabale-Ndoki Project, BP14537, Brazzaville, Republic of Congo, estokes@wcs.org **Richard W. Byrne** • Centre of Social Learning Cognitive Evolution & Scottish Private Research Group, School of Psychology, The University of St. Andrews, St. Andrews, Fife, KY16 9JP, Scotland (UK) rwb@st-andrews.ac.uk

Primates of Western Uganda, edited by Nicholas E. Newton-Fisher, Hugh Notman, James D. Paterson, and Vernon Reynolds. Springer, New York, 2006.

their survival, and the fact that such a large proportion of this community has sustained and survived major limb trauma suggests that these individuals are indeed able to compensate for their injuries.

A particularly important consequence of upper limb injuries would be feeding difficulties, likely to be specific to those foods that require a degree of manual skill in order to process. Studies on complex plant processing by gorillas and chimpanzees injured by snares (Stokes & Byrne, 2001; Byrne & Stokes, 2002) have shown them to be capable of implementing novel actions with their injured limb to execute feeding techniques that resemble those of able-bodied individuals in their overall structure. This was interpreted to mean that feeding techniques require learning, by program-level imitation, from the mother, who is normally able-bodied. In contrast, individual actions can be flexibly modified by individual learning, so compensation following injury is at the level of specific actions rather than overall plan. As a result of their ability to compensate, injured individuals were not found to show a significant reduction in feeding efficiency. The behavioral flexibility needed to compensate so effectively may be relatively unusual among animals. All simian primates are large-brained (Passingham, 1981), and there is now a growing body of evidence to suggest that environmental pressures have acted specifically on great apes to enhance their cognitive flexibility. Observations of skills shown when apes hunt (e.g., Teleki, 1973; Nishida, 1979; Boesch & Boesch, 1989), make tools (e.g., Goodall, 1964; Nishida 1973; Boesch & Boesch, 1983; McGrew 1992), manually process physically defended plants (Russon, 1998; Byrne *et al.*, 2001; Stokes & Byrne, 2001; Corp & Byrne, 2002;), or overcome locomotor problems associated with large body size (Povinelli & Cant, 1995), lend empirical support to the possibility of an unusual flexibility in behavior among great apes.

Arboreal fruit feeding by able-bodied chimpanzees has also been shown to demand skill (Stokes & Byrne, in review). While relatively simple to process, arboreal fruit presents a problem of a different sort for chimpanzees, one that is largely determined by environmental conditions such as quality and size of the fruit patch and the physical characteristics of the fruiting tree. These, in turn, are exacerbated by the large body size of chimpanzees and the social costs to them of foraging with conspecifics. In feeding on arboreal fruits, chimpanzees were found to show considerable behavioral flexibility in positioning themselves in relation to the constraints of the particular species of tree in which they were foraging (Stokes & Byrne, ibid). Injury is likely to exacerbate the problems of accessing arboreal food resources that are faced by able-bodied individuals, and preliminary studies have found that injured individuals exhibit significant differences in their positional behavior while feeding, compared with able-bodied individuals: injured chimpanzees preferring the canopy over the middle or lower portions of the tree, in which there are fewer secondary branches available for support (Reynolds *et al.*, 1996).

In this paper we systematically examine the effects of injury on arboreal fruit feeding by chimpanzees. We focus our attention on figs (*Ficus* sp.), which are a year-round staple food source for the Sonso community of chimpanzees, representing 50% of feeding time; of which 39% is taken up by two species, *Ficus mucuso* and *Ficus sur* (Stokes, 1999; see also Newton-Fisher, 1999a). Figs are a valuable food item, energy-rich with adequate protein, containing essential amino acids—in the form of dead fig wasps contained within the fig (Wrangham *et al.*, 1993). Figs are straightforward to process, and the key to efficient feeding is almost certainly maximizing food intake rate rather than special food preparation skills. In spite of largely asynchronous fruiting cycles, food patches are highly dispersed in space and time, and feeding competition within a single tree is likely to be high. Furthermore, within a particular fruiting tree, the size, orientation, and location of fruit varies considerably.

We present here new data on the fig-feeding behavior of injured individuals and compare with existing data on able-bodied chimpanzees (Stokes & Byrne, in review). We examine behavioral adaptations to injury through a descriptive analysis of postural and positional behavior and feeding technique (*sensu* Stokes & Byrne, 2001). In addition, we assess the extent to which injured individuals have compensated for injury, through a comparison of feeding efficiency with able-bodied individuals. Finally, we discuss the long-term implications for reproductive fitness.

METHODS

Study Site

The Budongo Forest Reserve lies in Western Uganda between 1°35′–1°55′ N and 31°18′–31°42′ E at an altitude of 1050 m. It is described as medium-altitude, semideciduous forest (Eggeling, 1947). Budongo Forest serves as the primary production forest in Uganda, and produced timber on a sustainable basis from the mid-1920s until the decline of the sawmill in the 1970s. The Sonso study site is found in the N (Nyakafunjo) 3 logging concession, and was last exploited between 1947 and 1952. It is composed of typical mixed-species forest, largely the result of previous forest management practices, and dominated by *Celtis mildbraedii*, *C. zenkeri*, *Khaya anthotheca*, *Chrysophyllum albidum*, and *Funtumia elastica* (Eggeling, 1947; Plumptre, 1996).

Study Community

The study was conducted between August 1997 and September 1998, at the beginning of which the study community consisted of 51 known individuals: 25 adults (13 males and 12 females), 8 subadults (3 males and 5 females), 11 juveniles (6 males and 5 females), and 8 infants (4 males and 4 females). Age classes were determined according to Goodall (1986). Eight individuals have permanent injuries to the hands, forelimbs, or feet. These comprise seven adults (three males and four females) and one subadult female. In all but one case, these injuries are the likely result of encounters with snares set on the forest floor. The one exception is an adult female with a congenital deformity to the foot (Waller, 1995). During the study a new injury was sustained by a juvenile female who lost her hand to a snare (personal observation). Certain adult females were irregularly observed on the periphery of the known community range, and are not considered here. Moreover, in order to set aside age-dependant variation in feeding skill (Corp & Byrne, 2002), juveniles and infants were not sampled. In total, 23 able-bodied individuals (15 adults [9 males and 6 females] and 8 subadults [4 males and 4 females]) and all 8 injured individuals were sampled.

Nature and Extent of Injuries

The nature and extent of injuries is highly idiosyncratic across the populations and described in detail elsewhere (Stokes, 1999; Stokes & Byrne, 2001). Table 1 provides a brief summary of each individual injury and classifies the nature of injury in three groups: injuries to (a) one upper limb, (b) both upper limbs, and (c) to one lower limb. However, it should be noted that these categories are largely based on morphology rather than functional limitations.

Feeding Tasks

We consider here two fig species, *F. sur* and *F. mucuso*, each of which illustrates a suite of relatively independent feeding tasks.

Fruits of *F. mucuso* are about the size of a golf ball and distributed in large drupes hung along wide horizontal branches. Processing is simple procurement,

Table 1. Description and classification of individual injuries

Category	Age–sex	Description
Injuries to one upper limb		
Kalema	Adult female	Right hand paralyzed. The hand is rigidly hooked at the wrist and the fingers are flexed and immobile. The whole hand is emaciated and wasted.
Muga	Adult male	Missing right hand from point distal to the wrist. Majority of the wrist joint is retained and capable of some flexion.
Kikunku	Adult male	Missing left forelimb from point midway up the forearm.
Kewaya	Subadult female	Right hand paralyzed. The wrist is extremely hooked and considerably stretched and twisted round the forearm. The hand is wasted and the fingers contorted so that the middle finger lies overlapping the forefinger. No voluntary movement of the hand.
Injuries to two upper limbs		
Zana	Adult female	Left hand is extremely wasted and the fingers are partially clawed and incapable of voluntary movement. Thumb has retained some function, but is incapable of flexion. Right wrist functions normally but the hand is missing digits 1–4. Thumb retains normal function.
Tinka	Adult male	Most of the muscles of the left wrist are paralyzed, the wrist is hooked and weakened and movement severely restricted. Digits 1–4 are in permanent flexion although the thumb has retained some function. Paralysis of the right wrist and hand is complete and voluntary movement impossible, although the digits can be passively extended.
Injuries to one lower limb		
Kigere	Adult female	Missing right foot from point just above the ankle
Banura	Adult female	Congenital deformity of left lower limb below the ankle, akin to "club foot." Foot enlarged and incapable of voluntary movement.

Description of injuries taken from Stokes (1999) and Stokes (2001), and from original drawings by Waller (1995).

but the relatively large size of the fruit appears to limit ingestion rate. To maximize fig intake, chimpanzees pick several figs from a particular patch and transfer them either to the other hand or to a foot while more fruits are accumulated. A chimpanzee may hold up to four fruits at any one time, which are subsequently retrieved to be processed individually. If displaced from a feeding patch, a chimpanzee will carry the accumulated fruit with them. Wadging is frequently observed and wadges are also transferred to another part of the body, while more fruits are accumulated. By retaining wadges, a chimpanzees can prolong and maximize nutrient extraction.

The fruits of *F. sur* are smaller than *F. mucuso* (~15 mm diameter), and thus food ingestion rate is less tightly constrained. Moreover, patches of figs are distributed vertically along the main trunk of the tree and thus limbs are required for postural support rather than for accumulating fruits. Chimpanzees exhibit short concentrated bouts of feeding, using a variety of feeding postures to access different parts of the tree. Processing is also simple procurement, but chimpanzees may elicit bimanual food processing in order to pull out-of-reach, fruit-bearing branches into range, although this is achieved at the expense of postural support.

Data Collection

Observations were made on the behavior of chimpanzees consuming both fig species. We used three different measures of feeding behavior: (a) position within the tree, (b) posture, and (c) structure of feeding technique.

Position

Scan sampling (Altmann, 1974) was used to record the activity and position of each individual in the feeding group every 15 min. The subject's position was distinguished as either canopy, apical branches of the tree, lower, main trunk of the tree and the first branch fork, and middle, all other parts of the tree.

Posture

Body posture was noted at the beginning of each feeding bout. We defined a feeding bout as beginning when an individual first touched the food, and ending when interrupted or terminated by switching to another activity, by moving away from the food item, or periods of inactivity of 10 s or more (see Appendix of Stokes & Byrne, 2001, for full glossary of terms). Feeding activities included procuring, manipulating, ingesting, and wadging food items.

We categorized posture on the basis of how many limbs were occupied in postural support. The following categories of posture (in order of decreasing stability) were distinguished: seated (SE), animal seated and feeding within arms reach; seated-reaching (SR), animal seated but with one hand in support while leaning to bring food item into range; one-arm support (1), animal suspended by one upper limb only, with remaining limbs available for feeding; bipedal support (2), animal bipedal and supporting with one upper limb while feeding with the other upper limb; and tripedal support (3), animal suspended by three limbs (upper or lower), with only one limb available for feeding.

Feeding Technique

Sequence sampling (Altmann, 1974) was used to record individual actions during feeding. A handful formed the basic unit of sequential analysis, comprising the sequence of events between picking a fruit and placing it in the mouth. For any one handful, processing was considered to be made up of sequences of individual elements (Byrne & Byrne, 1993). For each handful, we recorded the sequence of elements, and for each element the body part used: left or right hand, both hands together, left or right foot, or mouth. Any regularly used sequence of elements, co-coordinated so that the whole performance serves to process a handful of food, was defined post hoc as a technique (*sensu* Stokes & Byrne, 2001).

For *F. sur*, we further distinguished among techniques, according to whether the elements involved bimanual or monomanual co-ordination within a single handful. Techniques were also subdivided on the basis of postural support. In the case of *F. mucuso*, another part of the animal's body (usually an upper or lower limb) was frequently enlisted in holding a food item while another food item was being processed; this included both fruits and wadges, and we use the term "shelf" for this. Techniques were distinguished by the extent to which food items were shelved in a single handful (no shelf, fruit only, wadge only, fruit and wadge) and by the number of food items shelved during any one handful. More than one body part could be used as a shelf at a single time, and simultaneous use of limbs for shelving and processing were recorded during each handful.

Table 2. Summary of data collected

	Ficus mucuso		*Ficus sur*	
	Able-bodied	Injured	Able-bodied	Injured
Feeding posture				
N (individuals)[a]	16	8	23	7
N (bouts)	65	76	341	145
Median no. bouts/individual	4.5 (1–7)	5.5 (1–14)	15.0 (1–36)	14.0 (2–23)
Feeding position				
N (individuals)[a]	23	8	23	8
N (scans)	750	303	882	410
Median no. scans/individual	32.6 (4–53)	36.5 (18–68)	38.3 (12–141)	39.5 (1–58)
Feeding sequence				
N (individuals)[a,b]	13	8	15	7
N (handfuls)	290	361	985	865
Median no. handfuls/individual	26 (10–38)	41.5 (15–92)	58 (25–133)	123 (77–201)

[a] *F. sur* was more common than *F. mucuso* and more frequently fed on, thus more individuals were sampled on this food type.
[b] Only sampled individuals with 10 or more handfuls were included here for analysis of feeding sequence.

For analysis of feeding technique for both food types, we subdivided injured chimpanzees into those with injuries to one upper limb, those with injuries to both upper limbs, and those with injuries to lower limbs. Table 2 summarizes the sample size available for able-bodied and injured individuals for each feeding task and for each behavioral measure.

Feeding Efficiency

In addition to feeding behavior, feeding efficiency was measured by calculating rates of processing from data collected on a hand-held computer (Hewlett Packard 200LX). Key presses were used to record the time at which each handful of processed food was placed in the mouth, and intervals between successive key presses were used to measure the time taken to process a single handful in any given bout. The total bout length was also recorded. In the case of *F. mucuso*, the processing rate for each handful was divided by the mean number of fruits processed simultaneously (as recorded during observations of feeding technique), to indicate relative rates of food intake. However, the combination of two data sets reduced the total sample size in calculation of feeding efficiency for this food type.

Data Analysis

Goodness of fit tests were used to compare postural and positional patterns between able-bodied and injured individuals for each food type. Comparison of feeding techniques between able-bodied and injured individuals presents a number of statistical problems largely concerned with small sample sizes (see Stokes & Byrne, 2001, for discussion). To avoid these problems, variation in usage of a technique among the able-bodied population was used to estimate the likelihood of a deviant frequency occurring by chance in an injured individual. For the frequency of use of each technique, 95% confidence intervals were calculated from scores obtained from able-bodied individuals. Injured individuals whose score fell outside these intervals were considered to vary significantly in their usage of that technique.

RESULTS

Feeding Position and Posture

Injured individuals differ significantly from able-bodied individuals in their feeding position when feeding on *F. sur* ($X^2 = 16.841$, df $= 2$, $P < 0.0001$, Figure 1A). Injured individuals concentrate their feeding in the middle section of the tree (68%, compared to 55% by able-bodied individuals) and spend less time feeding in the canopy and in lower sections. In contrast, no significant difference in feeding position were found between injured and able-bodied individuals when feeding on *F. mucuso* ($X^2 = 0.388$, df $= 2$, $P = 0.824$, Figure 1B).

A significant difference was also found between able-bodied and injured individuals in their use of postures when feeding on *F. sur* ($X^2 = 39.425$, df $= 4$, $P < 0.0001$, Figure 2A). Injured individuals show a reduction in their use of different feeding postures and spend more time feeding from a seated posture (55%), than do able-bodied individuals (27%). As with feeding position, no significant difference in feeding posture was found between injured and able-bodied individuals when feeding on *F. mucuso* ($X^2 = 2.401$, df $= 2$, $P = 0.301$, Figure 2B).

When feeding on *F. sur*, bout length was significantly longer for injured individuals than for able-bodied individuals (mean bout length: injured $= 90 \pm 75$ s, able bodied $= 57 \pm 18$ s, t-test corrected for unequal variance $t = -2.261$, df $= 34$, $P < 0.05$). Injured individuals therefore relocate or change their

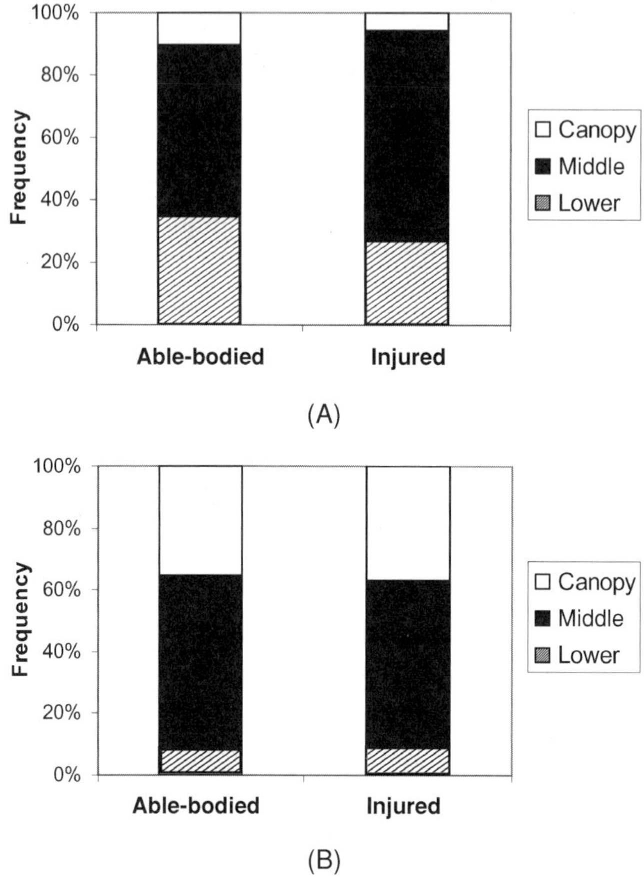

Figure 1. (A) Effect of injury on positional behavior when feeding on *F. sur*. (B). Effect of injury on positional behavior when feeding on *F. mucuso*.

feeding posture less frequently than able-bodied individuals, indicating reduced maneuverability in the tree. Once again, there was no significant difference in bout length between able-bodied and injured individuals when feeding on *F. mucuso* (mean bout length: injured = 159 ± 105 s, able-bodied = 203 ± 107 s, *t*-test $t = 1.070$, df = 26, $P = 0.294$).

Feeding Technique

Ficus sur

A total of six different techniques for processing *F. sur* have been identified for able-bodied individuals, five of which incorporate bimanual coordination. The

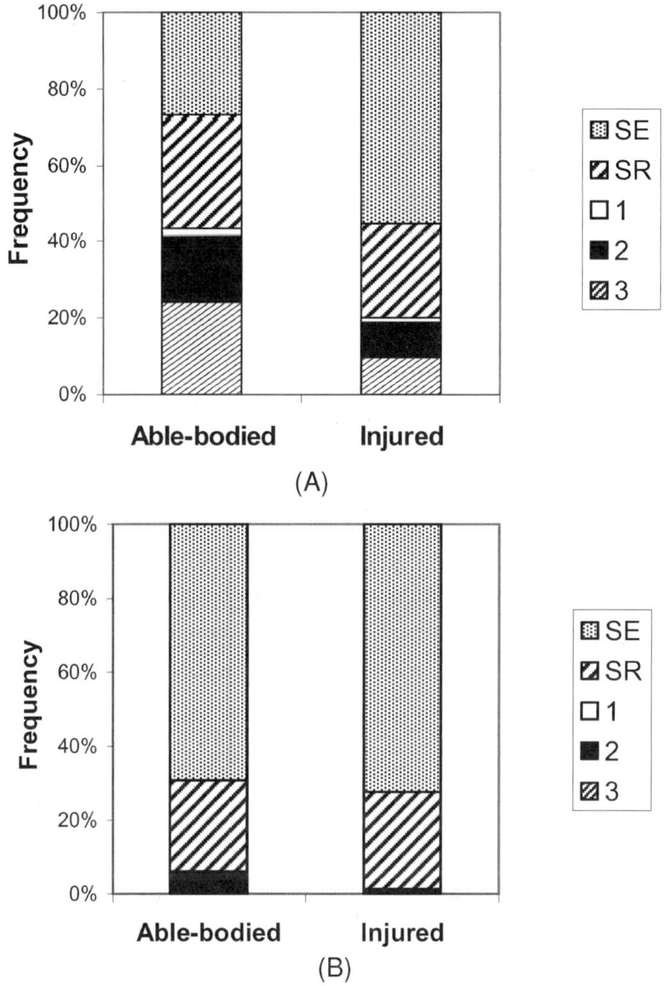

Figure 2. (A) Effect of injury on postural behavior when feeding on *F. sur*. See text for details: SE: seated; SR: seated-reaching; 1: one-arm support; 2: bipedal support; 3: tripedal support. (B) Effect of injury on postural behavior when feeding on *F. mucuso*. See text for details: SE: seated; SR: seated-reaching; 1: one-arm support; 2: bipedal support; 3: tripedal support.

frequency of use of each of monomanual and bimanual techniques by injured individuals is shown in Figure 3. Individuals with injuries to one and both upper limbs show a significant reduction in their use of bimanual techniques compared to able-bodied individuals. What is perhaps more surprising is that these individuals still incorporate bimanual processing into the majority of their techniques (mean 85% of techniques for all injured individuals pooled).

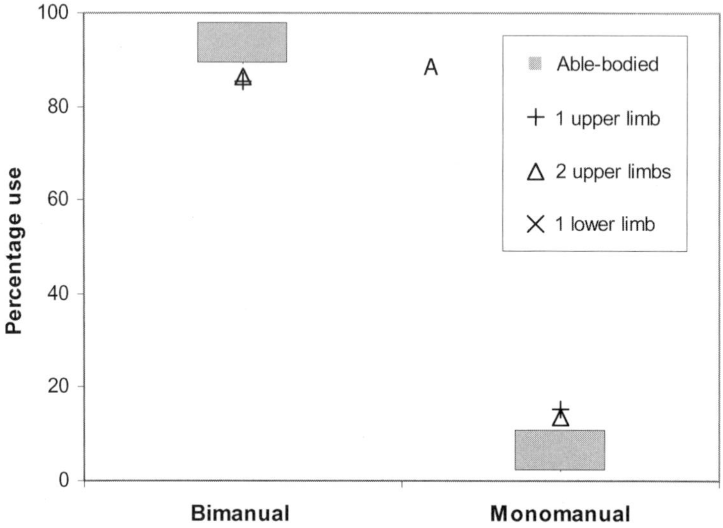

Figure 3. Effect of injury on the use of monomanual and bimanual techniques in processing fruits of *F. sur*. Injured individuals are divided into three categories according to nature and extent of injury. Grey bars represent 95% confidence intervals for frequency of use by able-bodied individuals. Only scores of injured individuals that fall outside these intervals are shown.

Able-bodied individuals rarely use their feet in bimanual food processing. In only 2.1% (0.5–3.7) of sequences were the feet used in combination with an upper limb, always to support a fig branch or bring a fig branch into range while the hand picked off individual fruits and almost exclusively (95%) when the other upper limb was tied up in postural support. Individuals with injuries to one or both upper limbs use their feet significantly more than able-bodied individuals (5.4%) and, perhaps more important, use their feet in food processing exclusively from a seated posture (100%), when no postural support is required. Furthermore, we observed injured individuals implementing novel elements with the injured limb to execute a bimanual feeding technique. In feeding on *F. sur*, injured chimpanzees used their injured upper limbs to pull a branch into range by the crook of the elbow and to support the branch between the forearm and the side of the body while the able limb picked off individual fruits (95% of all bimanual sequences).

Ficus mucuso

A total of 34 different sequences for processing *F. mucuso* have been identified for able-bodied individuals. We distinguished techniques for injured individuals

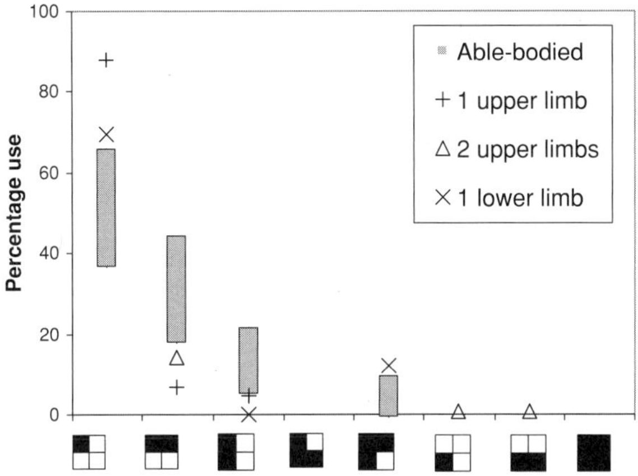

Figure 4. Effect of injury on limb coordination in processing fruits of *F. mucuso*. Each pattern represents the combination of upper and lower limbs used in feeding at any one time during a particular handful for all techniques. Left and right are used ambiguously. See figure for explanation of bars.

in processing fruits of *F. mucuso* and classified each according to the degree of shelf-use (no shelf, fruit shelved, wadge shelved, fruit and wadge shelved). The use of each group of techniques by injured individuals varies with the nature of injury. Individuals with one upper limb injury prefer to utilize a simple pick-and-eat process rather than techniques that shelve fruits (able-bodied: mean = 87.8%, 80.8–94.6; single upper limb injuries: 99.2%). However, individuals with injuries to both upper limbs shelve fruits and wadges at similar frequencies to able-bodied chimpanzees (fruit shelved: mean = 8.2%, 0.5–16.0; wadge shelved: 2.1%, –0.1–5.3; fruit and wadge shelved: 1.7%, –0.1–3.6).

Figure 4 shows the role of each limb in processing fruits of *F. mucuso*. Injured individuals differ in their use of a particular pattern of limb coordination according to injury types. Those individuals with injuries to one upper limb almost exclusively (88%) use their able upper limb in processing. In contrast, individuals with both upper limbs injured use their feet more, always to shelve fruits (1% of sequences for two individuals with both upper limbs injured, compared to zero occurrence for 23 able-bodied individuals).

Feeding Efficiency

No significant effect of injury on processing times was found for either *F. sur* or *F. mucuso* (*F. sur*: able-bodied individuals mean $= 6 \pm 1$ s/handful, injured individuals pooled mean $= 6 \pm 1$ s/handful, *t*-test $t = -0.871$, df $= 20$, $P = 0.39$;

F. mucuso: able bodied individuals mean = 22 ± 7 s/fruit, injured individuals pooled mean = 25 ± 8 s/handful, *t*-test $t = -0.713$, df = 11, $P = 0.491$). Sample sizes were too small to statistically examine feeding efficiency by injury type. In the case of *F. mucuso*, able-bodied individuals can process up to four fruits simultaneously, although 84% (±14) of handfuls are processed without shelving fruits (i.e., one fruit processed at a time). As expected, injury reduces the number of fruits that can be processed at the same time, and injured chimpanzees process a mean 94% (±6) of handfuls with only one fruit.

DISCUSSION

Injured individuals compensate remarkably well in processing complex foods (Stokes & Byrne, 2001). However, the analyses of this chapter have shown that in a feeding task that requires agility and maneuverability in order to rapidly procure fruits that are by themselves relatively simple to process, injured individuals show a marked deviation from their able-bodied counterparts. The breadth of feeding postures available to an injured chimpanzee is reduced, as is the number of limbs available for temporarily holding fruits before processing ("shelving") when feeding on *F. mucuso* and consequently the number of fruits that can be processed simultaneously.

In feeding on the leaves of *Broussonetia papyrifera*, injured chimpanzees retained the basic structure of the feeding technique observed in the able-bodied population, and compensated at the level of individual elements, where they employed novel use of the injured limb in order to execute each stage of the overall program (Stokes & Byrne, 2001; Byrne & Stokes, 2002). Fig processing does not require a complex program of actions, but instead presents a suite of dynamic challenges. In dealing with these challenges, injured individuals were able to modify their behavior to an extent, although no individual successfully compensated for reduced flexibility in postural behavior. Compensation was largely through the use of novel elements by the injured limb in bimanual processing. These novel elements require neither considerable muscle strength nor manual dexterity, and are likely to play an important role in enabling injured individuals to exploit bimanual processing techniques. Feet were also used more frequently than is observed in able-bodied individuals, both as a shelf for processing *F. mucuso* and to complement an upper limb in processing fruits of *F. sur*. For able-bodied chimpanzees feeding on *F. sur*, adapting to the constraints of a particular tree represents a trade-off between bimanual processing and postural support. Both strategies serve to effectively increase access to food patches, the former by bringing food into the range of the animal, and the latter by bringing the animal into the range of the food. The trade-off between the two strategies therefore provides an opportunity for injured individuals to compensate for their reduced maneuverability in the tree, by increasing their usage of bimanual techniques. However, in most cases, the frequency of use of bimanual techniques fell below even that of able-bodied individuals and was frequently at the expense of postural support. In general, therefore, injured chimpanzees showed a considerable reduction in their behavioral plasticity to environmental constraints.

Perhaps by their very nature, feeding tasks of this sort produce a great deal of idiosyncrasy in behavioral response. Here, behavioral variation within the injured population were observed in the preferred choice of technique and the nature of individual elements, but not in the structure of the technique itself, probably because the nature of the fruit largely determined what little needed to be done to process it. For injured chimpanzees, this variation was largely associated with the nature and extent of injury. In feeding on *F. mucuso*, chimpanzees with injury to a single upper limb appeared to "overcompensate" with the able limb and repertoires of feeding techniques were notably reduced as a result. In contrast, individuals with injuries to both upper limbs showed a more diverse repertoire of techniques, either through increasing the use of feet in bimanual processing and "shelving," or the use of novel elements with the injured limb. In the two cases of double upper limb injury examined here—Tinka and Zana—this was facilitated by the fact that both injured limbs retained some able functions, and that these functions were complementary across the two hands. For example, the right hand of Tinka retained a passive grasping and support function, while the left hand retained precision control of the thumb and index finger. In this way Tinka could pick an individual fruit with the left hand and place it into the passive grip of his right hand as a "shelf," before accumulating a second fruit. In this way he functioned as an able-bodied individual.

In measures of feeding efficiency, however, injured individuals did not show significantly lower processing rates than their able-bodied counterparts. Given that processing is a simple pick-and-eat process, at least for *F. sur*, this is perhaps not surprising. However, for *F. mucuso* feeding efficiency also takes into account the number of fruits that can be processed simultaneously, which we know to be lower for injured individuals, and thus it is perhaps at first glance surprising that this was not reflected in the results. It is worth noting here that able-bodied chimpanzees process the majority of handfuls of *F. mucuso* with a simple pick-and-eat sequence. Although shelving does occur, it is relatively rare and most likely a direct response to scramble competition within the tree. If levels of feeding competition are driving performance, then the disparity between able-bodied and injured individuals will most likely be observed at times of peak scramble competition over food. A similar argument can be made for *F. sur*, in that accessibility to food patches will be at a premium when levels of scramble competition in a tree are high. Those individuals who are able to make the most accurate judgments with regards to how to position themselves and adapt their feeding strategy, both with respect to the amount of food available and the presence of conspecifics are likely to have the highest success in feeding. Accordingly, the two males with the highest social ranking in the population (both able-bodied) recorded the two highest averages for the number of fruits of *F. mucuso* processed simultaneously. With the physical limitations imposed by injury, and limited capacity to compensate manually for loss of function, injured individuals have a higher risk of losing access to high quality food patches when scramble competition is high, and individual fitness will likely be compromised as a result.

Preliminary studies on injured individuals suggest that injury does reduce the social ranking of chimpanzees (Reynolds *et al.*, 1996). The effects of injury on feeding efficiency are therefore likely to be compounded by increasing time spent monitoring other individuals in the group at the expense of food intake. This phenomenon would be particularly prevalent when feeding on *F. sur*, as the large group size in the tree at any one time, coupled with the continual relocation and postural readjustment of able-bodied individuals around the feeding tree would suggest the need for continual reassessment of individuals' positions. Furthermore, the fact that injured individuals are unable to maneuver about the tree to the same extent as able-bodied chimpanzees would exacerbate the need for visual monitoring in keeping track of a conspecific's movements.

The long-term implications of injury on fig feeding may be mitigated to an extent by the prevalence of figs in the diet of Budongo chimpanzees. Figs are considered a staple, rather than fallback, food at Budongo and are consumed year-round (Newton-Fisher, 1999). From studies on activity budgets over a period of 14 months, injured chimpanzees were found to spend significantly more time feeding on *F. sur* than did able-bodied individuals (Stokes, 1999)—this flexibility in daily activity largely facilitated by the dynamics of a fission–fusion society. Therefore, while injury has a profound effect on positional and postural capabilities, the long-term negative implications for reproductive fitness may be buffered by the relative spatial and temporal abundance of figs in the Budongo Forest.

ACKNOWLEDGMENTS

This paper was originally presented at the 2001 Winter Meeting of the Primate Society of Great Britain. We thank Vernon Reynolds for the invitation to present this paper, and Jim Patterson for the invitation to contribute to this book. We thank the Uganda National Council for Science and Technology, the President's Office and the Ugandan Forestry Department for permission to work in Uganda, and the staff of the Budongo Forest Project, in particularly ES's field assistant James Kakura. This research was funded by a studentship from the Biological and Biotechnological Sciences Research Council.

CHAPTER SEVENTEEN

Deciphering Junglespeak

An Update on Chimpanzee Vocalizations, Referential Communication, and Their Relationship to Language

Hugh Notman

INTRODUCTION

Chimpanzees have long been favored animal subjects for researchers interested in studying the roots of human behavior, and particularly the origins of human communication. Language origin studies involving apes have typically focused on the animals' use of nonvocal, symbolic communication, such as gestures or keypads, in part because there are now recognized physiological constraints on apes' abilities for language-like vocal production. In tandem with the ape language studies, however, there has been a strong tradition of interest in determining the extent to which apes' natural vocalizations might also contain language-like qualities, such as referential calls.

In this paper I examine our current knowledge regarding referential communication in chimpanzees, focusing specifically on the species-typical, long-distance call, or pant hoot. Results from a number of studies have shown that pant hoots exhibit marked variation in their acoustic structures that is suggestive

---

Hugh Notman • Department of Anthropology, University of Calgary, 2500 University Drive NW, Calgary, Alberta, Canada.

Primates of Western Uganda, edited by Nicholas E. Newton-Fisher, Hugh Notman, James D. Paterson, and Vernon Reynolds. Springer, New York, 2006.

of a referential, or "semantic" communication system. However, recent analyses of Sonso chimpanzee vocalizations from the Budongo Forest, Uganda (Notman, 2003; Notman & Rendall, 2005), provide evidence to suggest this variation is not arbitrarily related to specific production contexts, as a referential model would predict. Rather, variation seems most salient at the level of the individual, and along acoustic features that are associated with physical activity and emotional arousal. These results suggest that pant hoots are not referential in the strictest sense of being arbitrary acoustic signals that symbolize external referents; however, it is likely that receivers infer details of eliciting contexts via a host of cues that both accompany and are integral to the acoustic property of received vocalizations. In addition, chimpanzees likely make inferences about a call's context and select an appropriate response by capitalizing on knowledge of, and experiences with, individuals in their social group and details of their ecological environment. I will discuss the implications for referential communication in chimpanzees in light of these recent studies, and then address the further implications these findings present for the evolution of language in humans.

Chimpanzees, Language, and the Scala Naturae

The continuity approach to the study of animal and human behavior has a long and varied history (Darwin, 1872; Preuss, 1995). The traditional method of ranking species as "lower" or "higher" on a relativistic scale according to biological, psychological, or behavioral complexity has provided generations of researchers with a convenient framework in which to organize biological systems and processes into hierarchies of continuity, from single-celled organisms at the "bottom" to *Homo sapiens* at the top. In the historical tradition of comparative psychology, for instance, this hierarchical scale is borne out as accumulations of cognitive capacities in successive taxa, so that "higher" organisms have cognitive functions (and presumed underlying neural structures) "tacked on" to existing abilities in lower taxa. This process reaches its apex in humans; our brains, it is often assumed, are ape brains equipped with extra features that have been selected to endow us with culture, language, arithmetic, and the ability to ponder our place in the universe.

There has perhaps been no component of behavior more subject to historical approaches of continuity between species as that of communication and language. Animal communication has traditionally been treated as symptomatic (expressions of emotion or motivation) at a minimum and iconic (representative of a behavioral state or as prescriptive directions, such as the honey bee "waggle dance") at best. However, the long-heralded fact of the apes' genetic propinquity with humans has contributed to our perception of them as linguistically intermediate between other animals and humans. Claims that chimpanzees could learn signs to refer to objects (Gardner & Gardner, 1969), create novel words by combining signs using basic syntax (Gardner & Gardner, 1975), communicate using artificial languages (Rumbuagh, 1977), and even understand novel spoken commands (Savage-Rumbaugh *et al.*, 1998) have convinced many scientists, as well as a mildly curious public that apes are, in fact, short hairy humans trapped in a monkey's body. While many of these earlier claims related to ape linguistic capacities have been refuted or even outright rejected (e.g., Chomsky, 1975; Terrace *et al.*, 1979; see Ristau & Robbins, 1982, for review), the effect has nonetheless been to emphasize the linguistic continuity between apes and humans.

Referential Communication: The "Holy Grail" of Language Origins?

The provocative if controversial interpretations of the ape language data ignited some interest in looking for possible vocal homologues in the wild calls of apes to the artificial lexicons used by language-trained subjects. To add further incentive to this endeavor, a new conceptual paradigm had emerged in the 1980s that reoriented many researchers' approach to studying and interpreting primate communication. This conceptual framework has variously been labeled the "referential" or "semantic" model of communication and was popularized by the early work of D. Cheney and R. Seyfarth on vervet monkey alarm calls (e.g., Seyfarth *et al.*, 1980; Cheney & Seyfarth, 1990). These researchers showed that vervet monkeys respond to played-back alarm call variants in a manner that was appropriate to the context in which each alarm call type was elicited. In other words, a "leopard alarm" elicited an escape response appropriate to the presence of a leopard (running into a tree), as did "eagle" and "snake" alarm calls (running for cover; scanning the ground, respectively). The important aspect of these observations is that the recorded calls by themselves were sufficient to elicit the correct response from listeners. Cheney and Seyfarth interpreted these observations to mean that vervet monkeys were responding to calls in a manner that suggested a shared representation of the call stimulus in the "minds" of the audience on hearing a specific call. In this regard, they argue, calls function in a manner at least analogous to human words in that there is an arbitrary relationship between the vocalization and the external entity to which that call is referring, and that this relationship is "understood" by receivers (Seyfarth *et al.*, 1980; Cheney & Seyfarth 1990).

This interpretation of call mechanism (shared mental representation of call stimulus) and accompanying function (to "inform" receivers about something in the environment) has subsequently been applied to many aspects of vocal behavior and to many different primate species (Hauser & Marler, 1993; Gouzoules *et al.*, 1995; Zuberbühler, 2000) as well as to some nonprimate species (Herman, 1986; Gyger *et al.*, 1987; Kaminski *et al.*, 2004). Its appropriateness to the latter, while exciting for many animal behaviorists, somewhat deflated the claims of a few primatologists who had already invoked some higher-order cognitive processing to account for their subjects' aptitude at semantic representation. In response, the mechanisms and functions described in the referential model for animal communication have been subject to much healthy debate and qualification (Hauser, 1996; Owren & Rendall, 2001; Cheney & Seyfarth, 2003; Hauser *et al.*, 2003), and the model is currently considered an appropriate explanatory framework in some signaling contexts. In particular, there is increasing recognition of the role that a species' ecology plays in shaping communication systems. For instance, we might expect selection to favor the production of referentially specific and unambiguous alarm calls when the costs of not doing so are high (e.g., in small, savannah-dwelling monkeys that associate in kin-based groups). However, the mechanism for how these calls are produced and represented in receivers is still uncertain.

THE CHIMPANZEE VOCAL LEXICON

Naturally, many researchers felt that chimpanzee vocalizations ought also to demonstrate some language parallels, given the apes' purported abilities with artificial languages, as well as the increasing corpus of evidence indicating that other "lower" animal species used vocalizations in a word-like manner. To be sure, there has been a mystique surrounding chimpanzee vocal communication—their calls are at once so obviously connected to emotion, as anyone who has worked with chimpanzees will attest, but they are also somehow opaque. Given what we know about chimpanzees' abilities with artificial languages (as well as what we know of their genetic relatedness to humans), there has been an underlying assumption that there is a "code" to be cracked in their calls that we have yet to decipher.

In fact, at first blush, the chimpanzee vocal repertoire is surprisingly simple, comprising primarily acoustic variations on grunts, barks, pants, and screams, as well as some softer vocalizations such as whimpers and "hoos." However, within and between each acoustic class is a pronounced variability that flows along individual, gender, arousal and perhaps even dialectal parameters (Marler & Hobbet, 1975; Mitani *et al.*, 1992; Clark Arcadi, 1996; Crockford *et al.*, 2004). While such variation serves, on the one hand, to confound acoustic distinctions along particular aspects of interest to researchers, it also presents tantalizing potential for the existence and examination of within-call type diversification that reflects the active modification of acoustic signals to broadcast information specific to the caller's environmental context.

Pant Hoots

One particular vocalization, the pant hoot, has been the focus of much attention. Pant hoots are the loud calls of chimpanzees typically used to communicate over large distances. Pant hoots are complex vocalizations, consisting of a series of different individual call units strung together in a sequence that can last between 3 and 23 s (Marler & Hobbet, 1975). The sequence can include up to four distinct phases: an introduction, a build-up, a climax, and a let-down (Figure 1). The introduction phase usually consists of a short series

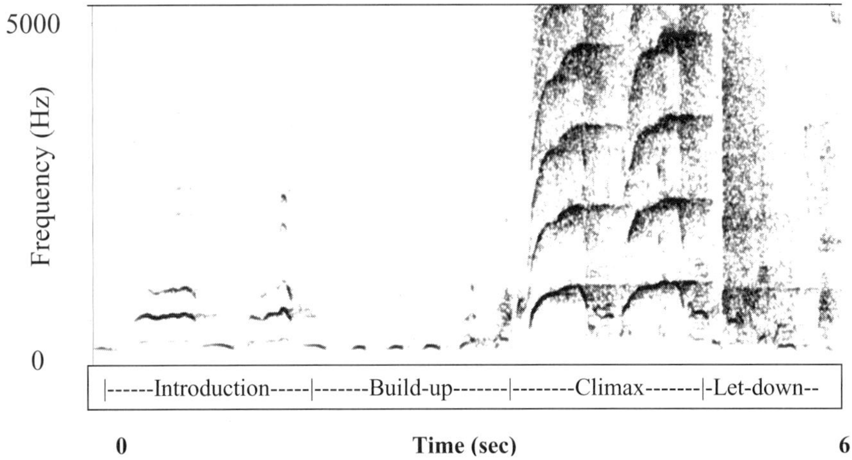

Figure 1. Spectrogram showing the four different phases of a typical pant hoot sequence. Spectrogram was created from a set of successive high-resolution (2000-point) FFTs produced with a Hanning window and 50% frame overlap.

of unmodulated, tonal elements that resemble the low-pitched "hoo" vocalization with a fundamental frequency (F0, which translates perceptually as the pitch) ranging from 200 to 600 Hz (Marler & Tenaza, 1977). The build-up phase consists of a series of shorter inhalation/exhalation, or "panting," elements that are low-pitched and slightly modulated and may become increasingly high-pitched as the call proceeds. The climax phase includes one or a series of long, high-frequency elements that may be tonal and frequency-modulated signals ranging from 500 to 1800 Hz in frequency and sound "wail-like," or they may be more broadband, "noisy" signals that sound "roar-like." Finally, the let-down phase is similar in acoustic structure to the build-up phase except that it tends to decrease in both amplitude and frequency over time. Let-down phases are also typically much shorter than the build-up phase. Although the typical pant hoot sequence contains all four phases, one or two phases can be omitted. Phases within the pant hoot can also grade into one another so that the distinction between them is not always clear.

The fact that pant hoots are used in distance communication means that all information contained within or accompanying the signal is typically received by listeners via the acoustic channel alone, as distance and habitat would limit or preclude the transmission of visual cues to call context. This, in addition to the fact that pant hoots are highly variable and produced in a variety of contexts, has made them ideal candidates for functionally referential signals, as current formulations of a referential communication system require that acoustic variation be specific to call context and be salient to conspecific receivers such that they can make inferences about the eliciting context and select a response appropriate to it.

Unfortunately, previous research regarding the function of pant hoots and the type of information they might broadcast has been inconclusive. Most researchers have emphasized the role of pant hoots in broadcasting information specific to social or ecological contexts. Because earlier observations of chimpanzees vocalizing tended to occur at or near feeding sites, some researchers proposed that pant hoots functioned primarily to announce the discovery and location of food to other community members (Reynolds & Reynolds, 1965; Goodall, 1968a; Wrangham, 1977; Clark & Wrangham, 1993). Subsequent observations, however, noted that pant hoots were produced and varied in their acoustic structure according to a variety of social contexts that appeared independent of feeding situations, such as prior to travel (Mitani & Brandt,

1994), as a function of the amount of time spent associating and calling with allies (Mitani & Nishida, 1994), and as advertisements of social status (Clark & Wrangham, 1994).

More recent studies have further pursued the possibility that chimpanzee vocalizations show functional diversification according to context. In an unpublished study, Uhlenbroek (1996) discriminated between three distinct acoustic variants of pant hoots, the "roar," "slow roar," and "wail" pant hoots, which were produced in the specific contexts of travel, arrival at food, and feeding, respectively. The most salient feature used to distinguish each variant was the relative tonality of the climax elements—roar pant hoots typically contained noisy, broadband climaxes that differed qualitatively from the more tonal, song-like climaxes of wail pant hoots. Slow roars were approximately intermediate in relative tonality between the other two. Uhlenbroek interprets these patterns of variation to mean that pant hoots exhibit selective diversification that functions in a referential manner by sending specific information to receivers about the activity of the caller.

DECIPHERING JUNGLESPEAK: WHAT DO PANT HOOTS REALLY TELL US?

In an effort to resolve some of these issues regarding the function of pant hoots and their relationship to a language-like communication system, Notman and Rendall (Notman, 2003; see Notman & Rendall, 2005, for methodological details and specific results) undertook a study of chimpanzee vocalizations from the Budongo Forest, Uganda.

In these studies, a large number of features were used to characterize the pattern of calling and detailed acoustic structure of chimpanzee pant hoots. To begin with, results showed that pant hoots are highly individually distinctive. Although we did not find evidence for additional differentiation in pant hoots according to many of the behavioral, social, and ecological dimensions considered, the analyses did reveal several statistically significant patterns. Thus, we found that pant hoots were more likely to be produced on arrival at an abundant food source. We also found that pant hoots produced while traveling along the ground in small parties prior to joining up with other community members were consistently different from all other pant hoots, varying reliably in the tonal quality, or "noisiness" of climax elements, the pitch of build-up elements, and in the presence of a let-down phase. Pant hoots that contained a let-down phase were also more likely to elicit pant hoots from others.

These patterns are generally consistent with and extend the results of previous studies of chimpanzee pant hoots. Several authors have described individual differences in pant hoots (Mitani & Brandt, 1994; Mitani *et al.*, 1996b), and many have also noted the variety of contexts in which pant hoots are produced, including while feeding, traveling, and joining up with other community members (Goodall, 1986; Clark & Wrangham, 1993; Mitani & Nishida, 1993; Mitani, 1994). The association between pant hoots and food sources, in particular, has also been emphasized previously, and some studies have documented an increased rate and likelihood of calling when food is especially abundant or of high quality (Wrangham, 1977; Hauser *et al.*, 1993). The variable inclusion of a let-down phase has also been reported previously and, as in our study, has been associated with details of food sources (Wrangham, 1977; Clark & Wrangham, 1993).

Several of the differences identified in Notman (2003) and Notman and Rendall (2005) in the more detailed spectral features of pant hoots are novel but some have been observed previously. Thus, consistent with these studies, Uhlenbroek (1996) noted that the climax elements of pant hoots produced by traveling chimpanzees were noisy and rough, lacking a clear harmonic structure, which made them "roar-like," whereas those produced while resting or feeding were more tonal and "wail-like."

Implications for Referential Communication

One interpretation of these patterns observed in pant hoots is that they reveal significant call variants or subtypes that reflect selective diversification in the structure of chimpanzee calls to support a system of referential communication about important features of the environment. However, if this were the case, we are led to question why the acoustic differences are not more obvious and more pervasive. Several studies by different researchers have failed to identify many consistent and obvious subtypes of the call that are associated with specific production contexts, and analyses using a large number of acoustic features and a similarly large number of potentially relevant behavioral, social, and ecological dimensions have revealed only a handful of significant effects, all of which are acoustically quite subtle. In one study, for instance, the tonal quality of climax elements, while the most salient acoustic feature to vary along behavioral contexts, only actually accounts for a fraction of the overall variation (Notman & Rendall, 2005).

A more parsimonious explanation is that the broad contextual usage of pant hoots might reflect the fact that they serve a generalized social function, such as monitoring and coordinating the activities and movements of dispersed social companions, that transcends the different contexts associated with production. On the surface of it, this seems at least plausible given the decidedly dispersed and fragmented but intensely social quality of chimpanzee society in which individual and kin identity, social status, and flexible alliance networks are paramount. This combination of social organization characteristics may create a vexing problem in coordinating a variety of critically important social relationships while often physically separated from one another. Under these circumstances, the primary function of pant hoots might be to clearly signal identity (and all the subsidiary social dimensions that flow from it, such as status, alliance relationships, etc.) in order to co-ordinate social activities and relationships at a distance. This proposal is, of course, very similar to some others offered for pant hoots (cf. Wrangham, 1977; Mitani & Nishida, 1993; Clark & Wrangham, 1994; Mitani, 1994). It is also supported by the fact that the most consistent finding from studies of pant hoots is that their structure is highly individually distinctive (Marler & Hobbett, 1975; Clark & Wrangham, 1994; Mitani *et al.*, 1996).

Social factors might thus be of overriding importance across the various contexts in which pant hoots are produced, and this may limit earlier attempts by researchers to identify clearly discrete subtypes of the call according to the many other contextual categories we ourselves recognize and define. The few subtle contextual differences that nevertheless emerge might then reflect rough correspondences that arise naturally from the way different contexts or associated behavioral activities affect vocal production. These could include effects on the pattern of calling or detailed acoustic features of the calls introduced by variation in caller arousal or motivation, variation in the caller's concurrent locomotor or other physical activities, or variation in the effort or exertion callers invest in vocalizing, or some combination of these factors.

For instance, Notman and Rendall (2005) found consistent acoustic differences associated with arousal (probability of calling, relative tonality of climax elements, variable inclusion of a "let-down" phase) in pant hoots produced when either traveling, arriving at a food source, or resting on the ground compared to when feeding or resting in a tree (Figure 2). One factor that unites the

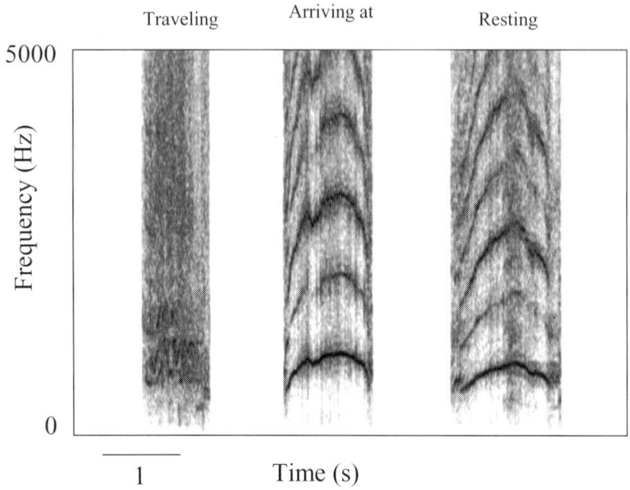

Figure 2. Spectrograms of three climax elements, one each taken from a pant hoot series produced while, traveling, arriving at a food tree, and resting in a tree. Spectrogram settings as per Figure 1.

first group of three disparate activities as well as the latter two and that also distinguishes between the two groups is the position of the callers: the first three activities all involve animals calling from positions on the ground, while the latter involve animals calling from an elevated position in a tree. In forest habitats, sound propagation is particularly influenced by the position of the sound source, favoring propagation of sounds produced from elevated positions over those produced on the forest floor. This effect is underscored by the additional finding we made in pant hoots that the pitch of build-up elements produced on the ground was significantly lower than in those produced in a tree, which is consistent with efforts to maximize propagation of the calls (Waser & Waser, 1977; Waser & Brown, 1986).

It's Not What You Say, It's How I Hear You

In addition, it is important to remember that the nuanced actualities of wild behavioral phenomena are likely not all captured within the rigid confines of empirical investigation. Far from received in an informational vacuum (as it has been presumed pant hoots must be if they are received at such distances from the caller), receivers likely have access to a host of informational cues that flesh out the situational contexts of the caller and which are completely independent of any intrinsic property of the call itself. Human observers can easily distinguish a ground-based from a canopy-based caller, can infer a moving chimpanzee's travel direction by the relative position of calls produced in a sequence, and can generally deduce that something of import is happening by the intensity and number of callers and duration of protracted bouts that might be emanating from a particular location, all without access to visual information. It is therefore reasonable to suspect that an adult chimpanzee familiar with its social and ecological environment can do the same.

One can imagine a scenario whereby a receiver can infer the necessary context of a call by using a host of cues that capitalize on various domains of individual social intelligence but that do not require either referential-like call properties or a linguistic capacity to represent them. For instance, suppose chimpanzee receiver A hears a pant hoot from location X within that community's home territory. Receiver A will likely be able to tell the identity of the caller, because he/she has learned the very salient vocal idiosyncrasies of all the community members (most perceptive human observers can do this by about 6 months of sustained observation). So, receiver A hears the caller "Bob." Receiver A knows from experience that Bob is a high-ranking male, and that location X is actually the site of a large fig tree that was monitored recently by A and assessed to have ripening fruits (receiver A likely knows the location and state of readiness of all the big food trees in his/her range). Receiver A then hears a number of other callers subsequent to Bob's initial pant hoot, some pant grunts (indicating submission to Bob), some screams and other signs of contest as individuals settle into their appropriate places at the food patch according to their relative rank. Receiver A can likely infer that "Bob is calling from location X where the big fig tree is—others are joining and also calling—there is some intense socializing; therefore, that big fig is in fruit and I will go/not go and partake."

Thus, it is possible that the selective forces that have shaped the acoustic structure of pant hoots are those that render individual calls perceptually salient to receivers, as well as those that effectively propagate the transmission of the signal. Superimposed onto this selective dimension are the constraints imposed by respiratory mechanisms that function to facilitate the expression of both. Into this mix are the further constraints imposed by phylogeny; indeed, all species of African great apes share a vocal repertoire that is largely composed of variations on hoots, pants, and screams (Mori, 1983). Given the distances and stationary heterogeneities comprising a forest habitat through which an acoustic signal must travel, pant hoots are perhaps limited in the degree to which they can broadcast contextual nuances via acoustic structure. As a result, it is the receiver who shoulders the onus of interpreting call context via accompanying contextual cues, in contrast to circumstances surrounding the selective diversification of some other call types in other species (such as vervet alarm calls) in which selection is operating on the sender to make the context of calls as perceptually salient and unambiguous as possible (Cheney & Seyfarth, 2003). In chimpanzee distance calls, then, selection for the integrity of acoustic properties that maximize signal propagation and of acoustic cues to identity are paramount in senders; in tandem, receivers are under selective pressure to parse call context by tapping into existing cognitive domains associated with the particulars of an individual's social and ecological environment.

IMPLICATIONS FOR HUMAN LANGUAGE

The search for a language-like communication system in the wild calls of chimpanzees has traditionally been motivated by several avenues of research that have emphasized continuity between humans and apes in many cognitive and behavioral domains. For chimpanzees, in particular, their purported facility with artificial languages has especially compelled endeavors to establish parallels with (or homologues to) their natural vocalizations. However, the "ape language" research has shown that, while chimpanzees are capable of matching symbols (whether manual signs, plastic chips, or keys on a computer) with objects, people, and actions, there are also important limitations that underscore the differences between ape communication and human language (see Bickerton, 1990; Shuttleworth, 1998; and Tomasello & Call, 1997, for reviews). One of many interesting observations that emerged from research on a particular ape (the bonobo, "Kanzi"), who was enculturated in a linguistic environment using a lexigram and by responding to spoken English, is that there is an apparent discrepancy between Kanzi's production and comprehension abilities. Kanzi is purportedly able to comprehend an impressive array of complex spoken English sentences, but he is unable (or unmotivated) to produce similar sentence types on a lexigram, and his linguistic output is restricted to short grammatical phrases that are predominantly requests for favored items or actions.

Arguably, there are epistemic challenges for researchers who evaluate the presence of a particular trait such as language in other species against the extent to which we posses it. A less anthropocentric approach might be to establish a set of "core properties" for target traits, such as language or culture, from which we may infer which properties are accessible to other species and, by extension, the "boundary conditions" that surround their respective abilities (Kako, 1999). The benefit to an approach that identifies constituent dimensions of behavioral domains in species is that we can infer the specific ecology-related evolutionary pressures that might have selected for the expression of certain properties. For example, it is possible that the comprehension bias evident in language-trained apes is rooted in perception-based mechanisms that have been selected in chimpanzees to parse acoustic information into meaningful contexts in the absence of calls that broadcast specific details about the environment.

One of the principal hallmarks of human language is its capacity to exchange or transfer detailed information. The social context that underlies an exchange is intrinsically cooperative; information transfer is an integral component to human pedagogy, group activity coordination (such as hunting and sport), social narrative, and social gossip. To be sure, language is also used in competitive contexts and for manipulation. However, the sharing of information integral to a language-based system to a language-based system may be a dimension of a larger human behavioural tendancy for cooperation.

Chimpanzees, by contrast, do not normally cooperate on a day-to-day basis. For the most part, male chimpanzees compete with one another for food and mates, and females compete for access to food. Where cooperation occurs it is largely inadvertent or does not require one individual sharing information with the other; cooperative hunting and territorial encounters generally consist of individuals behaving as independent, selfish agents whose collective activity increases the efficacy of an objective. Similarly, male alliances involve a cooperative effort against a third party, but this cooperation is largely based on mutually selfish objectives that need not be shared among participants in order for the desired effect to come about.

In fact, recent experimental evidence looking at social cognition in chimpanzees has shown that chimpanzees perform better at tasks in which subjects must attend to communicative and behavioral cues when the experimental paradigm incorporates a competitive, rather than cooperative, context (Hare *et al.*, 2000; Hare & Tomasello, 2004). This suggests that chimpanzees do not typically communicate information about their environment in a deliberate fashion. Indeed, the consensus arising from research regarding the function of loud calls dispels the notion that they deliberately broadcast information about food, in part because it is high-ranking animals that are more likely to call at food bonanzas, and these individuals are less likely to suffer the effects of feeding competition and may benefit from attracting allies or mates to their proximity.

If chimpanzee calling systems have been shaped by an ecology involving social competition, then the primary function of their vocalizations might be to directly affect the behavior of receivers by broadcasting information regarding the caller's identity and emotional state, which could be used by receivers to gauge a future course of action. At some stage during our evolution, humans likely began to socially cooperate to an extent that chimpanzees currently do not. Language may have facilitated the sharing of information, which was likely a critical component of efficient cooperation. Language may therefore be a dimension of a larger adaptive trait in humans that involves attending to others' mental states in order to cooperate with individuals within a social group—the so-called Theory of Mind. The outstanding issues in these areas would benefit from research that examines more closely the relationship between species ecology, cognition, and communication, and analyses the mechanisms that underlie the perception of acoustic signals in animals.

CONCLUSION

Much ink has been spilled debating the best approaches to studying animal origins of human language. Chimpanzees have been central to this debate; the "ape language" research has generated much productive theorizing in this arena, as has the issue of referential vocal signaling in chimpanzees and other primate species. The lack of any clear parallels between chimpanzee vocal communication and human language points to the importance of understanding species ecology in explaining the expression of certain behavioral traits, and urges us away from assuming that species who share a close phylogenetic relationship with humans are continuous in all other behavioral and cognitive domains.

CHAPTER EIGHTEEN

Instrumental Leaf Use by Chimpanzees of the Budongo Forest (Sonso Community)

Duane Quiatt

INTRODUCTION AND METHODOLOGY

Site, Subjects, Procedure

For 3 years, from mid-October 1996 to mid-November 1999, the resident research staff of the Budongo Forest Project in west-central Uganda recorded at my request ad hoc observations of leaf-assisted drinking (LAD) and other instrumental use of leaves by chimpanzees of the Sonso community (for a description and history of the Budongo Forest Reserve and its primate populations, see Reynolds & Reynolds, 1965, and Reynolds, 1992). Observations were entered on standard forms to speed on-site data entry and to help ensure consistency in subsequent comparison of numerous behavior records contributed over a long period of time by multiple observers. The data protocol also elicited, for

Duane Quiatt • Department of Anthropology, University of Colorado, Denver, P.O. Box 173364, Denver CO 80217.

Primates of Western Uganda, edited by Nicholas E. Newton-Fisher, Hugh Notman, James D. Paterson, and Vernon Reynolds. Springer, New York, 2006.

each observation of LAD (more commonly referred to as leaf sponging or leaf sponge drinking; see, e.g., Goodall, 1986; Nishida, 1987; McGrew, 1992; Quiatt, 1994; Boesch, 1996), basic information concerning the prevailing social and wider activity contexts. Finally, observers were encouraged to note in their own words features of the behavior they were reporting that struck them as "unusual or of particular interest." As seems not unreasonable in retrospect, that suggestion was treated with reserve; the most frequent response was "normal."

Leaf-Assisted Drinking

One hundred twenty-one instances of instrumental leaf use were recorded in the course of this study, 111 of which (Table 1) involved transport of water to mouth by dipping a leaf or, in 3 cases, a portion of fruit into the bole of a tree, a spring or standing water, or the Sonso River, for which this community of chimpanzees is named. On occasion, water was dipped or sponged from the bole of a fallen tree or log, but standing trees provided the great majority of source boles. Most but by no means all boles from which water was obtained were accessible from the ground. Leaves were recovered for examination after use in all but a few cases. Many had been modified, folded, and/or lightly crushed rather than chewed, increasing absorbency and/or adhesion of water to leaf surface and maybe releasing flavor elements, a possibility not to be ruled out. (The taste of medicinal plants and plant foods used by chimpanzees and other nonhuman primates has been investigated systematically by, most recently,

Table 1. Instances of Leaf-assisted drinking: 15 October 1996–15 November 1999

	Use context				
Age–sex class	Leaf dip in bole	Leaf dip in river	Leaf dip in spring	Fruit dip in river	Row subtotal
Adult male	53	4	1	–	58
Adult female	27	1	–	2	30
Subadult male	9	–	–	1	10
Subadult female	–	–	–	–	–
Juv/inf male	6	1	–	–	7
Juv/inf female	6	–	–	–	6
Total male	68	5	1	1	75
Total female	33	1	–	2	36
Combined total	101	6	1	3	111

Nishida *et al.* (2000), who provide a comprehensive review of prior studies.) Examination of the composition and flavor of water either in boles "sponged" by chimpanzees or in water procured by use of an Acalypha leaf sponge seems to have been limited so far to ad hoc tasting by a few field primatologists, for example and respectively, Michael Huffman and Vernon Reynolds (personal communications).

Unassisted Drinking

In addition to these 111 instances of leaf-assisted drinking (and 10 instances of instrumental leaf use that did not involve drinking, summarized in Table 2), 18 instances of UAD (UAD) were recorded. UAD most frequently consisted of drinking with mouth applied directly to river or spring (or, in one case, to water in a shallow tree bole); 5 instances involved transporting water from bole to mouth by means of fingers alone, without resort to leaf use. Geographic distribution of drinking events is shown in Figure 1; most observations of drinking behavior, whether LAD or UAD, were made within the perimeter of the Sonso community's high-use area, much of which was surveyed and roughly grid-mapped in the early 1990s (Figure 1). With four exceptions (excluded from analysis), observers' on-site records of drinking and other events identified the 100 × 100-m grid-block in which each observation was made (or, by extension, for events recorded outside the grid-trail system proper, an equivalent if approximate coordinate location), thus enabling comparison of the geographic distribution of LAD and UAD.

Table 2. Other (non-LAD) instances of leaf use: 15 October 1996–15 November 1999

	Use context					
Age–sex class	Eat leaf with meat	Clip/tear leaf	Wipe penis (M)	Wipe semen (F)	Leaf examination	Row subtotal
Adult male	3	2	–	–	1	6
Adult female	–	–	–	1	–	1
Subadult male	–	–	1	–	–	–
Subadult female	–	–	–	–	–	–
Juv/inf male	–	1	1	–	–	2
Juv/inf female	–	–	–	–	–	–
Total male	3	3	2	–	1	9
Total female	–	–	–	1	–	1
Combined total	3	3	2	1	1	10

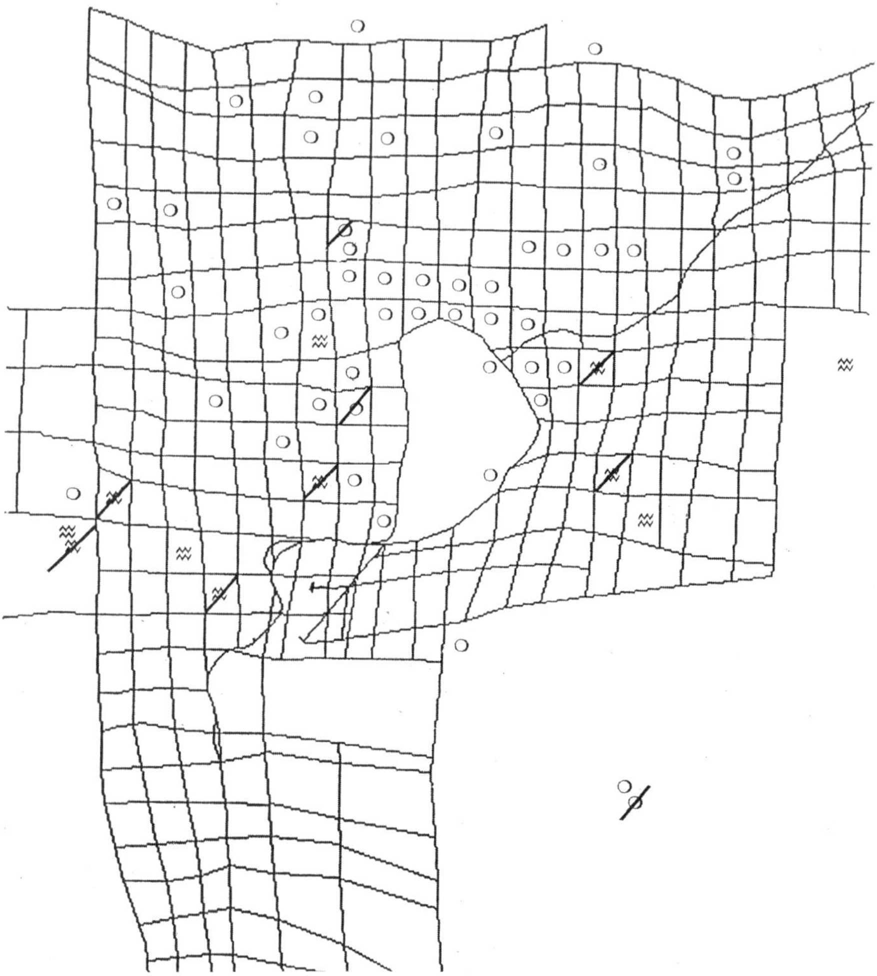

Figure 1. Local geography of leaf-assisted and unassisted drinking. Symbols on this grid-block map, of a high-use portion of the Sonso chimpanzees' range indicate in each case and for that particular area one or more bouts of drinking behavior, LAD or UAD from tree (or log) bole (open circles), or river (or spring) (double wavy lines), for a total of 129 bouts. UAD indicated by a diagonal strike though of the symbol for the location.

RESULTS

Leaf Use Within the Community

Although the sample of observations treated here is small (i.e., 139, of which 111 were of LAD, 18 of UAD, and 10 of instrumental leaf-use other than

LAD), it involves 38 different members of a chimpanzee community that numbered between 45 and 50 through the 3-year period of study. The sample includes individuals of every age–sex category except that of subadult female (Table 1). Several instances were recorded of mother and infant leaf sponging at the same source, with obvious implications for early learning if not manner of learning of this behavior. It seems likely that, had it been feasible to sample in greater depth or to extend sampling over a greater number of years, virtually all members of the community would have been recorded as engaging in LAD. Indeed, that LAD is widespread or universal and not sex-linked in its expression is clear from previous studies (see especially Whiten *et al.*, 1999).

The disproportionate representation of males over females in Table 1, in all but the Juvenile/Infant category, may be attributable in some degree to shyness on the part of female chimpanzees (subadult and young adult females in particular) in the presence of human observers, especially when those observers are male—and all but two of the cases of leaf use reported here were collected by male observers. Similarly, data collection could have been influenced (however slightly and for whatever reason) by observer bias, favoring observation and record of LAD by adults rather than by young animals and/or by adult males rather than by adult females. None of these tendencies—if indeed they did operate to influence in some degree the distribution by age and sex of LAD events reported in Table 1—can be applied to explanation of age/sex differences observed in the practice of other "cultural" behaviors noted and summarized in Table 2. Most of those, that is, clip/tear leaf, male wipe penis, female wipe semen (from vaginal area), are strongly indicated if not strictly determined by age and sex of actor involved. They are loaded with important and for the most part seemingly unambiguous social information relevant to consideration of reproductive role and status. Such behaviors seem, on the face of it, more interesting from an informational and communication standpoint. I will have more to say about them from that standpoint in the discussion that closes this chapter. However, this report has mainly to do with LAD, and I turn now to what is revealed when that subset of observations is sorted according to (1) species of leaf employed to dip or sponge water; (2) where, within the high-use area of the Sonso community's territorial range, different kinds of drinking behavior were observed; and (3) when cases were recorded, at what time of day and in which month over the 3 years of record.

Leaves Employed in Leaf-Assisted Drinking

All but a few of the leaves used by Sonso chimpanzees to transport water from source to mouth could be identified at least by genus (Table 3). Most were assignable to one or another species of *Acalypha*, a common shrub in the Budongo forest. Leaves used for water transport were described on data record forms as "hairy" in 84 cases, "smooth" in 23. Most of the leaves described in this study as "hairy" were assigned by contributors of data to *A. acrogyna*, *A. neptunica*, or—less specifically and with greater frequency—to *Acalypha* spp. or simply *Acalypha*. A problem arises in that contributors also assigned one-third (8) of the leaves described as "smooth" to *Acalypha* spp. or *Acalypha*. In Kenya, the leaves of both *A. caudata* and *A. Microstylis* are smooth or glabrous (M. Huffman, personal communication), while to the west, and similarly, of the *Acalypha* species listed in *Flora of West Tropical Africa*, just one, *A. grandis*, is represented as bearing leaves that are hairy or furry (i.e., densely tomentellous on the underside of the leaf). Because leaves in the present study were discarded after examination and identification, questions concerning heterogeneity of form at the species level cannot now be resolved. This does not affect the main conclusion that the great majority of leaves selected by Budongo Forest chimpanzees for LAD can be assigned to hairy-leaved species of the widespread genus *Acalypha*. Such selectivity of course makes good sense from a water transfer standpoint.

Table 3. Leaf species employed in leaf-assisted drinking

Genus/species	Cases
Acalypha spp.	78
Alchornea spp.	3
Broussonetia papyrifera	6
Cynometra alexandri	3
Desplatsia dewevrei	8
Ficus sur	3
Lasiodiscus mildbraedii[a]	3
Maesopsis eminii	2
Marantochloa leucantha	3
Unidentified	2
Total	111

[a] Leaves also employed and probably consumed in conjunction with the meat eating recorded in Table 2.

Geographic Distribution of Leaf-Assisted and Unassisted Drinking Behavior

Figure 1 maps the local geographic distribution of drinking events, leaf-assisted and unassisted, on a plan of the Budongo Forest Project's trail system, a loosely bounded grid net of forest paths, cut and maintained, that in their present extent comprehend an area roughly coextensive with the high-use sector of the Sonso chimpanzee community's range. In this system each constitutive grid block measures approximately 100 × 100 m. (There are of course exceptions to this generalization: "roughly" and "approximately" are terms that iron out differences of considerable magnitude between certain individual blocks in the grid pictured here: cf. those in Figure 1's far NW corner.) Vertical and horizontal gridlines represent for the most part trails cut prior to 1993, and maintained since by a permanent crew of trail cutters. Since 1993, secondary extension of these trails along established gridlines has proceeded intermittently, particularly in the southwest quadrant).

This map depicts schematically a substantial portion of the Sonso community's annual range, recently estimated at just under 7 km^2 (Newton-Fisher, 2003). Sonso chimpanzees have fed and nested regularly within or close to this grid-mapped area through many months of each year since studies were initiated in 1991. It is in every sense a high-use area for the community. Comparison of how behavior can vary within it—in particular, behavior associated with something as basic as water intake—should prove instructive. This study enables a first comparison of the geographic distribution of LAD by Sonso chimpanzees with that of UAD.

As Figure 1 reveals, LAD events were recorded with far greater frequency in the northern and north-central areas of the grid-map's main section, UAD events in the southern and south-central areas—taking center as the midway point of the east–west gridline that runs along the top of the large, irregularly shaped open area in the grid center. Situated in that "open" space are the Sonso Mill, now largely abandoned, and two clusters of houses originally constructed for workers and administrators of that now defunct sawmill. In late 1999 some of these houses were inhabited: one was in regular use as a school for local children, and the BFP was headquartered in two of the somewhat more "finished" administrators' dwellings. The presence of a small human community in a forest clearing may explain why instances of chimpanzee leaf use and drinking behavior are all but absent from this area of the grid map. What remains to be explained is why so many records of these and other chimpanzee activities are drawn from grid squares in close proximity to that cleared area and its human community. Part of the answer, but by no means all, lies in the distribution of *Broussonetia papyrifera*, a forest-edge tree species on which the Sonso chimpanzees rely for fall-back foods of one sort and another (leaves, flowers, fruits) over much of the year.

There is, however, another explanation for the geographic demarcation of LAD and UAD noted at the outset of the last paragraph. North of center, the area mapped in Figure 1 is relatively high and dry, with few marked changes in elevation. South of center, where the Sonso River (not represented here) runs from east to west across the area mapped, there is greater variability: in elevation, in forest cover, and in character of habitat. Although the river and its drainage system are not directly represented in Figure 1, the drainage area is roughly delineated by the concentration of grid map symbols that indicate UAD from river or spring. It should come as no surprise then to see that LAD from tree boles predominates to the north, UAD from river, spring, or standing water to the south.

There are, again not surprisingly, few instances (6) of UAD from a tree bole; more common are instances (10) of LAD from an open water source. Several records of drinking behavior describe two or three individuals (rarely more than three) utilizing the same source, bole or open water, simultaneously or in close succession. One three-bout LAD sequence involves three individuals dipping portions of *Desplatsia* fruit repeatedly into the Sonso River. I have treated these bouts as equivalent to "leaf-dipping" or "leaf-sponging" (in Table 3 they are included as three of the eight *Desplatsia* bouts) partly because such treatment may be warranted if portions thereof were not consumed, as appears to have been the case, or if the taste of the water or the flavor of the experience were not thereby enhanced, as is perhaps less likely to have been the case, but also and more importantly to emphasize the fact that such questions remain to be addressed systematically in connection with leaf sponging per se. I will return to this issue in closing, in connection with suggestions for future research.

Temporal Distribution of Observations Recorded

Drinking activity, as represented in observations of LAD and UAD over day-light hours (no nighttime observations were recorded) begins in the first or second hour after sunrise and peaks strongly around mid-morning (9–10 AM), with frequencies maintained at roughly the same level through the noon hour (12 noon–1 PM), then declines gradually and stops with construction of night nests at sunset. In short and as might be anticipated, drinking activity appears to coincide in a general way with feeding activities through the course of the day. It should be kept in mind however that many or most records of both LAD and UAD were made while the chimpanzees involved were on the move from one feeding area to another (or in late afternoon toward a nesting site), also that, as noted above, the majority of boles visited in this sampling of LAD activity turn out to be accessible from the ground. Together, these three observations—i.e., that LAD activity is broadly coincident with (1) feeding activity, more narrowly with (2) travel, and most narrowly though by no means exclusively with (3) terrestrial locomotion—suggest that connections between feeding, travel, and drinking behavior are likely to prove complex.

Sorting observations by day and month over the 37 months of record confirms only the obvious, that no observations of chimpanzee drinking behavior are recorded during periods in which chimpanzees are absent from the study area. Other patterns are not as easily interpreted.

DISCUSSION

Culture in Evolution

Whiten *et al.* (1999), examining "cultural" behavior reported from seven long-term chimpanzee study sites, sorted 65 candidate patterns into four categories: (a) patterns reported as absent at no long-term study site; (b) patterns not achieving habitual frequencies at any site; (c) patterns for which any absence can be explained by local ecological factors; and (d) patterns customary or habitual at some sites yet absent at others, with no ecological explanation. What I have termed LAD appears to be the same as Whiten *et al.*'s leaf-sponge (leaf mass used as sponge), one of seven candidate behavior patterns assigned to that report's Category A:

1. Investigatory probe (probe and sniff)
2. Play start (invite play holding stem in mouth)
3. Drag branch (drag large branch in display)
4. Leaf-sponge (leaf mass used as sponge)

5. Branch-clasp (clasp branch above, groom)
6. Branch-shake (to attract attention, court)
7. Buttress-beat (drum on buttress of tree)

Compare LAD/leaf-sponge with other candidate patterns recognized by Whiten *et al.* (and assigned to categories B and D above), patterns in which leaf use operations correlative with sponging (mopping, dabbing, wiping, cleaning) are accorded more specific designation, that is, in Whiten *et al.*'s terminology: B 20, Leaf-mop (leaves used to mop up insects); B 21, Leaf-wipe (food wiped from skull etc.); D 50, Leaf-napkin (leaves used to clean body); and D 51, Leaf dab (leaf dabbed on wound, examined).

The foundational significance of Whiten *et al.*'s comparison of chimpanzee behavior across several long-term study sites was immediately recognized (e.g., de Waal, 1999) and widely remarked. Geographically varying cultural traditions have since been reported for orangutans as well (van Schaik *et al.*, 2003). Whiten *et al.* apply the term "cultural tradition" exclusively to the site-specific behavior variants for which no ecological explanation can be assigned, that is to Category D in their assignment of candidate behaviors. There are sound methodological reasons for doing so. In brief, where systematic differences in behavior across sites cannot be explained as a consequence of genetic and/or ecological differences, social learning is a clear alternative; while, for a pattern of behavior that is customary and habitual at all sites, it is difficult to judge whether commonalities result from social transmission or simply reflect parallel acquisition of knowledge via nonsocial learning by individuals constrained by a common evolutionary heritage and similar life histories (Whiten *et al.*, 1999). I have no quarrel with this reasoning if it does not discourage investigation into patterns of social learning by nonhuman primates (and other animals) that appear to be both species-wide and species-limited. Such studies seem to me at least as important to an understanding of animal (including hominid) evolution and speciation. That consideration in mind, review of the data discussed here raises two afterthoughts in connection with Whiten *et al.*'s landmark study.

The first is that LAD (Leaf-sponging) has not always been considered a behavior customary or habitual among chimpanzees in communities throughout the species range. Nishida (1987), in the paper from which Whiten *et al.*, take their definition of culture as a "population-level characteristic," observed that while Gombe chimpanzees "occasionally . . . crush leaves and use them as sponges...for drinking from holes in trees, [...] this technique of "leaf-sponging" has never been seen at Mahale." Thus, what once interested Nishida as a cultural tradition limited in its distribution appears not so interesting from a Category D perspective. The same is likely to happen, for all the accumulated years of observation accrued to date at chimpanzee study sites, in the case of at least some of the behaviors currently assigned by Whiten *et al.* to Category D.

The second afterthought has to do with the instances of leaf use treated here that, unlike LAD, do not involve water intake (Table 2). Clip/tear leaf, Wipe penis, and Wipe semen (i.e., from vaginal area) interest me in particular, not so much in terms of their resemblance to Category D behaviors in Whiten *et al.*'s listing—Clip/tear leaf evidently equivalent to "Leaf-clip," and Wipe penis and Wipe semen to "Leaf-napkin" in that listing—but strictly from an informational standpoint. In this small sample of record, instances of non-LAD behaviors taken as a whole appear to be sex-linked although not in all cases age-linked. It seems likely that if sample size were increased, the same relation would obtain at least for these three varieties of non-LAD leaf use—the latter two of course sex-linked by definition. More to the point, perhaps, performance of any of these behaviors appears to interest companion chimpanzees (as well as human observers) at least as much as does LAD, if for different reasons: each carries social information—socio-sexual information—that is underlined by and may be embellished in performance. Social information embodied in a performance of LAD probably is insignificant in comparison, while information of a nonsocial character that is communicated by the act of dipping a leaf into a bole and bringing forth water may be in certain performance contexts (especially in periods when rains have been infrequent) of considerable significance to all who get the message, not exclusively or primarily those who are sexually active. The behaviors in Whiten *et al.*'s Category A (again, all but leaf-sponge) may be similar in respect to those discussed just above, signaling variously age and/or reproductive status and/or intent to initiate social or socio-sexual activity. A methodologically restricted definition of culture may be useful, indeed essential, to examination of within-species differences in socially learned behavior. For comparison of differences (and examination of similarities) in such behavior across species, in particular comparison extended beyond the close boundaries of our hominid lineage, less restrictive definitions of "culture" and "social knowledge" may prove more informative.

CONCLUSION

Future Research

Results of this survey appear to warrant a more intensive investigation into Sonso chimpanzees' LAD and UAD practices, with closer control over (1) observation schedules, to ensure comparable data records for every daylight hour through all seasons of the year; and (2) recording of behavior, to secure records of LAD and UAD for as many individuals as is practical and in detail adequate for at least a coarse-grained comparison of similarities and differences in individual practice. Primatologists seem in general to approve an assumption that dipping or sponging water from a tree bole, obtaining water in the same way from an open water source, and drinking water with mouth applied directly to the source are equivalent ways of satisfying a single primary physiological requirement. That assumption probably should be examined at least to the extent of (1) measuring or estimating water intake by each of these means; (2) ascertaining if possible whether *Acalypha* leaves are favored by Budongo chimpanzees (and chimpanzees in other forests?) strictly for properties important to water transport or, in part and perhaps as importantly, because they are abundant and/or distributed in convenient proximity with water sources; and (3) investigating through chemical analysis and taste assessment whether water samples obtained from tree boles at which LAD has been observed may not reveal traces of nutrient value and/or flavor character, either of which conceivably could encourage leaf sponging.

ACKNOWLEDGMENTS

I thank the Uganda Forest Department for permission to work in the forest, the Uganda National Council for Science and Technology for research permission, and the Budongo Forest Project for hospitality and research cooperation over extended periods beginning in 1993. For on-site encouragement, aid, and companionship, I thank its founder and director, Vernon Reynolds, a succession of field directors (in particular Andrew Plumptre, Christopher Bakuneeta, Jeremy Lindsell, and Fred Babweteera), and the resident research staff, trail crew, and maintenance and housekeeping personnel without exception. All who have worked at the Budongo Forest Project must be grateful for support funding contributed to that project over the years by the National Geographic Society, the ODA Forest Research Programme, and, more recently, the Norwegian

Agency for Development. I take particular pleasure in acknowledging gratitude to Trustees of the Boise Fund, Oxford University, for support applied in 1996 to the research reported here. Finally, and most particularly, I thank the following individuals for data contributed: Godfrey Jachan Biroch, James Kakura, Joseph Karamaji, Paula Pebsworth, Emma Stokes, and especially Zephyr Tuka Kiwede, Geresomu Muhumuza, and the late Tinka John.

SECTION FOUR

Conservation

Figure v. Blue monkey *Cercopithecus mitis* (above) and L'Hoest's monkey *Allochrocebus lhoesti* (below, photos T. Furuichi).

CHAPTER NINETEEN

A Survey of Prosimians in the National Parks and Forest Reserves of Uganda

Lesley Ambrose

INTRODUCTION

Little information has been available on the behavior, ecology, distribution, and conservation status of nocturnal primates in Uganda. This paper provides new data on five prosimian species in the region based on surveys conducted between June 2000 and June 2001.

The principal aims of the study were to identify which species were present and to access their distributions and relative densities for conservation purposes. It was also aimed to obtain more information on *Galago matschiei*, particularly its loud call repertoire, as this species remains unstudied. This chapter provides information on the identification of each species (morphological descriptions and loud call repertoires) as well as habitat use (locomotion and use of forest strata).

Lesley Ambrose • Nocturnal Private Research Group School of Social Sciences and Law Oxford Brookes University Oxford, OX3 OBP

Primates of Western Uganda, edited by Nicholas E. Newton-Fisher, Hugh Notman, James D. Paterson, and Vernon Reynolds. Springer, New York, 2006.

The results are discussed in terms of the factors affecting the distribution and current conservation status of prosimians in Uganda. These include the specific habitat requirement of each species with regard to their ecology but also relates to habitat loss. Areas of concern are highlighted, particularly in the southeast of the country.

The sites surveyed are representative of the various forest types found in Uganda, which reflect diverse climatic and altitudinal factors (Howard, 1991). The mountains of the Western Rift are covered by medium- and high-altitude forest and include Bwindi and Kibale National Parks. This region contains a number of endemic primates such as *Gorilla beringei*, *Procolobus badius tephrosceles*, and *Colobus angolensis ruwenzori* (Oates, 1996a). In the extreme southwest lies Mgahinga National Park, which is a small fraction of the Virunga Volcanic Region extending into D.R. Congo and Rwanda. The central part of the country is relatively flat, reaching elevations of no more than 1500 m, and contains relic forests such as Mabira and Sango Bay around Lake Victoria (Howard, 1991). In the extreme east a line of ancient volcanoes runs along the Kenyan border and includes Mount Elgon (Howard, 1991).

Previous surveys of nocturnal primates were conducted by Bearder and Harcourt (1989) at Semliki (20 survey hours), where *Galagoides thomasi* and *G. demidoff* were identified, and at Kibale (64 h) and Bwindi (15 h), where *G. thomasi* was found to occur sympatrically with *Galago matschiei*. Other records include Budongo Forest, where *G. thomasi* and *Perodicticus potto* (J. Kingdon, personal communication, V. Reynolds, personal communication) were identified. There is also a possible sighting of *G. matschiei* (J. Oates, personal communication) in this forest.

METHODOLOGY

The study took place from June to September 2000 (occasional surveys) and February to June 2001. Fourteen sites were visited over 83 survey nights with a total of 308 survey hours. Forest paths were walked systematically and the vegetation searched with Petzl zoom (4.5 volt) head torches to detect eyeshine. More detailed observations were facilitated by Maglite (6 volt) torches and Bushnell 8 × 42 Natureview binoculars. Galagos were also identified by their loud calls. Recordings were made using a Sony Professional tape recorder (frequency response 40 Hz–15 kHz) and Sennheiser MKE 300 directional microphone (150 Hz–17 kHz).

Ad libitum behavioral observations were made relating to approximate height in the forest strata on first encounter, locomotion, feeding, grooming, intraspecific interactions, and call context. Morphological descriptions were made relating to body size and pelage coloration. Records were also kept of other mammals encountered during surveys.

Chadonneret traps (Charles-Dominique & Bearder, 1979) were used at several sites where the length of survey made trapping a possibility. The traps were baited with fruit and placed at a height of approximately 2 m near sleeping trees or at locations where galagos were regularly observed.

Calls were displayed using Avisoft SASlab (Raimund Specht, Berlin). They are illustrated as a sonogram (frequencies in kilohertz against time in seconds), an oscillogram (amplitude against time in seconds), and as a sound spectrum (magnitude against frequency in kilohertz). The sampling frequency was 32,000 Hz and the resulting sonograms had an FFT length of 512 points and a frame size of 50%. A Hamming evaluation window was used with a bandwidth of 162 Hz. The frequency resolution was 63 Hz and the time resolution was 8 ms.

RESULTS

Galago matschiei

Distribution and Density

The Spectacled or Eastern needle-clawed galago was recorded at seven of the survey sites. It occurred from the Western Rift as far east as Mabira Forest near Jinja on the left bank of the Nile River and as far south as Echuya Forest near the Rwandan border. These galagos were also present in Mpanga Forest near Kampala and the small forest remnant of Zika near Entebbe, both of which are part of the Mpigi archipelago. At some locations (Maramagambo and Echuya), only one or two individuals were observed. The highest densities were recorded at Kibale (1.71 animals per hour, $n = 39$) and Bwindi (0.37 animals per hour, $n = 19$), but they occurred at fairly high densities at most locations. These galagos were sympatric with *Galagoides thomasi* at five sites (Kibale, Mabira, Bwindi, Mpanga, and Echuya) and with *G. demidoff* at three sites (Mabira, Bwindi, and Mpanga). This species was found in a broad range of forest types, including medium-altitude moist evergreen and tropical high forest. In parks and reserves with a broad altitudinal range, this species preferred the lower elevations of these forests.

Morphological Description and Loud Call Repertoire

Galago matschiei is a small galago (~200 g) with a dark brown dorsum and pale grey ventrum, the grey color extending up the sides of the face. Eye-rings are broad and black and a white nose stripe is often evident. The eyes are amber in color and the eye-shine by reflected torchlight is orange (more strongly orange in the Western Rift compared to sites in the east). The ears are black. The long tail is evenly furred and usually dark grey but sometimes brown. Tails are often held coiled when sitting and sometimes also when running. A mouse-sized juvenile had full adult coloration.

The principal loud call is a churr, a low frequency (<5 kHz) harmonic call that carries for some distance and is usually highly repeated. The context of this call is mate attraction and it comprised 82% of the vocal repertoire. Countercalling was recorded only at Kibale National Park (NP), where long sequences continued for several hours. Staccato yaps are given in the context of alarm. These have a broader frequency range (<10 kHz) and are given in short, rapid bursts of three to four units (phrase), with varying intervals between the phrases. Grunting sounds caused by rapid inhalation are heard at low levels of arousal. An unidentified grunt-chatter recorded in Kibale NP may be attributed to this species.

Behavior and Habitat Use

G. matschiei used the whole strata of the forest up to about 15 m, often moving rapidly between the understorey and the canopy (Figure 1). These galagos were frequently unconcerned at the presence of the observers and grooming was recorded on several occasions, the animals employing both the toothcomb and grooming claws. Several individuals were observed to chase and catch insects, and one appeared to be feeding on large red flowers in the canopy, running rapidly from one to the other. One observation was made of a galago entering a tree hollow. In Mabira Forest, a large, old *Prunus africanus* that was covered with numerous lianas was used as a sleeping site by a female and juvenile.

Galago senegalensis

Distribution and Density

The Senegal galago was identified at only three of the study sites. Single individuals were recorded at the forest margins at Pabidi and Mpanga, revealing a broad

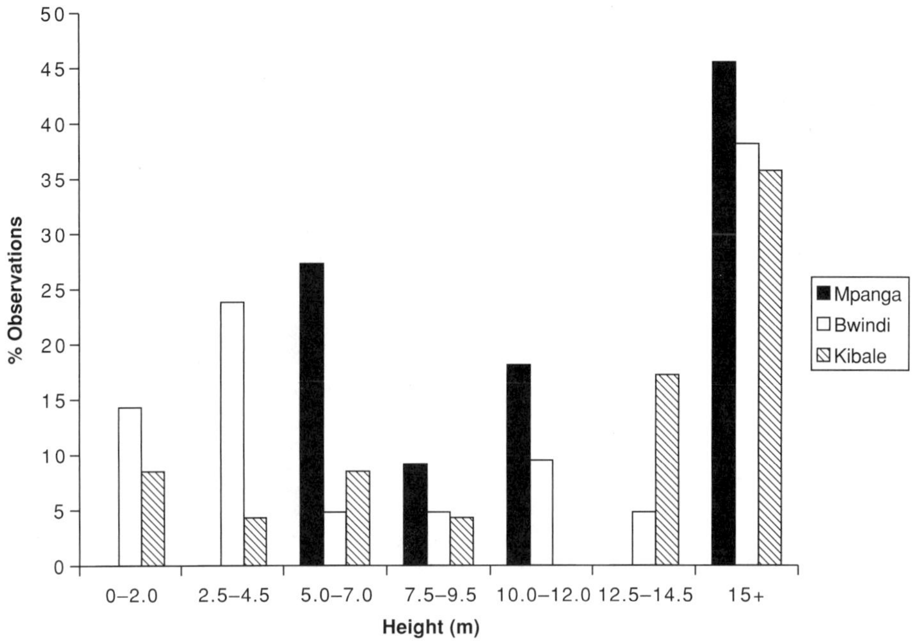

Figure 1. Use of forest strata by *Galago matschiei* in three Ugandan forests.

distribution across the country. Only on Mount Elgon were they recorded inside the forest, at elevations of 2600–2800 m. Here they were observed mainly in one part of the forest, where they occurred at densities of 0.67 animals per hour ($n = 41$). This species was parapatric with the other four prosimians at Mpanga and with *G. thomasi* at Pabidi.

Morphological Description and Loud Call Repertoire

G. senegalensis is a small galago (~200 g) with a grey dorsum and lighter grey ventrum. There is a pale nose stripe and pale grey around the snout and sides of the face. A yellow band may be observed along the flanks in more mature individuals, and this yellow coloration may become more extensive with age. The tail is darker grey and appears bushier toward the distal end.

The contact call is a soft honking sound reminiscent of a goose. This harmonic call is highly repeated, with regular intervals between the units. There is a broad frequency range, but the higher frequencies are lost with distance. This was the most frequently recorded call on Mount Elgon, comprising 95% of the loud call repertoire. Yaps varying in speed and intensity were recorded in the context of alarm.

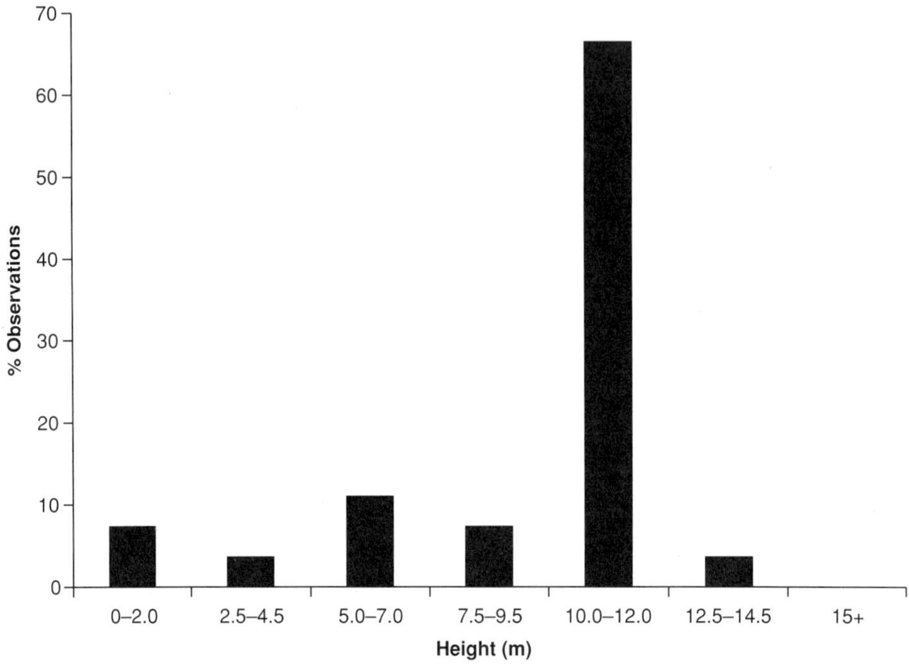

Figure 2. Use of forest strata by *Galago senegalensis* in Mount Elgon forest.

Behavior and Habitat Use

These galagos used the whole strata of the forest up to about 15 m, but the majority of observations were at 10–12 m (Figure 2). They rarely descended to the ground, but were observed on occasions to hop along branches on their hind limbs as in normal terrestrial locomotion. Associations of two to four individuals were observed on Mount Elgon, usually females with up to two offspring and in one instance a more peripheral individual was probably a male.

Galagoides thomasi

Distribution and Density

G. thomasi was the most widely distributed species occurring from Rubongo in the Murchison Falls area, as far south as Echuya on the Rwandan border and as far east as Mabira on the left bank of the Nile River. It was identified in seven locations and occurred at quite high densities at most sites (0.66

animals per hour at Kibale, $n = 15$). Only at Echuya and Bwindi were they recorded at low densities (0.1, $n = 2$, $n = 5$). This species was found sympatrically with *Galago matschiei* at five sites, with *Galagoides demidoff* at three sites and *Perodicticus potto* at four sites. It occurred in a broad range of forest types, including medium-altitude moist semideciduous and evergreen forest. At sites with a broad altitudinal range, it preferred the higher elevations of these forests.

Morphological Description and Loud Call Repertoire

A dwarf galago (~100 g) with a mid- to light-brown dorsum and slightly lighter brown ventrum. The eye-rings are variable, in some indistinct and in others broad giving the impression of a dark face mask. There is often a pale nose stripe. The tail is long but not bushy and the same color as the dorsum or slightly darker brown.

The contact call is a short (3–4 s) crescendo, which starts slowly but then rapidly increases in speed and pitch. This was the most commonly heard call type, comprising 60% of the loud call repertoire. Crescendos are usually given in sequences and often result in countercalling. These are heard at dusk as the animals leave their sleeping sites and at intervals throughout the night. Trills are used for gathering at dawn. This species was highly vocal in Mabira Forest, where periods of alarm calling lasted up to 30 min. Rapid yaps, wail, and shriek yaps were given in this context. Grunt yaps were heard at lower levels of arousal.

Behavior and Habitat Use

The majority of observations were above 10 m in the forest strata, but in Mabira Forest they frequently descended, sometimes as low as 3 m (Figure 3). These galagos moved around rapidly in the canopy and were difficult to observe.

Galagoides demidoff

Distribution and Density

Demidoff's galago was identified at three of the study sites: Bwindi in the Western Rift and Mabira and Mpanga Forest Reserves near Lake Victoria.

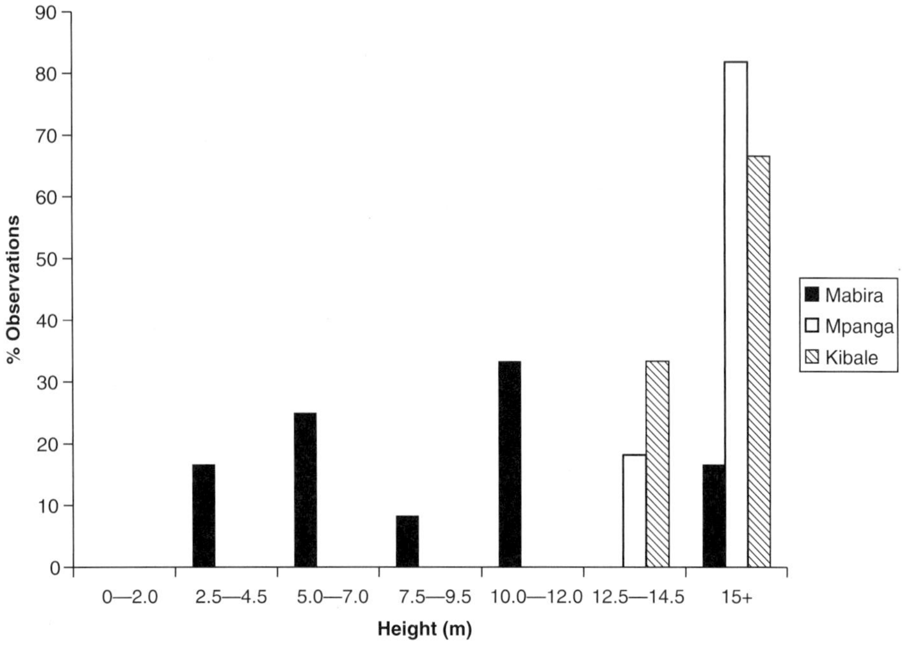

Figure 3. Use of forest strata by *Galagoides thomasi* in three Ugandan forests.

The highest encounter rate was at Mabira (0.57 animals per hour, $n = 26$). These galagos were rarely observed at Mpanga (0.16 animals per hour, $n = 5$). This species occurred sympatrically with *Galago matschiei*, *Galagoides thomasi*, and *Perodicticus potto* at all three sites. It was located in different forest types, including tropical high forest and medium-altitude moist evergreen forest. At Bwindi it was recorded up to elevations of about 1600 m.

Morphological Description and Loud Call Repertoire

It is a mouse-sized animal (~60 g) with variable pelage, even within populations. The dorsum is usually mid- to dark brown and the ventrum whitish. Eye-rings are variable and may be noticed in some individuals. The small, pointed upturned snout has a prominent pale nose stripe. The tail is sparsely haired and varies in color from brown to red.

The contact call is a crescendo that lasts for 4–5 seconds and rapidly increases in speed, pitch, and volume. It gives the impression of a lower frequency call compared to that of *G. thomasi* and is given only once or twice. This was the

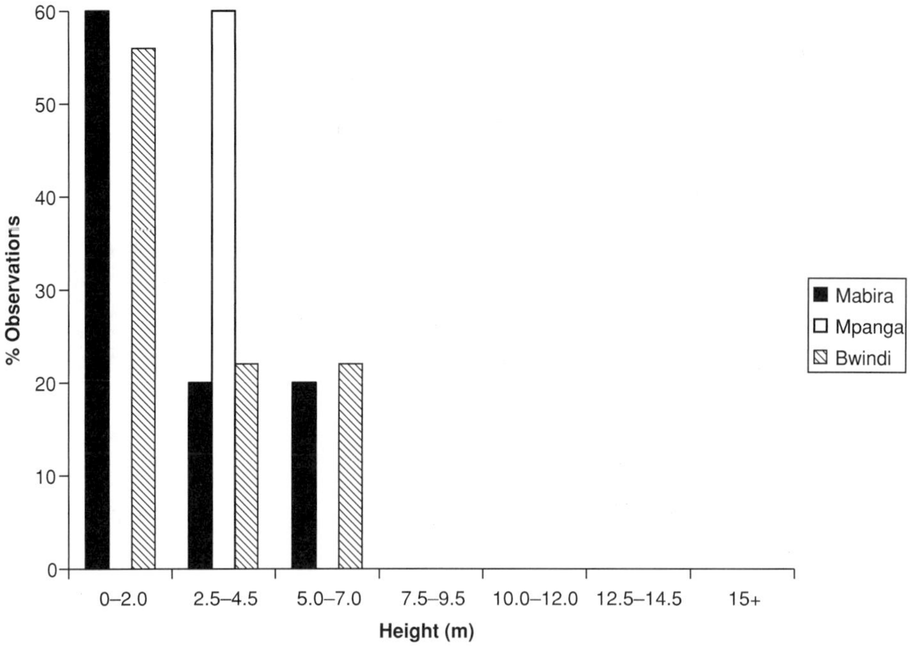

Figure 4. Use of forest strata by *Galagoides demidoff* in three Ugandan forests.

most frequently recorded call comprising 53% of the vocal profile. Rapid yaps in short phrases are given in the context of alarm and explosive buzzes for contact avoidance.

Behavior and Habitat Use

This species was encountered in the understorey, in dense tangles of vegetation below 7 m, with the majority of observations below 2 m (Figure 4). It made rapid darting movements along fine branches.

Perodicticus potto ibeanus

Distribution and Density

The potto was widely distributed throughout the region from Bwindi in the West as far north as Budongo Forest and extending across the River Nile into South Busoga Forest Reserve. It was identified at a total of seven study sites and occurred at low to moderate densities. The highest encounter rate was at Busingiro in Budongo Forest (0.48 animals per hour).

Morphological Description

Robust, medium-sized prosimian with a blunt snout, round face, and small ears. Two color types of potto were observed: a reddish brown dorsum with a slightly lighter brown ventrum, and a brown dorsum with a pale grey ventrum. Tail length was variable (~4–12 cm). One young individual was observed parked on a horizontal branch; it had adult coloration of the first type.

Behavior and Habitat Use

The majority of observations were above 10 m (Figure 5). In some forests such as Mpanga they were encountered as high as 20 m in the canopy. These animals moved slowly relying on crypsis, but if encountered at low elevations would ascend rapidly. At South Busoga they were observed to descend into low bushes to cross to a nearby tree. They were also encountered in exotic species such as *Eucalyptus* in this forest.

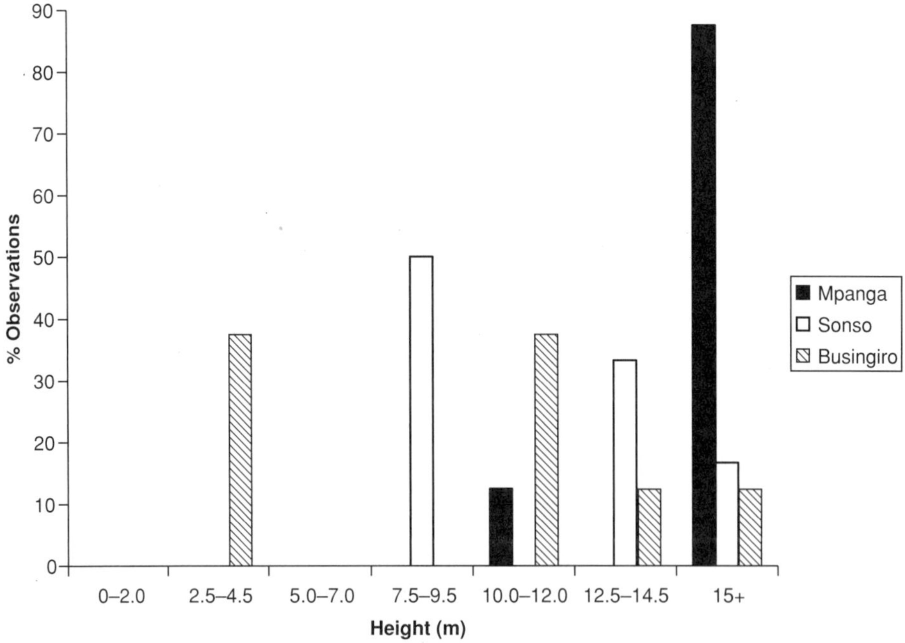

Figure 5. Use of forest strata by *Perodicticus potto ibeanus* in three Ugandan forests.

DISCUSSION

An overall assessment of distribution patterns will be followed by a more detailed examination of the distribution of each species in the context of behavior, habitat use, and food requirements. Conservation issues will then be discussed, particularly in regard to habitat loss and other pressures affecting prosimian densities and distributions.

Patterns of Distribution

The results reveal a range extension for *Galago matschiei*, *Galagoides thomasi*, and *G. demidoff*. All three species had formerly been identified in the Western Rift. *G. demidoff* was known to occur only in Semliki NP, while *Galago matschiei* and *Galagoides thomasi* were previously identified in Kibale and Bwindi NPs (Bearder & Harcourt, 1989). The forests of Uganda would once have been contiguous and there are no major faunal boundaries between the locations where these galagos were located. These distributions should not, therefore, be unexpected. The eastern extent of all three species is Mabira Forest on the left bank of the River Nile. It has not been possible to determine whether their range extends further eastwards. Apart from Mount Elgon, South Busoga Forest Reserve was the only site visited across the Nile and its apparent recent destruction is consistent with the destruction of other forests in eastern Uganda and western Kenya; little natural habitat remains in this region. *Galagoides thomasi* and *Galagoides demidoff* are two of the most widely distributed of all galago species, and major rivers such as the Niger, Sanaga, and Congo have not proved a barrier to their dispersal. It is therefore quite possible that these dwarf species may have extended further eastwards. The major factor limiting their dispersal would be the cessation of suitable forest habitat. It is not known if the range of *Galago matschiei* extends beyond the River Nile.

Galago matschiei

This species was found at all the sites visited in the Western Rift as well as Mpanga and Zika Forests, which are part of the Mpigi archipelago. There is at least one unconfirmed report of *G. matschiei* in Budongo Forest (J. Oates, personal communication) and an unidentified call recorded there during this study may have belonged to this species. It may well be present at extremely low densities. It is thought to be convergent with the needle-clawed galagos of

West-Central Africa. Its keeled nails terminate in claw-like points that enable it to move on broad supports. The diet comprises gums, probably from a limited number of tree and liana species, as well as insects and fruit. According to Kingdon (1971, 1997), gum is a dry season food while caterpillars and beetles are eaten in the rainy season. The presence of suitable food trees will therefore affect its distribution and density. In Kibale Forest, for example, there are a number of suitable gum-producing tree species such as *Albizia* (C. Turinawe, personal communication), which probably account for the higher densities. This species was not highly vocal and countercalling between familiar animals was not generally heard. The exception being in Kibale NP, where prolonged periods of contact calling may have been associated with mating activity. Locating and identifying this species was, therefore, more difficult. One new call was recorded but it could not positively be attributed to *G. matschiei*.

Galagoides thomasi

G. thomasi is present at Pabidi, a northeastern extension of Budongo Forest, and extends as far as the isolated forest of Rubongo in Murchison Falls National Park. This forest was once connected along the watercourses to the main Budongo Forest block. Both of these locations are within the loop of the River Nile. Two ambiguous records of galagos were made around the Sonso area in the southern part of Budongo Forest. Earlier reports have recorded *G. thomasi* at this location (J. Kingdon, personal communication). Its presence in the main part of the forest is quite likely, especially as it was identified at Pabidi. The fact that many survey hours at two sites in the reserve resulted in very little data, however, indicates that it now occurs at extremely low densities. An earlier report (A. Walker, personal communication) places *G. thomasi* as far south as Zika Forest near Entebbe. Its presence was not confirmed by the short survey conducted in this forest. *G. thomasi* prefers primary forest and in the Western Rift occurred at the higher altitudinal elevations of these forests. This is one of the most widely distributed and adaptable galago species and it seems to occur across Uganda wherever suitable forest remains.

Galagoides demidoff

G. demidoff was found at only three of the study sites and this is almost certainly due to lack of suitable habitat. Demidoff's galago occupies the fine-branch niche in the understorey and favors secondary growth, edge habitat, and river banks. This species is said to occur in Kibale NP (Weisenseel *et al.*, 1993), and particular attention was paid to grassland and roadsides at Kanyanchu during this survey. There were, however, no low dense tangles of vegetation normally associated with this species and it was not identified at this site. It may, of course, be present in other parts of the park. It occurred at quite high densities in Mabira Forest and certain parts of Bwindi NP but at lower densities in Mpanga Forest. At this latter site, the conditions were dark and no calling was heard for this species, the results might therefore be misleading. There is a possible identification in Minziro Forest in northern Tanzania, which is contiguous with Sango Bay, but it was not found at this location during the study. Bearder and Harcourt (1989) recorded this species in Semliki National Park.

Galago senegalensis

While the forest galagos occupy the main forest blocks it appears that the Senegal galago occurs on the forest margins and more marginal habitat. One was observed at the edge of Mpanga Forest and another near the forest edge at Pabidi in the Murchison Falls area. This species is known to occur around the shores of Lake Victoria. Its presence on Mount Elgon was previously unknown, although it was known to occur around the lower margins (Kingdon, 1997). Extensive surveys in the park revealed no other galago species. This almost certainly represents an example of competitive exclusion, *G. senegalensis* normally being out-competed by other galagos in forest habitat, and it has also been reported inside Mau Forest in Kenya (T. Butynski, personal communication). This species is normally associated with dry savannah woodland at lower elevations, where it may be frequently observed on the ground. It is a widely distributed species, ranging from Senegal to Kenya and Tanzania, and is therefore highly adaptable. It was found on Mount Elgon only in one main area of the forest between 2500 and 2700 m. The understorey was dense and the galagos remained almost exclusively in the canopy. *G. senegalensis braccatus* is the subspecies supposed to be represented in Uganda, but this short study was unable to confirm its subspecific status.

Perodicticus potto ibeanus

P. potto was found in the west in Kibale and Bwindi NPs and further east it occurred at quite high densities in Mpanga and Mabira Forest Reserves. Two color types were clearly observed at Busingiro and at several other locations. Hollister (1920, reported in Hill, 1953) suggests that this subspecies might be sexually dimorphic, with the males redder and the females greyer. This has not been reported in other populations of potto from West and Central Africa. This subspecies of potto may later be found to represent a distinct species. It was also identified at South Busoga on the right bank of the Nile River. This forest was largely destroyed and without habitat there appears to be no future for it at this location. *P. potto* occurs at least as far east as Kakamega Forest in Kenya (one night survey conducted by author in August 2000). Museum specimens exist from Mount Kenya (T. Butynski, personal communication).

Conservation

Many galago species are highly adaptable and are found in most forests in sub-Saharan Africa. Their presence or absence is therefore a good indicator of the health of a particular habitat. Galagos and pottos are not subject to hunting pressure in Uganda. The factors that affect their survival are habitat loss and disturbance. High and inaccessible areas such as those found in the Western Rift would appear to be the most secure but much lowland forest has been removed. In South Busoga Forest Reserve, for example, no natural forest could be found during the study period, some of it having been removed in the 6 months prior to the survey. This forest has been felled to make way for gardens and some of the area is being extensively planted with exotic species by the Saudi Arabian Marble Company. Pottos in this forest were found in isolated old forest trees as well as in exotic species such as *Eucalyptus*. Without suitable habitat, it is difficult to see how they can survive at this location. No galago species were recorded here. Southeastern Uganda is of particular concern for habitat loss; a similar destruction of forests has occurred in the highlands of Kenya. Museum specimens exist of galagos from southwestern Kenya, where no natural forest is found today. These include the silver galago *Otolemur monteiri*, one of the greater galago group. Other small, unidentified galagos from this region occur in collections; it is not known what may have been lost from this region.

Much illegal activity appears to continue in forest reserves such as pit-sawing and the laying of snares, which was witnessed in Budongo Forest Reserve. The virtual absence of galagos in the main part of this forest is highly unusual; the forest habitat appears suitable. The area has been subject to large-scale logging in the past but *Galagoides thomasi* has been identified in such forests in Gabon (Wickings *et al.*, 1995) and in Cameroon (Ambrose, 1999). The use of arboricides might have adversely affected smaller mammal species. *G. thomasi* is a highly adaptable species found widely across West and Central Africa; therefore, the low densities in Budongo Forest are difficult to explain.

Most of the sites surveyed had healthy populations of galagos, the highest densities being found in Kibale and Bwindi NPs in the west and Mabira and Mpanga Reserves near the Nile River and Lake Victoria. Mabira Forest is particularly important as it appears to mark the eastern extent of a number of West-Central African mammal species. Apart from the prosimian species mentioned, the Western tree hyrax *Dendrohyrax dorsalis* and the hammerhead fruit bat *Hypsignathus monstrosus* were also identified. All prosimian species appear secure in the region but their distribution is inevitably fragmented.

CONCLUSIONS

Species composition and distributions are now more clearly understood in an area that is transitional between the moist tropical forests of West-Central Africa and the drier regions to the east. *Galago matschiei* is the only known galago species to be endemic to this region. The three forest galagos and potto occur in suitable forest locations as far as the Nile River. *G. senegalensis* appears to be widely distributed in more marginal habitat; it also occurs in the north of the country (Kingdon, 1997). The actual presence or absence of each prosimian species is related to its ecology, particularly with regard to food and habitat use. Most forests support *Galago matschiei* and *Galagoides thomasi*, but *Galagoides demidoff* is restricted to forests with a suitable understorey. The major factors affecting their distribution are disturbance and habitat loss, which is particularly evident in lowland forests. None of these species is considered at risk of extinction in Uganda.

ACKNOWLEDGMENTS

This research was funded by the Margot Marsh Biodiversity Foundation through Conservation International and by the Nocturnal Primate Research Group, Oxford Brookes University. I would like to thank the Uganda National Council for Science and Technology, and the Uganda Wildlife Authority and the Forest Department for giving me permission to conduct research.

CHAPTER TWENTY

The Diets, Preferences, and Overlap of the Primate Community in the Budongo Forest Reserve, Uganda

Effects of Logging on Primate Diets

Andrew J. Plumptre

INTRODUCTION

The Budongo Forest Reserve in western Uganda forms the northernmost forest of the forests of the Albertine rift (Howard, 1991). Six species of diurnal primate are known to inhabit this forest: eastern chimpanzee *Pan troglodytes schweinfurthii*, black-and-white colobus *Colobus guereza*, redtail monkey *Cercopithecus ascanius*, blue monkey *Cercopithecus mitis*, olive baboon *Papio anubis*, and the vervet monkey *Cercopithecus aethiops*. The latter two species are generally found

---

Andrew J. Plumptre • Wildlife Conservation Society, Plot 802, Mitala Road, Kansanga, P.O. Box 7487, Kampala, Uganda.

Primates of Western Uganda, edited by Nicholas E. Newton-Fisher, Hugh Notman, James D. Paterson, and Vernon Reynolds. Springer, New York, 2006.

around the edges of the forest and did not form part of this study. Budongo has been Uganda's main forest for timber production and has been logged on a sustainable yield basis since the 1920s up to the present day although there has been much illegal logging in recent years. Research has shown that the forest management practices have changed the forest composition from a forest type dominated by *Cynometra alexandri* to a more mixed forest type where the diversity of tree species is higher (Plumptre, 1996). Plumptre and Reynolds (1994) showed that the densities of the *C. mitis*, *C. ascanius*, and *C. guereza* were significantly higher in logged compartments in the forest while there was no significant difference in the density of *P. troglodytes* between logged and unlogged compartments. This study also showed that *C. mitis* and *C. ascanius* were more abundant in mixed forest. At the time it was hypothesized that the increase in density could be attributed to the greater tree species diversity in the logged areas, particularly of trees bearing fruits edible for primates but the data at the time were not available to test this. Since this paper was produced, data were collected on the diets and food preferences of the primates in Budongo Forest. This hypothesis that variations in primate abundance in Budongo are due to dietary requirements is tested here using analyses of primate diets in different areas of the forest and measures of fruit production.

METHODOLOGY

Study Area

The Budongo Forest Reserve (latitude: 10°37′–20°00′ N; longitude: 31°22′–31°46′ E) covers 793 km^2 of moist semideciduous forest and grassland, 428 km^2 of which is forested (Howard, 1991). The Reserve lies at a mean altitude of 1100 m and is divided into 70 compartments for management purposes. Eight of these compartments were chosen as study sites, two unlogged (N15, K11–13) and six logged (B4, N3, N11, W21, B1, K4) at approximately 10-year intervals since 1950 (for map see Plumptre & Reynolds, 1994). Mean annual rainfall is 1392 mm (1993–1996) and the climate is characterized by a dry season of about 3 months from mid-December to mid-March (Newton-Fisher, 1999a,b,c). Temperatures vary little during the year with the highest maximum values, and lowest minimum values occur in the dry season. Temperatures rarely fall below 12°C at night.

Tree Species Enumeration and Fruit Production

During 1992, five 2-km transects were established in a stratified random manner (Plumptre & Reynolds, 1994) in each of eight compartments (B4, B1, N15, N3, N11, W21, K4, and K11–13). Every 50 m along these transects, all tree species of over 10 cm diameter at breast height (DBH) were identified in 7-m-radius circular plots (Plumptre, 1996) and their DBH at 1.3 m measured following Alder and Synnott (1992). Each plot was assigned one of five forest types: *Cynometra*, *Cynometra*-mixed, mixed, colonizing, and swamp. Every second plot at 100-m intervals was marked, trees numbered, and the measurement points of each tree painted. This enumeration enabled us to calculate the sum of the DBH values for different groups of trees producing different fruit types as a measure of potential fruit availability (see below).

In N15 (unlogged) and N3 (logged in 1950), the painted trees were used to obtain measures of phenology. These measures were concentrated in two sites because of the costs of collecting in all eight sites. Measures of availability of fruit are always fraught with problems and ideally should be based on the biomass of fruits available on the tree. Chapman *et al.* (1992) showed that DBH was a good predictor of fruit abundance (biomass) for particular tree species, and the sum of DBH values across different species have been used as a measure of availability (Chapman & Chapman, 1999). Others have calculated availability as the density of fruiting trees in a habitat (e.g., Sourd & Gautier-Hion, 1986) and some have multiplied tree basal area by a score of fruit production similar to that given above (Dasilva, 1989; Fairgrieve, 1995). Given the finding by Chapman *et al.* (1992), we decided to calculate two measures of availability: (1) the "potential fruit availability" calculated as the sum of the DBH values of trees and (2) the product of the DBH and whether the tree was fruiting or not. The second measure allows a seasonal analysis of fruit availability. To multiply DBH by a fruit score would have overweighted those trees producing much fruit if the correlations found by Chapman *et al.* (1992) also hold for the species in Budongo. We then summed these two measures for trees producing similar fruit types (small and fleshy [<2.5 cm long], large and fleshy [>2.5 cm long], and pods). Although these measures also have weaknesses, they have the advantage that there are fewer problems with interobserver error as only an identification of whether the tree was fruiting is required.

Measures of Primate Diet

Feeding Scans from Transects

Five 2-km transects were cut in each of the eight compartments using a stratified random sampling procedure to enable the primates to be censused during 1992 (see Plumptre & Reynolds, 1994). From March 1993 to February 1994 the same transects were walked from 7.30 to 11.00 am. Whenever a primate was seen, the observers watched it (provided it did not react to the presence of the observer by fleeing) and recorded the first food item it was seen to feed upon. Items recorded were young leaves, mature leaves, ripe fruit, unripe fruit, flowers, bark, and arthropods. For groups of primates this involved scans of the group, recording food items of all individuals that could be seen until all of them had been sampled once or until the group spotted the observer and fled. This method will bias the estimation of diets toward food items that are more easily visible. On the other hand it allows many more groups to be sampled and a larger area to be covered so that diets obtained will include a wider range of tree species parts eaten in each compartment.

Preference for certain food types was measured with Manly's alpha (Krebs, 1989) and calculated for fruits and leaves for each tree species consumed. Manly's alpha measures preference taking into account availability (as measured by sum DBH). Rare species are almost always preferred using this index because their availability is low and therefore it is not very meaningful to consider them.

Dawn to Dusk Follows

In compartments N3 (logged) and N15 (unlogged), a total of 17 groups of monkeys were habituated sufficiently to follow them. In both compartments, three groups of *C. mitis* and *C. guereza* as well as three *C. ascanius* groups in N3 and two *C. ascanius* groups in N15 were habituated. Some of the *C. mitis* groups were those followed by Fairgrieve (1995). In N3 the habituation of a *P. troglodytes* community had begun in 1990 and by December 1994 it was sufficiently far advanced that individual chimpanzees could be identified and followed (Newton-Fisher, 1999a,b,c). Only data from complete dawn to dusk follows were used in analyses for the chimpanzee data as individuals were sometimes lost when they left feeding trees and it was felt that diets might be biased if these were included. Primate groups or individual *P. troglodytes* were located at about 16.00 and followed for a period of 48 h from dawn to dusk. Each group of monkeys was followed for 2 days each month from October 1994 to January 1996. Individual *P. troglodytes* were followed on 2 days each week by four observers between December 1994 and December 1996.

Scan samples at 30-min intervals were made for all individuals visible and the following data recorded:

Activity. If feeding, the plant species and part. With habituated monkeys, it was possible to identify parts in more detail than in the study from the transects and so fruit was separated into whole fruit, flesh and seeds (discarding pericarp), and seeds only (discarding surrounding tissue and pericarp).

Dietary Overlap. This was calculated at the two different scales between the same species at different sites and between species at the same site. Horn's overlap index was used as the measure of overlap because it has been shown to have lower bias than other measures (Krebs, 1989). Monte Carlo simulations were made generating random numbers for the dietary intake of plant species and parts (only for those species and parts known to be consumed by the primates) to assess whether the dietary overlap values calculated were significantly higher or lower than might be expected from random sampling. This form of neutral sampling keeps plant parts that are not eaten as zeros but varies the proportion of plant parts that are eaten 100 times to obtain the lowest and highest 5% overlap values. If the true value falls within the lowest or highest five values, it is considered to be significantly different from a random value.

RESULTS

Fruit Availability

It was found that the variation in tree species composition of the forest was greatest between the west and east of Budongo (Plumptre, 1996) so that comparing compartments K11–13 in the east with B4 and B1 in the west was hampered by this effect. This was why it was decided to combine tree species into groups based on their fruit characters (see Methods). Table 1 gives the sum DBH values for the different fruit types for each compartment and also the density of the primates. In most compartments, trees that produce small fleshy fruits were the most abundant.

The results of the phenology analysis for the three fruit categories in compartments N15 and N3 for ripe and unripe fruit showed that trees producing

Table 1. Measures of the potential availability of fruit (sum DBH values) for different fruit types in each compartment per hectare and the density of each primate in each compartment (No. per km^2)

Compartment	N15	K11–13	B4	N3	N11	W21	B1	K4
Year logged	Unlogged	Unlogged	1941–1942	1947–1952	1960	1963–1964	1981–86	1988–1996
Small fleshy fruits	6,060	3,770	4,960	6,440	5,370	4,960	4,560	4,740
Large fleshy fruits	4,330	5,440	4,390	5,120	4,880	3,680	3,580	4,100
Pods	3,810	3,570	3,190	3,740	4,230	3,070	2,600	4,170
Total	**14,200**	**12,780**	**12,540**	**15,300**	**14,480**	**11,710**	**10,700**	**13,010**
C. mitis	31.2	7.7	61.6	57.5	45.1	45.3	86.8	33.2
C. ascanius	12.7	8.9	54.4	38.0	44.3	39.5	63.6	30.6
C. guereza	25.8	26.7	37.1	51.7	46.6	48.9	34.0	49.9
P. troglodytes	0.9	3.0	0.5	1.4	1.5	1.6	1.0	0.9

Census data are from Plumptre and Reynolds (1994).

small fleshy fruits were the most abundant but that production could fluctuate greatly between months and years. This fluctuation between boom and bust periods could be important for primates, with species preferring areas where food supply is relatively constant. In addition the patchiness of food availability will be important (Newton Fisher *et al.*, 2000). If there are areas with lots of fruit and areas with little, then primates will need larger home ranges to find sufficient food over the year and density will consequently be lower. The coefficient of variation (standard deviation and mean) between phenology scores in plots varied over the 4 years from January 1993 to December 1996 but was higher for 91% of months in N15 compared with N3 for small fleshy fruits and 71% of the time for large fleshy fruits. This shows that for much of the time for small fleshy fruits and large fleshy fruits, the variation between plots was greater in N15, the unlogged compartment. Therefore the food supply was more patchily distributed here.

Diets from Transect Scans in Compartments

Diets obtained from the transect scans determined some of the main food plants that provide food for these primates. Primate densities were correlated (Spearman rank correlations) with the sum DBH (potential food available) of the 10 most commonly consumed tree species that were found in all compartments in the forest, but few significant correlations were found (Table 2). *Colobus guereza* was positively correlated with *Celtis durandii*, and *Celtis mitis* and *Cercopithecus ascanius* were negatively correlated with *Cynometra alexandri* (Table 2). *Pan troglodytes* density was only correlated with the sum DBH of trees producing large fleshy fruits. Given the large number of correlations (64) in Table 2, it would be expected that three significant correlations would be obtained by chance (1 in 20 tests at $P = 0.05$). With only eight points to correlate, reducing the probability level to compensate for the number of correlations eliminates almost all pattern because a correlation coefficient of 1.00 is required for $P = 0.001$. Consequently the $P = 0.05$ level is given here to show that some patterns exist and that future work could investigate these further.

Correlations were calculated between the density of the three monkey species and the percentage of pods and fleshy, ripe, and unripe fruit in the diets obtained from scans along transects in each compartment. For *C. ascanius* and *C. mitis* there were significant correlations between the percentage of fleshy fruit (+ve), ripe fruit (+ve), and unripe fruit (−ve) in the diet and primate density. For

Table 2. *R* values and significance of Spearman rank correlations between primate density and (a) sum DBH of tree species and those producing certain fruit types in compartments and (b) percentage scans of food items.

	C. mitis	*C. ascanius*	*C. guereza*	*P. troglodytes*
(a) Tree density				
Cynometra alexandri	−0.93**	−0.82*	−0.14 ns	0.59 ns
Maesopsis eminii	0.71*	0.36 ns	0.52 ns	−0.24 ns
Celtis durandii	0.33 ns	−0.07 ns	0.76*	0.10 ns
Celtis mildbraedii	−0.02 ns	−0.61 ns	0.17 ns	−0.02 ns
Celtis zenkeri	0.52 ns	0.61 ns	0.38 ns	−0.21 ns
Alstonia boonei	0.13 ns	−0.38 ns	0.83*	0.12 ns
Albizia spp.	0.80*	0.88**	0.07 ns	−0.18 ns
Funtumia elastica	0.43 ns	0.21 ns	0.74 ns	0.04 ns
Croton macrostachys	0.79*	0.46 ns	0.60 ns	−0.16 ns
Ficus spp.	0.68 ns	0.31 ns	0.26 ns	−0.35 ns
Small fleshy fruits	0.24 ns	−0.07 ns	0.21 ns	−0.05 ns
Large fleshy fruits	−0.31 ns	−0.25 ns	−0.12 ns	0.88**
Pods	0.57 ns	0.46 ns	0.62 ns	0.04 ns
(b) Dietary scans				
Ripe fruit	0.74*	0.86*	0.19 ns	NA
Unripe fruit	−0.90**	−0.93**	−0.24 ns	NA
Fleshy fruit	0.74*	0.82*	0.29 ns	NA
Pods	−0.88**	−0.54 ns	−0.26 ns	NA

* $P < 0.05$. ** $P < 0.01$. NA = not able to calculate.

C. mitis the correlation between density and pods consumed was also significant (−ve) at $P = 0.01$ (Table 2). There were no significant correlations for these measurements and *C. guereza* density. It was not possible to obtain a measure of *P. troglodytes* diets in the eight compartments because sightings were too few.

Table 3 summarizes the preferences for the three monkey species in each compartment for the most common trees (top 15 species in each compartment) where a preference was shown. This table shows the importance of *Celtis durandii* in the diets of all these three primates with a preference shown for its fruit (and leaves in the case of the *C. guereza*) in almost all compartments. This is despite the fact that it is an abundant tree species in most areas and availability values are high. It is interesting to note, however, that many species may be preferred in some compartments but not others.

Diets from Dawn to Dusk Follows

Figures 1 through 7 show the percentage of scans of different plant parts consumed by primates in N15 (unlogged) and N3 (logged) throughout the year.

Table 3. Dietary preferences for each monkey species in each compartment

Tree	Part	N15	K11–13	B4	N3	N11	W21	B1	K4
No. of scans		450	50	332	823	362	576	393	192
C. mitis									
Aningeria altissima	Fruit	+							
Alstonia boonei	Leaf							+	
Albizia spp.	Leaf		+	+	+		+	+	+
Bosqueia phoberos	Fruit				+			+	
Broussonetia papyrifera	Fruit				+				
	Leaf								
Chrysophylum albidum	Fruit					+	+		+
	Leaf							+	+
Celtis durandii	Fruit	+	+	+	+	+	+	+	+
Celtis zenkeri	Fruit								+
Cleistopholis patens	Fruit						+		
Cordia millenii	Fruit							+	
C. alexandri	Leaf				+			+	
Erythrophleum suaveolens	Fruit								
	Leaf	+			+			+	
Ficus exasperata	Fruit				+			+	
Ficus sur	Fruit					+			
Funtumia elastica	Fruit		+				+	+	+
Maesopsis eminii	Fruit							+	+
Morus lactea	Fruit								
	Leaf								
Tapura fischeri	Fruit							+	+
C. ascanius									
Aningeria altissima	Fruit								
Alstonia boonei	Leaf								
Albizia spp.	Leaf			+	+		+	+	
Bosqueia phoberos	Fruit				+		+		
Broussonetia papyrifera	Fruit				+				
	Leaf				+				
Chrysophylum albidum	Fruit		+	+		+	+	+	
	Leaf	+		+	+	+	+		
Celtis durandii	Fruit	+	+		+	+	+	+	
Celtis zenkeri	Fruit	+							
Cleistopholis patens	Fruit	+				+	+		
Cordia millenii	Fruit								
Cynometra alexandri	Fruit								
	Leaf	+			+				
Erythrophleum suaveolens	Fruit								+
	Leaf								+
Ficus exasperata	Fruit				+			+	
Ficus sur	Fruit				+	+			
Funtumia elastica	Fruit	+							
Maesopsis eminii	Fruit	+				+			+

(*Cont.*)

Table 3. *(Continued)*

Tree	Part	N15	K11–13	B4	N3	N11	W21	B1	K4
C. guereza									
Aningeria altissima	Fruit			+					
Alstonia boonei	Leaf	+	+	+	+	+	+	+	+
Albizia spp.	Leaf				+			+	
Bosqueia phoberos	Fruit			+					
Broussonetia papyrifera	Fruit								
	Leaf								
Chrysophylum albidum	Fruit		+		+		+	+	
	Leaf							+	
Celtis durandii	Fruit	+	+	+	+	+	+	+	+
	Leaf	+	+	+	+	+	+	+	+
Celtis zenkeri	Fruit								
	Leaf					+			
Cleistopholis patens	Fruit					+	+		
Cynometra alexandri	Fruit				+	+		+	
	Leaf			+	+	+	+	+	+
Erythrophleum suaveolens	Fruit		+					+	
	Leaf								+
Ficus exasperata	Fruit				+				
Ficus sur	Fruit				+			+	
Funtumia elastica	Fruit								
Holoptelea grandis	Leaf	+		+		+	+		
Maesopsis eminii	Fruit			+	+	+	+	+	
	Leaf	+		+	+				

Only those tree species which were relatively abundant in all compartments and where a preference was shown are listed. Preferences were calculated using Manly's alpha and the sum DBH values were used as a measure of availability. The values of alpha are not given because they are dependent on the number of food items and hence are not comparable between compartments. + = preference.

These show that for most months there was more ripe fruit and more fruit in general being eaten in N3. Table 4 lists the 10 most commonly eaten food items for each species in each compartment and gives the total number of items eaten. The top 10 items constitute between 43 and 55% of the total diet for the monkey species while for *P. troglodytes* they form 70.7% of the diet. In Table 5, preference calculations (Manly's alpha) are given for the same species as in Table 3. These two tables show good agreement, which gives support to the method used on the transects as a way of collecting data on diets over a larger area.

Colobines are usually considered to be folivores or seed eaters (Davies & Oates, 1994; Davies *et al.*, 1999), but in Budongo Forest fruit eating (including ripe, unripe, and seeds) could form over 50% of the diet, and averaged 25.6% in N15 (Figure 3) and 36.3% in N3 (Figure 6). In N3 the four most abundantly eaten food items by *C. guereza* were fruit (Table 4). Seeds formed about

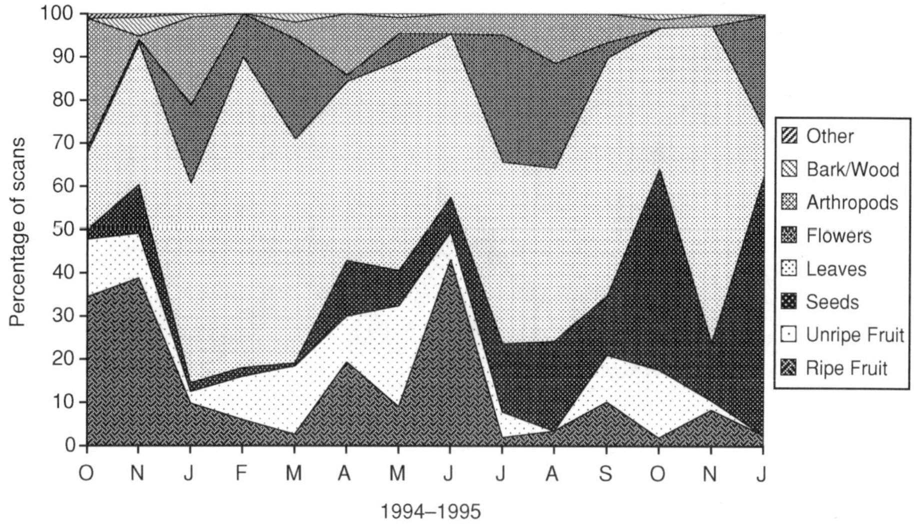

Figure 1. The diets of the *Cercopithecus ascanius* groups in compartment N15, unlogged. These diets are calculated from dawn-to-dusk follows of groups and averaged for the species.

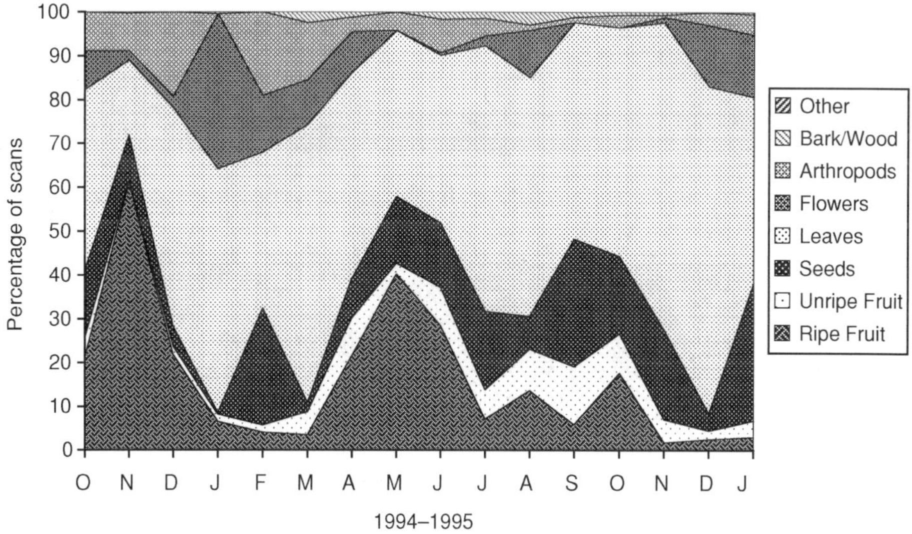

Figure 2. The diets of the *Cercopithecus mitis* groups in compartment N15, unlogged. These diets are calculated from dawn-to-dusk follows of groups and averaged for the species.

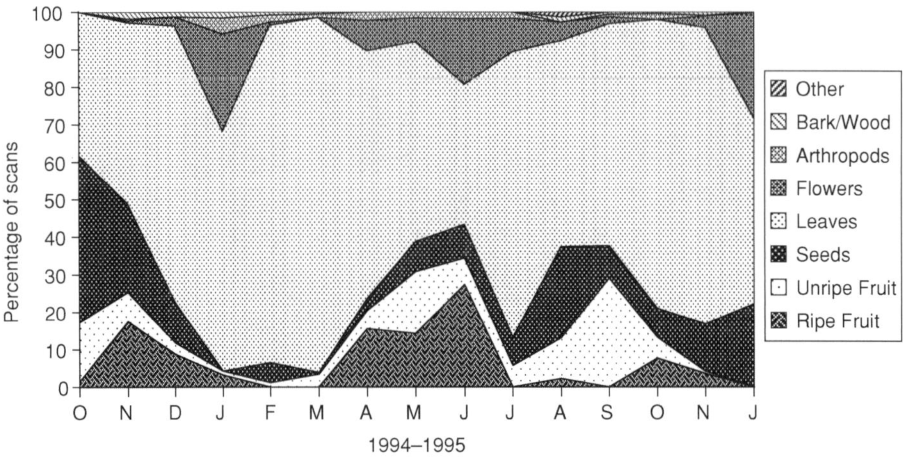

Figure 3. The diets of the *Colobus guereza* groups in compartment N15, unlogged. These diets are calculated from dawn-to-dusk follows of groups and averaged for the species.

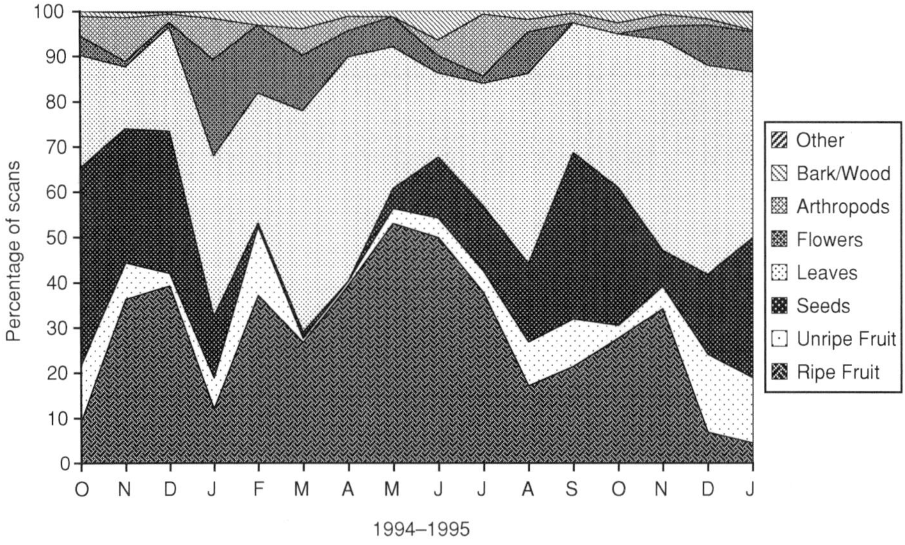

Figure 4. The diets of the *Cercopithecus ascanius* groups in compartment N3, logged. These diets are calculated from dawn-to-dusk follows of groups and averaged for the species.

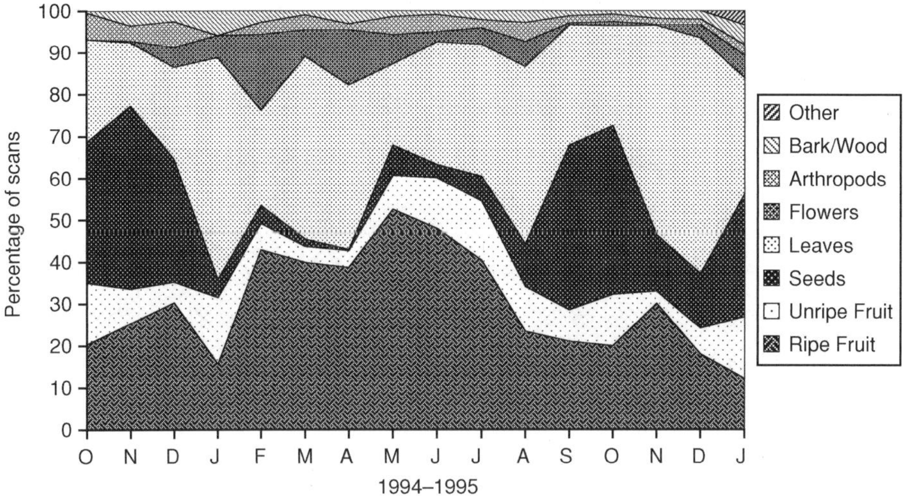

Figure 5. The diets of the *Cercopithecus mitis* groups in compartment N3, logged. These diets are calculated from dawn-to-dusk follows of groups and averaged for the species.

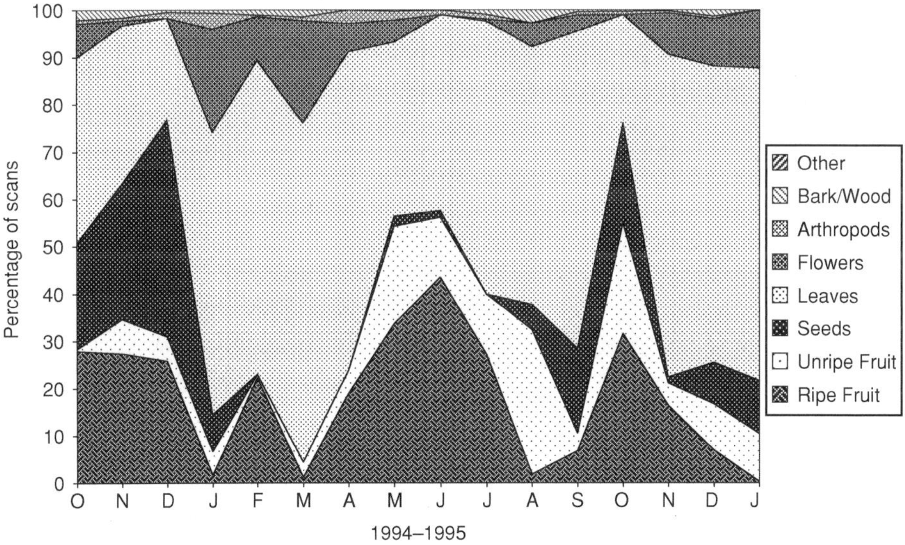

Figure 6. The diets of the *Colobus guereza* groups in compartment N3, logged. These diets are calculated from dawn-to-dusk follows of groups and averaged for the species.

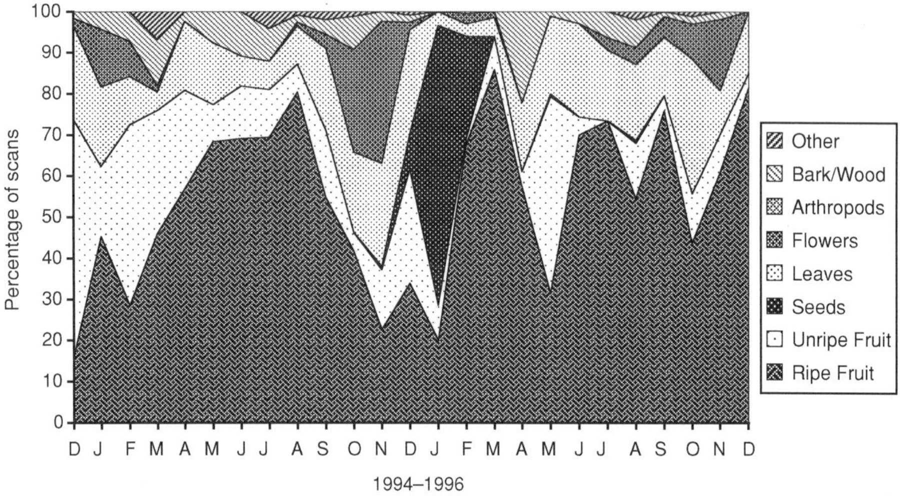

Figure 7. The diets of *Pan troglodytes* compartment N3, logged. A mean diet was calculated across the individuals followed.

30–50% of this fruit consumption (9.3% of the diet in N3 and 12.0% in N15). *Celtis durandii* is preferred by this primate for both fruit and leaves throughout Budongo (Tables 3 and 4) and forms a large percentage of its diet (31.4% in N3 and 14.6% in N15). This tree is abundant across Budongo and is in the top 10 trees by basal area for half of the compartments measured.

Dietary Overlap

Overlap Between the Same Species in Different Compartments

The overlap between diets of the same species in different compartments would have been low if based on species eaten because the tree species vary greatly between compartments. Consequently, overlap was calculated using the percentage intake of plant parts (buds, young leaves, mature leaves, ripe and unripe fruit, flowers, arthropods, and "other" [a category for minor dietary items such as fungi, resin, etc.]). The results showed that overlap was significantly higher than would be expected if these primates ate these plant parts at random for 78% of the 63 comparisons. All overlap values exceeded 89%, and 56 of the 63 comparisons were 95% or higher. Consequently the selection of plant parts was important for the diets of these monkeys. The compartments where overlap was lowest and not significantly different from random for *C. mitis* and *C. ascanius* (K4 and N15) are where these species were at lowest density. However, this

Table 4. The top 10 dietary items for each of the monkey species in N15 and N3 and for chimpanzees in N3

		N3				N15		
Species Scans	Part[a]	*C. mitis* 6,807	*C. ascanius* 6,004	*C. guereza* 4,910	*P. troglodytes* 4,318	*C. mitis* 4,394	*C. ascanius* 2,063	*C. guereza* 3,872
Maesopsis eminii	Rf	8.9	9.0	7.4	7.6	6.3	4.3	3.5
Celtis durandii	Rf	7.2	8.4	10.7	5.4	3.0	3.6	
Croton macrostachys	Rf	5.8	4.9					
Ficus sur	Rf	3.5	3.2		14.8			
Broussonetia sp.	Rf	3.7			9.1			
Ficus exasperata	Rf	2.6						
Celtis zenkeri	Rf					4.8		
Cynometra alexandri	Sd				4.2		6.0	
Celtis durandii	Ur	3.1	2.8	7.7			3.4	3.2
Celtis mildbraedii	Ur		2.7					
Bosqueia phoberos	Ur			7.4				
Ficus sur	Ur				7.6			
Alstonia boonei	Fl			2.6			3.5	3.7
Broussonetia sp.	Fl				4.8			
Celtis zenkeri	Yl		3.1			3.3	4.2	
Ficus sur	Yl		2.9					
Cynometra alexandri	Yl	2.7				11.1	7.7	6.4
Broussonetia sp.	Yl				8.9			
Celtis durandii	Yl			7.0				6.4
Alstonia boonei	Yl			3.7		3.4		7.8
Celtis mildbraedii	Yl				4.2	3.7	3.7	4.0
Celtis mildbraedii	Ml	3.0	2.6	4.0		4.6	6.2	7.3
Celtis durandii	Ml			6.0				5.0

(*Cont.*)

Table 4. *(Continued)*

		N3				N15		
Species Scans	Part[a]	*C. mitis* 6,807	*C. ascanius* 6,004	*C. guereza* 4,910	*P. troglodytes* 4,318	*C. mitis* 4,394	*C. ascanius* 2,063	*C. guereza* 3,872
Alstonia boonei	Ml			3.0				4.5
Cynometra alexandri	Ml					4.0		
Arthropods		2.9	3.8			5.8	6.5	
Khaya anthotheca	Bk				4.2			
1st 10 items (%)		43.4	43.4	59.5	70.8	50.0	49.1	51.8
No. items >1% of diet		33	33	27	19	22	23	30
Total number of items		152	128	112	98	152	116	111
Species of tree		58	49	47	62	63	48	48

The percentage of scans in which the species and part were observed being eaten are given for these 10 dietary items. Data were collected in dawn-to-dusk follows.
[a] Plant parts: rf = ripe fruit; ur = unripe fruit; fl = flower; yl = young leaf; ml = mature leaf, sd = seed, bk = bark.

Table 5. Manly's alpha preference values calculated for the same tree species as in Table 3 with a few others preferred by chimpanzees for the dawn-to-dusk follows data

		N3				N15		
Species	Part	*C. mitis*	*C. ascanius*	*C. guereza*	*P. trogodytes*	*C. mitis*	*C. ascanius*	*C. guereza*
Aningeria altissima	F							
Albizia spp.	L	+	+	+		+	+	+
Bosqueia phoberos	F	+	+	+				+
Broussonetia papyrifera	F	+	+	+	+			
	L	+	+	+	+			
Chrysophylum albidum	F							
	L	+					+	
Celtis durandii	F	+	+	+	+	+	+	+
	L		+	+		+	+	+
Celtis zenkeri	F	+	+			+	+	+
	L	+	+			+	+	+
Cleistopholis	F					+	+	
Cynometra alexandri	F				+		+	
	L	+	+	+		+	+	+
Erythrophleum suaveolens	F	+		+		+		+
	L	+	+	+		+	+	+
Ficus exasperata	F	+	+	+	+			
Ficus sur	F	+	+	+	+			
Ficus mucuso	F				+			
Funtumia elastica	F						+	
Holoptelea grandis	L	+	+	+		+	+	+
Maesopsis eminii	F	+	+	+	+	+	+	+
	L	+	+	+		+	+	+
Myrianthus holstii	F				+			

Part eaten is either fruit (F) or leaf (L); + = preference.

pattern did not hold for *C. guereza*, which had low overlap values in these compartments too but was at a relatively high density in K4.

Overlap Between Different Species in the Same Compartments

To compare diets between species in the same compartment, more detailed measures of the diets based on tree species and plant part were used (e.g., *Celtis durandii*, ripe fruit would differ from *C. durandii* flower) using transect scan data. Results showed that overlap between *C. mitis* and the other two species was always significantly higher than would be expected from neutral models (*C. mitis* vs. *C. ascanius*: 65–84% overlap; *C. mitis* vs. *C. guereza*: 43–66% overlap) but that overlap between *C. ascanius* and *C. guereza* (34–68% overlap) was only significantly higher about 50% of the time (in compartments K4, N3, and W21).

Overlap Between Adjacent Groups of the Same Species

The overlap in diet (plant parts of different tree species) between adjacent groups of the same species within a compartment were all higher than would be expected from random neutral models (*C. mitis*: 73–81% overlap; *C. ascanius*: 74–80%; *C. guereza*: 76–82%). This means that not only were similar tree species eaten but also similar parts of the species in similar proportions.

Overlap Between Species Whose Home Ranges Overlap Extensively

The dawn-to-dusk follows of the monkey groups allowed the calculation of home ranges of the various groups followed. Interestingly, in both N15 and N3, home range overlapped between species extensively, such that there was approximately one group of each of the three species for each region of the forest. For groups that occupied approximately the same home range, overlap was calculated on the basis of plant parts of each tree species (Plumptre *et al.*, 1997). Results show that all overlap values were significantly higher than expected (*C. mitis* vs. *C. ascanius*: 75–90%; *C. mitis* vs. *C. guereza*: 70–73%; *C. ascanius* vs. *C. guereza*: 69–76%). It is interesting to note that overlap between *C. mitis* and *C. ascanius* in compartment N3 was higher than values between the same species in adjacent groups.

Table 6. Dietary overlap (Horn's Index) based on detailed follows of primate groups but averaged for compartments N3 and N15

Species pairs	N3	N15
C. mitis vs. *C. ascanius*	**0.94**	**0.87**
C. mitis vs. *C. guereza*	**0.78**	**0.80**
C. ascanius vs. *C. guereza*	**0.79**	**0.78**
C. mitis vs. *P. troglodytes*	**0.50**	
C. ascanius vs. *P. troglodytes*	**0.45**	
C. guereza vs. *P. troglodytes*	**0.35**	

Overlap is calculated on the percentage intake of plant parts of different tree species, as well as arthropods, and other items. Where a value is higher than expected from random Monte Carlo analyses, it is marked in bold.

Overlap Between Species for Average Diets in a Compartment

This measure of overlap is similar to the measure calculated above but is based on the dawn-to-dusk follows of the monkeys rather than opportunistic scans from walking transects. It also allows a calculation of dietary overlap between the monkey species and chimpanzees (Table 6). All species pairings were significantly higher than expected, indicating that, despite very different diets in the case of chimpanzees (low overlap values), there was still selection for similar plant species and parts where they both did consume the same items.

DISCUSSION

Diets of the Budongo Primates

This study investigated the primate diets in Budongo at two different scales; one a coarser scale overview of the diets of three monkeys in eight compartments of the forest and the other, a more detailed study of the diets in two of these compartments including the diet of *Pan troglodytes*. Diets measured at both scales were within the same ball-park values, but the scans of unhabituated monkeys were very variable, depending on which site was considered. The preferences for dietary items were similar at both scales for the three monkeys (Tables 3 and 5). It appears therefore that diets observed from transects of many different groups can provide a reasonable measure of the dietary preference over a larger area than can be obtained by observing habituated groups. However, the diets were not as accurate and tended to miss certain food items such as arthropods.

Table 7 shows the relative proportions of the diets of these monkeys in comparison with the same species at other sites. This table shows that *C. mitis* and *C. ascanius* during this study ate more seeds than in other sites although Fairgrieve (1995) did not find this in his study in 1992–1993. The seeds that were primarily eaten in Budongo are those of *Funtumia elastica* (unripe), *Croton macrostachys* (ripe), *Celtis durandii* (ripe), *Bosqueia phoberos* (ripe and unripe), *Maesopsis eminii* (unripe), and *C. alexandri* (ripe). *C. alexandri* was very seasonal and develops unripe fruits between November and December, which ripen in January, which is when the large seed intake for *P. troglodytes* occurs (Figure 7). *F. elastica*, however, fruited throughout most of the year, and the unripe pods were eaten for the immature seeds and their plumes. This species was not usually a preferred food item though and may act as a fall-back food when nothing else can be found. Ripe and unripe whole fruit of the other tree species were consumed as well as the seeds alone and when the seeds are eaten they are squeezed out of the fleshy pulp into the mouth and the pulp is dropped to the forest floor. For some species, such as *C. durandii*, it is possible that these monkeys were storing seeds in cheek pouches to spit out later, as has been closely observed in *C. ascanius* in Kibale forest (Lambert, 1999), and that some of the pulp was dropped as part of this process. This would have made it appear that the pulp was being rejected in favor of the seed when it was not in fact the case. This is why Table 7 lists seeds as a subset of the fruit diet. Arthropod intake seems to be lower in Budongo in comparison with Kibale and Kakamega for these two monkeys, but for *C. guereza* it is slightly higher. Fruit intake is lower in Budongo if seed intake is partitioned out for the two cercopithecines, but if seed intake is included then the fruit intake is more similar to other studies. Fruit intake for *C. guereza* in logged forest was higher but the two unlogged forest compartments in Kibale and Budongo were about the same.

Fruit Availability

Measures of fruit availability in tropical rainforests are fraught with problems. Most use observations of certain tree species and usually pick 5–20 individuals to monitor. Then the density of these trees is calculated and multiplied by the mean abundance of fruits on the monitored trees to obtain a measure of availability. Usually mature individuals are selected for monitoring so that there is no allowance for variation in fruiting with tree size. Plumptre (1995a)

Table 7. Percentage contribution of plant parts to the diets of these primates in several study sites in East Africa

	YL	ML[a]	FL	FR[b]	SD	Arth.	Other	Reference
C. mitis								
Budongo Forest								
N3 logged	22.4	10.8	4.9	56.0	16.5	2.9	3.0	This study
N3 logged	26.2	0.6	4.6	57.3	1.5	8.6	2.7	Fairgrieve, 1995
N15 unlogged	33.8	13.1	7.4	38.0	14.8	5.9	1.8	This study
N15 unlogged	32.4	0.4	6.2	50.8	5.9	9.7	0.5	Fairgrieve, 1995
Range of 8 comp.	13–29	2–13	3–6	33–40	1–5	0–1	1–5	This study
Kibale Forest	13.7	6.8	12.5	45.1	2.4	19.8	2.1	Struhsaker, 1978
Kibale logged	42.3[c]		3.6	54.0	3.1	—	0.1	Wrangham *et al.*, 1998
Kibale unlogged	40.6[c]		4.2	55.3	7.1	—	0.0	Wrangham *et al.*, 1998
Kakamega Forest	16.6	2.3	3.7	57.1	2.5	16.8	3.5	Cords, 1986
Zomba Plateau	32.6[c]		10.2	51.9	—[b]	0.8	2.9	Beeson *et al.*, 1996
C. ascanius								
Budongo Forest								
N3 logged	24.8	9.4	6.1	53.3	18.3	3.8	2.6	This study
N15 unlogged	28.3	12.4	10.4	38.7	17.6	7.5	2.7	This study
Range of 8 comp.	12–20	2–10	0–16	61–73	1–5	0–1	0–4	This study
Kibale Forest	11.4	4.5	15.3	43.7	0.1	24.7	0.4	Struhsaker, 1978
Kibale logged		38.52	0.6	60.9	3.7	—	0.0	Wrangham *et al.*, 1998
Kibale unlogged		37.52	2.3	60.1	6.2	—	0.1	Wrangham *et al.*, 1998
Kakamega Forest	6.8	0.4	2.0	61.7	0.4	25.1	4.0	Cords, 1986
Kakamega Forest	8.9	0.3	1.4	55.0	—	29.5	4.9	Gathua, 1999
C. guereza								
Budongo Forest								
N3 logged	33.1	19.8	7.7	36.3	9.3	1.0	2.1	This study
N15 unlogged	36.9	24.8	8.9	25.6	12.0	1.2	2.6	This study
Range of 8 comp.	19–36	9–22	1–8	36–70	1–10	0	0–9	This study

(*Cont.*)

Table 7. *Continue.*

	YL	ML[a]	FL	FR[b]	SD	Arth.	Other	Reference
Kibale Forest	60.0	12.3	2.1	13.2	0	0	12.4	Struhsaker, 1978
Kakamega Forest	23.7	29.1	0.5	38.6	1.2	0	8.1	Fashing, 1999
Ituri Forest	29.7	28.0	2.9	24.6	22.0	0	14.5	Bocian, 1997
P. troglodytes								
Budongo Forest								
Sonso (incl. N3)	19.7[c]		8.8	64.5	—	0.0	7.0	Newton-Fisher, 1999a,b,c
N3	15.0	0.1	6.4	73.4	4.2	0.0	5.1	This study
Kibale Forest	9.3[c]		0.6	78.6	0.1	—	11.5	Wrangham *et al.*, 1998

Kibale Forest is in Uganda and Kakamega Forest is in Kenya. Where seed intake is known it has been added as an additional column but it is also combined with the fruit intake percentage data to allow comparison. Therefore total percentages sum to 100% + seed contribution. YL = young leaves; ML = mature leaves; FL = flowers; FR = Fruit; SD = seeds; Arth. = arthropods.

[a] In some studies there were unclassified as well as classified leaves. In this Table the unclassified leaves are included with mature leaves.

[b] Fruit includes consumption of seeds. Beeson *et al.* (1996) did not separate seeds and fruit.

[c] Wrangham *et al.* (1998), Beeson *et al.* (1996), and Newton-Fisher (1999a,b,c) did not separate young and mature leaves.

showed that some trees fail to fruit until they reach quite large sizes, sometimes up to 60–70 cm DBH. This study of phenology was more representative in that all trees were monitored in 100 or more plots sited in each study area in a stratified random manner. However, there were still problems in that there was no real method of combining tree species to form one measure of availability unless the biomass of fruit had been measured during the phenology (and even this would assume that the quality of the fruits was equal). Phenology measures are also affected by the fact that you see a tree at a certain point in time. Monkeys can strip a tree of its fruit within a few days and if you measure the tree just before they do and just after, you will get very different results. Where primate density is high, consumption of fruits will be heavy, so that fruits will be removed from trees more quickly than where primate density is low. Consequently it is theoretically possible that in areas of high primate density, less food will be found on the trees when being measured for phenology. Measures of potential availability such as sum DBH values can be used as an alternative method, but this method does not allow you to investigate seasonal and annual changes. Consequently, for seasonal availability I preferred to use a simple measure of whether the tree fruited or not and the sum of the DBH values for the monthly analyses in N15 and N3 and a simple measure of sum DBH for potential availability in all the compartments.

Apart from *P. troglodytes*, none of the primate densities showed any correlation with overall measures of potential fruit availability (Table 2(a)), and yet the two cercopithecines did show strong correlations with fruit intake in the diet (Table 2(b)). This result may be due to the problems with measuring availability, as outlined above, or it may be a function of variation in fruit quality between sites and that it is probably not valid to combine different species into fruit types. When comparing the data from the dawn-to-dusk follows in N15 and N3, it is clear that in the logged compartment there was a higher production of fruit throughout the year using this measure of availability. The variation between fruiting in plots is also less, indicating that the fruit is more evenly distributed. These two compartments are close and have similar tree species compositions so that the variation due to different tree species is lessened. Consequently the differences in the densities of monkeys between these two compartments was probably related to this more abundant and more even food production.

The dietary overlap analyses showed significantly higher overlap values than would be expected from random neutral models for most of the comparisons made. It would be expected that competition would lead to niche differentiation and low dietary overlap values. The fact that the overlap is higher than would be expected indicates how important the relative intake of plant species and their parts (and hence nutrient values) is for these primates. There is little indication therefore that competition is important in determining primate densities in these compartments at this scale of analysis, although for the two cercopithecines there may be increased competition for food where they occur at low density where overlap values become lower. It is possible that what is more important to these primates is whether competition is fierce during certain seasons when good food is scarce, and it is possible that dietary overlap during these "lean" times may drop. The data I have for each month though is probably not sufficient to validly calculate overlap because the sample sizes are low (only 2 days of observations per group).

Effects of Logging on Diets

Colobus monkey density was significantly correlated with the density and basal area of *C. durandii*, a species of tree whose leaves and fruits are preferred food items in every compartment studied for this primate (Table 3). This tree species also forms the greatest percentage contribution to the diet of this monkey (32.7% of colobus diets in N3, 17.3% in N15). *C. durandii* is not as good for timber as the other *Celtis* species (R. Plumptre, personal communication) and is unlikely to be harvested heavily even if the Uganda Forest Department promotes the use of more tree species for timber production. Consequently colobus are not likely to be at great risk from timber harvesting unless this tree species is targeted for removal in order to increase the stocking density of more desirable species. *C. durandii* was treated with arboricide during the 1950s and 1960s but has recolonised or regenerated in treated areas. In 1953 an experimental plot (RP100) was established in compartment N2 of Budongo to investigate the effects of different treatment programs on regeneration. Five replicates of four treatments and a control were established in a latin square design, each square of 100 × 100 m (only the central 60 × 60 m were measured to avoid edge effects). Rukundo (1996) remeasured the plot in 1994 and showed that where the treatments had occurred, *C. durandii* was more abundant (5–14% of trees) than in the untreated control plots (3% of trees). Consequently this tree seems to be a good colonizing species that can establish itself even where it has been selectively poisoned. This may explain why *C. guereza* often seems to survive in scrubby or isolated patches of forest throughout western Uganda where *C. durandii* also often occurs.

The two cercopithecines showed a strong correlation between fleshy fruit intake and density and are consequently likely to be affected by some aspect of food availability. To fully confirm this, populations need to be monitored over time to see how they fluctuate and to confirm that density has not been affected by some other factor such as a disease in some compartments. However, the preference analyses for these two species showed that fleshy fruits were often preferred dietary items, and the diet measures showed that they spent about 50% of their time feeding on these. Consequently, it is likely that food availability will be of some importance for determining densities of primate populations and that if logging affects this in one way or another then the primate populations will change. However, given the high dietary overlap values it is unlikely that food supply is limiting the primate populations except possibly where they are occurring at low densities. In Budongo, the removal of the monodominant *Cynometra* forest led to an increased diversity of tree species (Plumptre, 1996), and a consequent increase in the number of trees bearing fleshy fruits, particularly *C. durandii*. Consequently cercopithecine populations are higher in the logged and treated areas (Plumptre & Reynolds, 1994). The dawn-to-dusk follows in N3 and N15 showed that in the unlogged compartment N15, fruit intake was low for *C. mitis*, *C. ascanius*, and *C. guereza*. The values were lower than studies at most other sites, indicating that N15 is probably a sub-optimal habitat for these species, which would explain why they occur here at low densities.

P. troglodytes nest density correlated with the density of trees bearing large fleshy fruits. Large fleshy fruits formed 22.85% of the diet of the Sonso community, a much larger contribution than for the monkey species, and consequently this correlation makes sense. Of particular importance were *Chrysophyllum* species, *Uvariopsis congensis*, *Ficus mucuso*, *Ficus sansibarica* and *Ficus saussureana*. The latter three fig species and the *Uvariopsis* are not considered to be valuable timber species, but the *Chrysophyllum* trees are. In a separate study of chimpanzee diet in the Sonso community Newton-Fisher (1999a,b,c) found that four timber species were among the top seven species in the chimpanzees' diet, although two (*Celtis* species) are not valuable woods and not in high demand: *Khaya anthotheca*, *Maesopsis eminii*, *Celtis mildbraedii*, and *C. durandii*. Chimpanzees also were observed to consume significant amounts of dietary parts of three of these timber species in this study (Table 4).

Following this research the Budongo Forest Project has recommended to the Uganda Forest Department that some large *Chrysophyllum* trees are left in each logging concession to support the chimpanzee populations and also to provide seed for a future crop. Arboricide treatment of the forest in the past did target fig trees, because many species strangle other trees, and this will have had an effect on fig supply; however, fig trees seem to be able to recover quickly and are now more abundant in logged forest than unlogged forest (Earl, 1992). At present, a greater threat to chimpanzees in Budongo is probably the presence of large numbers of people in pit-sawing concessions. During 1996, numbers exceeding 100 men or more were common in and around two pit-sawing concessions. Most of these people were carrying out planks of timber to the loading bays. Human disturbance may cause chimpanzees to leave an area because of disturbance and noise rather than problems with food availability. If the disturbance is over a large area, it may lead to conflicts between neighboring chimpanzee communities, which can lead to the elimination of a chimpanzee community (Goodall, 1986). Pit-sawyers also set snares to catch duikers and bushpigs for meat and chimpanzees are accidentally trapped in these. At present, 25% of identified chimpanzees in Budongo forest have injuries to their limbs from snares, indicating that this may be a more significant determinant for this species rather than food availability (Plumptre *et al.*, 2003; see Reynolds, Chapter 22, this volume).

CONCLUSIONS

The forest management in Budongo over the past 60 years has reduced the dominance of *C. alexandri* and encouraged a more diverse and mixed forest. This has led to a greater density of trees bearing fleshy fruits, particularly *C. durandii*, which forms an important component of the diet for all the primates. Food availability is probably not the only factor determining primate densities in this forest though, as there is little indication that there is competition for food. It is possible that at the low density sites, food is a determining factor but that at high density sites social factors between primate groups may determine densities rather than food supply. Much of the change in the forest can be attributed to the arboricide treatment rather than the logging because it was not the logging that reduced the monodominance in the forest. In Uganda there has been little hunting of primates for meat because most Ugandans will not eat primate meat. More recently, with movements of people between the

Democratic Republic of Congo and Uganda, there are now some people who will hunt primates for meat; however, the level of hunting is still very low in comparison with Central Africa. Care must therefore be taken when extrapolating from these results to other logging sites in Africa. In Central and West Africa, the building of roads by logging companies opens up the forest and encourages hunters to penetrate deeper into the forest. Hunting, rather than the logging practices themselves, exacts a major toll on the wildlife, including primates, in these regions.

ACKNOWLEDGMENTS

I thank all the Ugandan field assistants who helped in the collection of these data: Nabert Mutungire, Geresomu Muhumuza, Julius Kyamanywa, Godfrey Biroch, Alfred Tolith, Edison Lokiri, and Stephen Hatari. The Forestry Research Programme of the Overseas Development Administration, USAID, National Geographic Society, and the Jane Goodall Institute provided financial support to the Budongo Forest Project, which enabled this study to be carried out. We are grateful to the Department of Biological Anthropology at Oxford University for providing a base for support and to Val DeNewton for her secretarial support. Nick Newton-Fisher kindly provided comments on a draft manuscript.

CHAPTER TWENTY-ONE

Behavioral Patterns of Colobus in Logged and Unlogged Forests

The Conservation Value of Harvested Forests

Colin A. Chapman, Michael D. Wasserman, and Thomas R. Gillespie

INTRODUCTION

Although tropical forests cover only 6% of Earth's arable surface, they account for nearly 50% of all known species (National Research Council, 1992). Despite this diversity, these systems are increasingly threatened, as exemplified by a recent Food and Agriculture Organization report (FAO, 1999) which indicates that tropical countries are losing 127,300 km^2 of forest annually; this does not consider the vast area being selectively logged (~55,000 km^2; FAO, 1990) or the forests degraded by fire (Laurance, 1999; Nepstad *et al.*, 1999;

Colin A. Chapman • Department of Anthropology and McGill School of Environment, McGill University, Montreal, Quebec, Canada, H3A 2T7 and Wildlife Conservation Society, 2300 Southern Boulevard, Bronx, NY 10460, USA **Michael D. Wasserman and Thomas R. Gillespie** • Department of Zoology, University of Florida, Gainesville, FL 32611, USA.

Primates of Western Uganda, edited by Nicholas E. Newton-Fisher, Hugh Notman, James D. Paterson, and Vernon Reynolds. Springer, New York, 2006.

Chapman & Peres, 2001). Furthermore, many tropical species are locally endemic or rare and patchily distributed (Struhsaker, 1975; Richards, 1996). Such restricted distributions predispose many species to increased risks of extinction (Terborgh, 1992) since they are less likely to be found inside national parks and reserves. As a result, conservation of many tropical forest species will depend on the capacity of disturbed forests to support their populations. Since all forms of extractive exploitation result in biological loss and ecosystem change, knowledge of how particular species are affected by extraction and an understanding of the mechanisms leading to this response are essential for developing sound conservation and management plans for disturbed forests. The most prevalent form of disturbed forest with conservation potential is selectively logged forest (Frumhoff, 1995; Struhsaker, 1997).

Primates are valuable subjects for examining the effects of logging. Most species are relatively easy to census and a great deal is known about their natural history. Furthermore, many primate species are endangered or threatened, making it critical that threats to their survival are better understood. Unfortunately, deriving generalizations about how primate populations respond to logging has proven to be difficult. Although a number of studies have examined the effects of selective logging on primate populations, the strength of the conclusions drawn is questionable. First, some studies were conducted soon after logging occurred (Plumptre & Reynolds, 1994; Bennett & Dahaban, 1995; Ganzhorn, 1995; Rao & van Schaik, 1997). Such studies may be inappropriate for examining the impact of logging on primate communities since habitat modification often lowers recruitment but does not usually kill primates (Struhsaker, 1997). Thus, in many cases, declines in primate populations will not appear until years after logging has occurred. Most other studies have not had data on primate abundance from before and after logging, and thus use neighboring unlogged sites to contrast to the logged site (for an exception, see Grieser Johns & Grieser Johns, 1995). This approach suffers by not taking into account natural variation in primate abundance within undisturbed forest (Johns, 1986; Chapman & Chapman, 1999). In addition, primate densities at many study sites are affected by confounding factors such as hunting (Wilkie *et al.*, 1992, 1998; Oates, 1996; Struhsaker, 1997; Rosenbaum *et al.*, 1998).

Primate studies have illustrated that variability in factors such as the intensity of timber extraction, associated hunting pressure, the nature of the habitat being studied, and even the methods used to conduct the study can lead to divergent results. As such, it is difficult to modify management plans based on existing studies. Two studies exemplify this divergence of results. Johns (1992) studied the effects of logging on animal populations in dipterocarp forests in Peninsular Malaysia, while Bennett and Dahaban (1995) addressed the same question in dipterocarp forests in Sabah. The intensity of logging was similar in the two regions. In Sabah, the logging produced an immediate 35–70% decline in the gibbon (*Hylobates muelleri*) and langur populations (*Presbytis* spp.; Bennett & Dahaban, 1995). In contrast, survival of the same genera in Peninsular Malaysia was much greater (10% decline to an increase of 74%; Johns, 1983). Bennett and Dahaban (1995) attribute the differences between their study and that of Johns (1983, 1992) to the nutrient-rich soils, initial high primate density, and virtual absence of hunting in Peninsular Malaysia, conditions quite different from those in Sabah.

As a second example, blue monkeys (*Cercopithecus mitis*) appeared to be severely impacted by logging at Kibale National Park, Uganda. Fifteen years after logging, areas had 20–30% fewer blue monkeys than unlogged areas (Skorupa, 1988), and this pattern continues to become more extreme to this day (Chapman *et al.*, 2000). In contrast, in Budongo Forest Reserve, Uganda, blue monkeys are 3.7 times more abundant in logged areas than in unlogged areas (Plumptre & Reynolds, 1994).

One might argue that examples such as these are simply exceptions to general trends, and, if good comparative methodologies were employed across a range of species and study sites, patterns would emerge. Johns and Skorupa (1987) attempted such a test with 37 primate species having populations from both undisturbed and disturbed habitats. They discovered that 44% of the variation in species' responses to moderate habitat disturbance could be accounted for by body size and dietary considerations. Body size and degree of frugivory were both negatively correlated with survival ability in degraded habitats. However, they concluded that while this relationship was statistically significant, it provided an insufficient basis to reliably predict how specific species would respond to disturbance. Unfortunately, it is these kinds of predictions that forest managers must make. The strongest suggestion that this comparative study was able to make was that large-bodied frugivores are the class of primates most vulnerable to habitat disturbance, and three examples were presented: *Ateles*, *Pan*, and *Pongo* (Johns & Skorupa, 1987). Unfortunately, if one looks closely at descriptions of even these three species, exceptions are evident. For example, an *Ateles geoffroyi* population has been described to be relatively abundant in a severely degraded area that was both intensively logged and grazed by cattle

(Chapman *et al.*, 1989). Similarly, *Pan troglodytes* groups are known to survive in areas that have been logged and almost totally converted to agriculture (Onderdonk & Chapman, 2000). These chimpanzees appear to survive by traveling between the few small remaining forest patches and by raiding crops planted by local farmers (Naughton-Treves, 1996).

As a result, predicting the effect of logging on particular primate species has proven difficult. On the basis of this perspective, the objective of this paper is to use a comparison of the behavior of the black-and-white (*Colobus guereza*) and red colobus (*Procolobus badius*) in logged and unlogged areas of Kibale National Park, Uganda, to provide insights into responses to logging. We selected these two species because they appear to respond differently to logging, while there are theoretical reasons to suggest that their biomass is determined by the same factors. The density of red colobus in a heavily logged area of Kibale was approximately half that of an adjacent unlogged area a decade after logging, while there were 5 times more black-and-white colobus in this logged area than in the unlogged area (Skorupa, 1988; Struhsaker, 1997; Chapman *et al.*, 2000). This difference exists despite the fact that these two colobus species are of similar size (red colobus and black-and-white colobus both weigh 8.8 kg; Struhsaker and Leland 1979), have similar locomotory patterns (Gebo & Chapman, 1995), have similar day ranges (red colobus 648 m/day, black-and-white colobus 535 m/day; Struhsaker & Oates, 1975), and have similar diets (Chapman *et al.*, 2000a,b). Furthermore, the two species respond differently to logging despite the fact that the biomass of both species can be predicted by the protein-to-fiber ratios of potential foods at sites within the park (Chapman *et al.*, 2000). Since the tree species that first colonize after logging have higher protein-to-fiber ratios than forest species, the response of the red colobus is not expected (Chapman *et al.*, 2005).

Here we contrast the activity budgets, travel costs, percentage of time in mixed-species groups, and diet between two groups of each species in unlogged forest and one group in the logged forest. We consider two groups in the unlogged forest to provide a better understanding of the natural variation that can exist in the variables considered. In the unlogged forest, we selected groups that varied in size, since many of the parameters we were contrasting are known to vary as a function of group size. This comparison is conducted approximately 30 years after the logging was completed. Most logging regimes call for some sort of rotation: the area is logged, it is left to recover for a specified period, often 30–50 years, and then it is logged again. We suggest that if logged areas are to be compatible with primate conservation, primate populations must recover from the initial disturbance, and their behavioral patterns should be similar prior to the time that the area would be scheduled to be reharvested. If a similar comparison was conducted at some time prior to this (i.e., only a few years after logging), it is likely that the findings would differ from those reported here.

METHODOLOGY

Study Site

Kibale National Park (795 km^2) is located in western Uganda (0°13′–0°41′ N and 30°19′–30°32′ E) near the foothills of the Ruwenzori Mountains (Struhsaker, 1975, 1997; Skorupa, 1988; Chapman *et al.*, 1997). The park consists of mature, midaltitude, moist semideciduous and evergreen forest (57%), grassland (15%), woodland (4%), lakes and wetlands (2%), colonizing forest (19%), and plantations of exotic trees (Chapman & Lambert, 2000). Mean annual rainfall in the region is 1749 mm (1990–2001, or 1547 mm for 1903–2001); the mean daily minimum temperature is 14.87°C, and the mean daily maximum temperature is 20.18°C (1990–2001). There are distinct wet and dry seasons that are bimodal in distribution. May to August and December to February tend to be drier than other months (Chapman *et al.*, 1999).

We contrasted the behavior of groups of both species living in areas that had been heavily logged to groups in an unlogged section of forest. The logged forest experienced heavy selective felling in 1969. The harvest averaged 21 m^3/ha or approximately 7.4 stems/ha (Skorupa, 1988; Struhsaker, 1997). Incidental damage in the area was high, and it is estimated that approximately 50% of all trees were destroyed by logging and incidental damage (Skorupa, 1988). The unlogged area had not been commercially harvested; however, prior to 1970, a few large stems (0.03–0.04 trees/ha) were removed by pit-sawyers. This extremely low level of extraction seems to have had very little impact on the structure and composition of the forest (Skorupa, 1988; Struhsaker, 1997).

Censuses conducted over a 17-year period have been used to quantify the long-term effects of high-intensity selective logging on the density of the five common primates in these areas (Chapman *et al.*, 2000). Red colobus populations were recovering in the heavily logged areas; however, the rate of increase appeared to be slow (0.005 groups/km^2 per year). Black-and-white colobus appeared to do well in some disturbed habitats and were found at higher group densities in the logged areas compared to the unlogged area. Group densities of blue monkeys (*Cercopithecus mitis*) and redtail monkeys (*C. ascanius*) in the heavily logged area were low and continued to decline decades after logging. There was no evidence of changes in mangabey (*Lophocebus albigena*) group density in the heavily logged area since the time of logging.

Behavioral Observations

Behavioral observations were made 4 days each month on two groups of each species in the unlogged forest between August 1998 and June 1999 (except March). The black-and-white colobus groups were of similar size (Group 1 = 9, Group 2 = 6), while the red colobus groups differed substantially in size (Group 1 = 48, Group 2 = 24). The groups are subsequently called the big and small groups. This sampling resulted in approximately 650 h of observations on black-and-white colobus, and 700 h of observations on red colobus. One group of each species was observed in the logged area for 5 days a month between July 1999 and May 2000 (except July and August, when they were watched for 4 days), producing approximately 500 h of observations for each species. The red colobus group contained 34 individuals, while the black-and-white group contained 7 individuals.

During each half hour, the observer was with the group, five point samples were made of different individuals. If the animal was feeding, the species and plant part (e.g., fruit, young leaf, and leaf petiole) were recorded. We made an effort to avoid repeatedly sampling particularly conspicuous animals by moving throughout the group when selecting subjects and by sampling animals that were both in clear view and those that were more hidden. At the end of each half hour, we estimated the distance the group had traveled and the group spread, and we determined if the group was in association with other primates. Distance traveled was estimated as the straight-line distance between the center of mass of the group in the previous 30-min sample to the current center of mass. Distance estimations were assisted by travel routes plotted on detailed trail maps in which trails were only 50–100 m apart. Given that it was sometime difficult to locate groups predawn, data was often not recorded until 0730 or 0800 h. As a result, daily distance traveled was calculated as the distance moved between 0800 and 1700 h. Days when follows were shorter than this (e.g., groups could not be found until after 0800 h) were excluded from the analysis. Group spread was estimated by having one observer frequently walk from the perceived center of mass of the group to where no additional animals could be seen. The observer would walk perpendicular to the direction of group travel. If the study group was within 50 m of another primate species, it was considered to be in a mixed-species association.

These behavioral observations were conducted by CC, Lauren Chapman, and a team of three Ugandan field assistants. The field assistants have worked with CC and LC since 1990 and knew the tree species and monkey age classes prior to the start of the project. The field assistants were supervised by CC and LC when they were in the field and in their absence by project directors. Detailed training was conducted prior to when a new observer started recording data. Subsequent to this training, interobserver reliability was assessed by having the new individual observe the activity of the same animal being watched by CC (all assistants and volunteers). The first 30 foraging bouts for new observers agreed with those of CC in >97% of the occasions (range = 97–100%, mean = 98.8%). In those instances where there were disagreements, they typically occurred with respect to assessing the stage of maturity (i.e., ripe versus unripe fruit or mature vs. young leaves, particularly for *Celtis africana*—a species with small leaves).

Food Tree Density

To evaluate if the groups in the logged area used foods that were less preferred than groups in the unlogged area, the density of food trees was estimated by establishing four permanent vegetation plots (200 by 10 m) in the home range of the study groups. Each tree with a diameter at breast height (DBH) >10 cm within 5 m of each side of the trail was individually marked with a numbered aluminum tag and measured (DBH).

Statistical Evaluations

We constructed monthly profiles of behavioral variables (i.e., activity budgets, proportion of time in mixed-species groups, distance traveled, group spread, and dietary profiles (i.e., % of scans eating different plant parts) and contrasted groups in logged and unlogged areas using one-way analysis of variance, with each month considered as an independent sample. Where appropriate we arcsine-square-root-transformed proportional data. In addition to this analysis, we attempted to control for seasonal variation by pairing samples by month and contrasting groups using a paired t-test. Since mixed-species association with a particular species was often rare and 4–5 days of observations were made each month, the sample size was not thought to be sufficient to accurately depict particular associations. Hence, for analysis of mixed-species associations, only the overall level of association is compared statistically (i.e., time in association with any primate).

RESULTS

For both species there were significant differences among groups in the amount of time spent feeding and traveling, while the amount of time resting did not differ (Figure 1, statistical comparisons in the figure legend). However, in all cases, the activity level of the groups in the logged area was intermediate between the two groups observed in the unlogged area. Thus, the differences in activity budgets cannot be attributed to the effects of logging. When we controlled for seasonal variation in activity pattern by pairing samples by month and contrasting groups using a paired t-test, for both species there were the same number of significant differences between the groups in the logged and unlogged areas as between the two groups in the unlogged area. This supports the previous analysis and suggests that observed differences in activity budgets cannot be attributed to the effects of logging.

The average distance traveled each day between 0800 and 1700 h by the different groups revealed similar, but not identical, patterns to the time spent traveling. For red colobus ($F = 21.65$, $P < 0.001$), the group in the logged forest (mean distance = 123 m) traveled less than the large group in the unlogged forest (232 m), and the small group in the unlogged forest (142 m) traveled shorter distances each day than the large group in the unlogged forest ($P < 0.001$ in all cases). However, there was no significant difference between the group in the logged forest and the small group. For black-and-white colobus ($F = 7.32$, $P = 0.001$), the large group in the unlogged forest (138 m) traveled farther than either of the other groups (unlogged small 83 m, $P = 0.01$, logged 83.5 m, $P = 0.003$).

Group spread differed among all groups for both species (black-and-white colobus, $F = 194.6$, $P < 0.001$; red colobus, $F = 254.03$, $P < 0.001$). The large groups in the unlogged area (black-and-white colobus 21.0 m, red colobus 63.5 m) were more spread out than small groups in the unlogged area (17.0 m, 47.8 m) or the groups in the logged area (12.0 m, 39.5 m). The small groups in

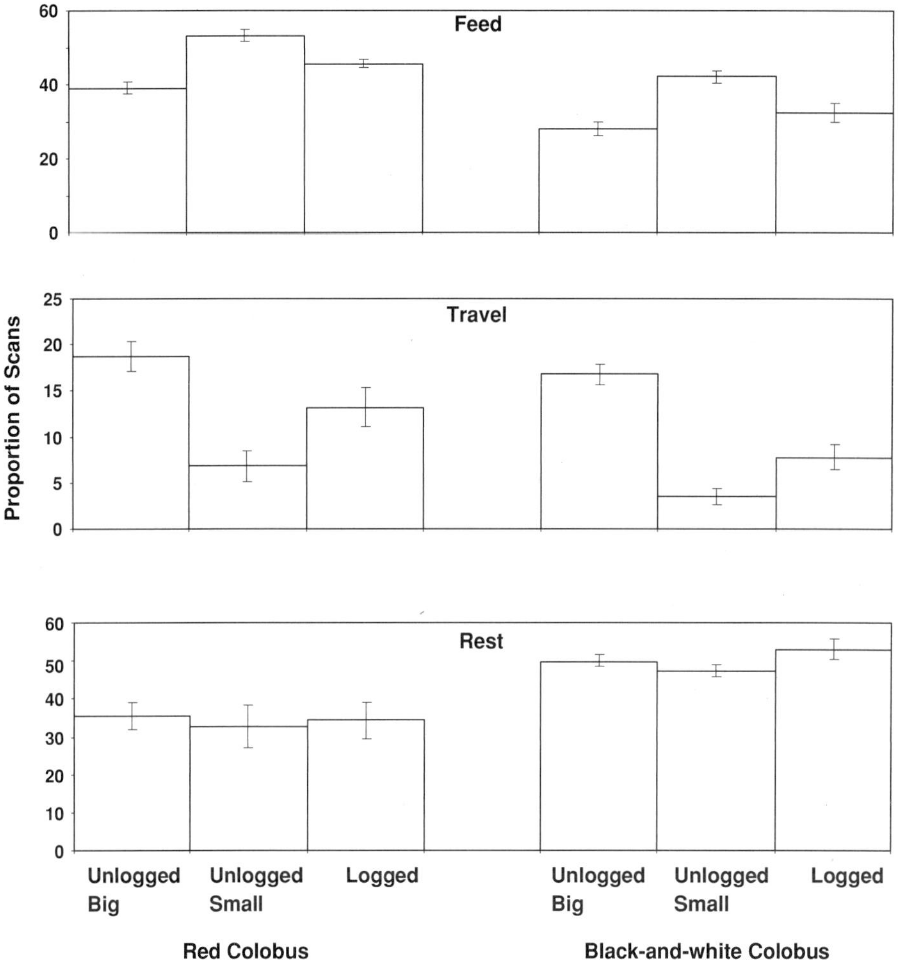

Figure 1. The average proportion of scans each month (+SE) that groups of red colobus and black-and-white colobus in Kibale National Park, Uganda, spent engaged in feeding, resting, and traveling. Black-and-white colobus: Feeding $F = 11.91$, $P < 0.001$, Scheffe post hoc tests logged versus unlogged big $P = 0.344$, logged versus unlogged small $P = 0.008$, unlogged big versus unlogged small $P \leq 0.001$; Travel $F = 30.49$, $P < 0.001$, Scheffe post hoc tests logged versus unlogged big P ¡ 0.001, logged versus unlogged small $P = 0.048$, unlogged big versus unlogged small P ¡ 0.001; Rest $F = 2.15$, $P = 0.137$. Red colobus: Feeding $F = 20.29$, $P < 0.001$, Scheffe post hoc tests logged versus unlogged big $P - 0.018$, logged versus unlogged small $P = 0.006$, unlogged big versus unlogged small $P \leq 0.001$; Travel $F = 9.64$, $P = 0.001$, Scheffe post hoc tests logged versus unlogged big $P = 0.121$, logged versus unlogged small $P = 0.072$, unlogged big versus unlogged small $P = 0.001$; Rest $F = 0.84$, $P = 0.442$.

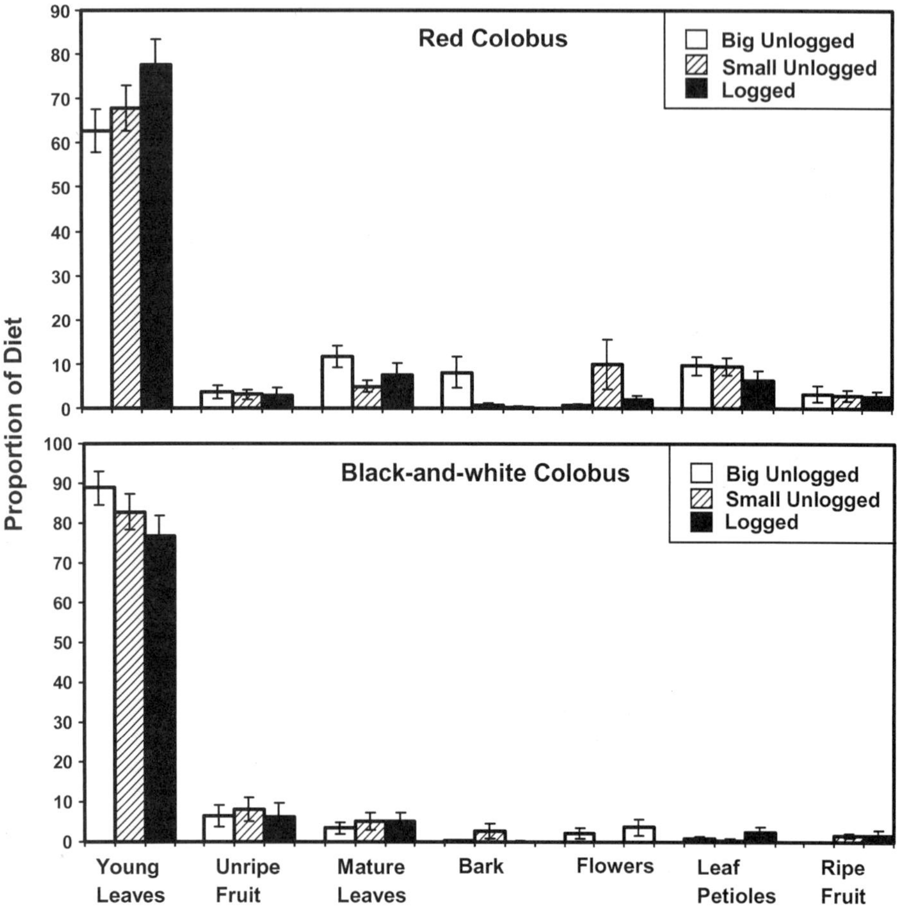

Figure 2. The average proportion of foraging scans each month (+SE) that groups of red colobus and black-and-white colobus in Kibale National Park, Uganda, spent eating different plant parts. Black-and-white colobus nonsignificant in all comparisons. Red colobus nonsignificant in all comparisons, except bark $F = 4.22$, $P = 0.023$.

the unlogged forest had a smaller group spread than the groups in the logged area (Scheffe, $P < 0.001$ for both species).

There were no differences in the diets of the different groups of black-and-white colobus in terms of the plant parts eaten ($P > 0.122$; Figure 2). All groups relied most heavily on young leaves (77–82% of all foraging scans). In contrast, there were differences in the diets of the red colobus among groups. The large group of red colobus in the unlogged area spent more time eating bark (primarily *Prunus africana*) than either of the other groups ($F = 4.218$, $P = 0.023$). Similarly, there was a marginal difference in the time spent eating young leaves between the large group in the unlogged area and the group in the logged area, with the group in the logged area spending more time eating young leaves. Since young leaves are nutritionally one of the best foods that red colobus eat (Chapman & Chapman, 2002), this difference does not suggest that the group in the logged area is eating poorer quality food. In contrast, this may represent a cost associated with large group size in red colobus (Gillespie & Chapman, 2001).

Given the heterogeneity in forest composition of tropical forests over short spatial scales, it is not surprising that the diets of the different groups differed with respect to the species and part consumed (Table 1). For all groups, *Celtis durandii* and *Celtis africana* young leaves were frequently eaten as these species were two of the top four most eaten items for all groups and the top two for 4 of the groups. *Celtis durandii* seems to be most important as it remained the top food for 5 of the 6 groups despite its density decreasing to one-third the unlogged value in the logged forest. Other than these two species, it is difficult to determine how logging altered the diet selection and preference of the animals. For example, while *Olea welwitschii* was one of the species targeted in the logging operation, its current density in the logged and unlogged areas are the same (1.3 individuals/ha). Despite the fact that its density is equivalent in both areas, the red colobus groups in the unlogged area never or rarely ate this tree's young leaves, while the group in the logged area fed on its young leaves extensively (7.8%, Table 1). The density of *Strombosia scheffleri* is also the same in the different areas, but red colobus fed on it less (2.7%) in the logged area than either group in the unlogged forest (5.3 and 4.6%). Therefore, for species that had identical availability in both forests, some species are used more frequently in the logged forest, while others are eaten more in the unlogged forest, making general trends unapparent. To further exemplify the inconsistent effect of logging on the colobus diets, the density of 11 food species decreased from unlogged to logged areas, 13 increased, and 10 stayed approximately the same.

When examining the diet richness (defined as the number of foods consisting of 1% or more of each colobus' diet), trends differ between the two colobus. For the black-and-white colobus, the small group had the richest diet with 16 foods, followed by the logged group, which fed on 14, and the big group fed on 11 foods. Similarly, the small red colobus group had the richest diet with 24 foods; however, in contrast to the black-and-white colobus, the big group fed on more foods (20) than the logged group (18). As for rare food feeding

Table 1. The proportion of all foraging scans that the red and black-and-white colobus of Kibale in logged and unlogged forests spent eating different species and parts

	BWC Big	BWC Small	BWC	RC Big	RC Small	RC	Unlogged	Logged
Species (Part)	Unlogged	Unlogged	Logged	Unlogged	Unlogged	Logged	Density/ha	Density/ha
Celtis durandii (YL)	31.6	26.3	20.6	18.3	5.3	12.3	47.5	16.3
Celtis africana (YL)	12.8	11.1	13.4	6.8	7.9	8.8	13.8	17.5
Albizia grandibracteata (YL)	3.5	5.2	12.7	1.1	3.0	5.5	1.3	2.5
Markhamia platycalyx (YL)	9.6	4.8	2.9	4.1	1.9	6.1	58.8	16.3
Dombeya mukole (YL)	7.0	0.4	2.8	7.1	2.2	4.4	5.0	6.3
Olea capensis (YL)	4.6	1.1	6.9	0.9	0.0	7.8	1.3	1.3
Strombosia scheffleri (YL)	2.8	2.8	0.0	5.3	4.6	2.7	11.3	11.3
Prunus africana (YL)	0.3	6.7	1.0	1.6	4.5	3.1	1.3	2.5
Parinari excelsa (YL)	0.4	2.4	1.9	6.4	7.7	5.5	1.3	2.5
Celtis durandii (UF)	0.0	3.3	7.1	1.2	1.3	3.6	47.5	16.3
Funtumia africana (YL)	1.3	4.2	0.7	2.9	4.6	2.7	45.0	15.0
Millettia dura (YL)	1.8	0.5	6.0	1.1	0.6	6.0	1.3	11.3
Trilepsium madagascariensis (YL)	0.4	0.4	0.3	7.4	3.9	2.1	48.8	23.8
Prunus africana (ML)	0.0	3.7	0.2	2.1	7.4	0.2	1.3	2.5
Funtumia africana (RF)	0.0	0.5	0.0	3.9	4.2	0.7	45.0	15.0

Ficus brachylepis (YL)	3.6	1.5	0.0	2.1	1.1	0.8	2.5	1.3
Markhamia platycalyx (Petiole)	0.7	0.4	0.3	4.6	1.1	1.7	58.8	16.3
Strombosia scheffleri (Petiole)	0.1	2.8	0.0	2.9	1.6	1.2	11.3	11.3
Diospyros abyssinica (YL)	2.0	1.5	0.2	0.0	0.1	2.9	26.3	35.0
Chrysophyllum spp. (YL)	0.1	2.7	0.0	1.3	1.9	0.0	1.3	1.3
Celtis durandii (ML)	0.8	0.5	1.8	0.0	0.6	0.9	47.5	16.3
Celtis durandii (FL)	0.9	0.0	1.7	1.0	0.0	0.9	47.5	16.3
Macaranga schweinfurthii (YL)	0.0	1.4	0.0	0.5	2.1	0.0	1.3	1.3
Prunus africana (RF)	0.0	0.8	0.0	0.1	2.7	0.0	1.3	2.5
Celtis africana (ML)	0.0	0.1	0.8	0.0	0.3	1.7	13.8	17.5
Prunus africana (Bark)	0.0	0.0	0.1	0.0	2.3	0.1	1.3	2.5
Mimusops bagshawei (YL)	0.0	0.0	0.0	0.3	1.7	0.5	1.3	1.3
Diospyros abyssinica (RF)	0.0	0.0	1.7	0.0	0.0	0.4	26.3	35.0
Funtumia africana (FL)	0.2	0.0	0.0	1.0	0.0	0.3	45.0	15.0
Balanites wilsoniana (YL)	0.0	0.3	0.0	0.0	1.1	0.0	1.3	1.3
Pancovia sp. (YL)	0.0	0.0	0.0	0.2	1.1	0.0	1.3	1.3
Fagaropsis angolensis (Petiole)	0.0	0.0	0.2	0.0	0.0	1.0	3.8	8.8
Spathodea campanulata (YL)	0.0	0.0	1.1	0.0	0.1	0.0	1.3	1.3
Pancovia sp. (Petiole)	0.0	0.0	0.0	0.0	1.1	0.0	1.3	1.3

(defined as a food that consists of 1% or more of the diet and is not fed on by the other two groups within that colobus species), the small group of red colobus in the unlogged area fed on seven rare foods, the group in the logged area fed on four, and the big group in the unlogged area fed on two rare foods. Within the black-and-white colobus groups, the small and logged groups fed on four rare foods, and the big group fed on no rare foods.

The average amount of time that the different groups spent in association with another primate species differed among black-and-white colobus groups ($F = 5.00$, $P = 0.014$), but not red colobus groups ($F = 2.06$, $P = 0.146$; Figure 3). In the unlogged area, the large group of black-and-white colobus spent more time in association than the small group (Scheffe = 0.014), but the

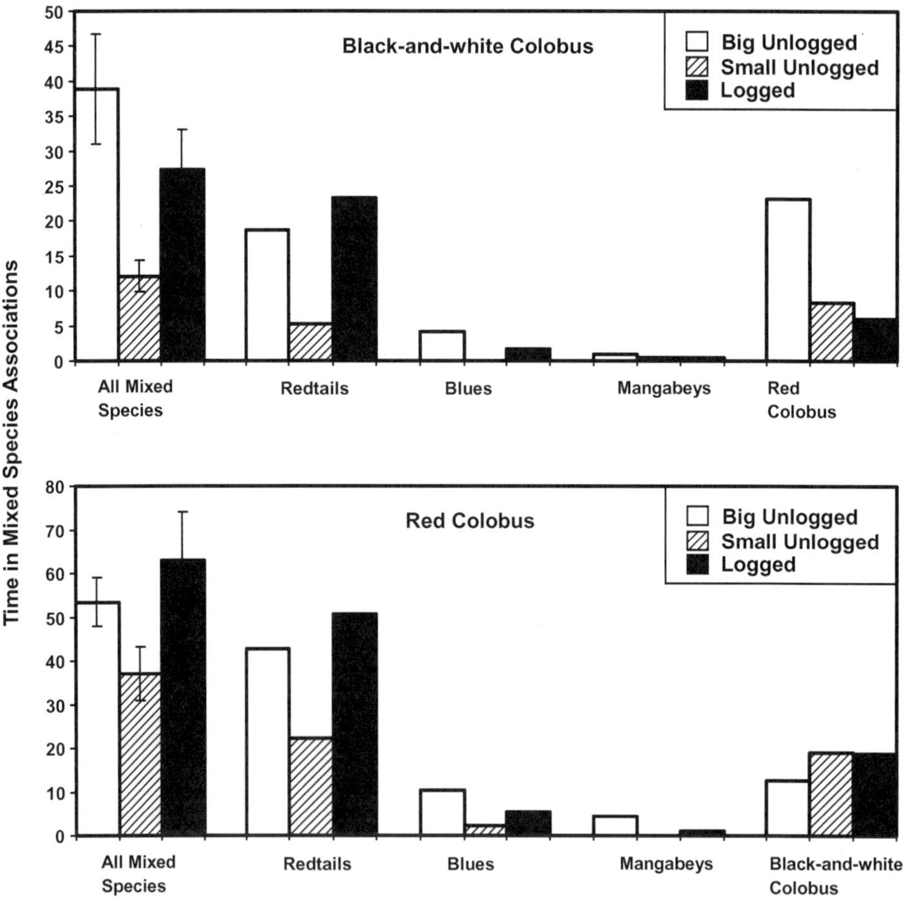

Figure 3. The time spent in mixed-species association by groups of red colobus and black-and-white colobus in Kibale National Park, Uganda.

time the other groups spent in association did not differ ($P > 0.228$). However, since the amount of time black-and-white colobus spend in association has been shown not to differ from random (Chapman & Chapman, 2000a), the biological significance of the difference found is questionable.

DISCUSSION

The differences in activity budgets, group spread, diet, mixed-species associations, and distance traveled do not suggest that the ecological differences between the logged and unlogged areas are creating large differences in the behavior of these two colobine species 30 years after logging. In contrast, the differences that were documented support the claim that group size is a strong determinant of many aspects of colobine behavior (Gillespie & Chapman, 2001). Large groups are thought to deplete food patches more rapidly, which alters foraging behavior and increases travel time (Janson & Goldsmith, 1995; Chapman & Chapman, 2000b).

Onderdonk and Chapman (2000) documented that the activity budget of a group of black-and-white colobus in a forest fragment that was an order of magnitude smaller than their typical home range was similar to a similarly sized group in the continuous forest. This study, in combination to the results obtained here, suggests that colobus groups in what one would expect to be stressful habitats (logged forest, fragments) are not spending more time traveling or feeding, as might be expected if they were forced to search harder for food or to eat lower quality foods. It may be that the densities and group sizes of the populations have been adjusted so that the animals are distributed between the logged and unlogged forest and among groups in a pattern that would represent an ideal free distribution (Fretwell & Lucas, 1970). Thus, where food resources are scarce, there are fewer animals to compete over those resources and those animals are in smaller groups. Therefore, the amount of resource available per individual is constant across habitats. In support of this, Teelen (1994) censused black-and-white colobus groups in the heavily logged and unlogged areas where we worked. She found that the average group size was 6.4 animals in the heavily logged area and 8.1 in the unlogged area. Struhsaker (1997) reported that the black-and-white colobus groups in lightly logged forest were 25% smaller than in the adjacent unlogged forest (7.9 vs. 10.5). Chapman *et al.* (2000) documented a 25% reduction in group size in red colobus between the heavily logged and unlogged areas of Kibale.

The smaller group size in degraded habitats is consistent with other studies of black-and-white colobus in forest fragments. Oates (1977) found smaller group size (modal size 7, $n = 2$) at Chobe, Uganda, a site consisting of patchy riparian forest, than in Kibale (modal size 9, $n = 7$). Similarly, Dunbar (1977) found that groups in Bole Valley, Ethiopia, in gully patches were smaller than groups in riverine forest, although this difference was not significant. In examining the relationship between habitat quality and black-and-white colobus group size across sites, Dunbar (1987) found that groups are significantly larger in more forested habitats.

The apparent relationship between colobine group size and habitat degradation suggests that the animals may be adjusting group size to the ecological conditions of the degraded habitat in such a fashion as to maintain relatively similar activity budget, travel costs, etc. It is generally held that group size is a trade-off between costs of reduced foraging efficiency and benefits of reduced predation risk (Chapman *et al.*, 1995; Janson & Goldsmith, 1995). As group size increases, reduced foraging efficiency of individuals requires them to travel farther to meet their nutritional requirements (the ecological constraints model; Chapman & Chapman, 2000b). For a group in a logged habitat, increasing foraging area may be too costly because it would involve traveling long distances between suitable feeding sites. Thus, the size of a group in degraded habitats may be constrained by the resources available.

The most recent survey in Kibale demonstrates that red colobus density is slightly higher in the unlogged forest than the heavily logged forest (unlogged = 5.5 groups/km^2, logged = 4.4 groups/km^2), while the density of black-and-white colobus is much higher in the heavily logged forest (unlogged = 2.0 groups/km^2, logged = 9.1 groups/km^2). The lower density of red colobus would presumably also reduce competition, possibly facilitating the maintenance of similar behavioral patterns in logged and unlogged forests. The fact that the black-and-white colobus are found at high density in the logged forest in comparison to the unlogged forest would increase competition and thus one would expect behavioral patterns to diverge. This may be offset by the reduction in group size; however, this explanation is somewhat unsatisfactory, as the reduction in group size between logged and unlogged areas is similar for these two species. Thus, the different response of the two colobine species to logging requires further investigation.

Primates have frequently been described forming mixed-species groups (Cords, 1987). These groups involve two or more species associating together and coordinating activities for a number of hours or days and these associations are often argued to have adaptive functions. Benefits of mixed-species associations are controversial, but studies in Kibale have emphasized the role they can play with respect to decreasing predation risk through increased detection of predators and predator defense (Struhsaker, 1981; Chapman & Chapman, 1996, 2000a). Similarly, research in Tai National Park, Ivory Coast, has experimentally demonstrated that these associations may function to reduce predation risk (Bshary & Nöe, 1997a,b; Nöe & Bshary, 1997). Playbacks of chimpanzee (*Pan troglodytes*) hoot recordings induced formation of new associations and extended the duration of existing associations.

In censuses, Struhsaker (1975, 1981, 1997) documented that red colobus were in associations with other primates in unlogged forest 65% of the time, but only in association in heavily logged forest 36% of the time, while black-and-white colobus associated with other primates in the unlogged area 37% of the time and in the logged area 29% of the time. The levels of association we documented for red colobus in the heavily logged forest was much higher (60%) and for the black-and-white the level was reduced (12%). However, we also documented large differences between groups within a habitat type in the percentage of time spent in association. For example, the big black-and-white colobus group spent 39% of its time in association, while the small group was with other primates only 27% of the time. This suggests that estimates of percentage of time in association from observations of single groups may not represent the population as a whole. The differences between Struhsaker's finding and those reported here may reflect the fact that our study was conducted many years after the logging, while the previous study was closer to the time of logging.

Despite this, it is somewhat surprising that the colobus could maintain high levels of mixed-species associations in the logged forest. Given the fact that primate density is much lower in the logged forest (Chapman *et al.*, 2000), this would suggest the animals are actively seeking out other monkeys to associate with. Since the density of food trees are lower, the costs of these associations will be elevated in the logged area (Chapman & Chapman, 2000a). That red colobus are associating more in the logged area suggests that the benefits of association may be great in this habitat. It would be consistent with these observations to suggest that red colobus are reacting to an increased risk of predation in logged forest by associating more when the costs are greater.

The fact that few differences were documented in the behavioral patterns of the two colobus species between logged and unlogged areas 30 years after logging suggests that mechanisms are present that facilitate their use of degraded forests. This mechanism may in part involve a reduction of group size to allow a decrease in feeding competition. These findings suggest that with respect to behavioral patterns, logged forest has considerable value for colobines. However, a number of issues remain to be resolved before the conservation value of logged forests are understood. For example, why is the density response to logging so different between these two colobus species? Why is the nature of forest recovery following logging in Kibale so slow (Struhsaker *et al.*, 1996; Chapman and Chapman 2004)? Why is the population growth rate of black-and-white colobus in the last decade much greater than that of red colobus (Chapman and Chapman 2004)? How will cascading consequences of logging affect the colobus? For example, if cercopithecine abundance continues to decline, a point may be reached where it is impossible to maintain the current level of mixed-species association, which may result in increased predation risk. Until such questions can be answered for colobines and many other species, caution should be used in following the lead of many conservation and development agencies that are suggesting that the conservation of many tropical forest species will depend on the capacity of disturbed forests to support their populations. If conservation funding is limited, it may be that conserving disturbed forests could be at the detriment of protected area management.

ACKNOWLEDGMENTS

Funding for the research was provided by the Wildlife Conservation Society (WCS) and the National Science Foundation (SBR 96-17664, SBR-990899). Permission was given to conduct this research from the Office of the President, Uganda, the National Research Council, the Uganda Wildlife Authority, and the Ugandan Forest Department. Lauren Chapman and Tom Struhsaker provided helpful comments and insights on this work.

CHAPTER TWENTY-TWO

Threats to, and Protection of, the Chimpanzees of the Budongo Forest Reserve

Vernon Reynolds

INTRODUCTION

In 1962, the Busingiro chimpanzee community at Budongo studied by my wife and myself (Reynolds, 1965; Reynolds & Reynolds, 1965) had, we estimated, between 60 and 80 members. We did not see any chimpanzees with snare or trap injuries. A recent survey of Budongo Forest (summarized in Plumptre *et al.*, 2003) found the mean population size of chimpanzees in Budongo Forest to be 584 individuals (95% confidence limits 356–723), thus confirming the earlier estimate of 570 (95% confidence limits 425–711) (Plumptre & Reynolds, 1996).

Sugiyama (1968:230) reported on one chimpanzee, Mkono (the word means "arm" in kiswahili), of the Busingiro community who had no right hand. Suzuki (1972) included a photograph of the same individual, Mkono, showing that

Vernon Reynolds • Emeritus, School of Anthropology, University of Oxford, 51 Banbury Road, Oxford, United Kingdom.

Primates of Western Uganda, edited by Nicholas E. Newton-Fisher, Hugh Notman, James D. Paterson, and Vernon Reynolds. Springer, New York, 2006.

his right hand and wrist were completely missing. This injury was likely to have been caused by an encounter with a trap, either while crop-raiding, or in a trap in the forest set for a buffalo.

Thus snare and trap injuries were comparatively rare 40 years ago. At that time, the traditional Bunyoro method of hunting was to go out communally from the village and hunt over the woodland savannah countryside using dogs and long nets, driving game into the nets. Snares were set in the forest but not the large numbers that are found today. In this chapter the main threats to the Budongo Forest chimpanzees are discussed, as are the ways they might be addressed in future.

THREATS

Snares and Traps

The people of Western Uganda are for the most part living at subsistence level, and meat is a luxury food for them. They grow a variety of crops (cassava [*Manihot esculenta*] is the staple) and do not often eat meat. If families have any domestic animals they have a goat or a few chickens, but very many do not have any at all. If they want to eat meat they have to buy it in the market. Beef, goat, chicken, and occasionally pork are all locally available, but beef costs 2000 Ugandan shillings a kilo (about 80 UK pence) and many families have only a few hundred shillings to spend in the course of a week.

As a result, heads of families and single men go into the forest and place snares on the trails that criss-cross the forest floor, hoping to catch a duiker or a wild pig. Snares are made of steel wire from inside car tires, bicycle brake cables, telephone lines, or electric cable. Nylon cord is also used.

Snare Types

Two snare types are used: one is attached to a sapling at the edge of the animal trail and hangs in a noose shape just above the ground placed so that a duiker or pig will put its head or foot into the snare as it walks along. The second, more effective, type is sprung: the noose is as above but the end of the snare is attached to a strong sapling that has been bent over. Both types are common in Budongo.

Traps

Traps are normally set for a different reason than snares, and in a different location. Whereas snares are set inside the forest to catch duikers, pigs, porcupines, rats, or guinea fowl, traps are set around the edges of farmers' fields to catch crop-raiding animals. The commonest form of trap is the leg-hold trap: a pair of toothed metal jaws is held open by a metal spring mechanism; when the animal steps between the open jaws the spring gets released and the trap snaps shut, gripping the ankle or foot of the victim. The trap is attached to a metal chain, which is tethered to the ground or a small tree. In the past, traps were set for buffaloes inside the forest but this is no longer the case.

Injuries from Snares and Traps

Chimpanzees walking along trails in the forest may put a hand or foot into a snare and then panic; if so, and if they succeed in getting away, the noose is embedded tightly around fingers, wrist, or ankle, often with consequent loss of function and sometimes of the extremities themselves. Chimpanzees engaging in crop raiding are likely to encounter traps and as a result get caught by them with severe damage to, or loss of, entire hands or feet.

Snare-Related Injuries to Sonso Chimpanzees. At the present time, of the 50 noninfant chimpanzees in the Sonso community, 14 chimpanzees are or have been affected by snare injuries (Table 1). Thus 14/50 (28%) of noninfant Sonso chimpanzees are suffering from snare injuries. Insofar as the Sonso range is representative, there may be 164 injured chimpanzees in the whole forest. There

Table 1. Snare injuries of chimpanzees in the Sonso community

Name	Age/sex	Injury
Tinka	AM	Both hands deformed (could be congenital but more likely the results of snare injuries to each wrist)
Zefa	AM	Missing two fingers on right hand but has removed snare
Bwoba	AM	Left hand was damaged but has recovered
Gashom	AM	Right hand was damaged but has recovered
Kalema	AF	Right hand—complete loss of function
Kigere	AF	Right foot lost
Zana	AF	Both hands severely damaged
Mukwano	AF	Right foot damaged
Kewaya	AF	Right hand severely damaged
Shida	SAF	Left hand lost, new snare on ankle
Wilma	AF	Right hand missing
Flora	AF	Right hand missing
Nora	JF	Snare lost and recovering
Kana	JF	Snare on ankle

AM = adult male; AF = adult female; SAF = subadult female; JF = juvenile female.

may be more, because in the Sonso range (see below) we remove snares from the forest on a daily basis. However, there may be less because the Sonso range is closer to human habitation than some of the ranges of other communities in the forest. If these factors balance out, then we may be looking at approximately the right figure.

Injuries in More Detail. Waller and Reynolds (2001) investigated the injuries of the Sonso chimpanzees. The injured limbs fell into two broad morphological categories: claw-hands and wrists, and missing digits and limb segments (Figure 1). Regarding how chimpanzees deal with snares once they are caught, it seems that individuals can sometimes chew through the wire. If the wire is multistranded, then each strand has to be individually bitten through. Once escaped from the snare site, the commonest situation is for the animal to trail the remaining wire, still tightly attached to the fingers or wrist, with loss of use of the affected hand. At this stage the animal is usually paying attention to the wire, licking the wound and occasionally trying to remove the snare. This may go on for days or weeks. In due course, either because the skin falls off or for other reasons, the snare disappears and is not seen any more. Injuries to younger animals may heal up completely, with no loss of function; this has happened in the cases of Zefa, Bwoba, and Gashom (see Table 1). In the case of adults this is less likely and the damage is normally permanent, and can have ramifying effects on an individual's behavior (see Munn, Chapter 15, this volume).

Deaths from Snares and Traps

Since the Budongo Forest Project (BFP) began in 1990, two individuals have been found dead in the forest near camp as a direct result of getting caught in snares. It is impossible, given this small sample size, to obtain an accurate estimate of how many chimpanzees lose their lives to snares in the forest each year. In the case of traps, there have been two cases of deaths from leg-hold traps, both in cultivated areas to the south of the main block of the Budongo Forest, one in Kasokwa Forest (Munn & Kalema, 1999/2000), and one nearby in Kasongoire Forest.

Snare Injuries at Other Sites

Kano (1984) described a number of snare-related injuries in the bonobos of Wamba, Zaire. Hashimoto (1999) has written about the injuries of the chimpanzees of the Kalinzu Forest: of 16 male chimpanzees identified, 2 had a wire

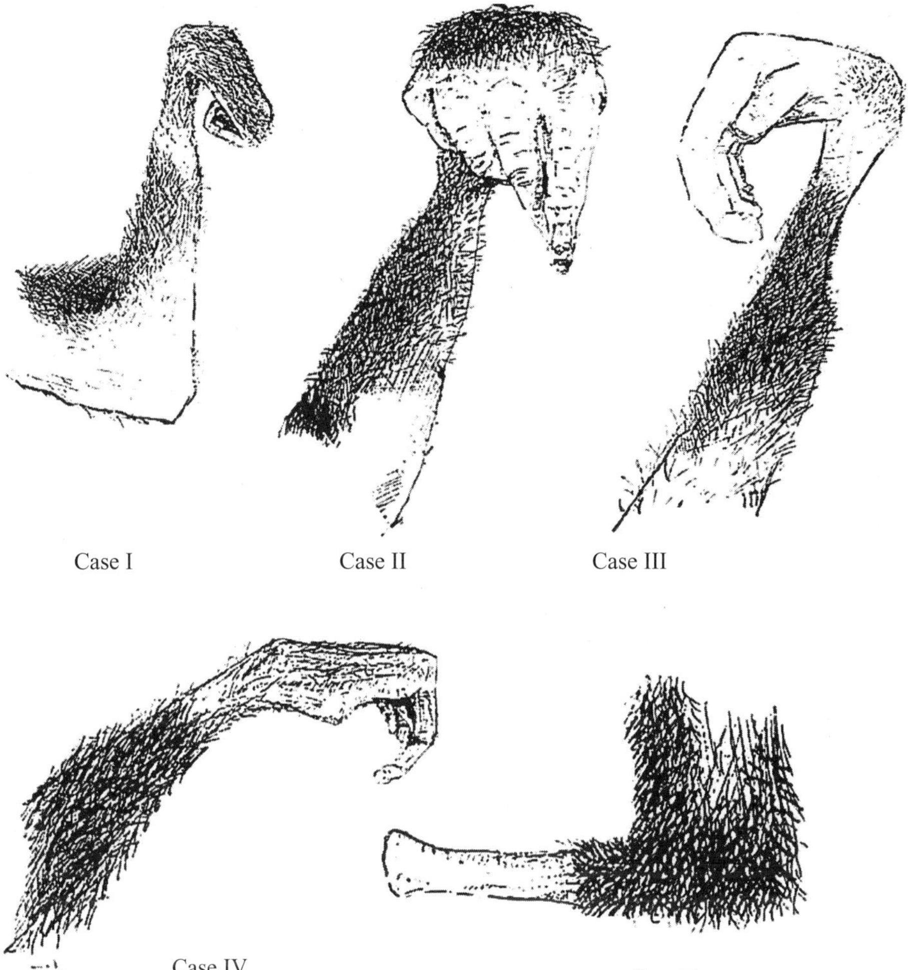

Figure 1. Cases of claw-hand and wrist (Cases I–III) and missing limb segments (Cases IV and V) (drawings by J. Waller, from Waller & Reynolds, 2001).

snare embedded in their hand, while another 7 had injuries on their limbs probably caused by snares, including loss of hand or foot, claw hand or wrist, and loss of digits.

Wrangham and Goldberg (1997) quantified the amount of snaring of chimpanzees in Kibale Forest. Over a period of 300 "snareable chimpanzee years," 11 snares were recorded (snareable chimpanzees were defined as chimpanzees aged 6 years or over). Of 55 snareable chimpanzees in the Kanyawara community at Kibale, 18 showed snare damage as follows—lost hand: 4, lame hand: 5, lame finger: 2, crippled toes: 1. A sex difference was found, with 9/19 (47%)

of males and 9/36 (25%) of females having snare damage. These proportions are horrific. We should note that Kibale Forest was not a National Park during most of the period when these snare injuries came about, it was a Forest Reserve until 1993 and a bushmeat trade was already operational there.

Contagious Diseases

There are a number of other threats to chimpanzees around Budongo Forest. These include diseases spread by human beings. The danger is particularly acute in the case of habituated chimpanzees, but all chimpanzees living near human habitations are at risk. There is one fully habituated community of chimpanzees in Budongo Forest: the Sonso community in the middle of the forest studied by BFP. Of the two ecotourism communities, one at Busingiro in the southwest has not been habituated, but the other, at Kaniyo-Pabidi in the northeast, has been partly habituated, as has the small community in the Kasokwa Forest Reserve to the south of Budongo.

Wallis and Lee (1999) and Woodford *et al.* (2002) are among those who have studied the risks of habituation, focusing on disease transmission. The great apes are susceptible to many human diseases that can cause severe morbidity or death. The commonest routes for transmission are respiratory and fecal–oral. A list of pathogens known to be transmissible between humans and apes can be found in Homsy (1999), copies of which may be obtained from the Senior Veterinarian, IGCP, c/o AWF, P.O. Box 48177, 00100 Nairobi, Kenya.

The danger increases the more habituation has been achieved. The respiratory route can be very dangerous. Five chimpanzees died of suspected pneumonia in the Kasakela community at Gombe National Park, Tanzania, in 1968 (Goodall, 1986) and a further nine died from a similar disease in 1987 (Wallis & Lee, 1999). In the neighboring Mitumba community a respiratory disease killed at least 11 chimpanzees in 1996. Similar events have occurred at Mahale Mountains National Park in Tanzania (Hosaka, 1995). It used to be the case that visitors to gorilla tourism sites were encouraged to hold a baby gorilla in their arms; a peak experience for any tourist, this was nonetheless a highly dangerous moment in the life of the infant gorilla. In 1988 no less than 81% of the gorillas in seven tourist and research groups in the Virunga Volcanoes region showed signs of influenza-like symptoms, 6 females died and 27 cases were treated with penicillin. As a result, a vaccination program was implemented and 65 gorillas were vaccinated using a dart-gun (Woodford *et al.*, 2002). A study by Kalema (1995b) compared two groups of gorillas at Bwindi Impenetrable Forest, one habituated tourist group and one unhabituated group. She examined fecal samples and found higher levels of gut parasites in the habituated group, probably because of contact with human excrement in the forest.

As awareness of this problem has grown, measures have now been taken to reduce the risks, and most sites where habituated apes can be seen have rules limiting the proximity of visitors to the apes. At Budongo, sick visitors, especially those with respiratory diseases such as the common cold, or fecal diseases such as diarrhea, are not allowed into the forest. Inside the forest, we adhere to the rule that proximity between humans and chimpanzees must never be less than 7 m. We also have a rule that all human feces must be buried. Finally, we limit to four the number of students working on our chimpanzees in the forest at any one time, to avoid causing them stress.

Thus the threats to chimpanzees at the present time come primarily from the human population. Direct threats are deaths and injuries caused by hunters setting snares, and people entering the forest for various purposes taking with them contagious pathogens. Indirect threats are forest fragmentation and degradation caused by the current influx of individuals seeking work, mainly at the Kinyara Sugar Works, who destroy forest in order to plant crops and build houses. These threats are formidably difficult to deal with and to date no proper solution has been found; the provision of live traps (see below) has led to the release of a small number of chimpanzees and must be regarded as a first step toward finding a solution. Probably, however, the limitations of this method make it unlikely to be successful on a wide scale. For a long-term solution it will be necessary to police forest reserves effectively, and to educate local people into an attitude whereby they will themselves value and want to conserve chimpanzees. We are far from either of these solutions at the present time.

PROTECTION MEASURES

Buffer Zones

A huge area of land to the south of Budongo Forest is today covered in sugar, owned by the Kinyara Sugar Works (KSW). In recent years, KSW has contracted hundreds of small farmers who own land adjoining its own land to grow sugar on their fields, for which there is a guaranteed payment each year at harvest time. This income is a tremendous incentive to farmers to turn their land over to sugarcane and most of them have done so. As a result, land right up to the forest edge is now planted with sugarcane. Consequently, there is more crop raiding than hitherto because of the lure of the sugarcane for primates. Baboons are the chief raiding species but chimpanzees also raid for sugarcane. I had always believed that farmers were more tolerant to chimpanzees than to baboons, and indeed many of them are. But not all, alas. Besides the two chimpanzees caught and killed by leg-hold traps to the south of Budongo referred to above, in 2003, a sugarcane guard speared to death one of the fine adult males in the Sonso community, Jambo, who was raiding crops. This individual had been known to staff and students of BFP since 1992 and it was a great hurt to all of us when he was killed. To date, despite the illegality of the killing of an endangered species, no court proceedings have been lodged against the guard by the relevant authorities. The BFP wants to ensure that all those who plant sugarcane right beside the forest understand their responsibilities toward Uganda's chimpanzees, and to this end we have an ongoing education project in the local schools and villages.

One way to reduce such needless loss of life by chimpanzees is to have buffer zones between the forest and the cropland. Such zones can be accompanied by electrified fencing, which is not hugely expensive for a large company such as Kinyara Sugar Works. They may be planted with crops chimpanzees do not normally eat such as chili peppers, yams, or cassava. Thorny fences can also be built in them. Housing can be placed in the buffer zone. Their main advantage is that they prevent farmers growing crops right up to the forest edge and thus encouraging chimpanzees (and other primate and nonprimate species) to come out and raid. In September 2000, the BFP discussed with the relevant authorities the idea of a buffer zone between Kinyara Sugar Works and the forest but the problem arose that the land concerned belonged to private farmers and they were free to grow sugarcane on it. There would have to be adequate compensation payments if they were to be prevented from growing sugarcane on it. With the death of Jambo in 2003 a new impetus has been given to the buffer zone idea and we have now seen the first steps toward the creation of something effective. In a recent development the Kinyara Sugar Works has agreed to leave a strip of unplanted land 12 m wide between the forest edge and the sugarcane fields of outgrowers, and has informed its outgrowers that it will not buy sugar grown within 12 m of the forest edge. We at BFP hope this new zone will be maintained and extended, and we shall see in due course if chimpanzees cross this strip.

Tree Corridors

In the present situation of population explosion around Budongo, with constant clearing of land at the forest edge and in forest outliers to the south of the forest, chimpanzees are being threatened in two ways. Their freedom of movement from the outliers to the main forest block has been severely curtailed and in some cases they are cut off. Since the food supply in the forest outliers is insufficient for their needs, they survive by crop raiding and suffer accordingly. When moving outside the forest outliers, they have to tread carefully and try to avoid contact with people and traps. There is currently no provision for their need to move around.

Tree corridors should be set up expressly so that chimpanzees can move between the existing forest fragments in safety. This has been achieved in Guinea, West Africa, between the Nimba Mountains and Bossou (Matsuzawa, 1998). This tree corridor was planted and is being maintained by local people, who are paid a small sum for this. It is a model Uganda would do well to study and follow.

The Budongo Forest Project has tried to initiate a tree corridor between the Kasokwa Forest Reserve and the main Budongo Forest block, but so far without success.

Snare Removal Project

The Budongo Forest Project began a snare removal project in January 2000, employing two local hunters from Nyakafunjo. In January they brought in 231 snares, in February they brought in 172 snares, in March they brought in 111 snares: 514 snares confiscated from the forest in 3 months. The types of snares are indicated above (Table 2).

The result of this project was that local hunters decided to take recrimi native action. They threatened the snare removers with injury and removed scientific equipment belonging to one of our PhD students from the forest,

Table 2. Types of snares removed from Budongo Forest, January–March 2000

Month, year	Thick wire	Thin wire	Cord or nylon	Fishing line or electric wire
January 2000	23	160	47	1
February 2000	7	147	14	4
March 2000	11	77	4	19
Total	41	384	65	24

prejudicing his project. The two staff were laid off snare removal duties and assigned to our education project. We started a program of weekly education meetings in villages and schools. This continues, in modified form, today, as part of our general education program. In July 2000 we restarted snare removal, recording the locations of snares with a GPS. During the 16 months from August 22, 2000, to December 13, 2001, 399 snares were found, fewer than in the first 3 months of 2000. However, our removers continued to work in and around the Sonso area only, and we do not know if hunters moved to join those in other areas, with consequently increased dangers for chimpanzees there. Since that time we have been removing snares daily and continuing the education program. There has not been a repeat of the removal of scientific equipment from the forest, nor have our snare removers been threatened again.

However, we are now faced with the growth of a small but significant bushmeat trade along the main Masindi–Butiaba road which runs to the south of Budongo Forest. Until the mid-1990s there was no organized sale of bushmeat; wild animals were occasionally sold in the local markets by the hunters themselves or their wives. Meat traders now set snares in the forest, collect the animals, and take them to particular places along the main road for sale. This is a new development and will inevitably lead to the depletion of the forest animals and the snaring of more chimpanzees, though to date I have not encountered any evidence that chimpanzees are being caught for bushmeat in or around Budongo Forest.

Plumptre *et al.* (2003) in their recent survey of the chimpanzees of Uganda found the greatest amount of hunting sign in Budongo Forest and its neighbor to the south, Bugoma Forest, of any of the western forests. Most of the sign was encountered around the Sonso area and to the east of Sonso, with an additional hot spot in the Busingiro area (Figure 2). These are areas where owing to the existence of trails made by legal and illegal loggers extracting mahogany, hunters (often the loggers themselves) are able to move freely in the forest. Our own trail system is also, undoubtedly, used by hunters.

Live-Trap Project

While we have made no more than limited progress with the snaring problem, we have been somewhat more successful in regard to the problem posed to chimpanzees by leg-hold traps. The Masindi District Wildlife Management

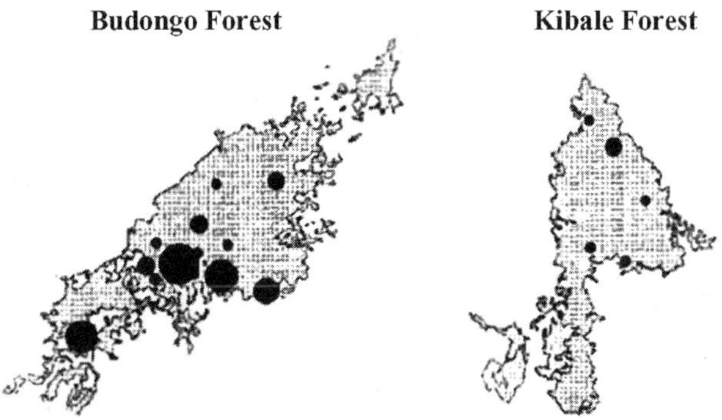

Figure 2. Map of hunting sign in Budongo and Kibale forests (courtesy of A. Plumptre, WCS, from Plumptre *et al.*, 2003).

Unit's director, Christopher Byarugaba, has developed a kind of trap that can be used to kill vermin and release chimpanzees. This he calls the "Christopher" trap and it catches whatever animal enters it alive; this animal can then be killed or released. The trap has a vertical door held up by stout wire that runs down to a baited toggle inside the trap; when the animal touches the bait the door comes down, trapping it. Having seen his traps in action, the Budongo Forest Project has supported his efforts in Budongo subcounty and several chimpanzees including two mother-infant pairs have been released (Figure 3). A condition of having such a trap is that leg-hold traps must no longer be used; if such traps are found the live trap is removed. This project has been successful and has shown farmers that we do understand their major problem, which is crop raiding by baboons and other species, notably wild pigs and porcupines.

Logging

Finally, mention needs to be made of logging and its primary effect on chimpanzees and all wildlife: habitat destruction. Since the 1930s, logging has been a feature of Budongo Forest, because this forest contained the country's biggest stock of mahoganies (one species of *Khaya* and three species of *Entrandrophragma*). Fortunately for Budongo and its wildlife, however, the logging, although practiced on a large scale by large companies, was selective, with valuable timber trees removed where they stood and dragged to nearby timber trails, rather than clear felling, which has never taken place in Budongo. In

Figure 3. Photo of chimpanzee mother and infant released from live trap (photo: C. Byarugaba).

some respects, selective felling has been beneficial for certain species of wildlife, such as monkeys (Plumptre & Reynolds, 1994; see Plumptre, Chapter 20, this volume).

However, in the 1980s and 1990s, logging ceased to be in the hands of a small number of large companies and was taken over by a large number of small privately hired logging gangs, mostly illegal, who moved quietly, often by night, removing valuable trees, pit-sawing them in the forest, and carrying the planks to waiting trucks. During this period the forest was heavily occupied by illegal pit-sawyers who practiced hunting as well as sawing, and the number of snares rose dramatically.

A second source of threat to chimpanzees constituted by logging is the gradual depletion of the forest by newly arrived immigrant farmers and laborers around forest edges and in forest outliers. Many of these outliers, which were formerly used by chimpanzees to move outside the forest, have now disappeared. And the forest edge is slowly being chipped away along the southern boundary as newly arrived immigrants make farms for themselves there. They quickly find that their fields are subject to crop raiding and take preventative measures, placing traps and snares around their fields on the forest edge. And they set guards to spear wildlife entering their fields.

CONCLUSIONS

The main conclusion of this chapter is that the chimpanzees of Budongo Forest Reserve face a growing threat to their survival, from a number of quarters. Increasing levels of hunting and an increase in the number of hunters looking for duiker and pig meat for their families are placing more and more snares in the forest, and traps around their fields, with the result that over a third of noninfant Sonso chimpanzees have sustained injuries, and the same proportion is likely to apply to the entire chimpanzee population in the forest. With increased numbers of people in the forest, the risk from contagious diseases is also increasing.

On the protection side, we have outlined a number of measures that we are taking at Sonso, notably snare removal and a live-trap project. Other measures that could and should be tried are buffer zones and tree corridors, the latter to give chimpanzees greater freedom of movement outside the main Budongo forest block.

We have also noted that there are weaknesses at the institutional level, with relevant authorities not taking sufficient action to protect the Budongo chimpanzees. It is to be hoped that this situation will change in the near future.

CHAPTER TWENTY-THREE

Gorillas Living on the Edge:

Literally and Figuratively

Michele L. Goldsmith, Joel Glick, and Evarist Ngabirano

INTRODUCTION

Owing to their dwindling numbers and endangered status, gorillas can be regarded figuratively as "living on the edge" of survival (e.g., Population Action International, 2000). Although numbers continue to decline in most of their range, two small populations in East Africa have actually increased in size. A 2003 gorilla census in the Virunga Volcanoes resulted in 380 individuals, representing a 17% increase since 1989 (as reported in Meder, 2004), while gorillas in Bwindi Impenetrable National Park increased in size from 290 individuals in 1997 (McNeilage *et al.*, 2001) to 320 individuals in 2002 (unpublished census data by Ugandan Wildlife Authority [UWA], International Gorilla Conservation Program [IGCP], and Institute of Tropical Forest Conservation [ITFC]). Although no direct evidence or data exist to explain these increases, it is the case that these areas experience committed protection by government officials, NGOs, and park rangers primarily for the maintenance and protection of

Michele L. Goldsmith • Emerson College, 120 Boylston St., Boston, MA 02116, USA **Joel Glick** • Mbeli Bai Study, Nouabale-Ndoki Project, Wildlife Conservation Society, BP 14537, Republic of Congo, Africa **Evarist Ngabirano** • Ecotourism and Behavioral Ecology of the Nkuringo Group, c/o Bwindi Impenetrable National Park, P.O. Box 862, Kabale, Uganda , East Africa

Primates of Western Uganda, edited by Nicholas E. Newton-Fisher, Hugh Notman, James D. Paterson, and Vernon Reynolds. Springer, New York, 2006.

populations for tourism enterprises (see Schmidt, 1999, for a story on the sacrifices made by some rangers).

Ecotourism, one of the fastest-growing global industries (Luzar *et al.*, 1998), was first implemented in East Africa in 1973 at Kahuzi Biega National Park, Democratic Republic of Congo (DRC), and was followed in 1979 with tourism in the Volcanoes National Park, Rwanda (Vedder & Weber, 1990). Tourism did not start in the Ugandan park of Bwindi Impenetrable until 1993. Ecotourism is often touted as this ape's saving grace (e.g., Weber & Vedder, 2001) and several researchers agree that it is vital for their survival (e.g., Harcourt, 1986; Vedder & Weber, 1990; Sholley, 1991; Stewart, 1991; Weber, 1993; McNeilage, 1996), while others take a more critical look and have become concerned with the potential consequences of this conservation tool (e.g., Frothmann *et al.*, 1996; Butynski & Kalina, 1998; Homsy, 1999; Goldsmith, 2004).

Butynski and Kalina (1998) present a thorough discussion of both the benefits and costs of gorilla tourism. Two main areas of concern with regard to the well-being of the gorillas are increased risk of disease transmission and changes in gorilla behavior. Studies on disease risks in great ape groups used for tourism emphasize the need for more precautions (Homsy, 1999; Wallis & Lee, 1999; Woodford *et al.*, 2002). Specific to mountain gorillas, studies have shown increases in endoparasitic loads and infections (Ashford *et al.*, 1996; Graczyk *et al.*, 1999; Nizeyi *et al.*, 1999, 2001; Sleeman *et al.*, 2000; Graczyk & Cranfield, 2001), multiple scabies outbreaks (Macfie, 1996; Kalema *et al.*, 1998; Meder, 2000; Graczyk *et al.*, 2001a), and complications with respiratory infections and measles (Sholley & Hastings, 1989; Hastings *et al.*, 1991). Other factors such as stress can also influence health (Hudson, 1992; Nizeyi, personal communication). With regard to behavioral changes, however, little research has been conducted, primarily because of tight restrictions of visitation to these groups and a policy of "no research" on tourist gorilla groups. This has changed recently as managers and government wildlife officials start recognizing the need for this information. The first such study conducted on a tourist group examined direct impacts of visitors in Volcanoes National Park (Steklis *et al.*, 2004). Among other findings, the investigators demonstrate that tourist gorilla groups spent significantly more time moving during tourist visits at the expense of feeding. A similar study has been conducted on the tourism groups in Buhoma in Bwindi (Muyambi, 2005).

In an attempt to reduce possible exposure to disease and influences on behavior, the minimum distance tourists are required to keep from the gorillas was increased from 5 to 7 m (following reports by Homsy, 1999, and Muyambi, 2005). However, tourism regulations, such as mandatory distance, are not always followed in either the Virungas (Stewart, 1992; McNeilage, 1996; Ströbele-Gregor, 2000) or Bwindi (Macfie, 1997; Schmitt, 1997, personal observation). In addition, an even less controlled situation exists regarding contact between gorillas and local people living in surrounding communities. In Bwindi, for example, gorilla groups used for tourism live close to the park edge, a decision made to reduce impacts in the park by providing most of its infrastructure along the boundary. The area surrounding Bwindi in Southwestern Uganda, however, maintains one of the highest densities of rural human habitation (about 200–300 people per square kilometer) and one of the highest population growth rates in the world (ranging from 1.7 to 4.5% in some districts) (Ministry of Planning and Economic Development, 1997). In most areas, agricultural land abuts the park boundary so if gorillas venture out or humans venture in there is a high likelihood that they will interact (either directly or indirectly), which can create problems regarding safety, health, crop raiding, and strained relations between local people and park officials.

A necessary prerequisite for great ape tourism (and research) is habituation. To "habituate" means to accustom by frequent repetition or prolonged exposure (*The American Heritage College Dictionary*, 1993). Primatologists often describe habituation as a process by which wild animals accept human observers as a neutral element in their environment (Tutin & Fernandez, 1991; Blom *et al.*, 2004). (However, whether observers can ever truly be neutral elements in any animal's environment has been challenged [Burghardt, 1992; Goldsmith, 2004].) Regardless, the result of habituation is that an animal's fear of humans is greatly reduced. So, in Bwindi, we have habituated gorilla groups not fearful of humans living along the edge of the forest.

This chapter presents preliminary findings of the impact that "living on the edge," literally, has on the behavioral ecology of a habituated group in Bwindi. Our aim is to document the extent to which this habituated group uses areas outside the park boundary, examine the factors that attract them to these areas, and determine preliminary impacts on behavioral ecology. Findings add to our knowledge base of Bwindi gorilla ecology and behavior, which is still poorly known, and in the long term will assist us in examining possible impacts of tourism. Data on habitat use are important in implementing management plans and in reducing human–animal conflict. Findings are also significant and timely as the Ugandan Wildlife Authority starts investigating the management of a new 350-m buffer zone around the park and as tourism begins on this habituated group.

METHODOLOGY

Study Site and Subjects

Bwindi Impenetrable National Park (331 km^2) in the southwestern section of Uganda was gazetted in 1991 and tourism started in the village of Buhoma in 1993. At present, there are four groups habituated for tourism and one habituated for research. These groups comprise at least 75 individuals, representing close to 25% of the entire Bwindi population. The focal population of this study is the Nkuringo group in the Nteko/Kashasha region. This site is a 6–7-h walk southeast of the main tourism site of Buhoma. Although the Nkuringo group had been habituated for tourism since 1998, preliminary (mock) tourism did not begin until early 2004. Tourism had been postponed because, according to the IGCP, "there is no infrastructure, the gorillas are spending too much time outside the park, and a clear tourism development plan has not been finished" (Meder, 2000:8). At the time of this study, the Nkuringo group was composed of 19 individuals, including 2 adult silverback males, 4 large blackback males, 6 adult females, 6 adolescents, and 1 juvenile.

General Methodology

Research on gorilla groups habituated and intended for tourism is politically and logistically difficult, since strict guidelines on visitation to these groups exist (limited to 1 h a day). This fact, along with a desire to compare Bwindi gorillas to that of unhabituated western lowland gorillas (Goldsmith, 2003), led investigators to use noninvasive methods that did not require observation (e.g., Goldsmith, 1996, 1999). Therefore, little pressure was placed on this group as direct contact was minimal. Visual observations were made, on occasion, during the 1 h allotted time and are referred to in the Discussion section.

The diet of the Nkuringo group was estimated by examining fresh feeding sites and dung along their trail, noting both species and the plant part eaten (e.g., leaf, pith, bark, flower, and fruit). Foods eaten were compiled on a daily and monthly basis for comparison with other variables.

The daily path length (DPL) of the group was measured as the distance they traveled between sleeping nest sites. Paths were followed by tracking fresh gorilla trails using a hand clicker to count paces. Occasionally a hip chain was used in conjunction to check for accuracy. Readings with a Garmin II global positioning system (GPS) were taken along the path at each nest site and at about every 100 steps along the trail. Researchers were sure to take into account when groups were chased back toward the park. "Chasing" occurred when UWA rangers and/or locals scared the gorillas with load noises from agricultural fields back into the park. We also noted when veterinarians visited the group and when intergroup encounters with "wild" gorillas occurred, as all three situations were thought to affect daily ranging patterns. DPLs were also mapped on a daily and monthly basis.

In addition to diet and ranging, we measured the cohesion of nesting individuals as an estimate of group spread. We were interested to see if this variable was affected by where the gorillas slept, hypothesizing they might be more cohesive when nesting in human-inhabitated areas. To control for differences in total nest count, since not all nests were always found, we measured cohesion as the mean area around each nest (cf. Williamson, 1988).

To establish how the gorillas were using their habitat, each nest site and path length (i.e., each step) were recorded as being in one of six different locations: WIP, well inside the park (>400 m from boundary); IP, inside the park (200–400 m from boundary); JIP, just inside the park (<200 m from boundary); JOP, just outside the park (<200 m from the boundary); OP, outside the park (200–400 m from boundary); WOP, well outside the park (>400 m from park boundary). For visual assistance in examining habitat use and determining home range, GPS data from nests and gorilla trails were downloaded into the software program IDRISI (1997) and then mapped in ArcView (1996).

RESULTS

Habitat Use by the Nkuringo Gorillas

Data were collected on the Nkuringo group for 18 months from July 2001 to December 2002. Data were sparse, however, for 2 months from August 2001 to September 2001 owing to illness of one of the investigators.

We examined 288 sleeping nest sites containing a total of 4170 individual nests. The Nkuringo group nested on 24 different mountains (three of which were only visited once). They slept most often on Bushaho Mountain (33%) followed by Miheno, Byahurumba, Rutebesa, and Nyamabuye Mountains (10,

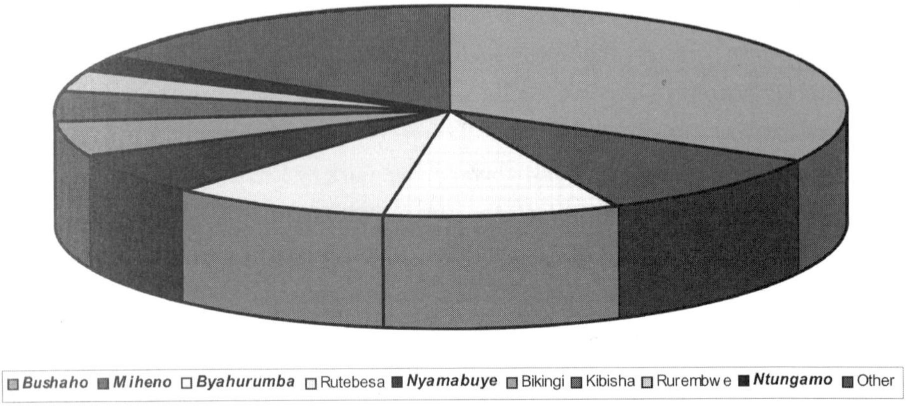

Figure 1. The proportion of nest sites found on 24 different mountains in and around Bwindi Impenetrable National Park during the study period. Bold and italicized names indicate mountains located outside of BINP.

10, 9, and 7%, respectively) (Figure 1). Of the top five mountains, all but one, Rutebesa, are found almost completely outside of the National Park. Nest sites ranged in altitude from 1500 to 2248 m above sea level, with an average of 1754 m (SD = 153 m). (However, altitude measurements using a Garmin II GPS should be viewed with caution.)

Table 1 presents the percentage of gorilla sleeping nest sites in each of the different locations. Overall, the group nested inside the park on only 69 of 288

Table 1. The frequency of sleeping nest sites and gorilla trails* within the different locations

Location	% Nest sites	No. of entrances[a]	% Gorilla trail
Inside the Park			
WIP, >400 m from boundary	8.2	27 (5.8)	10.0
IP, 200–400 m from boundary	5.0	41 (8.9)	13.3
JIP, <200 m from boundary	11.2	118 (25.5)	22.1
Outside the Park			
WOP, >400 m from boundary	15.3	30 (6.5)	6.9
OP, 200–400 m from boundary	22.7	83 (17.9)	15.5
JOP, <200 m from boundary	37.6	164 (35.4)	32.1
Total inside the park	24.4	186 (40.1)	45.4
Total outside the park	75.6	277 (59.9)	54.5

* Trail data include both complete and incomplete path lengths. [a]Gorillas could only enter each location once a day (percentages are in parentheses).

occasions (24%). Nest site frequency inside versus outside the park differed significantly even when chi-square expected values were set at 50% ($X^2 = 78.1$, $P < 0.0001$), with Nkuringo nesting most often (37.6%) within 200 m outside of the park (JOP) and least often (5%) within 200–400 m inside the park (IP).

Data from gorilla trails were collected over 281 days for a total of 189.13 km. Of the follows, 202 constituted complete DPLs, while 79 were partial path lengths, where trails were either lost, unclear, or incomplete because of work schedules. Complete DPLs averaged 737.4 m ($n = 202$), with a minimum of 36 m and a maximum of 4112 m traveled each day. Longest path lengths resulted from intergroup interactions with "wild" gorilla groups in the area, causing the Nkuringo group to move fast and far, with DPLs on August 12, 2002, of 2311.6 m and September 8, 2002, of 4112.2 m. Longer path lengths also resulted when the group was chased by park rangers (a mean of 1001 m for eight different occurrences, with a minimum of 713 m and a maximum of 1565 m). On one day, the group was chased and then had an interaction with a "wild" group (September 26, 2002), resulting in a path length of 2419.2 m. DPL averaged 704 m ($n = 191$) when these 11 values were removed from the analysis.

Using all complete and incomplete follows, Table 1 presents the total number of times each location was entered and the percentage of travel within each. The gorillas entered locations outside the park more often (60%) than those inside the park (40%) ($X^2 = 17.9$, $P < 0.001$) but the percentage of trails inside (45%) versus outside (55%) were not statistically significant ($X^2 = 2.8$, $P = 0.09$). However, the number of times each location was entered strongly predicted the percentage of trail within that location ($r^2_{[4]} = 0.95$; $P < 0.001$).

To be able to examine these data visually, a map was generated by ArcView (1996). Figure 2 represent GPS readings from sleeping nest sites and from those taken while following along gorilla trails. The estimated home range area for the Nkuringo group based on all GPS points collected during the length of study was approximated at 37 m^2 (width × length).

Like the location data, the GIS map demonstrates that the gorillas most often slept and foraged close to the park boundary. The gorillas also tended to use watershed areas when outside the park (mostly to sleep in; but this distinction is not evident from this map as it does not distinguish between nest sites and trail data). Note the heavy activity inside the park to the southwest. This area

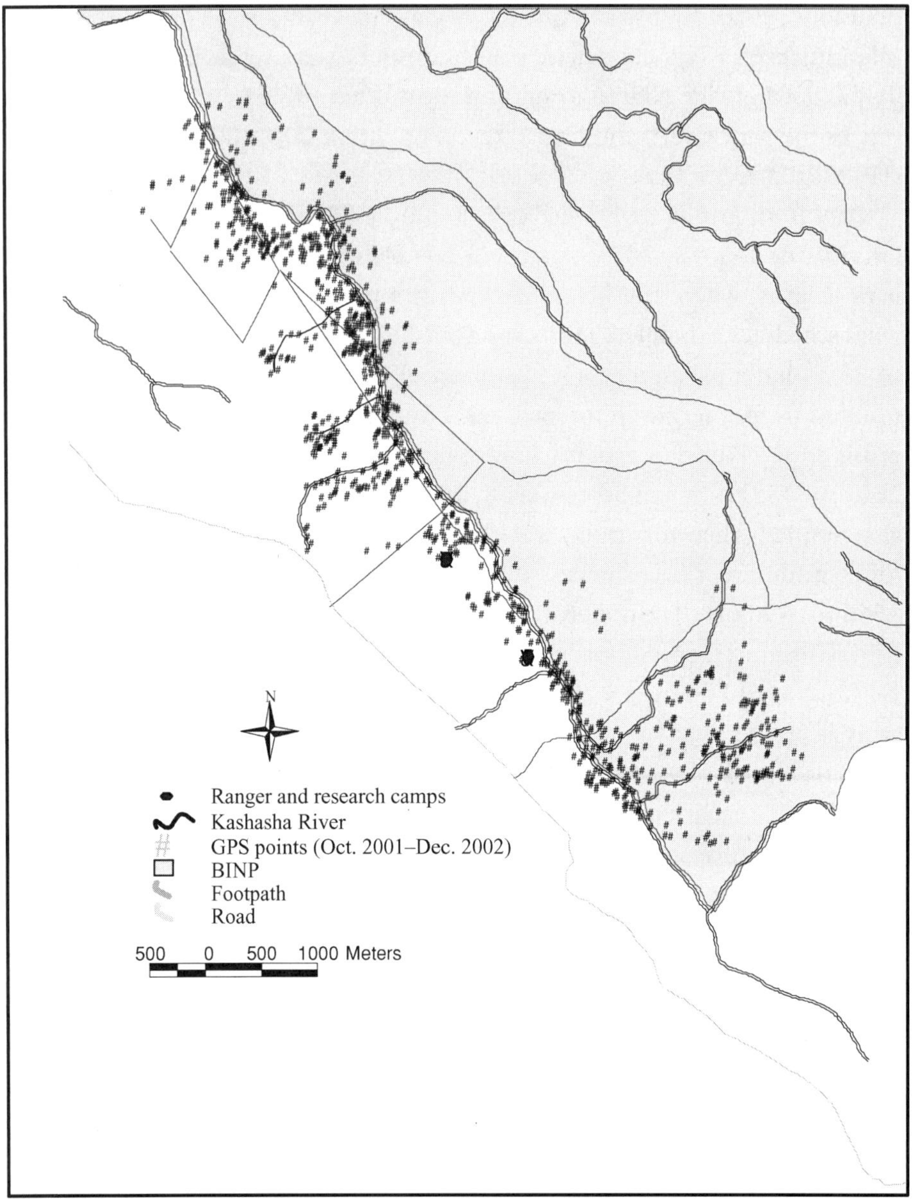

Figure 2. GIS map of the study area, with nest sites and path lengths of the Nkuringo group plotted.

was visited during high fruit availability months, which in this study spanned January 2002–March 2002.

Behavioral Correlates of Spending Time Outside the Park

Dietary Behavior

During the study period, the Nkuringo group ate the leaves, bark, pith, and flowers of 71 different species of plant—eating fruit from 18 of these species. When outside of the park, gorillas incorporated nonforest foods into their diet. They were recorded as destroying more than 842 (13%) sweet and plantain banana plants (matoke), 90 eucalyptus trees (1%), and smaller amounts of sweet potato and sugarcane. They also ate 928 wild banana plants (matembe) (15%) found outside the park boundary. This plant is not used as food by the local people but was an important attractant for the gorillas. Note that crop raiding numbers are underestimated, since data were not collected daily—only when events were observed by chance. Table 2 lists the most commonly eaten forest foods (regardless of feeding location) determined from 4519 feeding sites (after crop foods were removed).

Nest Cohesion

Of the 288 sleeping nest sites, six either had a large number of group members missing or the site was split into two separate sleeping areas. These cases were

Table 2. The most common noncrop foods eaten based on the total percentage of feeding sites found along trail ($n = 4519$)

Genus	Ruchiga Name	Total % feeding sites	% of sites composed of				
			Bark	Pith	Leaves	Fruit	Flowers
Myrianthus	Omufe	16.73	10.56		2.94	3.23	
Urera	Omushe	16.57	15.45	0.04	1.08		
Basella	Enderema	7.57			7.57		
Ficus	Omusene	7.72	7.48		0.20	0.04	
Pilea	Omumbya	6.04	4.8	0.89	0.33	0.02	
Cyathea	Ekigunju	5.27		5.27			
Mormodica	Omwihura	4.96			4.87	0.09	
Cola	Omulehe	4.05	2.88		0.75	0.42	
Brillantasia	Ebyunga	3.63	0.02	2.72	0.18		0.71
Afromomum	Amatahe	3.39		2.39		1.00	
Chrysophyllum	Omushayu	0.93				0.93	

Table 3. Mean area around each nest within a sleeping site (including standard deviation and sample size) for each location ($n = 282$)

Location	Mean	Standard deviation	Sample size (n)
Inside the Park			
WIP, >400 m from boundary	139.6	74.5	21
IP, 200–400 m from boundary	123.4	84.4	14
JIP, <200 m from boundary	95.5	72.3	53
Outside the Park			
WOP, >400 m from boundary	152.5	117.2	37
OP, 200–400 m from boundary	118.3	63.4	52
JOP, <200 m from boundary	108.7	75.2	105
Mean area inside	110.4	76.8	88
Mean area outside	121.0	83.1	194

removed when determining the mean area around each nest, which was not statistically different inside versus outside the park ($F_{[1,280]} = 1.1$, $P > 0.05$). When comparisons are made among all six locations, the only significant difference is found between the least cohesive nests in locations greater than 400 m outside the park (WOP) with the most cohesive sleeping nests just inside the park boundary (JIP) ($F_{[5,277]} = 2.74$, $P = 0.02$). There is a trend for sleeping sites to become less cohesive as gorillas nest farther from the park boundary, especially when outside the park. Table 3 gives the mean area around each nest—the larger the number the less cohesive.

Influences on Daily Path Length

To examine how gorilla travel was influenced by habitat use, all path lengths with greater than 90% of their travel inside versus outside the park were determined (Table 4). Of the 202 complete path lengths, 127 had 90% or more of their length either inside ($n = 26$) or outside ($n = 101$) the park (excluding chases, etc.). The mean distance traveled inside (692.9 m) did not differ significantly from the distance traveled outside the park (592.8 m) ($F_{[1,125]} = 1.47$, $P = 0.23$). We also considered each location separately and found that 93 path lengths actually had 90% or greater of their length within a specific location. Comparing across all locations ($F_{[5,86]} = 4.1$, $P = 0.002$) the longest path length, 200–400 m inside the park (IP), was significantly different from the three shortest distances (200–400 m OP; <200 m JIP; and <200 m JOP) (see Table 4).

Table 4. Mean daily path length for all path lengths with greater than 90% of their trail in one location

Location	Mean	Standard deviation	Sample size (n)
Inside the Park			
WIP, >400 m from boundary	721.7	226.0	7
IP, 200–400 m from boundary	1058.1	845.6	5
JIP, <200 m from boundary	469.1	318.3	15
Outside the Park			
WOP, >400 m from boundary	609.2	303.8	13
OP, 200–400m from boundary	409.8	257.4	24
JOP, <200 m from boundary	533.7	280.0	29
Mean DPL of all paths >90% inside	692.9	520.3	26
Mean DPL of all paths >90% outside	592.8	329.3	101

Monthly Summaries

Table 5 presents a number of monthly variables collected during the study period. High fruit months were determined as months when fruit comprised more than 15% of the food sites along gorilla trails. This included 5 of the 13 months with a mean monthly path length of 867.7 m. During the remaining 8 months, mean monthly path length was 747.6 m. A regression of the percentage of fruit sites on monthly path length found no relationship ($r^2_{[12]} = 0.01$, $P = 0.92$). Simple regressions demonstrate that mean monthly path length was not correlated with any of the monthly variables listed in Table 5, including total number of feeding sites, mean fruit species diversity in the diet, total monthly rainfall (in mm), average daily rainfall (in mm), or mean minimum and maximum temperatures (in Celsius).

DISCUSSION

The habituated Nkuringo gorilla group spent a substantial amount of their time nesting, feeding, and ranging outside of their home in Bwindi Impenetrable National Park. They frequented many mountains in the region, with four of the five most often visited being located almost completely outside the park boundary. They most often made sleeping nests and entered locations outside the park. The GIS map shows that most of their activity was just outside the park boundary but that the Nkuringo group also ventured far, sometimes greater than 1 km from the border. During at least two follows, the gorillas traveled (and nested) just below the main road in the town of Ntungamo. The gorillas

Table 5. Monthly average measures of daily path length, feeding sites, and climate for one full year

Month	n	Mean DPL	Total no. of food sites	% of sites with fruit	Mean fruit diversity	Total rainfall	Average daily rainfall	Average minimum temperature	Average maximum temperature
December 2001	13	865.2	36.7	6.6	1				
January 2002	10	634.1	28.8	20.3	0.9	155	5.01	14.35	22.06
February 2002	9	1107.3	38.0	56.8	1.11	92	3.29	14.21	24.21
March 2002	17	675.3	38.4	10	1	142.5	4.6	14.42	23.39
April 2002	15	570.8	32.5	26.4	1.67	287	10.25	15.57	21.86
May 2002	15	888.6	37.6	9.3	1.27	203.5	6.56	13.87	22.06
June 2002	15	953.3	29.9	9	0.87	35.5	1.18	14.23	25.07
July 2002	16	684.8	27.1	7.9	0.94	43.5	1.45	14.9	25.53
August 2002	15	844.9	54.7	7.3	1	127	4.54	14.68	24.68
September 2002	16	1211.7	57.4	16.5	1.31	127	4.23	14.73	23.47
October 2002	16	814.5	50.3	23.3	1.38	194.5	6.27	14.45	23.03
November 2002	11	616.7	53.5	12	1.36	310	10.33	14.07	23.5
December 2002	8	452.3	36.4	8.1	0.89				

Rainfall is measured in millimeters, while temperature is measured in degrees Celsius.

would have crossed over if it had not been for the ranger's efforts to chase them back toward the forest.

At the onset of the study, we had predicted that gorilla sleeping sites would be more cohesive outside of the park because of increased vulnerability of nesting in human-inhabited areas. There was, however, no significant difference in nest cohesion. In fact, nest sites were less cohesive (i.e., individual gorillas nested rather far from each other) when they were furthest from the park boundary, especially when outside the park. Nest site cohesion instead is most likely related to the availability and distribution of preferred nesting materials. In a recent study on topography (Goldsmith & Moles, in press), it was demonstrated that the area near the park boundary was relatively flat as it represents the river valley. Preliminary data from 3 km of transects demonstrate that vegetation was more dense outside the park (total density of 5172 stems) than inside the park (1904 stems). It may also be that as you move into or out of the park from the Kashasha River you travel up-slope, which may influence the availability and distribution of vegetation. We are presently examining the relationship between slope and the availability of food and nesting material to determine its influence on nest site construction and cohesion among nesting individuals.

As demonstrated by the GIS map, the gorillas often nested on or near streams, which, like Kashasha, are also found in flatter, valley areas with dense vegetation. These streams are important sources of fresh water for cooking, washing, and drinking by local people living in the area. Therefore, gorilla dung and urine may contaminate these sources and be problematic for the local people as well as all those that use the Kashasha River as these streams are major tributaries.

When examining a subset of the data where greater than 90% of any path length was inside or outside the park or within a specific location, we found that the total distance gorillas traveled did not significantly differ. The trend was for path lengths to be longest when inside the park, where food may be less densely distributed and available. It is also possible that longer path lengths result from increased effort when feeding on fruits, which was one reason the gorillas entered the forest. The GIS map shows most activity inside the park in the southwestern area of the range, where the gorillas traveled seasonally when fruits (mostly *Myrianthus*) were available. However, monthly measures do not demonstrate any relationship between the amount of fruit in the diet and average path length. In addition, Goldsmith and Moles (in press) examined the relationship between diet and DPL and found no relationship between the proportion of fruit sites and DPL. So it appears, in this area of Bwindi at least, that Nkuringo's DPL is not affected by frugivory. The trend for longer DPLs inside the park is consistent with the observation of each of the authors that gorillas tended to travel within the safety of the park and then come out across the river to enter agricultural fields. They would then spend time in this area (perhaps a few days) traveling relatively short distances and then reenter the park to travel to a different area. The number of feeding sites and the distance between them also appeared to be less when traveling inside the park, suggesting that overall gorilla foods are more scarce inside the park, increasing the group's foraging effort.

The Nkuringo group ate 71 plant species. Fruit was eaten from 18 of these species, similar to findings for the Kyagurilo group found at a higher elevation within Bwindi (Robbins & McNeilage, 2003). The gorillas ate a large proportion of their diet from only a few species, which tended to be found both inside and outside the park. When the Nkuringo gorillas were outside the park they incorporated nonforest foods by raiding crops in nearby fields. They predominately eat the pith of the banana plants, which ultimately destroyed the plant. Even though rangers and local people actively chased gorillas from these areas, crop raiding still remained a problem. Baboons also raided crops and were more difficult to scare off. Although they are not habituated they are quite courageous about coming out of the forest—very different from "wild" gorillas.

As a result of farming and crop raiding, three primate species share a common area: gorillas, baboons, and humans. We often found gorilla and human dung in close association with one another and saw evidence of human dung with a gorilla knuckle imprinted on it. To determine how each species influenced the health of the other, we conducted a preliminary analysis of parasite load on the baboons (Hope *et al.*, 2004) and found that they share many parasites with the gorillas. Most were of bovine origin, which is not surprising as cattle also graze within these same areas.

The gorillas had a relatively large home range of 37 km^2, which is similar to the largest of three home ranges measured (21.8, 21.1, and 40.1 km^2) for the Kyagurilo group (Robbins & McNeilage, 2003). Home range shape is also quite different, with the length and width of the Kyagurilo range being more similar to one other than the Nkuringo group's, in whose case the length was twice the width. The Kyagurilo gorillas are the only habituated group in Bwindi that do not live along the edge of the park, and their home range falls completely within (somewhat deep) the park boundary. The shape of the Nkuringo home range is definitely a result of the group's preferred use of areas along the park boundary and their seasonal use of fruit trees inside the park. A more detailed GIS map plotting GPS points by week and by month will help demonstrate this.

In Bwindi, tourist sites are located along the edge of the national park to reduce impact. As a result, gorillas habituated for tourism use areas outside of the park to a large extent. This is not only true for the Nkuringo group, as demonstrated in this study, but for Buhoma gorillas as well, who have been documented in areas outside the park boundary (personal observations, and Goldsmith, unpublished GPS data). The factors attracting them to these areas include access to crop foods, especially bananas. In addition, most of the area immediately outside the park is still heavily forested and also provides wild bananas, which is a preferred food. Areas outside the park have planted banana crops, wild bananas, popular forest foods, and a greater density of vegetation, so overall food abundance is most likely greatest outside the park. In contrast, inside the park, there are large areas of fern and open forest where gorilla foods are somewhat scarce and overall stem density was found to be low. Outside the park, forested areas, especially those near the river and around streams, provide ample vegetation for nesting and foraging. Although "wild" gorilla groups may also have their home range near the edge of the forest, they rarely use these areas (ranger interviews, personal observation). When they do come out of the park they tend to stay within meters of the boundary. For comparative purposes, a new phase of this study will include daily follows of at least one "wild" group in Nteko to document their potential use of areas outside the park.

Other Behaviors

Very little observational data are available on this group so we review some notes here. At the time of this study, the silverback Nkuringo was dominant over the much younger silverback Safari. When traveling or resting, the most common association was between Nkuringo and a young juvenile (Christmas) and his mother (Mama Christmas). Safari sometimes went off with his own small group, which usually included some of the blackbacks. The groups can split at times to distances greater than 400 m. This is also shown in their nesting patterns, where one silverback's nest is found surrounded by others from the group, whereas the other silverback's nest could be some 200 m away or more from the others (thus sleeping in two different areas). Usually, all the blackbacks are not accounted for at the nest sites as they tend to make their nests at the perimeters. This may represent some evidence that this group may split in the future.

During the 1-h visit, the gorillas were often hidden in the thick vegetation outside the park, which made observations difficult. However, we were able to observe some mating behavior and, interestingly, we noted a juvenile eating the feces of Nkuringo right after Nkuringo finished mating with one of the females. A very strange behavior not recorded for gorillas before. Both silverbacks were seen mating in the group and they appeared to hide this activity from each other. There have been aggressive encounters within the group during our brief observations, but the skirmishes are usually short-lived. Any aggressive behavior is usually between the two silverbacks or Nkuringo and one of the blackbacks. Nkuringo is very protective of the youngest juvenile. When the group does come into contact with a "wild" group, the habituated group is usually the one to flee, sometimes moving very far. After one of these interactions with a solitary male, an infant from the Nkuringo group was found dead. While there were no observable wounds on its body it is possible that her mother was scared and dropped her during the encounter.

Further Findings and Recommendations

Data collected on day and home ranges of the Nkuringo group and how habituation influences their habitat use should be useful in implementing tourism on this group in the future. In fact, these data have already been used by UWA and IGCP to determine where the tourist facilities should be set up to lessen the impact on the group. In addition, past study reports had recommended the implementation of a buffer zone, which has since been completed. Although these 350 extra meters is beneficial, it is important to note that this study shows the gorillas do range more than 1000 m outside the park boundary. Proper management of this zone, which is still in consideration after almost a year it has been in place, is crucial to keep the gorillas from exiting this area and once again entering agricultural fields. During my last field season in June–July 2005, the buffer zone was being used heavily by Nkuringo. As they used this area, they were allowed to feed on and destroy the crop plants (mostly bananas) that had been left behind by the farmers, who had moved out. The rangers made no effort to chase the group from these foods as they now saw this region as a continuation of the park. This is a very unfortunate situation. Whereas the gorillas were continually chased in the past, they were now free to eat in peace—possibly undoing all of the negative reinforcement of the previous 7 years. Once these plants within the buffer zone are gone, what is to stop them from leaving the zone, once again, to raid fields abutting the zone? It is recommended that management cut down and remove all crop plants within the buffer zone as soon as possible.

Health is also a major concern when it comes to tourism. Scabies has also been a huge problem in this area (Macfie, 1996; Kalema *et al.*, 1998; Meder, 2000; Graczyk *et al.*, 2001a). There have been numerous outbreaks during the past few years and veterinarians are often called in to treat the group. These treatments negatively impact the group as they result is extremely long DPLs, similar to those found when the group is chased or there is an intergroup encounter. Therefore, continued infections of scabies on individuals have a negative effect on all group members and needs to be controlled.

The reporting of health issues by the rangers is extremely helpful; however, their concerns are not always followed up on. In June 2004, a 5-month-old infant died and then her mother was found dead 1 month later. Neither body was collected for necropsy nor was a thorough investigation into the cause of death conducted. A few days after the female's death, four gorillas in the group were coughing. It is recommended that Bwindi managers devise a strict protocol for health reports and necropsy, as one contagious illness could devastate the entire population. Perhaps anatomical and tissue specimens can be preserved in the museum in Kampala so that a wealth of information on this species is not left to rot in the forest.

Finally, one last health threat is the presence and traveling of army personnel in the area. Since the massacre in Bwindi in 1999, the army escorts all visits to the gorillas. Over 100 soldiers are stationed in the Nteko area. During July 2004, we witnessed over 75 soldiers on their journey to Nteko as they walked right through the Nkuringo group, which was spread across the most popular footpath right outside the park boundary. We were fortunate to be there with rangers who were able to make the passage safe. Many, if not all of the men in the army, have never seen a gorilla and it is frightening to think what would have happened if one of the gorillas had charged. Each soldier was carrying a gun. We also found them sleeping, cooking, urinating, and defecating along the trail as we headed back to camp. We strongly recommend that all army personnel be trained as to what is proper behavior both inside and outside the park boundary.

Continued monitoring as a result of tourism may benefit the group by reducing risks of poaching. In Bwindi, this seems to be the case as there have been no recorded events since the killing of four gorillas from Kyagurilo in 1995.

During this study, traps used to capture duiker were found in certain areas inside the park. Rangers destroy these traps during patrols. In November–December 2002, direct poaching activities on the gorillas themselves increased. A group of poachers were found attacking a "wild" group of gorillas outside the park. While at the time the Nkuringo group was not near this area, it had been visited by them in the past. It was thought by management that poachers were trying to procure an infant gorilla, as a female was killed and her infant was missing in nearby Rwanda. The infant was later found and confiscated, which is why poachers might have now been in the area. A few of the poachers were caught during a weekly search by UWA and UPDF officers. For a time, all tourist gorilla groups were being monitored 24 h by UWA rangers. During December 2002, the Nkuringo group stayed closer to the park border, which could be a result of the rangers' constant monitoring of the group.

Within the Nkuringo area, conservation issues abound. When gorillas come out of the forest they threaten their own health as well as the health of the surrounding human population. As noted above, diseases can be transmitted from gorilla to human and vice versa by both direct and indirect contact. By collecting data on when and why gorillas come outside of the park, we can equip the local managers and community with information on how to reduce the occurrence. Written recommendations are reported to UWA annually. By examining the impact of certain conservation techniques on great ape populations, in this case ecotourism, we can better protect and manage greatly endangered populations. However, we must always keep in mind that the benefits of tourism to the gorillas must always outweigh the costs.

ACKNOWLEDGMENTS

We gratefully acknowledge the National Geographic Society for their financial support and the Ugandan Wildlife Authority (especially Aggrey Rwetsiba, John Makombo, John Bosco Nuwe, and Moses Olinga), the Ugandan National Council of Science and Technology, the Institute of Tropical Forest Conservation, and the International Gorilla Conservation Program (especially Helga Rainer and Stephen Asuma) for permission to conduct research within the park, and for logistical and field assistant support. In addition, we thank the many rangers in Nkuringo that we work with daily for sharing their knowledge, and for their patience.

CHAPTER TWENTY-FOUR

Local Ecological Perceptions of Chimpanzees and Forest Resources

A Case Study Near Budongo Forest, Uganda

Cristy Watkins

INTRODUCTION

Much of the world's biodiversity exists in developing nations, where the majority of people live in communities dependent on local natural resources. Uganda, for example, occupies only 2% of the world's area, yet is home to over 11 and 7% of the world's known bird and mammal species, respectively (EASD, State of the Environment Report, 1996).

Budongo Forest Reserve in western Uganda is one of the few remaining forests that is home to the endangered chimpanzee (Reynolds, 1965). While the chimpanzee population in Budongo is fairly stable, in Uganda alone, 78% of their original habitat has been lost due to human activity (Plumptre, 1996).

Cristy Watkins • School of Natural Resources and Environment, University of Michigan, USA.

Primates of Western Uganda, edited by Nicholas E. Newton-Fisher, Hugh Notman, James D. Paterson, and Vernon Reynolds. Springer, New York, 2006.

If this trend continues along its present trajectory, the impact on chimpanzee populations is likely to be significantly negative.

Local people around Budongo are dependent on the forest's resources. As a result, conflicts between chimpanzees and people are inevitable (Hill, 1993; Naughton-Treves, 1998). Here I seek to describe local human attitudes toward, and perceptions of chimpanzees and the natural resources they share. Ultimately, I hope that a better understanding of the cultural and psychological contexts within which human–wildlife conflicts occur will help mitigate the negative impacts of these conflicts and contribute to enhancing the effectiveness of both chimpanzee conservation efforts and forest resource management.

While it is essential to include local people in natural resource management, the implementation of successful conservation programs that incorporate local participation is nonetheless a challenging endeavor (Wild & Mutebi, 1997; Brosius *et al.*, 1998; Songorwa, 1999; Hill, 2000). Here I examine the role of "inclusion" in terms of individual perceptions, cultural views of ecological phenomena, and the factors motivating peoples' behavior toward wildlife and natural resources. How do people perceive chimpanzees and forest resources, and do they have any concern for resource or wildlife scarcity? How do social, economic, environmental, and demographic factors contribute to peoples' perspectives, and do attitudes and behavioral patterns shift according to perspective? If necessary, what incentives might motivate people to act in more conservation-oriented ways? More generally, how can conservation programs work in concert with local populations to best predict how people will respond to program goals?

One approach to addressing these questions is via the study of human behavioral ecology, which is rooted in evolutionary theory (see Barrett *et al.*, 2002). There is evidence that, in most conditions, people place more emphasis on short-term benefits to maximize immediate rewards, while discounting potential future costs (Smith, 1992; Hill, 1993; Alvard, 1998). Furthermore, social currencies such as reputation or status can be as important as physical resources (Buss, 1988). In part, HBE contends that the decisions humans make, and the actions stemming from them, are primarily influenced by short-term considerations that will maximize resource (material or other) acquisition. One goal of conservation programs, then, could be to address what people see as their immediate needs and to devise methods of fulfilling those needs that are both satisfactory and sustainable (Low, 1996a,b).

Human behavior is not cost-insensitive; behaviors that are highly costly are not likely to be widely adopted. Furthermore, what is seen as costly or beneficial will differ depending on, for example, one's status, wealth, gender, and cultural constraints. This suggests that to promote conservation-oriented behaviors, we have to find out what individuals in each situation (in this case, people living adjacent to Budongo Forest) deem costly and what beneficial. My analysis suggests that many deep-seated attitudes have their roots as "adaptations" to resource competition. These adaptations, in addition to proximate motivators, could guide our approaches to future conservation programs (Wilson *et al.*, 1998).

Community-Based Natural Resource Management

Historically, efforts to "conserve" natural spaces have involved foreign management and local exclusion (Matzke & Nabane, 1996). Community-based natural resource management (CBNRM) is now a widely used system that encourages the involvement of local people in decision making and long-term planning (Hill *et al.*, 1997; Brosius *et al.*, 1998; Jones, 1999; Mansuri & Vijayendra, 2003). It aims to minimize conflict between local resource users and governmental or conservation organizational powers, while protecting and managing wildlife and natural resources. CBNRM projects seek to understand and incorporate local cultural systems to determine if they are amenable to a particular conservation project, and to evaluate whether changes need to be made to existing projects (Uphoff & Langholz, 1998; Hitchcock, 2000). As noted above, however, many conservation projects have been unsuccessful, largely due to the inability to provide immediate and substantial benefits to local people (Parry & Campbell, 1992; Wild & Mutebi, 1997; De Boer & Baquete, 1998; Uphoff & Langholz, 1998; Songorwa, 1999; Infield & Namara, 2001). This failure suggests a need to better understand what drives peoples' perceptions and behaviors in order to establish successful management plans.

METHODOLOGY

The Study Site

The Budongo Forest Project (BFP) was created in 1990 to study local biodiversity (including a group of 46 habituated chimpanzees) and to understand the relationships between local people, resource use, and forest and wildlife conservation. Recent population increases in the areas surrounding Budongo (Marriott, 1999) make it likely that demands on forest resources will also increase. Therefore, studying these relationships is crucial for the continuing success of Budongo Forest.

Traditionally, the timber industry has been the major contributor to habitat loss in Budongo Forest (Paterson, 1991). The industry developed quickly during the twentieth century, creating permanent towns that required large quantities of fuel wood and building materials (Howard *et al.*, 1991). Despite the pressures of high commercial potential and human habitation nearby, Budongo has invariably been managed for both wildlife habitat and as a sustainable and productive resource for local communities (Reynolds, 1993). My research therefore seeks to understand how local human use and perception of the forest and chimpanzees have helped (or hindered) conservation and management strategies in Budongo.

Households at every economic level use forest resources (Scott, 1998). Fuel wood supplies make up 97% of the energy used in Uganda (Environmentally Sustainable Development Study, 1996), and each square kilometer of forest is estimated to support 150–350 people (Nowak, 1995). Although people living around Budongo cannot sell forest resources, they have free and legal access to firewood, wood for small building poles, and nontimber (and nonwildlife) products (Johnson, 1993). Village populations are growing, however, and more and more fuel wood and charcoal are being removed from the forest (Steven Khaukha, Uganda Forestry Sector, personal communication).

Nyakafunjo village (population = 381), 1 of 10 villages in Nyabyeya Parish, flanks the south side of Budongo Forest. Nyabyeya parish is a demographic composite of people from Zaire (Lendu, Okebu, and Logo), the West Nile Region (Lugbara, Alur), and various other Ugandan districts (Johnson, 1996a). The demographic history of the parish includes settlement by immigrant laborers during the booming timber industry of the 1950–1960s, as well as an immigrant influx in the 1980s from Zaire, Congo, and Rwanda due to civil unrest (Lauridsen, 1999). Thus, Nyabyeya is an ethnically diverse parish, with a history of forest employment.

Immigrants from Zaire (and other neighboring countries) may be responsible for increased illegal hunting of primates and other animals (Johnson, 1996a,b). Hunting signs are most intense at the southern end of the forest, close to Nyabyeya parish (Plumptre *et al.*, 2001). Furthermore, chimpanzees are often the accidental victims of snares illegally set for bush pigs and duikers (Waller, 1995; Stokes, 1999). Although recent censuses suggest that the BFP study population has been stable for the past few years (Plumptre, 1996; Plumptre *et al.*, 2001), these numbers represent a decline since the first studies in the 1960s (Reynolds, 1965; Reynolds & Reynolds, 1965; Sugiyama, 1968, 1969).

Data Collection

What guides individual perceptions of the forest and chimpanzees, and their subsequent approach to the utilization of natural resources (Figure 1)? I suggest that perceptions of chimpanzees and natural resources depend not only on social factors, such as occupation, duration of residence in the village, level of education, frequency and type of interactions with chimpanzees, but also on specific life history traits, such as respondent age, gender, and ethnicity. I conducted semistructured interviews using a designated questionnaire with both closed- and open-ended questions about people's perceptions of the forest, the chimpanzees, and the future of resource abundance. Field data were collected June 8, 2001–August 5, 2001. With the help of a translator, I completed 53 individual interviews (23 women, 30 men). Data were analyzed in SPSS v. 10, using multiple regression to control for potentially confounding effects.

RESULTS

Perception of Chimpanzees

People viewed chimpanzees as an important part of the forest and worthy of conservation (Table 1). Eighty six percent of respondents said that chimpanzees were not pests. People generally viewed chimpanzees as nondestructive, despite stealing the occasional corncob or sugarcane stalk. Multiple respondents noted that, "according to the law of chimpanzees," they take one helping for themselves and if they have a "friend" or a "wife," they will take two. Respondents maintained that it was better to allow the chimpanzee to eat the bottom of a corn stalk, because when it was through, it would leave the useful top part (ear) still intact. No variables affected the perception of chimpanzees as a neutral forest species.

People thought chimpanzees were extremely clever: 54.7% of respondents gave chimpanzees the highest "cleverness" rating of 5. People also thought chimpanzees had "good character," and were friendly, well-behaved, and considerate. This is in stark contrast to people's views of baboons, which were perceived as equally clever, but vicious and greedy. People noted that many

Objective 1: To understand the local perception of chimpanzees in the forest:

1) What is a chimpanzee?
2) How clever are chimpanzees? 1-5
3) Are chimpanzees a pest?
4) How would you describe your interactions with chimpanzees?
5) Are chimpanzees an important part of the forest? If yes, how?
6) What are some of the ways that you help to protect chimpanzees?

Predictions:

1) Higher education promotes the knowledge that chimpanzees are related to humans.
2) Frequent interaction will increase the likelihood that chimpanzees are related to humans.
3) Frequent entry into the forest results in the belief that chimpanzees can live 50-100 years or longer.
4) Farmers, craftsmen and sawmill employees see chimpanzees as pests, due to resource conflict.
5) Regardless of age, occupation, language, and duration in the village, most will not be willing to give up any money to help save a chimpanzee.

Objective 2: To understand any underlying cultural ecological philosophy

1) Who is responsible for the forest? For chimpanzees? Who *should* be responsible?
2) Are you satisfied with how you are allowed to use the forest? Why or why not?
3) What does conservation mean? Do you conserve resources? Which ones?
4) Are there rules regarding how to use the forest? If yes, where did you learn them?

Predictions:

1) The longer a person has lived in the village, the more likely they are to view locals as the party responsible for the forest and chimpanzees.
2) Sawmill workers and craftsmen believe that locals are responsible for the forest.
3) With time to build a tie to the forest, younger people who have lived in the village a long time will believe that locals are responsible for the forest.
4) Those people who believe that the forest belongs to locals may not be satisfied with their resource use because they feel it is their forest, so they should have full access.
5) Those resources that are most valued are also the resources that are conserved.

Objective 3: To assess the local knowledge of the forest in terms of its ecology and as a finite resource

1) How long will the forest last?
2) What would you do if the forest ran out? What would your children do?
3) Do you worry about the availability of resources in the future? Why or why not?

Predictions:

1) Because the forest is used on a daily basis, everyone should have an answer for what they would do if it ran out.
2) Those people not worried that the forest will run out, will not worry about future resource availability.
3) Younger people should be more worried about resource availability than older people.

Objective 4: To determine the proximate motivators desired by locals

1) What types of trade-off's would you want for limited use of the forest? (Money, land ownership, alternative resources, or other)
2) Rank the jobs according to which is the most economically advantageous (Forest resource use, farming, formal business, craftsman)
3) Which would you rather grow: Tobacco, which makes less money but is not eaten by chimpanzees, or sugarcane, which makes more money but is eaten by chimpanzees.

Predictions:

1) Farmers choose land.
2) Older people want money.
3) Younger women want land (food production for family).
4) Younger men want money (to gain status, wealth)
5) Farmers choose tobacco over sugarcane, because the costs are lower than those for sugarcane.

Figure 1. Questionnaire excerpt, showing questions and predictions.

Table 1. Study population descriptive statistics

Age	N	%	Language group	n	%
<25 years	9	16.9	Sudanic	21	39.6
26–45 years	17	32.1	Bantu	11	20.8
46–65 years	12	22.7	Nilotic	15	28.3
>65 years	9	16.9	Nilo-Hamitic	5	9.4
Occupation	*N*	%	**Village residence**	*n*	%
Farming	30	56.6	<5 years	16	30.2
Craftsman	6	11.3	6–10 years	8	15.1
Sawmill	5	9.4	11–20 years	8	15.1
Other[a]	11	20.8	>20 years	20	37.7
None	1	1.8			
Education	*n*	%			
None	19	35.8			
Primary	26	49.1			
Secondary	8	15.1			

Fifty-three individuals, 30 males and 23 females.
[a] Other included tailor, cook, housewife, teacher, and businessman.

chimpanzee behaviors were similar to human behavior, like food sharing, grooming, and maternal behaviors (Table 2). Some respondents said that they often stop and watch the chimpanzees in the trees around their homes.

Ninety-six percent of respondents believed that chimpanzees were an important part of the forest. Tourism and research were the most frequently given

Table 2. Differences and similarities between chimpanzees and humans

	Similarities	Differences
1	Similar facial structure	Different language/do not talk
2	Five fingers	Walk differently
3	Considerate and peaceful	They are wild
4	Come close to people, occasionally interact	They are hairy
5	Smart (i.e., do not take a lot of food because they know that people need to eat it too)	
6	Food share	
7	Protective of infants	
8	Thinks before acting	
9	Similar life-span	
10	Captive chimps able to learn how to drive and shake hands	
11	Wild chimps can steal children and take care of them	

reasons for this importance (34%). Chimpanzees were also viewed as important for seed dispersal (19%), their nondestructive nature (17%), and other reasons, such as their beautification of the forest, or their similarity to humans (19%).

Some respondents stated that wild pigs and baboons avoid their crops when a chimpanzee is nearby. Although people said that they would rather chimpanzees not eat their crops at all, they also thought that chimpanzees can actually serve as crop-guarders, and the small helpings of food that they take for themselves is worth the protection from more destructive wildlife. All respondents said that chimpanzees should be saved, while only six people (11%) believed that baboons should be protected, and only if they could be controlled and kept in the forest.

Most people did not understand that chimpanzees as a species were in danger of becoming extinct in the wild (62%). Almost half of respondents did not believe that chimpanzees were threatened in any way. Twenty-eight percent listed snares and human-induced harm, and 13.2% suggested drought, lack of food, and illness as potential threats. Eleven percent of respondents did not know if chimpanzees were threatened or not.

Stakeholder Responsibility and Ownership

Although there was variation among age, gender, and occupation groups, half of the respondents felt that the forest is the government's responsibility (50%). Thirty percent of respondents felt that the government was also responsible for the chimpanzees. Only 19% of respondents felt that local people were responsible for the forest, although almost 30% felt that locals were responsible for chimpanzees. Generally, whoever respondents believed owned the forest (whether it was locals, the government, God, or BFP) was indeed the actor they believed should own the forest (Pearson chi-square, $\chi^2 = 98.764$, df = 36, $P < 0.001$). No one younger than 25 believed that the forest belonged to local people. Only 12% of those 26–45, 17% of those 46–65, and 11% of those over 65 believed that the forest belonged to locals.

The majority of respondents were satisfied with how they were allowed to use the forest. None of the variables affected this perception. Eighty-nine percent of respondents believed that there were rules about how the forest should be used, and 58.5% stated that they learned these rules from the government or forestry department. The two rules mentioned most frequently were "license needed to cut timber" and "cannot kill animals."

Eighty-five percent of respondents felt that they were conserving resources ("conservation" was translated as "to protect and manage"). In order of importance, the conserved resources were identified as trees (including wood for building and firewood) and wildlife. Seven percent of respondents felt that they did not need to conserve because animals could take care of themselves, or that conservation was the government's responsibility. Seventy-three percent of Nyakafunjo villagers, however, did not know what the English word "conservation" meant. Those respondents who knew the word defined it as simply "protecting the forest," "leaving it as it is," "using it sustainably," or, "good to see animals in forest."

The Fate of Natural Resources in Budongo and the Extent of Personal Worry

The majority of respondents felt that the forest would last forever (77%). When asked what they would do if the forest did indeed "run out," there were mixed answers. Some people expressed deep worry, while others felt that resource availability was not a problem or not their concern. Of those who voiced substantial worry about future resource availability (37%), the majority of them tended to fear "death and destruction" if the forest was depleted (70%). Many of the respondents who were not worried about resource availability did not believe that the forest would run out (38%).

Women worried about resource availability more than men: 65% of women and 40% of men worried very much, and conversely 26% of women and 40% of men worried little or not at all (Likelihood ratio, $\chi^2 = 14.352$, df $= 5$, $P = 0.014$). Age was also a factor in the level of worry: 88% of those aged 26–45 worried "a lot," in comparison to those over 65, the majority of whom did not worry at all (67%).

When asked what their children would do if the forest ran out, 15% of all respondents argued that it would never happen, 45% stated that "the world would end and everyone would suffer," and 37% thought that their children would be better educated, know how to take care of the forest, and plant trees. One man replied, however, "How should I know what my children will do? My father did not know where I would end up! My children would just find something else to do." What people thought they would do if the forest ran out correlated with what they thought their children would do (Pearson chi-square, $\chi^2 = 98.753$, df $= 16$, $P < 0.001$).

Motivating Incentives and Desired Trade-offs

If access to forest resources were to be reduced, land ownership was the most desired trade-off: 60% of respondents desired land, 22% desired money, 8% desired cheaper fuel alternatives, and 6% said that nothing was worth losing access to forest. There was a substantial difference between the desired incentives of men and women: 47% of men and 78% of women wanted land. Conversely, 30% of men and 13% of women wanted money. There were also some discrepancies among age categories. Only 10% of those younger than 25 and 30% of those aged 26–45 chose money. Neither occupation nor level of education made a difference in desired incentive.

Understanding the result-driven perceptions of people in terms of their daily investments provides insight into what trade-offs they would accept and why. I predicted that people would plant crops that were the most immediately beneficial with the least amount of work. Sugarcane is a growing industry throughout the parish, but while the crop has its economic and societal benefits, there are also maintenance costs. Sugarcane is more labor intensive than tobacco, there is a longer wait to receive profits, and chimpanzees enjoy eating it. However, 64% would choose to plant sugarcane, and only 20% would choose tobacco.

DISCUSSION

Four significant points arise from this research. First, Nyakafunjo villagers see chimpanzees as a neutral, and sometimes beneficial, part of the forest. Unlike many cultures (Burton; 2002; Cormier; 2002; Lizarralde, 2002), no symbolic or religious connection to primates (nor any other wildlife) existed in Nyakafunjo. That no mythological, spiritual, or historical stories involving chimpanzees were shared may simply be a result of the short history of the village. There are, however, critical connections in terms of benefits (tourism, crop protecting, and seed dispersal) and costs (forest resource competition, crop raiding).

While chimpanzees are not considered pests, and some people believe that they act to deter other wildlife from crop raiding, my observations differed. The baboons in Budongo Forest Project's camp did not appear to be affected by the chimpanzees that occasionally came to feed in the surrounding fig trees. Perhaps the interspecific "fear and avoidance" between chimpanzees and baboons that appears to aid in crop protection may decrease with time and familiarity. Future studies need to explore this incidental crop-guarding behavior.

We also need to understand the impacts of the growing sugarcane industry on chimpanzee behavior and subsequent changes in human perception. Chimpanzees tend to visit sugarcane crops more often than other crops, and they eat it more readily. If people invest in sugarcane crops, chimpanzees could become costly crop raiders as opposed to beneficial crop guarders.

Nyakafunjo villagers are well aware of the lucrative business of tourism as well as the potential to profit from researchers. Tourism and research certainly have the potential to provide money and job opportunities to locals (Wunder, 2000; but see Stem *et al.*, 2003). The ecotourism projects in Budongo are far from Nyakafunjo, however, and generate only a small income increase for the 28 employed locals (Langoya & Long, 1997). Still, the desire to benefit from whatever tourism exists is exhibited by Nyakafunjo residents. It is still unclear whether or not a monetary benefit would indeed change behaviors and attitudes of people in Nyakafunjo.

Second, multiple variables influence conservation ethos and behavior. It is not surprising that the term "conservation" as we interpret it in industrial societies is an ideal that perhaps does not readily find its analog in subsistence societies, where basic needs are often barely met (e.g., Hough, 1988; Berkes, 2004). If a conservation ethic were present in Nyakafunjo, one might expect people to say that the forest and its wildlife belonged to locals, should belong to locals, and that locals are responsible for its care (Redford, 1990; Oelschlanger, 1991; Pimbert & Pretty, 1995). Nyakafunjo residents, however, see the government as owning and being responsible for the forest, and they are generally content with that. One explanation for this perception is that Nyakafunjo is a relatively young village, and at its conception the forest department was already established as the governing body. Thus, the new, ethnically diverse residents (even with their shared economic interests) have not developed a traditional ecological philosophy of their own; the government, instead, provided the "ethics" of resource use.

This fact does, of course, have serious implications for fostering a sense of stewardship in residents toward natural resources. Particular attention needs to be paid to the most recent of residents, those from traditional hunting cultures. While the "older" rural migrants' lack of forest culture has apparently not had a detrimental effect on the forest and its wildlife, newer migrants from traditional hunting cultures, such as those from the Congo area, obviously pose a more direct threat. Indeed, it has been recognized that the persistence of hunting in Budongo must be met with clear disincentives (Plumptre *et al.*, 1997).

Social currencies may also provide motivation for conservation. I argue that there is a greater perceived benefit in being part of a legal market enterprise outside of the forest, rather than an illegal one, such as hunting or illegally cutting timber. In this case, aspiring for immediate benefits drives interests away from the forest and, subsequently, the chimpanzees.

Furthermore, the use of the forest for food resources and other subsistence items is minimal and even undesired in parts of Uganda because it symbolizes a "poor man's lifestyle" (Johnson, 1993; Banana & Turino-habwe, 1997). The more people are influenced by the demands and standards of the "outside world," the less they will rely on the forest. If, however, it is perceived that there are greater benefits to be harnessed within (or on the border) rather than outside of the forest, it is likely that forest resource use, legal or otherwise, will increase. Indeed, relative success may be more important than absolute success (Penn, 2003) and people may strive to acquire and display more than others, whatever "it" is and in whatever amount (Wilson *et al.*, 1998). If relative success is perceived to be enhanced by using forest resources, it can be predicted that people will increase their use of those resources.

People have mixed perceptions about future resource availability. That 50% of respondents worried very much about future resource availability illustrates a strong dependency on resources. Most of these people were in the 26–45-year age group, which suggests the particular importance of resources for people in their prime family-bearing/rearing years. Although it might be predicted that young men will be more risky than older men (given their propensity to compete with other males for access to females), the conditions may be such that younger men are required to help care for their offspring, and thus have more to lose than older men (Wilson *et al.*, 1996). A future study could test the more specific prediction that worry about and interest in resource availability will depend on the size and makeup of the family.

It is still unclear whether parents are more or less optimistic about resource availability in their own lives versus the lives of their children. Rogers (1991) argues that the inclination to conserve resources on behalf of one's children is related to the security of inheritance. In Uganda, inheritance is strongly patrilineal and land ownership is rarely obtained by women (Tripp, 2004). A future study could explore the effects of inheritance and land security, or lack thereof, on one's propensity to conserve. If parents are indeed more optimistic about resource availability in future in their children's lives than they are about their own, perhaps they will be supportive of conservation education programs (which can be much less costly to the parent than the conservation of resources themselves) that teach ecology and sustainability.

Third, land and money are the most desired incentives. Conservation programs that are not sensitive to respective cultural, economic, and political contexts are not likely to be effective (Hutton & Leader-Williams, 2003). Therefore, an understanding of specific incentives is crucial to effective local participation. For the people of Nyakafunjo, land is the most desired incentive, but patterns of variation in age and gender groups are evident. Men see money as the better incentive while women would opt for land. This finding complements other studies that characterize women as "home-makers," who are heavily dependent on food and other resources essential to the care of their family, and men as "providers," whose interests include income-generating activities (Quinn, *et al.*, 2003; Turton, 2000). Similar to their propensity to worry more about resource availability than men, women are (unconsciously) partial to less risky but more reliable and direct means of resource utilization (Wilson *et al.*, 1996, 1998).

Evolutionary theory predicts that individuals behave in ways that maximize their reproductive success, or proxies for such success (Buss, 1988). Furthermore, differences in reproductive interests are frequently correlated with differential resource valuation and use (Turke & Betzig, 1985; Low & Heinen, 1993; Wilson *et al.*, 1998). In Nyakafunjo, then, men view money as a proxy for success (and a way to "show off" their assets) and women perceive land as a proxy for success (and a means by which to procure resources that directly benefit their family).

That people would rather grow sugarcane than tobacco (even though more people grow the latter) illustrates the desire to become part of a burgeoning economic industry. It is interesting that this interest was found in both men and women. Given the economic appeal of sugarcane, men would be predicted to be more interested than women. However, in this case, sugarcane is also a food source, providing the incentive for women to pursue it as well. Thus, the growth of the sugarcane industry may have serious consequences for chimpanzee conservation.

While pure economics can explain much of the rationale behind the desire to plant a more lucrative crop, evolutionary theory predicts that if that crop is too risky and the social benefits of growing it are not secure, they may opt for the less lucrative one (Alvard, 1998). The prestige of being included in this new industry, however, is a tantalizing incentive. The perceived benefit is greater than the perceived cost. This study, however, only reveals what people "intend" or "desire" to do, not what they are actually doing: only one household at the time of this study had a sugarcane crop.

Fourth, anomalies are important. Behavioral ecology predicts resource-degrading behavior when future benefits are unforeseeable. This study hails one exception to this theory. "Frank" was the only man in the entire village with a pine tree farm on his land plot. He had invested in 150 seedlings after a forest department officer came to the village suggesting that people buy pine tree seedlings and start a nursery of their own so that one day Budongo Forest will not have to be used. That was 4 years ago, and at the time of this study, he was still maintaining the trees.

While Daly and Cobb (1989) recognize that the judicious decisions made by one individual may not be the sensible policy for the entire group, Low and Heinen (1993) point out that it is a coalition of individuals willing to impose current costs on themselves for the sake of a long-term benefit that can create change. Thus, individuals choosing to make rational, long-term, altruistic decisions should not be ignored. People like Frank may have the power to convince other individuals into behaving in a similar manner (Ruttan & Borgerhoff-Mulder, 1999). Gaining social currency (such as status, prestige, and pride) may be the impetus that gets individuals motivated or even, as Ruttan and Borgerhoff-Mulder (1999) suggest, coerced into minimizing resource degradation.

CONCLUSION

This study illustrates how both proximate cultural influences and evolved behavioral and attitudinal tendencies lead to the ecological philosophies and perceptions of a single Ugandan village. I have shown that a government-dominated ecological philosophy may have an effect on the local perception of the forest and its usage. In the case of Budongo Forest, it appears that Nyakafunjo villagers support conservation power in the hands of the government and of conservation-oriented programs the like Budongo Forest Project.

I have also shown that there are clear differences in gender valuations and perceptions of the environment, availability of resources, and desired incentives. I have suggested the importance of future research on social, cultural, and family unit structure and their influence on values and perceptions.

Finally, the often human-like behaviors of chimpanzees and the awareness of the benefits to be acquired through tourism and foreign research are factors that may promote the protection of chimpanzees. Perhaps even more crucial, however, is the need for continued research on the impacts of the sugarcane industry and illegal hunting on chimpanzees. Although it appears that the perceived benefits of an outside market economy may have protected chimpanzees in the past, an increase in the sugarcane industry will likely produce increased pressure on forest boundaries as well as the chimpanzees. Taking advantage of the positive perception of chimpanzees and the potential to mold a community conservation ethic by offering desired incentives could lead to continuing chimpanzee stability in Budongo Forest.

REFERENCES

Abbiw, D. K. (1990). *Useful plants of Ghana.* Intermediate Technology Publications and Royal Botanic Gardens, Kew.

Achoka, I. (1993). *Home range, group size and group composition of mountain gorilla (*Gorilla gorilla beringei*) in the Bwindi Impenetrable National Park, south-western Uganda.* MSc thesis, Makerere University.

Adesina, G. A., Adesogan, E. K., Okorie, D. A., Taylor, D. A. H., & Styles, B. T. (1971). The limonoid chemistry of the genus *Khaya* (Meliaceae). *Phytochemistry, 10,* 1845–1853.

Adesina, S. K. (1983). Chemical examination of *Khaya ivorensis* and *Khaya senegalensis. Fitoterapia, 54,* 141–143.

Akah, P. A., Orisakwe, O. E., Gamaniel, K. S., & Shittu, A. (1998). Evaluation of Nigerian tradional medicines, II: Effects of some Nigerian folk remedies on peptic ulcer. *Journal of Ethnopharmacology, 62,* 123–127.

Ake-Assi, Y. A. (1992). *Contribution au recensement des espèces végétales utilisées traditionnellement sur le plan zootechnique et vétérinaire en Afrique de l'Ouest.* Thèse de doctorat (Sc. Vétérinaires), Lyon, Université Claude Bernard, 220 pp.

Akman, M., Erden, H., Bener, F., Liu, J., & Bahceci, M. (2002). Can luteal phase estradiol levels predict the pregnancy outcome in in vitro fertilization cycles of good responders whose excess embyros yield blastocysts? *Fertility & Sterility, 77,* 638–639.

Alcorn, J. B. (1991). Ethics, economies and conservation. In M. L. Oldfield & J. B. Alcorn (eds.), *Biodiversity: Culture, conservation and ecodevelopment.* Westview Press, San Francisco, pp. 317–349.

Alder, D., & Synnott, T. (1992). Permanent sample plot techniques for mixed tropical forest. *Tropical Forestry Papers, 5,* 1–124. Oxford Forestry Institute, Oxford, UK.

Alicata, J. E. (1935). Early developmental stages of nematodes occurring in swine. *USDA Technical Bulletin,* No. 489.

Allison, A. C. (1982). Co-evolution between hosts and infectious disease agents and its effect on virulence. In R. M. Anderson & R. M. May (eds.), *Population biology of infectious diseases.* Springer-Verlag, Berlin, pp. 245–267.

Altmann, J. (1974). Observational study of behaviour: Sampling methods. *Behaviour, 9,* 227–267.

Altmann, S. A. (1991). Diets of yearling female primates (*Papio cynocephalus*) predict lifetime fitness. *Proceedings of the National Academy of Sciences, 88*, 420–423.

Altmann, S. A. (1998). *Foraging for survival: Yearling baboons in Africa.* The University of Chicago Press, Chicago.

Altschul, S. V. R. (1973). *Drugs and foods from little-known plants.* Harvard University Press, Cambridge, MA.

Alvard, M. S. (1998). Evolutionary ecology and resource conservation. *Evolutionary Anthropology, 7*, 62–74.

Amberson, J. M., & Schwarz, E. (1952). *Ternidens deminutus* Railliet and Henry, a nematode parasite of man and primates. *Annals of Tropical Medicine and Parasitology, 46*, 227–237.

Ambrose, L. (1999). *Species diversity in West and Central African galagos (Primates, Galagonidae): The use of acoustic analysis.* Unpublished PhD thesis, Oxford Brookes University.

Anderson, R. C. (2000). *Nematode parasites of vertebrates: Their development and transmission* (2nd ed.). CAB International, Oxon.

Anderson, R. A., Bryden, N. A., Polansky, M. M., & Thorp, J. W. (1991). Effects of carbohydrate loading and underwater exercise on circulating cortisol, insulin and urinary losses of chromium and zinc. *European Journal of Applied Physiology & Occupational Physiology, 63*, 146–150.

Anderson D. P., Nordheim, E. V., Boesch, C., & Moermond, T. C. (2002). Factors influencing fission-fusion grouping in chimpanzees in the Tai National Park, Cote d'Ivoire. In C. Boesch, G. Hohmann, & L. F. Marchant (eds.), *Behavioural diversity in chimpanzees and bonobos.* Cambridge University Press, Cambridge, pp. 90–101.

Andrews, P., & van Couvering, J. A. H. (1975). Palaeoenvironments in the East African Miocene. In F. S. Szalay (ed.), *Approaches to primate paleobiology* (Contributions to Primatology, 5). Karger, Basel, pp. 62–103.

Apter, D., & Vihko, R. (1985). Premenarcheal endocrine changes in relation to age at menarche. *Clinical Endocrinology, 22*, 753–760.

ArcView (1996). *Using ArcView GIS.* Environmental Systems Research Institute, Inc. Redlands, CA.

Asaka, Y., Kubota, T., & Kulkarni, A. B. (1977). Studies on bitter principle from *Vernonia anthelmintica. Phytochemistry, 16*, 1838–1839.

Ashford, R. W., Reid, G. D. F., & Butynski, T. M. (1990). The intestinal faunas of man and mountain gorillas in a shared habitat. *Annals of Tropical Medicine and Parasitology, 84*, 337–340.

Ashford, R. W., Barnish, G., & Viney, M. E. (1992). *Strongyloides fuelleborni* Kellyi infection and disease in Papua New Guinea. *Parasitology Today, 8*, 314–318.

Ashford, R. W., Lawson, H., Butynski, T. M., & Reid, G. D. F. (1996). Patterns of intestinal parasitism in the mountain gorilla *Gorilla gorilla* in the Bwindi Impenetrable Forest, Uganda. *Journal of Zoology (London), 239*, 507–514.

Ashford, R. W., Reid, G. D. F., & Wrangham, R. W. (2000). Intestinal parasites of the chimpanzees *Pan troglodytes* in Kibale Forest, Uganda. *Annals of Tropical Medicine and Parasitology, 94*, 173–179.

Audebert, F., Cassone, J., Kerboeuf, D., & Durette-Desset, M. C. (2003). Development of *Trichostrongylus colubriformis* and *Trichostrongylus vitrinus*, parasites of ruminants in the rabbit and comparison with *Trichostrongylus retortaeformis. Parasitology Research, 90*, 57–63.

Aufreiter, S., Mahaney, W. C., Milner, M. W., Huffman, M. A. Hancock, R. G. V., Wink, M., & Reich, M. (2001). Mineralogical and chemical interactions of soils eaten by chimpanzees of the Mahale Mountains and Gombe Stream National Parks, Tanzania. *Journal of Chemical Ecology, 27*, 285–311.

Aujard, F., & Perret, M. (1998). Age-related effects on reproductive function and sexual competition in the male prosimian primate, *Microcebus Murinus Physiology & Behavior, 64*(4), 513–519.

Aveling, C. (1984). Notes on the golden monkey, *Cercopithecus mitis kandti* of the Virunga Volcanos, Rwanda. *African Journal of Ecology, 22*, 63–64.

Ayensu, E. S. (1978). Medicinal plants of West Africa. Reference Publications, Inc.

Bain, O., Moisson, P., Huerre, M., Landsound-Soukate, J., & Tutin, C. (1995). Filariae from a wild gorilla in Gabon with description of a new species of Mansonella. *Parasite, 2*, 315–322.

Balagizi Karhagomba, I., & Ntumba Kayembe, F. (1998). Plantes utilisées dans le traitement des helminthoses gastro-intestinales des petits ruminants dans le groupement d'Irhambi-Katana (Région du Bushi, Province du Sud-Kivu, Rép. Dém. du Congo). *Recherches Africaines,* 1, 90–99.

Banana, A. Y., & Turino-habwe, G. P. (1997). A socio-economic analysis of forest foods consumption in Hoima and Masindi Districts of Uganda. *African Crop Science Conference Proceedings, 3*, 1435–1442.

Bandyopadhyay, B. C., & Poddar, M. K. (1998). Dietary protein-induced change in mammalian corticosterone status (index of immune response) during aging. *Mechanisms of Ageing & Development, 103*, 57–68.

Baranga, D. (1983). Changes in chemical composition of foods parts in the diet of colobus monkeys. *Ecology, 64*, 668–673.

Barrett, L., Dunbar, R., & Lycett, J. (2002). *Human evolutionary psychology.* Palgrave, Hampshire, England, 434 pp.

Barton, R. A. (2000). Socioecology of baboons: The interaction of male and female strategies. In P. M. Kappeler (ed.), *Primate males: Causes and consequences of variation in group composition*. Cambridge University Press, Cambridge, pp. 97–107.

Bearder, S. K., Honess, P. E., & Ambrose, L. (1995). Species diversity among galagos with special reference to mate recognition. In L. Alterman, G. A. Doyle, & M. K. Izard (eds.), *Creatures of the dark: The nocturnal prosimians*. Plenum, New York, pp. 331–352.

Bearder, S. K., & Harcourt, C. S. (1989). *A field study of prosimian primates in Uganda*. Unpublished report to the Wenner-Gren Foundation for Anthropological Research.

Beeson, M. (1989). Seasonal dietary stress in a forest monkey (*Cercopithecus mitis*). *Oecologia, 78*, 565–570.

Beeson, M., Tame, S., Keeming, E., & Lea, S. E. G. (1996). Food habits of guenons (*Cercopithecus* spp.) in afro-montane forest. *African Journal of Ecology, 34*, 202–210.

Béguin, P., & Aubert, J. (1994). The biological degredation of cellulose. *FEMS Microbiology Review*, *13*, 25–58.

Bennett, E. L., & Dahaban, Z. (1995). Wildlife responses to disturbances in Sarawak and their implications for forest management. In R. B. Primack & T. E. Lovejoy (eds.), *Ecology, conservation, and management of Southeast Asian rainforests*. Yale University Press, New Haven, pp. 66–86.

Bercovitch, F. B. (1992). Re-examining the relationship between rank and reproduction in male primates. *Animal Behaviour, 44*, 1168–1170.

Bercovitch, F. B. (1994). Female choice in savanna baboons [Abstract]. *American Journal of Primatology, 33*, 195.

Bercovitch, F. B. (1995). Female cooperation, consortship maintenance, and male mating success in savanna baboons. *Animal Behaviour, 50*, 137–149.

Berkes, F. (2004). Rethinking community-based conservation. *Conservation Biology, 18*, 621–630.

Bernstein, I. S., Rose, R. M., & Gordon, T. P. (1977). Behavioural and hormonal responses of male rhesus monkeys introduced to females in the breeding and non-breeding seasons. *Animal Behaviour, 25*, 609–14.

Beroza, G. A., Barclay, W. P., Phillips, T. M., Foerner, J. J., & Donawick, W. J. (1983). Cecal perforation and peritonitis associated with *Anoplocephala perfoliata* infection in three horses. *Journal of the American Veterinary Medical Association, 183*, 804–806.

Beuning, K. R. M., Talbot, M. R., & Kelts, K. (1997). A revised 30,000-year paleoclimatic and paleohydrologic history of Lake Albert, East Africa. *Palaeogeography, Palaeoclimatology, Palaeoecology, 136*, 259–279.

Bhat, M. N., & Manickam, R. (1998). Corproculture and demonstration of third stage larvae of *Murshidia* sp. in elephants (*Elephas maximus*). *Indian Journal of Veterinary Medicine,* 1140–1142.

Bibby, C. J., Collas, M. J., Heath, C. H., Timboden, T. H., Johnston, A. J., Stattersfield, J. A., & Thirgood, S. J. (1992). *Putting biodiversity on the map: Priority areas for global conservation.* International Council for Bird Preservation, Cambridge, U.K.

Bickerton, D. (1990). *Language and species.* Chicago University Press, Chicago.

Bitariho, R. (1999). *The abundance, distribution and use of montane bamboo in Bwindi Impenetrable and Mgahinga Gorilla National Parks, S.W. Uganda.* MSc dissertation, Makerere University, Kampala, Uganda.

Bligh, M. E., Douglass, L. W., & Castonguay, T. W. (1993). Corticosterone modulation of dietary selection patterns. *Physiology & Behavior, 53,* 975–982.

Bligh-Tynan, M. E., Bhagwat, S. A., & Castonguay, T. W. (1993). The effects of chronic cold exposure on diurnal corticosterone and aldosterone rhythms in Sprague-Dawley rats. *Physiology & Behavior, 54,* 363–367.

Blom, A., Cipoletta, C., Brunsting, A. M. H., & Prins, H. H. T. (2004). Behavioral responses of gorillas to habituation in the Dzanga–Ndoki National Park, Central African Republic. *International Journal of Primatology, 25,* 179–196.

Bocian, C. M. (1997). *Niche separation of black-and-white colobus monkeys* (Colobus angolensis *and* C. guereza) *in the Ituri Forest.* PhD thesis, City University of New York, New York.

Boesch, C. (1991). The effects of leopard predation on grouping patterns in forest chimpanzees. *Behaviour, 117,* 220–242.

Boesch, C. (1996a). The emergence of cultures among wild chimpanzees. *Proceedings of the British Academy of Sciencs, 88,* 251–268.

Boesch, C. (1996b). Social grouping in Tai chimpanzees. In W. C. McGrew, L. F. Marchant, & T. Nishida (eds.), *Great ape societies.* Cambridge University Press, Cambridge, pp. 99–113.

Boesch, C., & Boesch, H. (1983). Optimisation of nut-cracking with natural hammers by wild chimpanzees. *Behaviour, 26,* 265–286.

Boesch, C., & Boesch, H. (1989). Hunting behaviour of wild chimpanzees in the Tai National Park. *American Journal of Physical Anthropology, 78,* 547–573.

Boesch, C., & Boesch-Achermann, H. (2000). *The chimpanzees of the Tai Forest: Behavioural ecology and evolution.* Oxford University Press, Oxford.

Boinski, S. (1987). Habitat use by squirrel monkeys (*Saimiri oerstedi*) in Costa Rica. *Folia Primatologica, 49,* 151–167.

Boinski, S., Treves, A., & Chapman, C. A. (2000). A critical evaluation of the influence of predators: Effects on group travel. In S. Boinski & P. A. Garber (eds.), *On the move: How and why animals travel in groups.* University of Chicago Press, Chicago.

Boreham, R. E., McCowen, M. J., Ryan, A. E., Allworth, A. M., & Robson, J. B. M. (1995). Human trichostrongyliasis in Queensland. *Pathology, 27,* 182–185.

Bouquet, A. (1969). Féticheurs et médecines traditionnelles du Congo (Brazzaville). *Mémoire O.R.S.T.O.M*

Bouquet, A., & Debray, M. (1974). Plantes médicinales de la Côte d'Ivoire. Travaux et Documents de l' O.R.S.T.O.M., Paris, no. 32, 232 pp.

Bouquet, A., Cave, A., & Paris, R. (1971). Plantes Medicinales du Congo–Brazzaville (III). *Plantes Medicinales et phytotherapie, V*, 154–158.

Boutin, S. (1990). Food supplementation experiments with terrestrial vertebrates: Patterns, problems and the future. *Canadian Journal of Zoology, 68*, 203–220.

Bower, C. A., & Wilcox, L. V. (1965). Soluble salts, In C. A. Black (ed.), *Methods of soil analysis, part 2.* American Society of Agronomy, Madison, Wisconsin, pp. 933–951.

Brack, M. (1987). *Agents transmissible from simians to man.* Springer-Verlag, Berlin.

Brosius, J. P., Lowenhaupt Tsing, A., & Zerner, C. (1998). Representing communities: Histories and politics of community-based natural resource management. *Society & Natural Resources, 11*, 157–168.

Brown, R. C., & Girardeau, M. H. F. (1977). Trans-mammary passage of *Strongyloides* sp. in larvae in human host. *American Journal of Tropical Medicine and Hygiene*, 26, 215–219.

Brownlow, A. R., Plumptre, A., Reynolds, V., & Ward, R. (2001). Sources of variation in the nesting behaviour of chimpanzees (*Pan troglodytes schweinfurthii*) in the Budongo Forest, Uganda. *American Journal of Primatology, 55*, 49–55.

Brugière, D. (1998). Population size of the black colobus monkey *Colobus satanas* and the impact of logging in the Lopé Reserve, Central Gabon. *Biological Conservation, 86*, 15–20.

Bruorton, M. R., & Perrin, M. R. (1991). Comparative gut morphometrics of vervet (*Cercopithecus aethiops*) and samango (*C. mitis erthrarchus*) monkeys. *Zeitschrift Fur Saufertierkune, –International Journal of Mammalian Biology, 56*, 65–71.

Bruorton, M. R., Davis, C. L., & Perrin, M. R. (1991). Gut microflora of vervet and samango monkeys in relation to diet. *Applied and Environmental Microbiology, 57*, 573–578.

Bshary, R., & Nöe, R. (1997a). Anti-predation behavior of red colobus monkeys in the presence of chimpanzees. *Behavioral Ecology and Sociobiology, 41*, 321–333.

Bshary, R., & Nöe, R. (1997b). Red colobus and Diana monkeys provide mutual protection against predators. *Animal Behavior, 54*, 1461–1474.

Burger, A. E., & Millar, R. P. (1980). Seasonal changes of sexual and territorial behaviour and plasma testosterone levels in male Lesser Sheathbills (*Chionis minor*). *Zeitschrift Fuer Tierpsychologie, 52*, 397–406.

Burghardt, G. (1992). Human–bear bonding in research on black bear behavior. In H. Davis & D. Balfour (eds.), *The inevitable bond: Examining scientist–animal interactions.* Cambridge University Press, Cambridge, pp. 365–382.

Burton, F. (2002). Monkey King in China: Basis for a conservation policy. In A. Fuentes & L. Wolfe (eds.), *Primates face to face: The conservation implications of human–nonhuman primate interconnections.* Cambridge University Press, Cambridge, pp. 137–162.

Buss, D. (1988). The evolution of human intrasexual competition: Tactics of mate attraction. *Journal of Personality and Social Psychology, 54*, 616–628.

Butynski, T. M. (1984). *Ecological survey of the Impenetrable (Bwindi) Forest, Uganda, and recommendations for its conservation and management.* Report to the Government of Uganda.

Butynski, T. M. (1990). Comparative ecology of blue monkeys (*Cercopithecus mitis*) in high and low density subpopulations. *Ecological Monographs, 60*, 1–26.

Butynski, T. M., & Kalina, J. (1998). Gorilla tourism: A critical look. In E. J. Milner-Gulland & R. Mace, (eds.), *Conservation of biological resources.* Blackwell Press, Oxford, pp. 294–313.

Byrne, R. W., & Byrne, J. M. E. (1993). Complex leaf-gathering skills of mountain gorillas (*Gorilla g. beringei*): Variability and standardization. *American Journal of Primatology, 31*, 241–261.

Byrne, R., Corp, N., & Byrne, J. (2001). Estimating the complexity of animal behaviour: How mountain gorillas eat thistles. *Behaviour, 138*, 525–557.

Byrne, R. W., & Stokes, E. J. (2002). Effects of manual disability on feeding skills in gorillas and chimpanzees. *International Journal of Primatology, 23*, 539–554.

Cahusac, A. B. (1958). Impenetrable C.F.R Forest Types (Intepretation Report No. 6). Reference No. KI/15. Published as Appendix G in *Working Plan for the Impenetrable C.F.R* by G. L. Leggat & H. H. Osmaston, 1961, and modified as CARE-DTC map 1997.

Calvert, J. (1985). Food selection by western gorillas (*G. g. gorilla*) in relation to food chemistry. *Oecologia, 65*, 236–246.

Campana-Rouget, Y. (1959). Un nouveau Murshidia (Strongylidae, Nematoda) parasite accidentel du gorille. *Bulletin de l' Institute Royal des Sciences Naturelles de Belgique, 40*, 1–8.

Carroll, R. W. (1986). Status of the lowland gorillas (*Gorilla g. gorilla*) and other wildlife in the Dzanga-Sangha region of southwestern Central African Republic. *Primate Conservatiion*, 7, 38–41.

Casimir, M. J. (1975). Feeding ecology and nutrition of an eastern gorilla group in Mt. Kahuzi region (Republique-du-Zaire). *Folia Primatologica, 24*, 81–136.

Casmir, J. M., & Butenandt, E. (1973). Migration and core area shifting in relation to some ecological factors in the mountain gorilla group (*Gorilla gorilla beringei*) in the Kahizi Region (Republique du Zaire). *Folia Primatologica, 24*, 81–136.

Cassone, J., Vuong, P. N., & Durette-Desset, M. C. (1992). Cycle biologique de *Paralibyostrongylus hebrinicutus* (Nematoda: Strongylidae). *Annales de Parasitologie Humaine et Comparee, 2*, 33–41.

Caton, J. M. (1999). Digestive strategy of the Asian colobine genus *Trachypithecus. Primates, 40*, 311–326.

Chabaud, A. G., & Rousselot, R. (1956). Un nouveau spiruridae du gorille Chitwoodspirura gorille Chitwoodspirura wehri. *Bulletin de la Societe de Pathologie Exotique, 51*, 467–472.

Chabaud, A. G, & Lariviere, M. (1958). Sur les Oesophagostomes parasites de l'homme. *Bulletin de la Societe de Pathologie Exotique, 3*, 384–393.

Chapman, C. A. (1988). Patterns of foraging and range use by three species of neotropical primates. *Primates, 29*, 177–194.

Chapman, C. A., & Wrangham, R. W. (1993). Range use of the forest chimpanzees of Kibale: Implications for the understanding of chimpanzee social organization. *American Journal of Primatology, 31*, 263–273.

Chapman, C. A., & Chapman, L. J. (1996). Mixed-species primate groups in the Kibale Forest: Ecological constraints on association. *International Journal of Primatology, 17*, 31–50.

Chapman, C. A., & Chapman, L. J. (1999). Implications of small scale variation in ecological conditions for the diet and density of red colobus monkeys. *Primates, 40*, 215–232.

Chapman, C. A., & Chapman, L. J. (2000a). Interdemic variation in mixed-species association patterns: Common diurnal primates of Kibale National Park, Uganda. *Behavioral Ecology and Sociobiology, 47*, 129–139.

Chapman, C. A., & Chapman, L. J. (2000b). Determinants of group size in social primates: The importance of travel costs. In S. Boinski & P. A. Garber (eds.), *On the move: How and why animals travel in groups*, University of Chicago Press, Chicago, pp. 24–42.

Chapman, C. A., & Lambert, J. E. (2000). Habitat alteration and the conservation of African primates: A case study of Kibale National Park, Uganda. *American Journal of Primatology, 50*, 169–186.

Chapman, C. A., & Peres, C. (2001). Primate conservation in the new millennium: The role of scientists. *Evolutionary Anthropology, 10*, 16–33.

Chapman, C. A., & Chapman, L. J. (2002). Foraging challenges of red colobus monkeys: Influence of nutrients and secondary compounds. *Comparative Biochemistry and Physiology, 133*, 861–875.

Chapman, C. A., & Chapman, L. J. (2004). Unfavorable successional pathways and the conservation value of logged tropical forest. *Biodiversity and Conservation, 13*, 2089–2015.

Chapman, C. A., Chapman, L. J., & Glander, K. E. (1989). Primate populations in northwestern Costa Rica: Potential for recovery. *Primate Conservation, 10*, 37–44.

Chapman, C. A., Chapman, L. J., Wrangham, R. W., Hunt, K., Gebo, D., & Gardner, L. (1992). Estimators of fruit abundance of tropical trees. *Biotropica, 24*, 527–531.

Chapman, C. A., Struhsaker, T. T., & Lambert, J. E. (2005). Thirty years of research in Kibale national park, Uganda, reveals a complex picture for conservation. *International Journal of Primatology, 26*, 539–555.

Chapman C. A., White F. J., & Wrangham R. W. (1993). Defining subgroup size in fission-fusion societies. *Folia Primatologica, 61*, 31–34.

Chapman C. A., White F. J., & Wrangham, R. W. (1994a). Party size in chimpanzees and bonobos: A reevaluation of theory based on two similarly forested sites. In R. W. Wrangham, W. C. McGrew, F. B. M. de Waal, & P. G. Heltne (eds.), *Chimpanzee cultures.* Harvard University Press, Cambridge, pp. 41–58.

Chapman C. A., Wrangham, R. W., & Chapman, L. J. (1994b). Indices of habitat-wide fruit abundance in tropical forests. *Biotropica, 26*, 160–171.

Chapman, C. A., Wrangham, R. W., & Chapman, L. J. (1995). Ecological constraints on group size: An analysis of spider monkey and chimpanzee subgroups. *Behavioral Ecology and Sociobiology, 36*, 59–70.

Chapman, C. A., Chapman, L. J., Wrangham, R., Isabirye-Basuta, G., & Ben-David, K. (1997). Spatial and temporal variability in the structure of a tropical forest. *African Journal of Ecology, 35*, 287–302.

Chapman, C., Gautier-Hion, A., Oates, J., & Onderdonk, D. (1999a). African primate communities: Determinants of structure and threats to survival. In J. Fleagle, C. Janson, & K. Reed (eds.), *Primate communities*, Cambridge University Press, Cambridge, pp. 1–37.

Chapman, L. J., Chapman, C. A., Brazeau, D., McGlaughlin, B., & Jordan, M. (1999b). Papyrus swamps and faunal diversification: Geographical variation among populations of the African cyprinid *Barbus neumayeri*. *Journal of Fish Biology, 54*, 310–327.

Chapman, C. A., Balcomb, S. R., Chapman, L. J., Gillespie, T., Oates, J. F., Skorupa, J. P., & Struhsaker, T. T. (2000). Long-term effects of logging on African primate communities: A 28 year comparison from Kibale National Park, Uganda. *Conservation Biology, 14*, 207–217.

Chapman, C. A., Chapman, L. J., Bjorndal, K. A., & Onderdonk, D. A. (2002). Application of protein-to-fiber ratios to predict colobine abundance on different spatial scales. *International Journal of Primatology, 23*, 283–310.

Chapman, C. A., Chapman, L. J., Rode, K. D., Hauck, E. M., & McDowell, L. R. (2003). Variation in the nutritional value of primate foods: Among trees, time periods, and areas. *International Journal of Primatology, 24*, 317–333.

Chapman, C. A., Chapman, L. J., Naughton-Treves, L., Lawes, M. J., & McDowell, L. R. (2004). Predicting folivorous primate abundance: Validation of a nutritional model. *American Journal of Primatology, 62*, 55–69.

Charles-Dominique, P., & Bearder, S. K. (1979). Field studies of Lorisid behavior: Methodological aspects. In G. A. Doyle & R. D. Martin (eds.), *The study of prosimian behavior*. Academic Press, London, pp. 567–629.

Cheney, D. L., & Seyfarth, R. M. (1990). *How monkeys see the world: Inside the mind of another species.* University of Chicago Press, Chicago.

Chivers, D. J., & Hladik, C. M. (1980). Morphology of the gastrointestinal tract in primates: Comparisons with other mammals in relation to diet. *Journal of Morphology, 116*, 337–386.

Chomsky, N. (1975). *Reflections on language*. Pantheon, New York.

Choo, G. M., Waterman, P. G., McKey, D. B., & Gartlan, J. S. (1981). A simple enzyme assay for dry matter digestibility and its value in studying food selection by generalist herbivores. *Oecologia, 49*, 170–178.

Clark Arcadi, A. (1996). Phrase structure in wild chimpanzee pant hoots: Patterns of production and interpopulation variability. *American Journal of Primatology, 39*, 159–178.

Clark, A. P., & Wrangham, R. W. (1993). Acoustic analysis of chimpanzee pant hoots: Do chimpanzees have an acoustically distinct food arrival pant hoot? *American Journal of Primatology, 31*, 99–109.

Clark, A. P., & Wrangham, R. W. (1994). Chimpanzee arrival pant-hoots: Do they signify food or status? *International Journal of Primatology, 15*, 185–205.

Clore, J. N., Estep, H., Ross-Clunis, H., & Watlington, C. O. (1988). Adrenocorticotropin and cortisol-induced changes in urinary sodium and potassium excretion in man: Effects of spironolactone and RU 486. *Journal of Clinical Endocrinology & Metabolism, 67*, 824–831.

Clutton-Brock, T. H. (1974). Primate social organisation and ecology. *Nature, 250*, 539–542.

Clutton-Brock, T. H. (1975). Feeding behaviour of red colobus and black and white colobus in East Africa. *Folia Primatologica, 23*, 165–207.

Clutton-Brock, T. H., & Harvey, P. H. (1977). Primate ecology and social organization. *Journal of Zoology, 183*, 1–39.

Coe, C. L., Connolly, A. C., Kraemer, H. C., & Levine, S. (1979). Reproductive development and behavior of captive female chimpanzees. *Primates, 20*, 571–582.

Colyn, M. (1991). L'importance zoogéographique du basin du Fleuve Zaire pour la speciation: la cas des Primates Simiens. Koninklijk Museum voor Midden-afrika, Tervuren. *Annalen, Zoologische Wetenschappen, 264*, 1–250.

Cords, M. (1986). Interspecific and intraspecific variation in diet of two forest guenons, *Cercopithecus ascanius* and *C. mitis*. *Journal of Animal Ecology, 55*, 811–827.

Cords, M. (1987). *Mixed-species association of* Cercopithecus *monkeys in the Kakamega Forest Kenya*. University of California Publications 117.

Conklin-Brittain, N. L., Wrangham, R. W., & Hunt, K. D. (1998). Dietary response of chimpanzees and cercopithecines to seasonal variation in fruit abundance, II: Macronutrients. *International Journal of Primatology, 19*, 971–998.

Conklin-Brittain, N. L., Dierenfeld, E. S., Wrangham, R. W., Norconk, M., & Silver, S. C. (1999). Chemical protein analysis: A comparison of Kjeldahl crude protein and total ninhydrin protein using wild, tropical vegetation. *Journal of Chemical Ecology, 25*, 2601–2622.

Cooper, J. E. (1996). Protocol for collection and examination of faecal samples from the mountain gorilla *Gorilla gorilla beringei*. *African Primates, 2*, 24–25.

Cork, S. J. (1994). Digestive constraints on dietary scope in small and moderately-sized mammalian folivores in relation to chemical defenses in temperate and tropical forests. In R. T. Palo & C. T. Robbins (eds.), *Plant defenses against mammalian herbivory*. CRC Press, Boca Raton, pp. 133–166.

Cork, S. J. (1996). Optimal digestive strategies for arboreal herbivorous mammals in contrasting forest types: Why koalas and colobines are different. *Australian Journal of Ecology, 21*, 10–20.

Cormier, L. (2002). Monkey as food, monkey as child: Guajá symbolic cannibalism. In A. Fuentes & L. Wolfe (eds.), *Primates face to face: The conservation implications of human–nonhuman primate interconnections*. Cambridge University Press, Cambridge, pp. 63–84.

Corp, N., & Byrne, R. (2002a). Leaf processing of wild chimpanzees: Physically defended leaves reveal complex manual skills. *Ethology, 108*, 1–24.

Corp, N., & Byrne, R. (2002b). The ontogeny of manual skill in wild chimpanzees: Evidence from feeding on the fruit of *Saba florida*. *Behaviour, 139*, 137–168.

Courtenay, J. (1987). Post-partum amenorrhoea, birth intervals and reproductive potential in captive chimpanzees. *Primates, 28*, 543–546.

Cousins, D., & Huffman, M. A. (2002). Medicinal properties in the diet of gorillas: an ethnopharmacological evaluation. *African Study Monographs, 23*, 65–89.

Cowlishaw, G., & Dunbar, R. I. M. (2000). *Primate conservation biology*. The University of Chicago Press, Chicago.

Cranfield, M., Gaffikin, L., Sleeman, J., & Rooney, M. (2002). The mountain gorilla and conservation medicine. In A. A. Aguirre, R. S. Ostfeld, G. M. Tabor, C. House, & M. C. Pearl (eds.), *Conservation medicine: Ecological health in process*. Oxford University Press, New York, pp. 282–296.

Crockford, C., Herbinger, I., Vigilant, L., & Boesch, C. (2004). Wild chimpanzees produce group-specific calls: A case for vocal learning? *Ethology, 110*, 221–243.

Crompton, D. W. T. (1999). How much human helminthiasis is there in the world? *Journal of Parasitology, 48*, 285–375.

Crompton, D. W. T., & Nesheim, M. C. (2002). Nutritional impact of intestinal helminthiasis during the human life cycle. *Annual Review of Nutrition, 22*, 35–39.

Cronk, L. (1995). Is there a role for culture in human behavioral ecology? *Evolution and Human Behavior, 16*, 181–205.

Curtin, S. H. (1976). Niche separation in sympatric Malaysian leaf monkeys (*Presbytis obscura* and *Presbytis melalophos*). *Yearbook of Physical Anthropology, 20*, 421–439.

Czekala, N. M., Gallusser, S., Meier, J. E., & Lasley, B. L. (1986). The development and application of an enzyme immunoassay for urinary estrone conjugates. *Zoo Biology, 5*, 1–6.

Czekala, N. M., Shideler, S. E., & Lasley, B. L. (1988). Comparisons of female reproductive hormone patterns in the hominoids. In J. H. Schwartz (ed.), *Orang-utan biology*. Oxford University Press, New York, pp. 117–122.

Dahl, J. F., Nadler, R. D., & Collins, D. C. (1991). Monitoring the ovarian cycles of *Pan troglodytes* and *P. paniscus*: A comparative approach. *American Journal of Primatology, 24*, 195–209.

Daly, H., & Cobb, J. (1989). *For the common good: Redirecting the economy toward community, the environment, and a sustainable future*. Beacon Press, Boston.

Dandelot, P. (1959). Note sur la classification des Cercopithèques du groupe aethiops. *Mammalia, 23*, 357–368.

Darwin, C. (1872). *The expressions of the emotions in man and animals*. Murray, London.

DaSilva, G. (1989). *The ecology of the western black and white colobus (*Colobus polykomos ploykomos *Zimmermann 1780) on a riverine island in south-eastern Sierra Leone*. Unpublished DPhil thesis, University of Oxford.

DaSilva, G. L. (1992). The western black-and-white colobus as a low energy strategist: Activity budgets, energy expenditure and energy intake. *Journal of Animal Ecology, 61*, 79–91.

DaSilva, G. L. (1994). Diet of *Colobus polycomos* on Tiwai Island: Selection of food in relation to its seasonal abundance and nutritional quality. *International Journal of Primatology, 15*, 655–680.

Davidian, M., & Giltinan, D. M. (1995). *Nonlinear models for repeated measurement data*. Chapman & Hall, London.

Davies, A. G. (1991). Seed-eating by red leaf monkeys (*Presbytis rubicunda*) in dipterocarp forest of northern Borneo. *International Journal of Primatology, 12*, 119–144.

Davies, A. G. (1994). Colobine populations. In A. G. Davies & J. F. Oates (eds.), *Colobine monkeys: Their ecology, behaviour and evolution*. Cambridge University Press, Cambridge, pp. 285–310.

Davies, A. G., & Richards, P. (1991). *Rain forest in Mende Life.* A Report to ESCOR, UK Overseas Development Administration.

Davies, A. G., & Oates, J. F. (1994). *Colobine monkeys: Their ecology, behaviour and evolution.* Cambridge University Press, Cambridge.

Davies, A. G., Oates, J. F., & DaSilva, G. L. (1999). Patterns of frugivory in three west African colobine monkeys. *International Journal of Primatology, 20*, 327–357.

Day, P. (1965). Particle fractionation and particle size analysis. In C. A. Black (ed.), *Methods of soil analysis.* American Society of Agronomy, Madison, Wisconsin, pp. 545–567.

De Boer, W., & Baquete, D. (1998). Natural resource use, crop damage and attitudes of the local population around the Maputo Elephant Reserve, Mozambique. *Environmental Conservation, 25*, 208–218.

De Vicente, F., Rodriguez-Perez, M. C., & Gomez-Jarabo, G. (1991). The effects of protein malnutrition and cortisol treatment on motor activity of rats. *Behavioural Processes, 25*, 1–14.

DeVore, I., & Washburn, S. L. (1963). Baboon ecology and human evolution. *Viking Fund Publications in Anthropology, 36*, 335–367.

DeVore, I., & Hall, K. R. L. (1965). Baboon ecology. In I. DeVore (ed.), *Primate behaviour: Field studies of monkeys and apes.* Holt, New York.

De Waal, F. B. M. (1999). Cultural primatology comes of age. *Nature, 399*, 635–636.

Deblauwe, I., Dupain, J., Nguenang, G. M., Werdenich, D., & Van Elsacker, L. (2003). Insectivory by *Gorilla gorilla gorilla. International Journal of Primatology, 24*, 493–502.

Defour, G. (1994). Plantes médicinales traditionnelles au Kivu (République du Zaïre). Documentation du sous-réseau Prélude.

Denegri, G., Bernadina, W., Perez-Serrano, J., & Rodriguez-Caabeiro, F. (1998). Anoplocephalid cestodes of veterinary and medical significance: A review. *Folia Parasitologica, 45*, 1–8.

Dewalt, S. J., Schnitzer, S. A., & Denslow, J. S. (2000). Density and diversity of lianas along a chronosequence in a central Panamanian lowland forest. *Journal of Tropical Ecology, 16*, 1–19.

Dierenfild, E. S., Koontz, F. W., & Goldstein, R. S. (1992). Feed intake, digestion and passage of the proboscis monkey (*Nasalis larvatus*) in captivity. *Primates, 33*, 399–405.

Doran, D. M. (1993). Sex differences in adult chimpanzee positional behaviour: The influence of body size on locomotion and posture. *American Journal of Physical Anthropology, 91*, 99–115.

Doran, D. M. (1997). Influence of seasonality on activity patterns, feeding behavior, ranging, and grouping patterns in Tai chimpanzees. *International Journal of Primatology, 18*, 183–206.

Doran, D., & McNeilage, A. (1998). Gorilla ecology and behavior. *Evolutionary Anthropology, 6*, 120–131.

Doran, D. M., McNeilage, A., Greer, D., Bocian, C., Mehlman, P., & Shah, N. (2002). Western lowland gorilla diet and resource availability: New evidence, cross-site comparisons, and reflections on indirect sampling methods. *American Journal of Primatology, 58*, 91–116.

Doran, D. M., Jungers, W. L., Sugiyama, Y., Fleagle, J. G., & Heesy, C. P. (2002). Multivariate and phylogenetic approaches to understanding chimpanzee and bonobo behavioral diversity. In C. Boesch, G. Hohmann, & L. F. Marchant (eds.), *Behavioural diversity in chimpanzees and bonobos*. Cambridge University Press, Cambridge, pp. 14–34.

Doran-Sheehy, D. M., Greer, D., Mongo, P., & Schwindt D. (2004). Impact of ecological and social factors on ranging in western gorillas. *American Journal of Primatology, 64*, 207–222.

Dudley, J. P., Ginsberg, J. R., Plumptre, A. J., Hart, J. A., & Campos, L. C. (2002). Effects of war and civil strife on wildlife and wildlife habitats. *Conservation Biology, 16*, 319–329.

Dunbar, R. I. M. (1987). Habitat quality, population dynamics and group composition in colobus monkeys (*Colobus guereza*). *International Journal of Primatology, 8*, 299–330.

Dunbar, R. I. M. (1988). *Primate social systems: Studies in behavioral adaptation*. Croom Helm, London.

Dunbar, R. I. M., & Dunbar, E. P. (1974). Ecology and popualtion dynamics of *Colobus guereza* in Ethiopia. *Folia Primatologica, 21*, 188–208.

Dupain, J., van Elsaker, L., Nell, C., Garcia, P., Ponce, F., & Huffman, M. A. (2002). *Oesophagostomum* infections and evidence for leaf swallowing in bonobos (*Pan paniscus*): Indication for self-medicative behavior? *International Journal of Primatology, 23*, 1053–1062.

Durette-Desset, M. C. (1974). Keys to the genera of the Superfamily Trichostrongyloidea. No. 10. In R. C. Anderson & A. G. Chabaud (eds.), *CIH keys to the nematode parasites of vertebrates*. CAB, Slough, England, pp. 1–86.

Durette-Desset, M. C., Chaubaud, A. G., Ashford, R. W., Butynski, T., & Reid, G. D. F. (1992). Two new species of the Trichostrongylidae (Nematoda: Tichostrongyloidea), parasitic in *Gorilla gorilla beringei* in Uganda. *Systematic Parasitology, 23*, 159–166.

Earl, D. E. (1992). Wise management of natural tropical forest for timber production, tourism and wildlife. In F. R. Miller & K. L. Adam (eds.), *Wise management of tropical forests*. Oxford Forestry Institute, Oxford University Press, Oxford, pp. 211–213.

Eberhard, M. L., Kovacs-Nace, E., Blotkamp, J., Verwij, J. J., Asigri, V. A. A., & Polderman, A. M. (2001). Experimental *Oesophagostomum bifurcum* in monkeys. *Journal of Helminthology, 75*, 51–56.

Edroma, E. L., Rosen, N., & Miller, P. S. (1997). *Conserving the Chimpanzees of Uganda: Population and Habitat Viability Assessment for* Pan troglodytes schweinfurthii. IUCN/SSC Conservation Breeding Specialist Group, Apple Valley, Minnesota.

Eggeling, W. J. (1947). Observations of the ecology of the Budongo Rain Forest, Uganda. *Journal of Ecology, 34*, 20–87.

Eilenberger, U. (1998). *Individual, group specific and ecological influence factors on the status of endoparasites of wild eastern lowland gorillas: A multidisciplinary approach.* Dissertation presented to Ausdem Institut fur Parasitologie und Tropenveterinarmedizin des Fachbereichs Veterinarmedizin der Freien Universitat Berlin. Journal Nr. 2140, pp. 135.

Eissa, M. K., Obhrai, M. S., Docker, M. F., Lynch, S. S., Sawers, R. S., & Newton, R. R. (1986). Follicular growth and endocrine profiles in spontaneous and induced conception cycles. *Fertility and Sterility, 53*, 81–87.

Ellison, P. T. (1993). Understanding natural variation in human ovarian function. In: *Galton Institute Symposium on Reproductive Decisions: Biological and Social Perspectives*, London.

Ellison, P. T. (1996). Developmental influences on adult ovarian hormonal function. *American Journal of Human Biology, 8*, 725–734.

Ellison, P. T., Peacock, N. R., & Lager, C. (1989). Ecology and ovarian function among Lese women of the Ituri Forest, Zaire. *American Journal of Physical Anthropology, 78*, 519–526.

Ellison, P. T., Lipson, S. F., O'Rourke, M. T., Bentley, G. R., Harrigan, A., Panter-Brick, C., & Vitzthum, V. (1993a). Population variation in ovarian function. *The Lancet, 342*, 433–434.

Ellison, P. T., Panter-Brick, C., Lipson, S. F., & O'Rourke, M. T. (1993b). The ecological context of human ovarian function. *Human Reproduction, 8*, 2248–2258.

Empowerment for African Sustainable Development. (1996). *State of the environment report, Uganda*. Available at: http://easd.org.za/download.asp?path=Uganda_soe_1996.pdf&Is=File.

Environmentally Sustainable Development (ESD)(1995). A study of woody biomass derived energy supplies, final report. Foresty Department, Ministry of Natural Resources, Uganda.

Etkin, N. L., & Ross, P. J. (1982). Food as medicine and medicine as food: An adaptative framework for the interpretation of plant utiliszation among the Hausa of Northern Nigeria. *Social Science & Medicine, 16*, 1559–1573.

Ewing, H. E. (1927). Descriptions of three new species of sucking lice, together with a key to some related species of the genus *Polyplax*. *Proceedings of the Entomological Society of Washington, 29*, 118–121.

Ewing, L. L. (1982). Seasonal variation in primate fertility with an emphasis on the male. *American Journal of Primatology, 1*(Suppl.), 145–160.

Fairgrieve, C. (1995). *The comparative ecology of blue monkeys (*Cercopithecus mitis stuhlmannii*) in logged and unlogged forest, Budongo Forest Reserve, Uganda*. Unpublished PhD thesis, University of Edinburgh.

Fain, A. (1957). L'acaraise pulmonaire chez la chimpanzé et le gorilla par des acariens du genre *Pneumonyssus. Banks. Revue de Zoologie et Botanique. Africaines, 56*, 234–242.

Falconer, J. (1991). *The major significance of minor forest products: The local use and value of the forests in the West African humid forest zone*. Food and Agriculture Organization of the United Nations, Rome, 1990.

FAO. (1990). *Forest resources assessment 1990—Tropical countries*, FAO Forestry Paper 112, Rome.

FAO. (1999). *State of the World's Forests*. Rome: Food and Agriculture Organization of the United Nations.

Food and Agriculture Organization. (1990). *Forest resources assessment 1990: Tropical countries*. Food and Agriculture Organization of the United Nations, Forestry Paper 112, Rome.

Food and Agriculture Organization. (1999). *State of the world's forests*. Food and Agriculture Organization of the United Nations, Rome.

Fashing, P. J. (1999). *The behavioural ecology of an African colobine monkey: Diet, range use, and patterns of intergroup aggression in eastern black and white colobus monkeys (*Colobus guereza). Unpublished PhD thesis, Columbia University, New York.

Fedigan L. M. (1991). History of the Arashiyama West Japanese Macaques in Texas. In Fedigan, L. M., and Asquith, P. J. (Eds.), *The monkeys of Arashiyama*. State University of New York Press, pp. 54–73.

Feigl, F., & Anger, V. (1966). Replacement of bensidine by copper ethylacetoacetate and tetra-base as spot-test reagent for hydrogen cyanide and cyanogens. *Analyst, 91*, 282–284.

File, S. K., McGrew, W. C., & Tutin, C. E. G. (1976). The intestinal parasites of a community of feral chimpanzees, *Pan troglodytes schweinfurthii*. *Journal of Parasitology, 62*, 259–261.

Flynn, R. J. (1973). *Parasites of laboratory animals*. Iowa State University Press, Ames.

Foley, W. J. (1992). Nitrogen and energy retention and acid-base status in the common ringtail possom (*Pseudocheirus peregrinus*): Evidence of the effects of absorbed allelochemicals. *Physiological Zoology, 65*, 403–421.

Fong, H. H. S., Tin-Wa, M., & Farnsworth, N. R. Unpublished lab guide, Dept. of Pharmacognosy and Pharmacology, College of Pharmacy, University of Illinois.

Fontenot, M. B., Stavisky, R. C., Seraphin, S. B., & Whitten, P. L. (2001). Aging primates: Patterns of dehydroepiandrosterone sulfate excretion in chimpanzees (*Pan troglodytes*). *American Journal of Primatology, 54*, 72.

Fossey, D. (1974). Observations on the home range of one group of mountain gorillas. *Animal Behaviour, 22*, 568–581.

Fossey, D. (1983). *Gorillas in the mist.* Houghton Mifflin, New York.

Fossey, D., & Harcourt, A. H. (1977). Feeding ecology of free-ranging mountain gorilla (*Gorilla gorilla beringei*). In T. H. Clutton-Brock (ed.), *Primate ecology.* Academic Press, London, pp. 415–447.

Francis, G., Kerem, Z., Makkar, H. P. S., & Becker, K. (2002). The biological action of saponins in animal systems: A review. *British Journal of Nutrition, 88*, 587–605.

Fretwell, S. D., & Lucas, H. L. (1970). On territorial behaviour and other factors influencing habitat distribution in birds. *Acta Biotheoretica, 19*, 16–36.

Frothmann, D. L., Burks, K. D., & Maples, T. L. (1996). Letter to the editor: African great ape ecotourism considered. *African Primates, 2*, 52–54.

Frumhoff, P. C. (1995). Conserving wildlife in tropical forests managed for timber. *Bioscience, 45*, 456–464.

Fujita, S., Mitsunaga, F., Sugiura, H., & Shimizu, K. (2001). Measurement of urinary and fecal steroid metabolites during the ovarian cycle in captive and wild Japanese macaques, *Macaca fuscata. American Journal of Primatology, 53*, 167–176.

Furuichi, T. (1987). Sexual swelling, receptivity, and grouping of wild pygmy chimpanzee females at Wamba, Zaïre. *Primates, 28*, 309–318.

Furuichi, T., & Hashimoto, C. (2002). Why female bonobos have a lower copulation rate during estrus than chimpanzees. In C. Boesch, G. Hohmann, & L. F. Marchant (eds.), *Behavioral diversity in chimpanzees and bonobos.* Cambridge University Press, Cambridge, pp. 156–167.

Gallaher, R. N., Weldon, C. O., & Futral, J. G. (1975). An aluminum block digester for plant and soil analysis. *Soil Science Society of America Proceedings, 39*, 803–806.

Ganas, J., & Robbins, M. M. (2004). Intrapopulation differences in ant eating in the mountain gorillas of Bwindi Impenetrable National Park, Uganda. *Primates, 45*, 275–278.

Ganas, J., Robbins, M. M., Nkurunungi, J. B., Kaplin, B. A., & McNeilage, A. J. (2004). Dietary variability of mountain gorillas in Bwindi Impenetrable National Park, Uganda. *International Journal of Primatology, 25*, 1043–1072.

Ganzhorn, J. U. (1988). Food partitioning among Malagasy primates. *Oecologia, 75*, 436–450.

Ganzhorn, J. U. (1992). Leaf chemistry and the biomass of folivorous primates in tropical forests. *Oecologia, 91*, 540–547.

Ganzhorn, J. U. (1995). Low-level forest disturbance effects on primary production, leaf chemistry and lemur populations. *Ecology, 76*, 2084–2096.

Ganzhorn, J. U. (2002). Distribution of a folivorous lemur in relation to seasonally varying food resources: Integrating quantitative and qualitative aspects of food characteristics. *Oecologia, 131*, 427–435.

Gardner, B. T., & Gardner, R. A. (1975). Evidence for sentence constituents in the early utterances of child and chimpanzee. *Journal ofExperimental Psychology, 104*, 244–267.

Garin, Y., Tutin, C. E. G., Fernandez, M., & Goussard, B. (1982). A new intestinal parasitic entodiniomorph ciliate from wild lowland gorillas (*Gorilla gorilla gorilla*) in Gabon? *Journal of Medical Primatology, 11*, 186–190.

Garner, K. J., & Ryder, O. A. (1996). Mitochondrial DNA diversity in gorillas. *Molecular and Phylogenetic Evolution, 6*, 39–48.

Gasquet, M., Bamba, D., Babadjamian, A., Balansard, G., Timon-David, P., & Metzger, J. (1985). Action amoebicide et anthelminthique du vernolide et de l'hydroxybernolide isoles des feuilles de Vernonia colorata (Willd.) Drake. *European Journal of Medicinal Chemistry, 2*, 111–115.

Gathua, J. M. (1999). *Intraspecific variation in foraging patterns of redtail monkeys (*Cercopithecus ascanius*) in the Kakamega Forest, Kenya.* Unpublished PhD thesis, Columbia University, New York.

Gaulin, S. J. C. (1979). A Jarman/Bell model of primate feeding niches. *Human Ecology, 7*, 1–20.

Gautier-Hion, A. (1988). The diet and dietary habits of forest guenons. In A. Gautier-Hion, F. Bourliere, & J-P. Gautier (eds.), *A primate radiation: Evolutionary biology of the African guenons.* Cambridge University Press, Cambridge, pp. 257–283.

Gebo, D. L., & Sargis, E. J. (1994). Terrestrial adaptations in the postcranial skeletons of guenons. *American Journal of Physical Anthropology, 93*, 341–371.

Gebo, D. L., & Chapman, C. A. (1995). Habitat, annual, and seasonal effects on positional behavior in red colobus monkeys. *American Journal of Physical Anthropology, 96*, 73–82.

Gerwing, J. J., & Farias, D. L. (2000). Integrating liana abundance and forest stature into an estimate of total aboveground biomass for an eastern Amazoniain forest. *Journal of Tropical Ecology, 16*, 327–335.

Ghiglieri, M. P. (1984). *The chimpanzees of Kibale forest: A field study of ecology and social structure.* Columbia University Press, New York.

Ghiglieri, M. P. (1988). *East of the mountains of the Moon: Chimpanzee society in the African rainforest.* The Free Press, New York.

Gillespie, T. R., & Chapman, C. A. (2001). Determinants of groups size in the red colobus monkey (*Procolobus badius*): An evaluation of the generality of the ecological-constraints model. *Behavioural Ecology and Sociobiology, 50*, 329–338.

Giner-Chavez, B., Van Soest, P. J., Robertson, J. B., Lascano, C., Reed, J. D., & Pell, A. N. (1997). A method for isolating condensed tannins from crude plant extracts with trivalent ytterbium. *Journal of the Science of Food and Agriculture, 74*, 359–368.

Glander, K. E. (1975). Habitat description and resource utilization: A preliminary report on mantled howler monkey ecology. In R. H. Tuttle (ed.), *Socioecology and Psychology of Primates*. Moulton, The Hague, pp. 37–57.

Glander, K. E. (1982). The impact of plant secondary compounds on primate feeding behavior. *Yearbook of Physical Anthropology, 25*, 1–18.

Glander, K. E., Wright, P. C., Seigler, D. S., Randrianasolo, V., & Randrianasolo, B. (1989). Consumption of cyanogenic bamboo by a newly discovered species of bamboo lemur. *American Journal of Primatology, 19*, 119–124.

Goering, H. K., & van Soest, P. J. (1970). *Forage fiber analysis*. Agricultural Handbook No. 379. Agricultural Research Service, U.S. Dept. of Agriculture.

Goldman, B. D., & Nelson, R. J. (1993). Melatonin and seasonality in mammals. In H. S. Yu & R. J. Reiter (eds.), *Melatonin: Biosynthesis, physiological effects, and clinical applications*. CRC Press, Boca Raton, FL, pp. 225–252.

Goldman, J., Wajchenberg, B. L., Liberman, B., Nery, M., Achando, S., & Germek, O. A. (1985). Contrast analysis for the evaluation of the circadian rhythms of plasma cortisol, androstenedione, and testosterone in normal men and the possible influence of meals. *Journal of Clinical Endocrinology & Metabolism, 60*, 164–167.

Goldsmith, M. L. (1996). *Ecological influences on the ranging and grouping behavior of western lowland gorillas at Bai Hokòu, Central African Republic*. PhD dissertation, University at Stony Brook, New York.

Goldsmith, M. L. (1999). Ecological constraints on the foraging effort of western gorillas (*Gorilla gorilla gorilla*) at Bai Hokou, Central African Republic. *International Journal of Primatology, 20*, 1–23.

Goldsmith, M. L. (2003). Comparative behavioral ecology of a lowland and highland gorilla population: Where do Bwindi gorillas fit? In A. B. Taylor & M. L. Goldsmith (eds.), *Gorilla biology: A multidisciplinary perspective*. Cambridge University Press, Cambridge, pp. 358–384.

Goldsmith, M. L. (2004). Habituating primates for field study: Ethical considerations for African great apes. In T. Turner (ed.), *Biological anthropology and ethics: From repatriation to genetic identity*. State University of New York Press, New York, pp. 49–64.

Goldsmith, M. L., & Moles, H. (2003). Does topography affect the foraging effort of mountain gorillas in Bwindi Impenetrable National Park, Uganda? *American Journal of Physical Anthropology* (Suppl. 20), 102 (abstract).

Goldsmith, M. L., & Moles, H. (in press). Topography affects the foraging effort of mountain gorillas in Bwindi Impenetrable National Park Uganda. *American Journal of Primatology*.

Goldsmith, M. L., Nkurunungi, J. B., & Stanford, C. B. (1999). Gorilla behavioral ecology: Effects of altitudinal changes on highland/lowland populations. *American Journal of Physical Anthropology* (Suppl.) (abstract).

Goodall, J. (1964). Tool-using and aimed throwing in a community of free-living chimpanzees. *Nature, 201*, 1264–1266.

Goodall, J. (1968a). Expressive movements and communication in free-ranging chimpanzees: A preliminary report. In P. Jay (ed.), *Primates: Studies in adaptation and variability*. Holt, Rinehart and Winston, New York.

Goodall, J. (1968b). The behaviour of free-living chimpanzees in the Gombe Stream Reserve. *Animal Behaviour Monographs, 1*, 161–311.

Goodall, A. (1977). Feeding and ranging behaviour of a mountain gorilla group (*Gorilla gorilla beringei*) in the Tshibinda-Kahuzi Region (Zaire). In T. H. Clutton-Brock (ed.), *Primate ecology*. Academic Press, London, pp. 415–447.

Goodall, J. (1986). *The chimpanzees of Gombe: Patterns of behaviour*. Harvard University Press, Cambridge, MA.

Goodman, M., Porter, C. A., Czelusniak, J., Page, S. L., Schneider, H., Shoshani, J., Gunnell, G., & Groves., C. P. (1998). Toward a phylogenetic classification of primates based on DNA evidence complemented by fossil evidence. *Molecular Phylogenetics and Evolution, 9*, 585–598.

Gordon, T. P., Bernstein, I. S., & Rose, R. M. (1978). Social and seasonal influences on testosterone secretion in the male rhesus monkey. *Physiology & Behavior, 21*(4), 623–627.

Goussard, B., Collet, J. Y., Garin, Y., Tutin, C. E. G., & Fernandez, M. (1983). The intestinal entodiniomorph ciliates of wild lowland gorillas (*Gorilla gorilla gorilla*) in Gabon, West Africa. *Journal of Medical Primatology, 12*, 239–249.

Gouzoules, H., Gouzoules, S., & Ashley, J. (1995). Representational signaling in non-human primate vocal communication. In E. Zimmerman, J. D. Newman, & U. Jürgens (eds.), *Current topics in primate vocal communication*. Plenum Press, New York, pp. 235–252.

Graber, M., & Gevrey, J. P. (1981). Parasites internes des primates de la République Démocratique du Congo (d'aprés la collection Cassard-Chambron 1956–1960) Rôle pathogène-diagnostic-prophylaxie. *Revue d Elevage et de Medecine Veterinaire des Pays Tropicaux, 34*, 27–41.

Graczyk, T. K., & Cranfield, M. R. (2001). Coprophagy and intestinal parasites: Implications to human-habituated mountain gorillas (*Gorilla gorilla beringei*). *Recent Research Developments in Microbiology, 5*, 285–303.

Graczyk, T. K., Lowenstein, L. J., & Cranfield, M. R. (1999). *Capillaria hepatica* (Nematoda) infections in human habituated mountain gorillas (*Gorilla gorilla beringei*) of the Parc National de Volcans, Rwanda. *Journal of Parasitolology, 85*, 1168–1170.

Graczyk, T. K., Mudakikwa, A. B., Cranfield, M. R., & Eilenberger, U. (2001a). Hyperkeratotic mange caused by *Sarcoptes scabiei* (Acariform Sarcoptidae) in juvenile human-habituated mountain gorillas (*Gorilla gorilla beringei*). *Parasitology Research, 87*, 1024–1028.

Graczyk, T. K., DaSilva, A. J., Cranfield, M. R., Nizeyi, J. B., Kalema, G. R. N. N., & Pieniazek, N. J. (2001b). *Cryptosporidium parvum* Genotype 2 infections in free-ranging mountain gorillas (*Gorilla gorilla berengei*) of the Bwindi Impenetrable National Park, Uganda. *Parasitology Research, 87*, 368–370.

Graczyk, T. K., Nizeyi, J. B., Ssebide, B., Andrew Thompson, R. C., Read, C. T., & Cranfield, M. R. (2002a). Anthropozoonotic *Giardia duodenalis* Genotype (Assemblage A) infections in habitats of free-ranging human-habituated gorillas, Uganda. *Journal of Parasitology, 88*, 905–909.

Graczyk, T. K., Nizeyi, J. B., da Silva, A. J., Moura, I. N. S., Pieniazek, N. J., Cranfield, M. R., & Lindquist, H. D. A. (2002b). A single genotype of *Encephalitozoon intestinalis* infects free-ranging gorillas and people sharing their habitats in Uganda. *Parsitology Research, 88*, 926–931.

Grady, J. P., Yeager, C. H., Esra, G. N., & Thomas, W. (1982). Ultrasonic evaluation of echinococcosis in four lowland gorillas. *Journal of American Veterinery Medical Association, 181*, 1348–1350.

Graham, C. E. (1970). Reproductive physiology of the chimpanzee. In *The Chimpanzee.* Karger, New York, pp. 183–220.

Graham, C. E. (1973). Chimpanzee endometrium and sexual swelling during menstrual cycle or hormone administration. *Folia Primatologica,* 19, 458–468.

Graham, C. E., Gould, K. G., Wright, K., & Collins, D. C. (1978). Luteal estrogen secretion and decidualisation in the chimpanzee. In D. J. Chivers & E. H. R. Ford (eds.), *Recent advances in primatology, Vol IV: Medicine.* Academic Press, New York, pp. 209–211.

Grant, J. K., & Beastall, B. H. (1983). *Clinical biochemistry of steroid hormones: Methods and applications.* Elsevier, New York.

Grieser Johns, A., & Grieser Johns, B. (1995). Tropical forest primates and logging: Long-term co-existence? *Oryx, 29*, 205–211.

Groves, C. P. (1978). Phylogenetic and population systematics of the mangabeys (Primates: Cercopithecoidea). *Primates, 19*, 1–34.

Groves, C. (2001). *Primate taxonomy.* Smithsonian Institution Press, Washington.

Groves, C. (2005). Geographic variation within Eastern Chimpanzees (*Pan troglodytes* cf. *schweinfurthii* Giglioli, 1872). *Australasian Primatology, 17*, 19–46.

Grubb, P. (1978). Patterns of speciation in African mammals. *Bulletin of the Carnegie Museum of Natural History, 6*, 152–167.

Grubb, P. (2001). Endemism in African rain forest mammals. In W. Weber, L. J. T. White, A. Vedder & L. Naughton-Treves (eds.), *African rain forest ecology and conservation: An interdisciplinary perspective.* Yale University Press, New Haven, pp. 88–100.

Grubb, P., Butynski, T. M., Oates, J. F., Bearder, S. K., Disotell, T. R., Groves, C. P., & Struhsaker, T. T. (2003). Assessment of the diversity of African primates. *International Journal of Primatology, 24*, 1301–1357.

Guerrera, W., Sleeman, J. M., Jasper, S. B., Pace, L. B., Ichinose, T. Y., & Reif, J. S. (2003). Medical survey of the local human population to determine possible health risks to the mountain gorillas of Bwindi Impenetrable Forest National Park, Uganda. *International Journal of Primatology, 24*, 197–207.

Gupta, M. B., Nath, R., Srivastava, K., Shanker, K., Kishor, K., & Bhargava, K. P. (1980). Anti-inflammatory and anti-pyretic activities of β-sitosterol. *Planta Medica, 39*, 157–163.

Gyger, M., Marler, P., & Pickert, R. (1987). Semantics of an avian alarm call system: The male domestic fowl, *Gallus domesticus. Behaviour, 102*, 15–40.

Hackney, A. C., & Hodgdon, J. A. (1991). Norwegian military field exercises in the arctic: Endocrine and metabolic responses. *Arctic Medical Research, 50* (Suppl. 6), 137–141.

Hall, M. B., Hoover, W. H., Jennings, J. P., & Miller-Webster, T. K. (1999). A method for partitioning neutral detergent-soluble carbohydrates. *Journal of the Science of Food and Agriculture, 79*, 2079–2089.

Hambleton, L. G. (1977). Semiautomated method for simultaneous determination of phosphorus, calcium, and crude protein in animal feeds. *Journal of the Association of Official Agricultural Chemists, 60*, 845–852.

Hames, R. (1987). Game conservation or efficient hunting? In B. McCay & J. Acheson (eds.), *The question of the commons.* University of Arizona Press, Tucson, pp. 97–102.

Hamilton, A. C. (1988). Guenon evolution and forest history. In A. Gautier-Hion, F. Bourliere, & J-P. Gautier (eds.), *A primate radiation: Evolutionary biology of the African guenons.* Cambridge University Press, Cambridge, pp. 13–34.

Hamilton, W. J. I., Buskirk, R. E., & Buskirk, W. H. (1978). Omnivory and utilization of food resources by chacma baboons, *Papio ursinus. American Naturalist, 112*, 911–924.

Hamilton, A. C., Taylor, D., & Vogel, J. C. (1986). Early forest clearance and environmental degradation in south-west Uganda. *Nature, 320*, 164–167.

Hamilton, A., Cunningham, A., Byarugaba, D., & Kayanja, F. (2000). Conservation in a region of political instability: Bwindi Impenetrable Forest, Uganda. *Conservation Biology, 14*, 1722–1725.

Hancock, R. G. V. (1984). On the source of clay used for Cologne Roman pottery. *Archaeometry, 26*, 210–217.

Harada, Y., & Mori, O. (1955). A new method for culturing hookworm. *Yonago Acta Medica, 1*, 177–179.

Harborne, J. B. (1991). Recent advances in the ecological chemistry of plant terpenoids. In J. B. Harborne & F. A. Tomas-Barberan (eds.), *Ecological chemistry and biochemistry of plant terpenoids.* Academic Press, London, pp. 399–426.

Harborne, J. B. (1993). *Introduction to ecological biochemistry* (4th ed.). Academic Press, New York.

Harcourt, A. H. (1986). Gorilla conservation: Anatomy of a campaign. In K. Benirschke (ed.), *Primates: The road to self sustaining populations.* Springer-Verlag, New York, pp. 31–46.

Harcourt, A. H., & Stewart, K. G. (1978). Coprophagy by wild mountain gorilla. *East African Wildlife Journal, 16*, 223–225.

Harcourt, A. H., & Harcourt, S. A. (1984). Insectivory by gorillas. *Folia Primatologica, 43*, 229–233.

Harding, R. S. O. (1981). An order of omnivores: non human primate diets in the wild. In R. S. O. Harding & G. Teleki (eds.), *Omnivorous primates: Gathering and hunting in human evolution.* Columbia University Press, New York.

Hare, B., Call, J., Agnetta, B., & Tomasello, M. (2000). Chimpanzees know what conspecifics do and do not see. *Animal Behavior, 59*, 771–785.

Hare, B., & Tomasello, M. (2004). Chimpanzees are more skilful in competitive than in cooperative cognitive tasks. *Animal Behavior, 68*, 571–581.

Harrison, M. J. S. (1986). Feeding ecology of black colobus, *Colobus satanas*, in central Gabon. In J. G. Else & P. C. Lee (eds.), *Primate ecology and conservation.* Cambridge University Press, Cambridge, pp. 31–37.

Hart, B. L. (1990). Behavioural adaptations to pathogens and parasites: Five strategies. *Neuroscience and Biobehavioral Reviews, 14*, 273–294.

Hartwell, J. L. (1967–1971). Plants used against cancer. A survey. *Lloydia*, 30–34.

Hasegawa, T. (1992). Mesu nitotteno rankon—chimpanzee to nihonzaru o hikakushite. In Y. Ito (ed.), *Doubutsushaka niokeru kyoudou to kougeki.* Tokai University Press, Tokyo.

Hasegawa, T., & Hiraiwa-Hasegawa, M. (1983). Opportunistic and restrictive matings among wild chimpanzees in the Mahale Mountains, Tanzania. *Journal of Ethology, 1*, 75–85.

Hasegawa, T., & Hiraiwa-Hasegawa, M. (1990). Sperm competition and mating behavior. In T. Nishida (ed.), *The chimpanzees of the Mahale Mountains: Sexual and life history strategies.* University of Tokyo Press, Tokyo, pp. 115–132.

Hashimoto, C. (1995). Population census of the chimpanzees in the Kalinzu Forest, Uganda: Comparison between methods with nest counts. *Primates, 36*, 477–488.

Hashimoto, C. (1999). Snare injuries of chimpanzees in the Kalinzu Forest, Uganda. *Pan Africa News, 6*, 20–22.

Hashimoto, C., Furuichi, T., Tashiro, Y., & Kimura, D. (1999). Vegetation of the Kalinzu Forest, Uganda: Ordination of forest types using principal component analysis. *African Study Monographs, 20*, 229–239.

Hashimoto, C., Furuichi, T., & Tashiro, Y. (2001). What factors affeect the size of chimpanzee parties in the Kalinzu Forest, Uganda? Examination of fruit abundance and number of estrous females. *International Journal of Primatology, 22*, 947–959.

Hashimoto, C., Suzuki, S., Takenoshita, Y., Yamagiwa, J., Basabose, A. K., & Furuichi, T. (2003). How fruit abundance affects the chimpanzee party size: A comparison between four study sites. *Primates, 44*, 77–81.

Hasler, R. (1995). *Political ecologies of scale: The multi-tiered co-management of Zimbabwean Wildlife Resources,* International Institute for Envoirnment and Development, London.

Hastings, B. E., Condiotti, M., Sholley, C., Kenny, D., & Foster, J. W. (1988). Clinical signs of disease in wild mountain gorillas. *Proceedings of the Meeting of the American Association of Zoo Veterinarians/American Association of Wildlife Veterinarians,* 107.

Hastings, B. E., Kenny, D., Lowenstine, L. J., & Foster, J. W. (1991). Mountain gorillas and measles: Ontogeny of a wildlife vaccination program. *Proceedings of AAZA Meetings*, 198–205.

Hastings, B. E., Gibbons, L. M., & Williams, J. (1992). Parasites of free ranging mountain gorillas: Survey and epidemiological factors. *Proceedings of the Meeting of the American Association of Zoo Veterinarians/American Association of Wildlife Veterinarians,* 301–302.

Hauser, M. D. (1996). *The evolution of communication.* MIT Press, Cambridge, MA.

Hauser, M. D., & Marler, P. (1993). Food associated calls in rhesus macaques (*Macaca mulatta*), I: Socioecological factors influencing call production. *Behavioural Ecology, 4*, 194–205.

Hauser, M. D., Teixidor, P., Field, L., & Flaherty, R. (1993). Food-elicited calls in chimpanzees: Effects of food quantity and divisibility. *Animal Behaviour*, *45*, 817–819.

Hegarty, E. E. (1990). Leaf life-span and leafing phenology of lianas and associated trees during a rainforest succession. *Journal of Ecology, 78*, 300–312.

Hegarty, E. E. (1991). Vine-host interactions. In: Putz, F. E., & Mooney, H. A. (eds.), *The biology of vines.* Cambridge University Press, Cambridge, United Kingdom, pp. 357–376.

Hegarty, E. E., & Caballé, G. (1991). Distribution and abundance of vines in forest communities. In F. E. Putz & H. A. Mooney (eds.), *The biology of vines*. Cambridge University Press, Cambridge, pp. 263–282.

Heine, B., & König C. (1988). Plant concepts and plant use. An ethnobotanical survey of the semi-arid and arid lands of East Africa. Part 2: Plants of the So (Uganda). Band Kolner Beiträge zur Entwicklungsländerforschung/Cologne Development Studies. Verlag breitenbach Publishers, Saarbrücken, Fort Lauderdale,p. 142.

Heistermann, M., Finke, M., & Hodges, J. K. (1995). Assessment of female reproductive status in captive-housed hanuman langurs (*Presbytis entellus*) by measurement of urinary and fecal steroid excretion patterns. *American Journal of Primatology, 37*, 275–284.

Heistermann, M., Möle, U., Vervaecke, H., van Elsacker, L., Hodges, J. K. (1996). Application of urinary and fecal steroid measurements for monitoring ovarian function and pregnancy in the bonobo (*Pan paniscus*) and evaluation of perineal swelling patterns in relation to endocrine events. *Biology of Reproduction, 55*, 844–853.

Hennig, J., Laschefski, U., Becker, H., Rammsayer, T., & Netter, P. (1993). Immune cell and cortisol responses to physically and pharmacologically induced lowering of body core temperature. *Neuropsychobiology, 28*, 82–86.

Hercberg, S., Chauliac, M., Galan, P., Devanlay, M., Zohoun, I., Agboton, Y., Soustre, Y., Bories, C., Christides, J. P., Potier de Courcy, G., Masse-Raimbault, A. M., & Dupin, H. (1986). Relationship between anaemia, iron and folacin deficiency, haemoglobinopathies and parasitic infection. *Human Nutrition: Clinical Nutrition, 40C*, 371–379.

Herman, L. M. (1986). Cognition and language competencies of bottlenosed dolphins. In R. J. Schusterman, J. Thomas, & F. G. Wood (eds.), *Dolphin cognition and behavior: A comparative approach*. Erlbaum, Hillsdale, NJ, pp. 221–251.

Hermanussen, M., Jensen, F., Hirsch, N., Friedel, K., Kroger, B., Lang, R., Just, S., Ulmer, J., Schaff, M., Ahnert, P., Heyne, K., & Zeisel, H. J. (1995). Acute and chronic effects of winter swimming on LH, FSH, prolactin, growth hormone, TSH, cortisol, serum glucose and insulin. *Arctic Medical Research, 54*, 45–51.

Hess, J. D., & Resko, J. A. (1973). The effects of progesterone on the patterns of testosterone and estradiol concentrations in the systemic plasma of the female rhesus monkey during the intermenstrual period. *Endocrinology, 92*, 446–453.

Hill, C. (2002). Primate conservation and local communities—Ethical issues and debates. *American Anthropologist, 104*, 1184–1194.

Hill, K. (1993). Life history theory and evolutionary anthropology. *Evolutionary Anthropology, 2*, 76–88.

Hill, K., Padwe, J., Bejyvagi, C., Bepurangi, A., Jakugi, F., Tykuarangi, R., & Tykuarangi, T. (1997). Monitoring hunting impact on large vertebrates in the

Mbaracayu Reserve, Paraguay, using native research assistants. *Conservation Biology, 11*, 1339–1353.

Hill, K., Boesch, C., Goodall, J., Pusey, A., Williams, J., & Wrangham, R. (2001). Mortality rates among wild chimpanzees. *Journal of Human Evolution, 40*, 437–450.

Hill, R. A., & Dunbar, R. I. M. (2002). Climatic determinants of diet and foraging behaviour in baboons. *Evolutionary Ecology, 16*, 579–593.

Hill, W. C. O. (1953). *Primates: Comparative Anatomy and Taxonomy, Vol. 1: Strepssirhini.* Edinburgh University Press, Edinburgh.

Hiraiwa-Hasegawa, M. (1990). Maternal investment before weaning. In T. Nishida (ed.), *The chimpanzees of the Mahale Mountains: Sexual and life history strategies.* University of Tokyo Press, Tokyo, pp. 257–266.

Hiraiwa-Hasegawa, M., Byrne, R. W., Takasaki, H., & Byrne, J. M. E. (1986). Aggression towards large carnivores by wild chimpanzees of Mahale Mountains National Park, Tanzania. *Folia Primatologica, 47*, 8–13.

Hitchcock, R. (2000). Traditional African wildlife utilization: subsistence hunting, poaching, and sustainable use. In H. Prins, J. Grootenhuis, & T. Dolan (eds.), *Wildlife conservation by sustainable use.* Kluwer Academic, Dordrecht, pp. 389–415.

Hitchcock, R. (2002). 'We are the first people': Land, natural resources and identity in the Central Kalahari, Botswana. *Journal of Southern African Studies, 28*, 797–824.

Hladik, C. M. (1975). Ecology, diet and social patterning in Old and New World monkeys. In R. H. Tuttle (ed.), *Socioecology and psychology of primates.* Mouton, Paris, pp. 3–35.

Hladik, C. M., & Gueguen, L. (1974). Géophagie et nutrition minérale chez les Primates Sauvages. Comptes Rendus des Seances de'l' Academie des Sciences, pp. 1393–1396.

Hladik, C. M. (1977). Chimpanzees of Gabon and Chimpanzees of Gombe: Some comparative data on the diet. In TH Cluton-Brock (ed.), Primate Ecology. London: Academic Press, pp. 324–353.

Hoffman, J. R., Falk, B., Radom-Isaac, S., Weinstein, Y., Magazanik, A., Wang, Y., & Yarom. Y. (1997). The effect of environmental temperature on testosterone and cortisol responses to high intensity, intermittent exercise in humans. *European Journal of Applied Physiology & Occupational Physiology, 75*, 83–87.

Hollister, N. (1920). Two new East African primates. *Smithsonian Miscellaneous Collection, 72*, 1.

Holmes, J. C., & Zohar, S. (1990). Pathology and host behavior. In C. J. Barnard & J. M. Behnke (eds.), *Parasitism and host behavior.* Taylor & Francis, London, pp. 34–63.

Homsy, J. (1999). *Ape tourism and human diseases: How close should we get? A critical review of the rules and regulations governing park management and tourism for wild mountain gorillas* (Gorilla gorilla beringei). Unpublished report of the Consultancy for the International Gorilla Conservation Programme, Nairobi, Kenya.

Hooge, P. N., Eichenlaub, W. M., & Solomon, E. K. (2000). *Using GIS to analyze animal movements in the marine environment.* Unpublished manuscript.

Hope, K., Goldsmith, M., & Graczyk, T. (2004). Parasitic health of olive baboons in Bwindi Impenetrable National Park, Uganda. *Veterinary Parasitology, 122*, 165–170.

Horwitz, W. (ed). (1970). *Official methods of analysis of the Association of Official Analytical Chemists* (11th ed.). Association of Official Analytical Chemists, Washington, DC.

Horwitz, W. (2000). *Official methods of analysis. 4.5.01. Crude fat or ether extract in animal feed. 41.1.33. Polymers and oxidation products of heated vegetable oils.* (17th ed.), Association of Officical Analytical Chemists, Washington, DC.

Hosaka, K. (1995). Epidemics and wild chimpanzee study groups. *Pan Africa News, 2*, 1–4.

Hough, J. (1988). Obstacles to effective management of conflicts between national parks and surrounding human communities in developing countries. *Environmental Conservation, 15*, 129–136.

Howard, P., Butono, F., Kayondo-Jjemba, P., & Muhumuza, C. (1991). Integrating forest conservation into district development: A case study. In P. Howard (ed.), *Nature conservation in Uganda's tropical forest reserves.* IUCN Forest Conservation Programme. IUCN Gland, Switzerland, and Cambridge, U.K.

Howard, P. C. (1991). *Nature conservation in Uganda's Tropical Forest Reserves.* IUCN, Gland, Switzerland.

Howell, D. C. (1997). *Statistical methods for psychology* (4th ed.). Duxbury Press, New York.

Hudson, H. R. (1992). The relationship between stress and disease in orphaned gorillas and its significance for gorilla tourism. *Gorilla Conservation News, 6*, 8–10.

Huffman, M. A. (1997). Current evidence for self-medication in primates: A multidisciplinary perspective. *Yearbook of Physical Anthropology, 40*, 171–200.

Huffman, M. A. (2003). Animal self-medication and ethno-medicine: Exploration and exploitation of the medicinal properties of plants. *Proceedings of the Nutritional Society, 62*, 371–381.

Huffman, M. A. (in press). Primate self-medication. In C. Campbell, A. Fuentes, K. MacKinnon, M. Panger, & S. Bearder (eds.), *Primates in perspective.* University of Oxford Press, Oxford.

Huffman, M. A., & Seifu, M. (1989). Observations on the illness and consumption of a possibly medicinal plant *Vernonia amygdalina* (DEL.), by a wild chimpanzee in the Mahale Mountains National Park, Tanzania. *Primates, 30*, 51–63.

Huffman, M. A., & Wrangham, R. W. (1994). Diversity of medicinal plants use by chimpanzees in the wild. In R. W. Wrangham, W. C. McGrew, F. B. de Wall, &

P. G. Heltne (eds.), *Chimpanzee cultures.* Harvard University Press, Cambridge, MA, pp. 129–148.

Huffman, M. A., & Caton, J. M. (2001). Self-induced increase of gut motility and the control of parasitic infections in wild chimpanzees. *International Journal of Primatology, 22*, 329–346.

Huffman, S., Chowdhury, A., & Allen, H. L. N. (1987). Suckling patterns and postpartum amenorrhea in Bangladesh. *Journal of Biosocial Science, 19*, 171–179.

Huffman, M. A., Gotoh, S., Izutsu, D., Koshimizu, K., & Kalunde, M. S. (1993). Further observations on the use of *Vernonia amygdalina* by a wild chimpanzee, its possible effect of parasite load, and its phytochemistry. *African Study Monographs, 14*, 227–240.

Huffman, M. A., Page, J. E., Sukhdeo, M. V. K., Gotoh, S., Kalunde, M. S., Chandrasiri, T., & Towers, G. H. N. (1996). Leaf-swallowing by chimpanzees, a behavioral adaptation for the control of strongyle nematode infections. *International Journal of Primatology, 17*, 475–503.

Huffman, M. A., Gotoh, S., Turner, L. A., Hamai, M., & Yoshida, K. (1997). Seasonal trends in intestinal nematode infection and medicinal plant use among chimpanzees in the Mahale Mountains, Tanzania. *Primates, 38*, 111–125.

Huffman, M. A., Ohigashi H., Kawanaka, M., Page, J. E., Kirby G. C., Gasquet, M., Murakami, A., & Koshimizu, K. (1998). African great ape self-medication: A new paradigm for treating parasite disease with natural medicines. In Y. Ebizuka (ed.), *Towards natural medicine research in the 21st century.* Elsevier Science B.V., Amsterdam, pp. 113–123.

Hunt, C. D., Johnson, P. E., Herbel, J., & Mullen, L. K. (1992). Effects of dietary zinc depletion on seminal volume and zinc loss, serum testosterone concentrations, and sperm morphology in young men. *American Journal of Clinical Nutrition, 56*, 148–157.

Hutton, J., & Leader-Williams, N. (2003). Sustainable use and incentive-driven conservation: Realigning human and conservation interests. *Oryx, 37*, 215–226.

Ibrahim, K. E., & Abu-samra, M. T. (1985). A severe outbreak of sarcoptic mange amongst goats naturally infected with a sheep strain of *Sarcoptes scabiei. Revue d Elevage et de Medecine Veterinaire des Pays Tropicause, 38*, 258–265.

Ichikawa, M. (1987). A preliminary report on ethnobotany of the Suiei Dorobo in Northern Kenya. *African Study Monographs, 7*, 1–52.

Ichikawa, M./AFlora (1998). Center for African Area Studies, Kyoto University (2005). Available at: http://130.54.103.36/aflora.nsf

IDRISI (1997). *Users Guide, Version 2.0* by J. Ronald Eastman. IDRISIS producation, Clark University, Massachusetts.

Imai, S., Ikeda, S., Collet, J. Y., & Bonhomme, A. (1991). Entodiniomorphid ciliates from the wild lowland gorilla with the description of a new genus and three new species. *European Journal of Protistology, 26,* 270–278.

Infield, M., & Namara, A. (2001). Community attitudes and behavior towards conservation: An assessment of a community conservation program around Lake Mburo National Park, Uganda. *Oryx, 35,* 48–60.

International Commission on Zoological Nomenclature. (2002). Opinion 1995 (Case 3004): Lorisidae Gray, 1821, Galagidae Gray, 1825 and Indriidae Burnett, 1828 (Mammalia, Primates); conserved as the correct original spellings. *Bulletin of Zoological Nomenclature, 59,* 65–67.

Isabirye-Basuta, G. (1998). Food competition among individuals in a free-ranging chimpanzee community in Kibale Forest, Uganda. *Behaviour, 105,* 135–147.

Isbell, L. A. (1983). Daily ranging behaviour of red colobus (*Colobus badius tephrosceles*) in Kibale Forest, Uganda. *Folia primatologia, 41,* 234–48.

Iwu, M. M. (1993). *Handbook of African medicinal plants.* CRC Press, London.

Janson, C. H., & Emmons, L. E. (1990). Ecological structure of the non-flying mammal community of the Cocha Cashu Biological Station, Manu National park, Peru. In A. Gentry (ed.), *Four neotropical rain forests.* Yale University Press, New Haven, pp. 314–338.

Janson, C. H., & Goldsmith, M. L. (1995). Predicting group size in primates: Foraging costs and predation risks. *Behavioural Ecology, 6,* 326–336.

Janson, C. H., & Chapman, C. A. (2000). Primate resources and the determination of primate community structure. In J. G. Fleagle, C. H. Janson, & K. Reed (eds.), *Primate communities.* Cambridge University Press, Cambridge, England, pp. 237–267.

Janzen, D. H. (1978). Complications in interpreting the chemical defenses of trees against tropical arboreal plant-eating vertebrates. In. G. G. Montegomery (ed.), *The ecology of arboreal folivores.* Smithsonian Institute Press, Washington, DC, pp. 73–84.

Jeanrenaud, S. (1991). *A study of forest use, agricultural practices and perception of the rain forest. Etinde Rainforest, South West Cameroon.* The conservation-development interface for ODA.

Jenkins, P. D. (1987). *Catalogue of primates in the British Museum (Natural History) and elsewhere in the British Isles. Part IV: Suborder Strepsirrhini, including the subfossil Madagascan lemurs and family Tarsiidae.* British Museum (Natural History), London.

Jensen-Seaman, M. I., & Kidd, K. K. (2001). Mitochondrial DNA variation and biogeography of eastern gorillas. *Molecular Ecology, 10,* 2241–2247.

Jisaka, M., Kawanaka, M., Sugiyama, H., Takegawa, K., Huffman, M. A., Ohigashi, H., & Koshimizu, K. (1992). Antischistomsomal activities of sesquiterpene lactones and steroid glucosides from *Vernonia amygdalina*, possibly used by wild chimpanzees against parasite-related diseases. *Bioscience Biotechnology and Biochemistry, 56*, 845–846.

Jisaka, M., Ohigashi, H., Takegawa, K., Huffman, M. A., & Koshimizu, K. (1993). Antitumor and antimicrobial activities of bitter sesquiterpene lactones of Vernonia amygdalina, a possible medicinal plant used by wild chimpanzees. *Bioscience Biotechnology and Biochemistry, 57*, 833–834.

Johns, A. D. (1983). Tropical forest primates and logging: Can they co-exist? *Oryx, 17*, 114–118.

Johns, A. D. (1986). Effects of selective logging on the behavioural ecology of West Malaysian primates. *Ecology, 67*, 684–694.

Johns, A. D. (1988). Effects of "selective" timber extraction on rain forest structure and composition and some consequences for frugivores and folivores. *Biotropica,* 20, 31–37.

Johns, A. D. (1992). Vertebrate responses to selective logging: Implications for the design of logging systems. *Philosophical TransActions of the Royal Society, London B, 335*, 437–442.

Johns, A. D., & Skorupa, J. P. (1987). Responses of rain forest primates to habitat disturbance: A review. *International Journal of Primatology,* 8, 157–191.

Johns, T. (1990). *With bitter herbs shall they eat it: Chemical ecology and the origin of human diet and medicine.* University of Arizona Press, Tucson.

Johns, T., & Duquette, M. (1991). Detoxification and mineral supplementation as functions of geophagy. *American Journal of Clinical Nutrition, 53*, 448–456.

Johnson, K. (1993). *Local use of Budongo's forest products.* Msc thesis, Oxford University.

Johnson, K. (1996a). Hunting in the Budongo Forest, Uganda. *SWARA, 19*, 24–26.

Johnson, K. (1996b). Local attitudes towards the Budongo Forest. *Indigenous Knowledge and Development Monitor, 4*, 31.

Johnson, T. C., Kelts. K., & Odada, E. (2000). The Holocene history of Lake Victoria. *Ambio, 29*, 2–11.

Jones, D. A. (1998). Why are so many food plants cyanogenic? *Phytochemistry, 47*, 155–162.

Jones, B. (1999). Community-based natural resource management in Botswana and Namibia: An inventory and preliminary analysis of progress. *Evaluating Eden Series, Discussion paper no. 6.* International Institute for Environment and Development (IIED), London, UK.

Jones, C., & Sabater-Pi, J. (1971). Comparative ecology of *Gorilla gorilla* (Savage and Wyman) and *Pan troglodytes* (Blumembach) in Rio Muni, West Africa. *Bibliotheca Primatologica, 13*, 1–95.

Kako, E. (1999). Elements of syntax in the systems of three language-trained animals. *Animal Learning and Behaviour, 27*, 1–14.

Kalema G. (1995). Epidemiology of the intestinal parasite burden of mountain gorillas, Gorilla gorilla beringei, in Bwindi Impenetrable National Park, Southwest Uganda. *Newsletter of the Zebra Foundation for Veterinary Zoological Education,* 19–34.

Kalema, G., Kock, R., & Macfie, E. (1998). An outbreak of sarcoptic mange in free-ranging mountain gorillas in Bwindi-Impenetrable National Park, South Western Uganda. In C. K. Baer (ed.), *Proceedings of the American Association of Zoo Veterinarians/American Association of Wildlife Veterinarians Joint Conference.* American Association of Zoo Veterinarians, Philadelphia, p. 438.

Kalema-Zikusoka, G., Kock, R. A., & Macfie, E. J. (2002). Scabies in free ranging mountain gorillas (*Gorilla beringei beringei*) in Bwindi Impenetrable National Park, Uganda. *The Veterinary Record, 150,* 12–15.

Kalina, J. (1991). *Mgahinga Gorilla National Park: Reference for management.* MGNP Management Plan Annex 1. Unpublished report, CARE, Uganda.

Kaminski, J., Call, J., & Fischer, J. (2004). Word learning in a domestic dog: Evidence for "Fast Mapping". *Science,* 1682–1683.

Kano, T. (1984). Observations of physical abnormalities among the wild bonobos (*Pan paniscus*) of Wamba, Zaire. *American Journal of Physical Anthropology, 63,* 1–11.

Kaplin, B. A. (2001). Ranging behavior of two species of guenons (*Cercopithecus l'hoesti* and *C. mitis doggetti*) in the Nyungwe Forest Reserve, Rwanda. *International Journal of Primatology, 22,* 521–548.

Katende, A. B., Birnie, A., & Tengnäs, B. O. (1995). *Useful trees and shrubs for Uganda, Identification, propagation and management for agricultural and pastoral communities.* Regional Soil Conservation Unit.

Kaur, T., & Huffman, M. A. (2004). Descriptive urological record of chimpanzees (*Pan troglodytes*) in the wild and limitations associated with using multi-reagent dipstick test strips. *Journal of Medical Primatology, 33,* 187–196.

Kay, R. N. B., & Davies, A. G. (1994). Digestive physiology. In A. G. Davies & J. F. Oates (eds.), *Colobine monkeys: Their ecology, behaviour and evolution.* Cambridge University Press, Cambridge, pp. 229–259.

Kayonga, A., & Habiyaremye, F. X. (1987). Médecine traditionnelle et plantes médicinales rwandaises. Contribution aux études ethnobotaniques de la flore rwandaise. Préfecture de Gisenyi. Univ. Nat. Rwanda, Centre universitaire de recherche sur la pharmacopée et la médecine traditionnelle, CURPHAMETRA, inédit, p. 121.

Kelly, T. R., Sleeman, J. M., & Wrangham, R. W. (2004). Urinalysis in free-living chimpanzees (*Pan troglodytes schweinfurthii*) in Uganda. *The Veterinary Record, 154,* 729–730.

Kesner, J. S., Knecht, E. A., & Krieg, E. F., Jr. (1995). Stability of urinary female reproductive hormones stored under various conditions. *Reproductive Toxicology, 9,* 239–244.

Ketch, L. A. (1998). *Microbiological investigations of geophagy in chimpanzees.* MSc thesis (Botany), University of Toronto.

Ketch, L. A., Malloch, D., Mahaney, W. C., & Huffman, M. A. (2001). Chemistry, microbiology and clay mineralogy of soils eaten by chimpanzees (*Pan troglodytes schweinfurthii*) in the Mahale Mountains National Park, Tanzania. *Soil Biology and Biochemistry, 33*, 199–203.

Khera, S. (1999). *Ascaris: The intestinal roundworm.* New Dehli: Shree Books.

Kim, K. C., & Emerson, K. C. (1968). Descriptions of Pediculidae (Anoplura) from great apes (Primates, Pongidae). *Journal of Parasitology, 54*, 690–695.

Kingdon, J. S. (1971). *East African mammals: An atlas of evolution in Africa* (Vol. 1). Academic Press, London.

Kingdon, J. S. (1997). *The Kingdon field guide to African mammals.* Academic Press, London.

Kiwede, Z. (2000). A live birth by a primiparous female chimpanzee at the Budongo Forest. *Pan African News, 7*, 6.

Kling, O. R., & Westfahl, P. K. (1978). Steroid changes during the menstrual cycle of the baboon (*Papio cynocephalus*) and human. *Biology of Reproduction, 18*, 392–400.

Kloos, H., & McCullough, F. S. (1987). Plants with recognized molluscicidal activity. In K. E. Mott (ed.), *Plant molluscicides.* Wiley, New York, pp. 45–108.

Knott, C. D. (1997a). The effects of changes in food availability on diet, activity and hormonal patterns in wild Bornean orangutans. *American Journal of Physical Anthropology Supplement, 24,* 145 (abstract).

Knott, C. D. (1997b). Field collection and preservation of urine in orangutans and chimpanzees. *Tropical Biodiversity, 4*, 95–102.

Knott, C. D. (1997c). Interactions between energy balance, hormonal patterns and mating behavior in wild Bornean orangutans (*Pongo pygmaeus*). *American Journal of Primatology, 42*, 124 (abstract).

Knott, C. D. (1999). *Reproductive physiological and behavioral responses of orangutans in Borneo to fluctuations in food availability.* Department of Anthropology, Harvard University, Cambridge, MA.

Knott, C. D. (2001). Ape models of female reproductive ecology. In P. T. Ellison (ed.), *Reproductive ecology and human evolution.* Aldine, Chicago

Koenig, A. (2000). Competitive regimes in forest-dwelling hanuman langur females (*Semnopithecus entellus*). *Behavioural Ecology and Sociobiology, 98*, 93–109.

Koenig, A., Beise, J., Chalise, M. K., & Ganzhorn, J. U. (1998). When females should contest for food—testing a hypotheses about resource density, distribution, size, and quality with Hanuman langurs (*Presbytis entellus*). *Behavioural Ecology and Sociobiology*, 42, 225–237.

Koenig, A., & Borries, C. (2001). Socioecology of Hanuman langurs: The story of their success. *Evolutionary Anthropology, 10*, 122–137.

Kokwaro, J. O. (1976). *Medicinal plants of East Africa*. General Printers Ltd, Nairobi.

Korstjens, A. H., Sterck, E. H. M., & Noë, R. (2002). How adaptive or phylogenetically inert is primate social behaviour? A test with two sympatric colobines. *Behaviour, 139*, 203–225.

Krafft, M. (1990). *Führer zu den Virunga Vulkanen*. Ferdinand-Enke Verlag, Stuttgart.

Khraisha, S. (1990). Comparative study of serum insulin, glucose, growth hormone and cortisol of students at 794.7 mm Hg (Dead Sea level) and 697.5 mm Hg (Amman) barometric pressures. *Aviation Space & Environmental Medicine, 61*, 145–147.

Krebs, C. J. (1989). *Ecological methodology*. Harper Collins, New York.

Krebs, C. J. (1999). *Ecological methodology* (2nd ed.). Addison Wesley Longman, Menlo Park, CA.

Krebs, J. R., & Davies, N. B. (1993). *An introduction to behavioural ecology*. Blackwell Science Limited, Oxford, pp. 31–38.

Kreis, V. H. A. (1955). Ibid XVIII Das genus *Probstmayria*. Ransom 1907. *Schweiz. Archiv. Tierheik., 97*, 422–433.

Krepel, H. P. (1994). *Oesophagostomum bifurcum* infection in man. A study on the taxonomy, diagnosis, epidemiology and drug treatment of *Oesophagostomum bifurcum* in northern Togo and Ghana.

Krief, S. (2004). Effets prophylactiques et thérapeutiques de plantes ingérées par les chimpanzés: la notion d' "automédication" chez les chimpanzés. *Primatologie, 6*, 171–191.

Krief, S., Bories, C., & Hladik, C. M. (2003). Résultats des examens parasitologiques de selles pratiqués sur une population de chimpanzés sauvages (*Pan troglodytes schweinfurthii*) d'Ouganda. *Bulletin de la Société de Pathologie Exotique, 96*, 80–82.

Krief, S., Hladik, C. M., & Haxaire, C. (2005a). Ethnomedicinal and bioactive properties of plants ingested by wild chimpanzees in Uganda. *Journal of Ethnopharmacology, 101*(1–3), 1–15 .

Krief, S., Huffman, M. A., Sévenet, T., Guillot, J., Bories, C., Hladik, C. M., & Wrangham, R. W. (2005b). Non-invasive monitoring of the health condition of wild chimpanzees (*Pan troglodytes schweinfurthii*) in the Kibale National Park, Uganda.*International Journal of Primatology, 26*, 467–490.

Krief, S., Thoison, O., Sevenet, T., Wrangham, R. W., & Lavaud, C. (2005c). Triterpenoid saponin anthranilates from Albizia grandibracteata leaves ingested by primates in Uganda. *Journal of Natural Products, 68*, 897–903.

Krief, S., Huffman, M., Sévenet, T., Hladik, C. M., Grellier, P., Loiseau, P., & Wrangham, R. W. (in press). Bioactive properties of plants ingested by chimpanzees

(*Pan troglodytes schweinfurthii*) in the Kibale National Park, Uganda. *American Journal of Primatology.*

Krishnamani, R., & Mahaney C. W. (2000). Geophagy among primates; adaptive significance and ecological consequences. *Animal Behaviour, 59*, 899–915.

Kupchan, S. M., Hemingway, R. J., Karin, A., & Werner, D. (1969). Tumor Inhibitors XLVII, vernodalin and vernomygdin, two new cytotoxic sesquiterpene lactones from *Vernonia amygdalina* Del. *Journal of Organic Chemistry, 34*, 3908–3911.

Kuroda, S. (1992). Ecological interspecies relationships between gorillas and chimpanzees in the Ndoki-Nouabale reserve, northern Congo. In: N. Itoigawa, Y. Sugiyama, G. P. Sackett, & R. K. R. Thompson (eds.),*Topics in primatology, Vol. 2: Behavior, ecology and conservation.* University of Tokyo Press, Tokyo, pp. 385–394.

Kuroda, S., Nishihara, T., Suzuki, S., & Oko, R. A. (1996). Sympatric chimpanzees and gorillas in the Ndoki Forest, Congo. In W. C. McGrew, L. Marchant, & T. Nishida (eds.), *Great ape societies.* Cambridge University Press, Cambridge, pp. 71–81.

Kutzner, H. J. (1981). The family Streptomycetaceae. In M. P. Starr, H. Stolp, H. G. Trüper, A. Balows, & H. G. Schlegel (eds.), *The prokaryotes: A handbook on habitats, isolation and identification of bacteria, Vol. II.* Springer, New York, pp. 2028–2090.

Lado-Abeal, J., Lukyanenko, Y. O., Swamy, S., Hermida, R. C., Hutson, J. C., & Norman, R. L. (1999). Short-term leptin infusion does not affect circulating levels of LH, testosterone or cortisol in food-restricted pubertal male rhesus macaques. *Clinical Endocrinology, 15*(1), 41–51.

Lambert, J. E. (1998). Primate digestion: Interactions among anatomy, physiology, and feeding ecology. *Evolutionary Anthropology, 7*, 8–20.

Lambert, J. E. (1999). Seed handling in chimpanzees (*Pan troglodytes*) and redtail monkeys (*Cercopithecus ascanius*): Implications for understanding hominoid and Cercopithecine fruit-processing strategies and seed dispersal. *American Journal of Physical Anthropology, 109*, 365–386.

Lambert, J. E. (2005). Competition, predation, and the evolutionary significance of the cercopithecine cheek pouch: The case of *Cercopithecus* and *Lophocebus. American Journal of Physical Anthropology, 126*, 183–192.

Lane, C. (1914). Bursate nematodes from the Indian elephant. *Indian Journal of Medical Research, 2*, 381–398.

Lane, C. (1923). Some Strongylata. *Parasitology*, 348–364.

Landsoud-Soukate, J., Tutin, C. E. G., & Fernandez, M. (1995). Intestinal parasites of sympatric gorillas and chimpanzees in the Lopé Réserve, Gabon. *Annals of Tropical Medicine and Parasitology, 89*, 73–79.

Langdale-Brown, I., Osmaston, H. A., & Wilson, J. G. (1964). *The vegetation of Uganda and its bearing on land use.* Government of Uganda, Govt. Printer, Entebbe.

Langoya, C., & Long, C. (1997). *Local communities and ecotourism development in Budongo Forest Reserve, Uganda.* Rural Development Forestry Network, Network Paper 22e-i, winter 97/98, 1–13.

Laurance, W. F. (1999). Reflections on the tropical deforestation crisis. *Biological Conservation, 91*, 109–117.

Lauridsen, M. (1999). *Workers in a forest: Understanding the complexity of incorporating local people in modern management. A case study of the Nyabyeya Parish, in western Uganda.* Msc thesis, University of Copenhagen.

Lawes, M. J. (1990). The distribution of the samango monkey (*Cercopithecus mitis erythrarchus*) Peters, 1852 and *Cercopithecus mitis labiatus* I. Geoffroy, 1843) and forest history in southern Africa. *Journal of Biogeography, 17*, 669–680.

Lawes, M. J. (1991). Diet of samago monkeys (*Cercopithecus mitis erythrarchus*) in the Cape Vidal dune forest, South Africa. *Journal of Zoology (London), 224*, 149–173.

Lawes, M. J., Henzi, S. P., & Perrin, M. R. P. (1990). Diet and feeding behaviour of samango monkeys (*Cercopithecus mitis labiatus*) in Ngoye forest, South Africa. *Folia Primatologica, 54*, 57–69.

Leakey, M. (1988). Fossil evidence for the evolution of the guenons. In A. Gautier-Hion, F. Bourliere, & J-P. Gautier (eds.), *A primate radiation: Evolutionary biology of the African guenons.* Cambridge University Press, Cambridge, pp. 7–12.

Leceta, J., & Zapata, A. (1985). Seasonal changes in the thymus and spleen of the turtle, Mauremys caspica. A morphometrical, light microscopical study. *Developmental & Comparative Immunology, 9*, 653–668.

Leigh, S. R. (1994). Relations between captive and noncaptive weights in anthropoid primates. *Zoo Biology, 13*, 21–43.

Lemmon, P. E. (1956). A spherical densioimeter for estimating forest overstory density. *Forest Science, 2*, 314–320.

Lemmon, P. E. (1957). A new instrument for measuring forest overstory density. *Journal of Forestry, 55*, 667–668.

Lernould, J-M. (1988). Classification and geographical distribution of guenons: A review. In A. Gautier-Hion, F. Bourliere, & J-P. Gautier (eds.), *A primate radiation: Evolutionary biology of the African guenons.* Cambridge University Press, Cambridge, pp. 54–78.

Levine, N. D. (1980). *Nematode parasites of domestic animals and man* (2nd ed.). Burgess, Minneapolis, 477 pp.

Lewicka, S., Nowicki, M., & Vecsei, P. (1998). Effect of sodium restriction on urinary excretion of cortisol and its metabolites in humans. *Steroids, 63*, 401–405.

Li, H., Nakajima, S., Chen, J., Todd, H., Overstreet, J. W., & Lasley, B. L. (2001). Differences in hormonal characteristics of conceptive versus nonconceptive cycles. *Fertility & Sterility, 75*, 549–553.

Lilly, A. A., Mehlmann, P. T., & Doran, D. (2002). Intestinal parasites in gorillas, chimpanzees, and humans at Mondika Research Site, Dzanga-Ndoki National Park, Central African Republic. *International Journal of Primatology, 23*, 555–573.

Linden, E. (2002). The wife beaters of Kibale. *Time Magazine.*

Lipson, S. F. (2001). Metabolism, maturation, and ovarian function. In P. T. Ellison (ed.), *Reproductive ecology and human evolution.* Aldine de Gruyter, New York, pp. 235–248.

Lipson, S. F., Ellison, P. T. (1996). Comparison of salivary steroid profiles in naturally occurring conception and non-conception cycles. *Human Reproduction, 11*, 2090–2096.

Lizarralde, M. (2002). Ethnoecology of monkeys among the Barí of Venezuela: perception, use and conservation. In A. Fuentes & L. Wolfe (eds.), *Primates face to face: The conservation implications of human–nonhuman primate interconnections.* Cambridge University Press, Cambridge, pp. 85–100.

Lohiya, N. K., Sharma, R. S., Manivannan, B., & Anand Kumar, T. C. (1998). Reproductive exocrine and endocrine profiles and their seasonality in male langur monkeys (*Presbytis entellus entellus*). *Journal of Medical Primatology, 27*, 15–20.

Low, B. (1996a). Behavioral ecology of conservation in traditional societies. *Human Nature, 7*, 353–379.

Low, B. (1996b). Men, women and sustainability. *Population and Environment, 18*, 111–141.

Low, B., & Heinen, J. (1993). Population, resources, and environment: implications of human behavioral ecology for conservation. *Population and Environment, 15*, 7–41.

Lowenstine, L. J. (1990). Long-distance pathology, or will a mountain gorilla fit in the diplomatic pouch? *Proceedings of the American Association of Zoo Veterinarians*, 178–185.

Luzar, E. J., Diagne, A., Gan, C. E. C., & Henning, B. R. (1998). Profiling the nature-based tourist: A multinomial logit approach. *Journal of Travel Research, 37*, 48–55.

Mace, G. M., & Harvey, P. H. (1983). Energetic constraints on home range size. *American Naturalist, 121*(1).

Macfie, E. (1991). The Volcanoes Veterninary Center. *Gorilla Conservation News, 5*, 21.

Macfie, E. (1996). Case report on scabies infection in Bwindi gorillas. *J. Berggorilla and Regenwald Direkthilfe, 13*, 19–20.

Macfie, E. (1997). Gorilla tourism in Uganda. *Journal of Berggorilla and Regenwald Direkthilfe, 15*, 16–17.

Macleod, M. M. (2000). *The reproductive strategies of samango monkeys* (Cercopithecus mitis erythrarchus). PhD dissertation, School of Life Sciences, University of Surrey, Roe Hampton, 273 pp.

Magliocca, F., & Gautier-Hion, A. (2002). Mineral content as a basis for food selection by western lowland gorillas in a forest clearing. *American Journal of Primatology, 57*, 67–77.

Mahaney, W. C. (1993). Scanning Electron Microscopy of Earth Mined and Eaten by Mountain Gorillas in the Virunga Mountains, Rwanda. *Primates, 34*, 311–319.

Mahaney, W. C. (1999). Paleoclimate and paleonutrition? Paleozoopharmacognosy: A timely connection. *Frankfurter Geowissenchaftliche Arbeiten,* Festschrift for Wolfgang Andres, *25*, 123–134.

Mahaney, W. C., & Krishnamani, R. (2003). Understanding geophagy in animals: standard procedures for sampling soils. *Journal of Chemical Ecology, 29*, 1477–1499.

Mahaney, W. C., Watts, D. P., & Hancock, R. G. V. (1990). Geophagia by mountain gorillas (*Gorilla gorilla beringei*) in the Virunga Mountains, Rwanda. *Primates, 31*, 113–120.

Mahaney, W. C., Hancock, R. G. V., & Inoue, M. (1993). Geochemistry and clay mineralogy of soils eaten by Japanese macaques. *Primates, 34*, 85–91.

Mahaney, W. C., Aufreiter, S., & Hancock, R. G. V. (1995a). Mountain gorilla geophagy: A possible seasonal behaviour for dealing with the effects of dietary changes. *International Journal of Primatology, 16*, 475–488.

Mahaney, W. C., Stambolic, A., Knezevich, M., Hancock, R. G. V., Aufreiter, S.,Sanmugadas, K., Kessler, M. J., & Grynpas, M. D. (1995b). Geophagy amongst rhesus macaques on Cayo Santiago, Puerto Rico. *Primates, 36*, 323–333.

Mahaney, W. C., Stambolic, A., Milner, M. W., Russon, A., Hancock, R. G. V., & Aufreiter, S. (1996a). *Geochemistry of soils eaten by orangutans in Indonesia*, SLOWPOKE Reactor Facility, Annual Report.

Mahaney, W. C., Hancock, R. G. V., Aufreiter, S., & Huffman, M. A. (1996b). Geochemistry and clay mineralogy of termite mound soil and the role of geophagy in chimpanzees of the Mahale Mountains, Tanzania. *Primates, 37*, 121–134.

Mahaney, W. C., Milner, M. W., Sanmugadas, K., Hancock, R. G. V., Aufreiter, S., Wrangham, R., & Pier, H. W. (1997). Analysis of geophagy soils in Kibale forest, Uganda. *Primates, 38*, 159–176.

Mahaney, W. C., Zippin, J., Milner, M. W., Huffman, M. A., Hancock, R. G. V., Aufreiter, S., Wink, M., Malloch, D., & Ketch, L. (1998). The geochemistry and mineralogy of termite mound soils eaten by chimpanzees of the Mahale Mountains, western Tanzania, Geological Society of America, Toronto. *Abstracts with Program, 30*, 222–223.

Mahaney, W. C., Zippin, J., Milner, M. W., Sanmugdas, K., Hancock, R. G. V., Aufreiter, S., Campbell, S., Huffman, M. A., Wink, M., Malloch, D., & Kalm, V. (1999). Chemistry, mineralogy and microbiology of termite mound soil eaten by chimpanzees of the Mahale Mountains, Western Tanzania. *Journal of Tropical Ecology, 15*, 565–588.

Mahaney, W. C., Milner, M. W., Aufreiter, S., Hancock, R. G. V., Wrangham, R., & Campbell, S. (in press). Notes on geophagy soils eaten by chimpanzees of the Kanyawara community in the Kibale Forest, Uganda. *International Journal of Primatology.*

Maisels, F., Gautier-Hion, A., & Gautier-Hion, J. P. (1994). Diets of two sympatric colobines in Zaire: More evidence on seed-eating in forests on poor soils. *International Journal of Primatology, 15*, 681–701.

Mansuri, G., & Vijayendra, R. (2003). *Evaluating community-based and community-driven development: A critical review of the evidence.* United States: Development Research Group, World Bank, Washington, DC.

Marler, P. (1969). *Colobus guereza*: Territorality and group composition. *Science, 163*, 93–95.

Marler, P., & Hobbett, L. (1975). Individuality in a long range vocalization of wild chimpanzees. *Zeitschrift fur Tierpsychologie, 38*, 97–109.

Marler, P., & Tenaza, R. (1977). Signaling behaviour of apes with special reference to vocalization. In T. Seborek (ed.), *How animals communicate.* Indiana Press, Bloomington.

Markham, K. R. (1982). *Techniques of flavonoid identification.* Academic Press. London.

Markham, R. (1995). Doing it naturally: Reproduction in captive orangutans (*Pongo pygmaus*). In R. D. Nadler, B. M. F. Galdikas, L. K. Shuran, & N. Rosen (eds.), *The neglected ape.* Plenum Press, New York, pp. 273–278.

Marriot, H. (1999). Microdemography of a farming population of Western Uganda: A preliminary analysis investigating population pressure on the Budongo Forest Reserve. *Social Biology and Human Affairs, 64*, 1–11.

Marsh, C. W. (1981). Diet choice among red colobus (*Colobus badius rufomitratus*) on the Tana River, Kenya. *Folia Primatologica, 35*, 147–178.

Marsh, C. W., & Wilson, W. L. (1981). Effects of natural habitat disturbance on the abundance of Malaysian primates. *Malaysian Journal of Applied Biology, 10*, 227–249.

Matsumoto-Oda, A. (1999a). Female choice in the opportunistic mating of wild chimpanzees (*Pan troglodytes schweinfurthii*) at Mahale. *Behavioral Ecology and Sociobiology, 46*, 258–266.

Matsumoto-Oda, A. (1999b). Mahale chimpanzees: Grouping patterns and cycling females. *American Journal of Primatology, 47*, 197–207.

Matsumoto-Oda, A., & Kasagual, M. B. (2000). Preliminary study of feeding competition between baboons and chimpanzees in the Mahale Mountains National Park, Tanzania. *African Study Monographs, 21*, 147–157.

Matsuzawa, T. (1998). Green passage plan (tree-planting project) and environmental education using documentary videos at Bossou: A progress report. *Pan Africa News, 5*, 18–20.

Matzke, G., & Nabane, N. (1996). Outcomes of a community controlled wildlife program in a Zambezi Valley community. *Human Ecology, 24*, 65–85.

Maughn, J. E., & Stanford, C. B. (in press). Ground-nesting by chimpanzees in Bwindi Impenetrable National Park, Uganda. In R. Tuttle and C. Litchfield (eds.), *Developments in primatology: Progress & prospects.* Plenum Press, New York.

McArthur, J. W., Beitins, I. Z., Gorman, A., Collins, D. C., Preedy, J. R. K., & Graham, C. E. (1981). The interrelationship between sex skin swelling and the urinary excretion of LH, estrone, and pregnanediol by the cycling female chimpanzee. *American Journal of Primatology, 1*, 265–270.

McCallum, H., & Dobson, A. (1995). Detecting disease and parasite threats to endangered species and ecosystems. *Trends in Evolution and Evolution, 10*, 190–194.

McCullagh, P., & Nelder, J. A. (1989). *Generalized linear models* (2nd ed.). Chapman & Hall, London.

McGrew, W. C. (1992). *Chimpanzee material culture: Implications for human evolution.* Cambridge University Press, Cambridge.

McGrew, W. C., Tutin, C. E. G., Collins, D. A., & File, S. K. (1989). Intestinal parasites of sympatric *Pan troglodytes* and *Papio* spp. at two sites: Gombe (Tanzania) and Mt. Assirik (Senegal). *American Journal of Primatology, 17*, 147–155.

McKey, D. B. (1978). Soils, vegetation, and seed-eating by black colobus. In G. G. Montgomery (ed.), *The ecology of arboreal folivores.* Smithsonian Press, Washington, DC, pp. 423–438.

McKey, D. B., Gartlan, S. J., Waterman, P. G., & Choo, G. M. (1981). Food selection by black colobus monkeys (*Colobus satanas*) in relation to plant chemistry. *Biological Journal of the Linnaean Society, 16*, 115–146.

McNeilage, A. (1995). *Mountain gorillas in the Virunga Volcanoes: Ecology and carrying capacity.* Unpublished PhD thesis, University of Bristol.

McNeilage, A. (1996). Ecotourism and mountain gorillas in the Virunga Volcanoes. In V. J. Taylor & N. Dunstone (eds.), *The exploitation of mammal populations.* Chapman & Hall, London.

McNeilage, A. (2001). Diet and habitat use of two mountain gorilla groups in contrasting habitats in the Virungas. In M. M. Robbins, P. Sicotte, & K. J. Stewart (eds.), *Mountain gorillas: Three decades of research at Karisoke.* Cambridge University Press, Cambridge, pp. 265–292.

McNeilage, A., Plumptre, A. J., Brock-Doyle, A., & Vedder, A. (1998). *Bwindi Impenetrable National Park, Uganda: gorrila and large mammal census, 1997.* Wildlife Conservation Society Working Papers, New York.

McNeilage, A., Plumptre, A., Brock-Doyle, A., & Vedder, A. (2001). Bwindi Impenetrable National Park, Uganda: Gorilla census, 1997. *Oryx, 5*, 39–47.

Meder, A. (2000). Scabies again. *J. Berggorilla and Regenwald Direkthilfe, 21*, 8–9.

Meder, A. (2004). Virunga Gorilla Census. *J. Berggorilla and Regenwald Direkthilfe, 28*, 5.

Mehlman, P. T., Higley, J. D., Fernald, B. J., Sallee, F. R., Suomi, S. J., & Linnoila, M. (1997). CSF 5-HIAA, testosterone, and sociosexual behaviors in free-ranging male rhesus macaques in the mating season. *Psychiatry Research, 72*(2), 89–102.

Ministry of Planning and Economic Development. (1997). The Republic of Uganda 1997 Statistical Abstract, Uganda, pp. 18–19.

Mitani, J. C., & Rodman, P. S. (1979). Territoriality: The relation of ranging pattern and home range size to defendability with an analysis of territoriality among primate species. *Behavioural Ecology and Sociobiology, 5*, 241–251.

Mellanby, K. (1944). The development of symptoms, parasitic infections and immunity in human scabies. *Parasitology, 35*, 197–206.

Mesfin, T., & Obsa, T. (1994). Ethiopian traditional veterinary practices and their possible contribution to animal production and management. *Revue scientifique et technique de'l' office international de epizooties, 13*, 417–424.

Messner, E. J., & Wrangham, R. W. (1996). In vitro testing of biological activity of *Rubia cordifolia* leaves on primates Strongyloides species. *Primates, 37*, 105–108.

Meyer, E. A. (1994). Giardia as an organism. In R. C. A Thompson, J. A. Reynoldson, & A. J. Lymbery (eds.), *Giardia: From molecules to disease*. CAB International, Oxon, pp. 3–14.

Milton, K. (1979). Factors influencing leaf choice by howler monkeys: A test of some hypotheses of food selection by generalist herbivorous. *American Naturalist, 114*, 362–378.

Milton, K. (1980). *The foraging strategy of howler monkeys: A study of primate economics*. Columbia University Press, New York.

Milton, K. (1981a). Food choice and digestive strategies of two sympatric primate species. *American Naturalist, 117*, 496–505.

Milton, K. (1981b). Distribution patterns of tropical plant foods as a stimulus to primate mental development. *American Anthropologist, 83*, 534–548.

Milton, K., & May, M. (1976). Body weight, diet and home range area in primates. *Nature, 259*, 459–462.

Milton, K., & Dintzis, F. R. (1981). Nitrogen-to-protein conversion factors for tropical plant samples. *Biotropica, 12*, 177–181.

Mitani, J. C. (1994). Ethological studies of chimpanzee vocal behaviour. In R. W. Wrangham, W. C. McGrew, F. B. M. de Waal, & P. G. Heltne (eds.),*Chimpanzee cultures*. Harvard University Press, Cambridge, MA.

Mitani, J. C., & Rodman, P. S. (1979). Territoriality: The relation of ranging pattern and home range size to defendability with an analysis of territoriality among primate species. *Behavioural Ecology and Sociobiology, 5*, 241–251.

Mitani, J. C., & Nishida, T. (1993). Contexts and social correlates of long-distance calling by male chimpanzees. *Animal Behaviour, 45*, 735–746.

Mitani, J. C., & Brandt, K. L. (1994). Social factors influence the acoustic variability in the long distance calls of male chimpanzees. *Ethology, 96*, 233–252.

Mitani, J. C., & Watts, D. P. (1999). Demographic influences on the hunting behavior of chimpanzees. *American Journal of Physical Anthropology, 109*, 439–454.

Mitani, J. C., Hasegawa, T., Gros-Louis, J., Marler, P., & Byrne, R. (1992). Dialects in wild chimpanzees? *American Journal of Primatology, 27*, 233–243.

Mitani, J. C., Gros-Louis, J., & Richards, A. F. (1996a). Sexual dimorphism, the operational sex ratio, and the intensity of male competition in polygynous primates. *American Naturalist, 147*, 966–980.

Mitani, J. C., Gros-Louis, J., & Macedonia, J. (1996b). Selection for acoustic individuality within the vocal repertoire of wild chimpanzees. *International Journalof Primatology, 17*, 569–583.

Mitani J. C., Watts, D. P., Pepper, J. W., & Merriwether, D. A. (2002). Demographic and social constraints on male chimpanzee behaviour. *Animal Behaviour, 64*, 727–737.

Moeller, W. (2000). *Chimp meat in western Uganda.* Unpublished report, Ugandan Wildlife Education Centre, Uganda.

Mori, A. (1983). Comparison of the communicative vocalizations and behaviours of group ranging in eastern gorillas, chimpanzees and pygmy chimpanzees. *Primates, 24*, 486–500.

Mortelmens, J., Vercruysse, J., & Kageruka, P. (1970). Three pathogenic intestinal protozoa of anthropoid apes: *Entamoeba histolytica*, *Balantidium coli*, and *Tronglodytella abrassarti*. *Proceedings of the Third International Congress of Primatologists, Zurich, 2*, 187–191.

Mturi, F. A. (1993). Ecology of the zanzibar red colobus monkey, *Colobus badius kirkii* (Gray, 1868), in comparison with other red colobines. In J. C. Lovett & S. K. Wasser (eds.), *Biogeography and ecology of the rain forests of eastern Africa*. Cambridge University Press, Cambridge, pp. 243–266.

Mudadkikwa, A. B., Cranfield, M. R., Sleeman, J. M., & Elenberger, U. (2001). Clinical medicine, preventative health care and research on mountain gorillas in the Virunga Volcanos region. In M. M. Robbins, P. Sicotte, & K. J. Stewart (eds.), *Mountain gorillas: Three decades of research at Karisoke*. Cambridge University Press, Cambridge, pp. 342–360.

Mudakikwa, A. B., Sleeman, J., Foster, J. Meader, L. L., & Patton, S. (1998). An indicator of human impact: Gastrointestinal parasites of mountain gorillas (*Gorilla gorilla beringei*) from the Virunga Volcanoes Region, Central Africa. *Proceedings of the Joint Meeting of the American Association of Zoo Veterinarians and the American Association of Wild Veterinarians*, 436–437.

Mullin, W. J., & Wolynetz, M. S. (1995). Effect of milling procedure on the measurement of dietary fiber by a gravimetric method. *J. AOAC Int., 78*, 83–87.

Muller, M. N. (2002). Agonistic relations among Kanyawara chimpanzees. In C. Boesch, G. Hohmann, & L. F. Marchant (eds.), *Behavioural diversity in chimpanzees and bonobos.* Cambridge University Press, Cambridge, pp. 112–124.

Muller, M. N., & Wrangham, R. W. (2004). Dominance, aggression and testosterone in wild chimpanzees: A test of the 'challenge hypothesis.' *Animal Behaviour, 67*, 113–123.

Munn, J. (2003). *The impact of injuries on free-living chimpanzees: How injury affects the social behaviour, mother-dependant behaviour and the locomotion of adult female chimpanzees from the Sonso community, Budongo Forest Reserve, Uganda.* MPhil thesis, Australian National University, Canberra.

Munn, J., & Kalema, G. (1999–2000). Death of a chimpanzee *Pan troglodytes schweinfurthii* in a trap in Kasokwa Forest Reserve, Uganda. *African Primates, 4*, 58–61.

Muyambi, F. (2005). The impact of tourism on the behaviour of mountain gorillas. *J. Berggorilla and Regenwald Direkthilfe, 30*, 14–15.

Mwanzia, J. M., Kock, R. A., Wambua, J. M., Kock, N., & Jarret, O. (1995). An outbreak of sarcoptic mange in free living cheetah (*Acinonyx jubatus*) in the Mara region of Kenya. *Proceedings of the American Association of Zoo Veterinarians/Wildlife Disease Association/American Association of Wildlife Veterinarians Joint Conference.* East Lansing, MI, August 12–17, pp. 105–114.

Nadler, R. D., Graham, C. E., Collins, D. C., & Gould, K. G. (1979). Plasma gonadotropins, prolactin, gonadal steroids and genital swelling during the menstrual cycle of lowland gorillas. *Endocrinology, 105*, 290–296.

Nadler, R. D., Collins, D. C., & Blank, M. S. (1984). Luteinizing hormone and gonadal steroid levels during the menstrual cycles of orangutans. *Journal of Medical Primatology, 13*, 305–314.

Nadler, R. D., Graham, C. E., Gosselin, R. E., & Collins, D. C. (1985). Serum levels of gonadotropins and gonadal steroids, including testosterone, during the menstrual cycles of the chimpanzee (*Pan troglodytes*). *American Journal of Primatology, 9*, 273–284.

Nagaya, M., & Widmaier, E. P. (1993). Twenty-four hour profiles of glucose, corticosterone and adrenocorticotropic hormone during the first postnatal day in rats. *Biology of the Neonate, 64*, 261–268.

National Research Council. (1992). *Conserving biodiversity: A research agenda for development agencies.* National Academy Press, Washington, DC.

National Research Council. (2003). *Nutrient requirements of nonhuman primates.* The National Academies Press, Washington, DC.

Naughton-Treves, L. (1997). Farming around the forest edge: Vulnerable places and people around Kibale National Park, Uganda. *The Geographical Review, 87*, 27–46.

Naughton-Treves, L. (1998). Predicting patterns of crop damage by wildlife around Kibale National Park. *Conservation Biology, 12*, 156–168.

Ndayitwayeko, A., & Ntungwanayo, V. (1978). Contribution à l'étude de plantes médicinales dans la région de Mugamba. (Commune Mugamba). Mémoire de licence, Univ. Burundi, Fac. Sc., p. 129.

Nepstad, D. C., Veríssimo, A., Alencar, A., Nobre, C., Lima, E. Lefebvre, P., Schlesinger, P., Potter, C., Moutinho, P., Mendoza, E. Cochrane, M., & Brooks, V. (1999). Large-scale impoverishment of Amazonian forests by logging and fire. *Nature, 398*, 505–508.

Newton, P. N., & Dunbar, R. I. M. (1994). Colobine monkey society. In A. G. Davies & J. F. Oates (eds.), *Colobine monkeys: Their ecology, behaviour and evolution.* Cambridge University Press, Cambridge, pp. 311–346.

Newton-Fisher, N. E. (1997). *Tactical behaviour and decision making in wild chimpanzees.* PhD thesis, University of Cambridge.

Newton-Fisher, N. E. (1999a). Association by male chimpanzees: A social tactic? *Behaviour, 136*, 705–730.

Newton-Fisher, N. E. (1999b). The diet of chimpanzees in the Budongo Forest Reserve, Uganda. *African Journal of Ecology, 37*, 344–354.

Newton-Fisher, N. E. (1999c). Termite eating and food sharing by male chimpanzees in the Budongo Forest, Uganda. *African Journal of Ecology, 37*, 369–371.

Newton-Fisher, N. E. (2003). The home range of the Sonso community of chimpanzees from the Budongo Forest, Uganda. *African Journal of Ecology, 41*, 150–156.

Newton-Fisher, N. E. (in press). Wild female chimpanzees form coalitions against male aggression. *International Journal of Primatology.*

Newton-Fisher, N. E., Plumptre, A. J., & Reynolds, V. (2000). Food supply and chimpanzee (*Pan troglodytes schweinfurthii*) party size in the Budongo Forest Reserve, Uganda. *International Journal of Primatology, 21*, 613–628.

Ngabirano, I. (2005). *Changing the role of African militaries in conservation and sustainable development.* MPS project paper, Cornell University, Ithace, New York, UK.

Ngounou, F. N., Lontsi, D., & Sondegam, B. L. (1988). Myrinathic acid: A new triterpenoid from *Myrianthus arboreus. Phytochemistry, 27*, 301–303.

Nieuwenhuijsen, K., de Neef, K. J., van der Werff ten Bosch, J. J., & Slob, A. K. (1987). Testosterone, testis size, seasonality, and behavior in group-living stumptail macaques *(Macaca arctoides) Hormones & Behavior, 21*(2), 153–169.

Nishida, T. (1973). Ant gathering behaviour by the use of tools among wild chimpanzees of the Mahale Mountains. *Journal of Human Evolution, 2*, 357–370.

Nishida, T. (1979). Predatory behavior among wild chimpanzees of the Mahale Mountains. *Primates, 20*, 1–20.

Nishida, T. (1987). Local traditions and cultural transmission. In B. B. Smuts, D. L. Cheney, R. M. Seyfarth, R. W. Wrangham, & T. T. Struhsaker (eds.), *Primate societies*. University of Chicago Press, Chicago, pp. 462–474.

Nishida, T. (1997). Sexual behavior of adult male chimpanzees of the Mahale Mountains National Park, Tanzania. *Primates, 38*, 379–398.

Nishida, T., & Uehara, S. (1983). Natural diet of chimpanzees (*Pan troglodytes* schweinfurthii). Long-term record from the Mahale mountains, Tanzania. *African Study Monographs, 3*, 109–130.

Nishida, T., Ohigashi, H., & Koshimizu, K. (2000). Tastes of chimpanzee plant foods. *Current Anthropology, 41*, 431–438.

Nishida, T., Corp, N., Hamai, M., Hasegawa, T., Hiraiwa-Hasegawa, M., Hosaka, K., Hunt, K. D., Itoh, N., Kawanaka, K., Matsumoto-Oda, A., Mitani, J. C., Nakamura, M., Norikoshi, K., Sakamaki, T., Turner, L., Uehara, S., & Zamma, K. (2003). Demography, female life history, and reproductive profiles among the chimpanzees of Mahale. *American Journal of Primatology, 59*, 99–121.

Nishihara, T. (1992). A preliminary report on the feeding habits of western lowland gorillas (*Gorilla gorilla gorilla*) in the Ndoki Forest, Congo. In N. Itoigawa, Y. Sugiyama, G. P. Sackett, & R. K. R. Thompson (eds.), *Topics in primatology, Vol. 2: Behavior, ecology and conservation*. University of Tokyo Press, Japan, pp. 225–240.

Nishihara, T. (1995). Feeding ecology of western lowland gorillas in the Nouabale-Ndoki National Park, Congo. *Primates, 36*, 151–168.

Nishihara, T., & Kuroda, S. (1991). Soil-scratching behavior by western lowland gorillas. *Folia Primatologica, 57*, 48–51.

Nizeyi, J. B., Mwebe, R., Nanteza, A., Cranfield, M. R., Kalema, G. R. N. N., & Graczyk, T. K. (1999). *Cryptosporidium* sp and *Giardia* sp infections in mountain gorillas (*Gorilla gorilla beringei*) of the Bwindi Impenetrable National Park, Uganda. *Journal of Parasitology, 85*, 1084–1088.

Nizeyi, J. B., Innocent, R. B., Erume, J., Kalema, G. R. N. N., Cranfield, M. R., &Graczyk, T. K. (2001). Campylobacteriosis, salmonellosis and shigellosis in free-ranging human-habituated mountain gorillas of Uganda. *Journal of Wildlife Diseases, 37*, 239–244.

Nizeyi, J. B., Cranfield, M. R., & Graczyk, T. K. (2002a). Cattle near the Bwindi Impenetrable National Park, Uganda, as a reservoir of *Cryptosporidium parvum* and *Giardia duodenalis* for local community and free-ranging gorillas. *Parasitology Research, 88*, 380–385.

Nizeyi, J. B., Sebunya, D., DaSilva, A. J., Cranfield, M. R., Pieniazek, N. J., & Graczyk, T. K. (2002b). Cryptosporidiosis in people sharing habitats with free-ranging mountain gorillas (*Gorilla gorilla beringei*), Uganda. *American Journal of Tropical Medicine and Hygiene, 66*, 442–444.

Nkurunungi, J. B. (1999). A survey of the gastro-intestinal helminths of the wild mountain gorilla (*Gorilla gorilla beringei* Matschie) and man in Bwindi Impenetrable National Park, Southwestern Uganda. In *Proceedings of the Ecological Monitoring Programme Workshop: Research as an important tool of ecological monitoring in Bwindi Impenetrable National Park, Uganda*, Bwindi Impenetrable National Park,December 9–11, 1999.

Nkurunungi, J. B. (2003). *Feeding and ranging ecology of gorillas (Gorilla gorilla beringei) in Bwindi Impenetrable National Park.* PhD dissertation, Kampala, Uganda, Makerere University.

Nkurunungi, J. B. (2005). *The distribution of fruit and non-fruit resources and its influence on the feeding ecology of the gorillas in the Bwindi Impenetrable National Park, Uganda.* Unpublished PhD thesis, Makerere University.

Noda, R., & Yamada, H. (1964). On two species of nematodes, *Necator gorillae* sp. nov. (Ancylostomidae) and Chitwoodspirura wheri. Chabaud and Rousselot, 1956. *Bulletine of the University of Osaka Prefecture Ser. B, 15*, 176–180.

Nöe, R., & Bshary, R. (1997). The formation of red colobus-diana monkey associations under predation pressure from chimpanzees. *Proceedings of the Royal Society, London B, 264*, 253–259.

Norton, G. W., Rhine, R. J., Wynn, G. W., & Wynn, R. D. (1987). Baboon diet: A five-year study of stability and variability in plant feeding and habitat of yellow baboons (*Papio cynocephalus*) of Mikumi National Park, Kenya. *Folia Primatologica, 48*, 79–120.

Notman, H. (2003). *The meaning, structure and function of chimpanzee pant hoots from the Budongo Forest, Uganda.* PhD thesis, University of Calgary.

Notman, H., & Rendall, D. (2005). Contextual variation in chimpanzee pant hoots and its implications for referential communication. *Animal Behaviour, 70*, 177–190.

Nowak, R. (1995). Endangered species: Uganda enlists locals in the battle to save the gorillas. *Science, 267*, 1761–1762.

Nutter, F. B., & Whittier, C. A. (2000). Occupational health programs for primate field researchers: Improving human health care benefits nonhuman primates. *Proceedings of a joint multi-national conference "The Apes: Challenges for the 21st Century."* Brookfield Zoo, pp. 244–248.

Nyakabwa, M., & Gapusi, R. (1990). Plantes médicinales utilisées chez les Banyamulenge de Fizi au Sud-Kivu (Zaïre). *African Studies Monograph, 11*, 101–114.

Nybelin, O. (1924). *Anoplocephala gorillae* n. sp. *Arkiv for Zoologi, 19B*, 1–3.

Nziza, J. (2003). *Prevalence of* Sarcoptes scabiei *infection in human population and dogs living around Bwindi National Impenetrable Park in Kisoro.* Report to Makerere

University Faculty of Veterinary Medicine Wild Animal and Resource Management Department, Kampala, Uganda.

Oates, J. F. (1977). The *C. guereza* and its food. In T. H. Clutton-Brock (ed.), *Primate Ecology*. Academic Press, London, pp. 275–321.

Oates, J. F. (1994). The natural history of African colobines. In A. G. Davies &J. F. Oates (eds.), *Colobine monkeys: Their ecology, behaviour and evolution*. Cambridge University Press, Cambridge, pp. 75–128.

Oates, J. F. (1996a). *Action plan for African primate conservation: 1986–1990.* IUCN/SSC Primate Specialist Group, New York.

Oates, J. F. (1996b). Habitat alteration, hunting and the conservation of folivorous primates in African forests. *Australian Journal of Ecology, 21*, 1–9.

Oates, J. F., & Davies, A. G. (1994). Conclusions: past, present and future of the colobines. In A. G. Davies & J. F. Oates (eds.), *Colobine monkeys: Their ecology, behaviour and evolution*. Cambridge University Press, Cambridge, pp. 347–358.

Oates, J. F., Swain, T., & Zantoska, J. (1977). Secondary compounds and food selection by colobus monkeys. *Biochemical Systematics and Ecology, 5*, 317–321.

Oates, J. F., Waterman, P. G., & Choo, G. M. (1980). Food selection by the south Indian leaf monkey, *Presbytis johnii*, in relation to leaf chemistry. *Oecologia, 45*, 45–56.

Oates, J. F., Whitesides, G. H., Davies, A. G., Waterman, P. G., Green, S. M., Dasilva,G. L., & Mole, S. (1990). Determinants of variation in tropical forest primate biomass: new evidence from West Africa. *Ecology, 71*, 328–343.

Oates, J. F., Davies, A. G., & Delson, E. (1994). The diversity of living colobines. In A. G. Davies & J. F. Oates (eds.), *Colobine monkeys: Their ecology, behaviour and evolution*. Cambridge University Press, Cambridge, pp. 45–73.

Obminski, Z. (1998). Changes in the free (unbound) fraction of testosterone in serum in vitro as affected by pH and temperature. *Experimental & Clinical Endocrinology & Diabetes, 106*, 85–88.

Obminski, Z., & Stupnicki, R. (1996). Effect of temperature and pH on the magnitude of the free fraction of cortisol in serum. *Experimental & Clinical Endocrinology & Diabetes, 104*, 350–352.

Odyek, O., Makanga, B., & Byaruhanga, M. A. (1990). Toxicity of the seed extract of Khaya grandifoliola on some pulmonate fresh water snails. *Fitoterapia, 61*, 50–53.

Oelschlanger, M. (1991). *The idea of wilderness: From prehistory to the age of ecology.* Yale University Press, New Haven.

Ohigashi, H., Jisaka, M., Takagaki, T., Nozaki, H., Tada, T., Huffman, M. A., Nishida, T., Kaji, M., & Koshimizu, K. (1991). A bitter principle and a related steroid glucoside of vernonia amygdalina, a possible medicinal plant for wild chimpanzees. *Agricultural and Biological Chemistry, 55*, 1201–1203.

Ohigashi, H., Huffman, M. A., Izutsu, D., Koshimizu, K., Kawanaka, M., Sugiyama, H., Kirby, G. C., Warhurst, D. C., Allen, D., Wright, C. W., Phillipson, J. D.,

Timmon-David, P., Delnas, F., Elias, R., & Balansard, G. (1994). Toward the chemical ecology of medicinal plant-use in chimpanzees: The case of *Vernonia amygdalina* Del. a plant used by wild chimpanzees possibly for parasite-related diseases. *Journal of Chemical Ecology, 20*, 541–553.

Ojinnaka, C. M., Okogun, J. I., & Okorie, D. A. (1980). Triterpene acids from *Myrianthus arboreus. Phytochemistry, 19*, 2482–2483.

Onderdonk, D. A., & Chapman, C. A. (2000). Coping with forest fragmentation: The primates of Kibale National Park, Uganda. *International Journal of Primatology, 21*, 587–611.

Osmaston, H. A. (1959). *Working plan for the Kibale and Itwara forests.* Uganda Forest Department, Entebbe, Uganda.

Overdorff, D. J., & Strait S. G. (1996). *Do prosimian primates function as seed dispersers in Madagascar?* Paper presented at XVIth Congress of the International Primatological Society.

Owren, M. J., & Rendall, D. (2001). Sound on the rebound: Bringing form and function back to the forefront in understanding nonhuman primate vocal signalling.*Evolutionary Anthropology, 10*, 58–71.

Paciepnik, O. (1976). Intestinal parasites in monkeys at the Wroclaw Zoo. *Parasitology News, 22*, 289–296.

Page, J. E., Huffman, M. A., Smith, V., & Towers, G. H. N. (1997). Chemical basis for medicinal consumption of *Aspilia* (Asteraceae) leaves by chimpanzees: A re-analysis. *Journal of Chemical Ecology, 23*, 2211–2225.

Palmer, B., Jones, R. J., Wina, E., & Tangendjaja, B. (2000). The effect of sample drying conditions on estimates of condensed tannin and fiber content, dry matter digestibility, nitrogen digestibility and PEG binding of *Calliandra calothyrsus. Animal Feed Science and Technology, 87*, 29–40.

Palmer, W. L., Cowan, R. L., & Ammann, A. P. (1976). Effect of inoculum source on *in-vitro* digestion of deer foods. *Journal of Wildlife ManageMent, 40*, 301–307.

Palmquist, D. L., & Jenkins, T. C. (1980). Fat in lactation rations: Review. *Journal of Dairy Science, 63*, 1–14.

Palmquist, D. L., & Jenkins, T. C. (2003). Challenges with fats and fatty acid methods. *Journal of Animal Science, 81*, 3250–3254.

Panter-Brick, C., Lotstein, D. S., & Ellison, P. T. (1993). Seasonality of reproductive function and weight loss in rural Nepali women. *Human Reproduction, 8*, 684–690.

Parry, D., & Campbell, B. (1992). Attitudes of rural communities to animal wildlife and its utilization in Chobe Enclave and Mababe Depression, Botswana. *Environmental Conservation, 19*, 249–252.

Passingham, R. (1981). Primate specializations in brain and intelligence. *Symposia of the Zoological Society of London, 46*, 361–388.

Paterson, J. D. (1973). Ecologically differentiated patterns of aggressive and sexual behavior in two troops of Ugandan baboons, *Papio anubis. American Journal of Physical Anthropology, 38*, 641–647.

Paterson, J. D. (1976). *Varying ecology and differential adaptation of the Ugandan Baboon,* Papio cynocephalus anubis, *with special reference to forest conditions and behavioral models for early hominids.* PhD thesis, University of Toronto.

Paterson, J. (1991). The ecology and history of Uganda's Budongo Forest. *Forest Conservation History, 35*, 179–186.

Pehrson, A., & Faber, W. E. (1994). Individual variation of *in-vitro* dry matter digestibility in moose. *Journal of Range Management, 47*, 392–394.

Penn, D. (2003). The evolutionary roots of our environmental problems: Toward a Darwinian ecology, *The Quarterly Review of Biology, 78*, 275–301.

Penry, D. B. (1993). Digestive constraints on diet selection. In R. N. Hughes (ed.), *Diet selection: An interdisciplinary approach to foraging behaviour.* Blackwell Scientific, Oxford, pp. 32–55.

Pepper J. W., Mitani J. C., & Watts D. P. (1999). General gregariousness and specific social preferences among wild chimpanzees. *International Journal of Primatology, 20*, 613–632.

Perret, M. (1985). Influence of social factors on seasonal variations in plasma testosterone levels of *Microcebus murinus. Zeitschrift Fuer Tierpsychologie, 69*, 265–280.

Peters, C. R., & O'Brien, E. M. (1981). The early hominid plant-food niche: Insights from an analysis of plant exploitation by *Homo*, *Pan* and *Papio* in Eastern and Southern Africa. *Current Anthropology, 22*, 127–140.

Pezzlo, M. (1988). Detection of urinary tract infections by rapid methods. *Clinical Microbiology Review*, 268–280.

Pfaller, M., Ringenberg, B., Rames, L., Hegeman, J., & Koontz, F. (1987). The usefulness of screening tests for pyuria in combination with culture in the diagnosis of urinary tract infection. *Diagnostic Microbiology and Infectious Disease, 6*, 207–215.

Phillips-Conroy, J. E. (1986). Baboons, diet, and disease: Food plant selection and schistosomiasis. In D. Taub & F. King (eds.), *Current perspectives in primate social dynamics.* Nostrand Reinhold, New York, pp. 287–304.

Pietras, R. J., & Wenzel, B. M. (1974). Effects of androgens on body weight, feeding, and courtship behavior in the pigeon. *Hormones & Behavior, 5*, 289–302.

Pimbert, M., & Pretty J. (1995). *Parks, people and professionals: Putting "participation" into protected area management*, UNRISD Discussion Paper No. 57.

Plumptre, A. J. (1995). The chemical composition of montane plants and its influence on the diet of large mammalian herbivores in the Parc National Des Volcans, Rwanda. *Journal of Zoology, 235*, 232–337.

Plumptre, A. J. (1996). Changes following 60 years of selective timber harvesting in the Budongo Forest Reserve, Uganda. *Forest Ecology and Management, 89*, 101–113.

Plumptre, A. J. (2000). Monitoring mammal populations with line transect techniques in African forests. *Journal of Applied Ecology, 37*, 356–368.

Plumptre, A. J., & Reynolds, V. (1994). The effects of selective logging on the primate populations in the Budongo Forest Reserve, Uganda. *Journal of Applied Ecology, 31*, 631–641.

Plumptre, A. J., & Reynolds, V. (1996). Censusing chimpanzees in the Budongo Forest, Uganda. Internat. *Journal of Primatology, 17*, 85–99.

Plumptre, A. J., & Williamson, E. A. (2001). Conservation-oriented research in the Virunga region. In M. M. Robbins P. Sicotte, & K. J. Stewart (eds.), *Mountain gorillas: Three decades of research at Karisoke*. Cambridge University Press, Cambridge, pp. 362–389.

Plumptre, A. J., Reynolds, V., & Bakuneeta, C. (1994). *The contribution of fruit eating primates to seed dispersal and natural regeneration after selective logging*. Final Report of Overseas Development Association Project R4738.

Plumptre A. J., Reynolds, V., & Bakuneeta, C. (1997). *The effects of selective logging monodominant tropical forests on biodiversity*. Final report of Overseas Development Association project, number R6057.

Plumptre, A. J, Mugime, S., Cox, D., & Montgomery, C. (2001). *Chimpanzee and large mammal survey of Budongo Forest Reserve and Kibale National Park*. Report to WCS on surveys in Western Uganda.

Plumptre A. J., Cox, D., & Mugume, S. (2003). *The status of chimpanzees in Uganda*. Albertine Rift Technical Report Series No. 2, Wildlife Conservation Society.

Polygenis-Bigendako, M. J. (1990). *Recherches ethnopharmacognosiques sur les plantes utilisées en médecine traditionnelle au Burundi occidental*. Thèse de Docteur en sciences, Univ. libre de Bruxelles, Fac. Sc., Lab. de Botanique systématique et de Phytosociologie, p. 352.

Pomeroy, D. E. (1976). Some effects of mound-building termites on soils in Uganda. *Journal of Soil Science, 27*, 377–394.

Popovich, D. G., Jenkins, D. J. A., Kendall, C. W. C., Dierenfeld, E. S., Carroll, R. W., Tariq, N., & Vidgen, E. (1997). The western lowland gorilla diet has implications for the health of humans and other hominoids. *Journal of Nutrition, 127*, 2000–2005.

Poppenwimer, C. J. (1999–2000). Encounter in Uganda between chimpanzees *Pan troglodytes* and a leopard *Panthera pardus*. *African Primates, 4*, 75–76.

Population Action International. (2000). Relative scarcity: Apes on the edge. *Nature's Place: Human Population and the Future of Biological Diversity*. Available at: http://www.actionbioscience.org/biodiversity/pai.html#Primer.

Porter, L. J., Hrstich, L. N., & Chan, B. G. (1986). The conversion of procyanidins and prodelphinidins to cyaniding and delphinidin. *Phytochemistry, 25*, 223–230.

Povinelli, D., & Cant, J. (1995). Arboreal clambering and the evolution of self-conception. *Quarterly Review of Biology, 70*, 393–421.

Premvati (1958). Studies on *Strongyloides* of primates I. Morphology and life history of *Strongyloides fulleborni* von Linstow. *Canadian Journal of Zoology, 36*, 65–77.

Preuss, T. M. (1995). The argument from animals to humans in cognitive neuroscience. In M. Gazzaniga (ed.), *The cognitive neurosciences.* MIT Press, Cambridge, MA, pp. 1227–1241.

Pulliam, H. R. (1975). Diet optimization with nutrient constraints. *American Naturalist, 109*, 765–768.

Pusey, A. E. (1980). Inbreeding avoidance in chimpanzees. *Animal Behaviour, 28*, 543–552.

Pusey, A. (1998). Scabies in chimpanzees of Gombe National Park, Tanzania. *Newsletter 1, European Association of Zoo and Wildlife Veterinarians.*

Pusey, A. E., Williams, J., & Goodall, J. (1997). The influence of dominance rank on the reproductive success of female chimpanzees. *Science, 277*, 828–831.

Putz, F. E. (1983). Liana biomass and leaf area of a "Tierra Firme" forest in the Rio Negro basin, Venezuela. *Biotropica, 15*, 185–189.

Quiatt, D. (1994). Leaf sponge drinking by a Budongo Forest chimpanzee. *American Journal of Primatology, 33*, 236.

Quiatt, D. (1996). *Budongo Forest chimpanzees: Behavioural accommodations to physical disabilities.* Paper presented at the XVth Congress of the International Primatological Society/XIXth Conference of the American Society of Primatologists.

Quiatt, D., Reynolds, V., & Stokes, E. J. (2002). Snare injuries to chimpanzees (*Pan troglodytes*) at 10 study sites cross East and West Africa. *African Journal of Ecology, 40*, 303–305.

Quinn, C., Huby, M., Kiwasila, H., & Lovett, J. (2003). Local perceptions of risk to livelihood in semi-arid Tanzania. *Journal of Environmental Management, 68*, 111–119.

Ransom, T. W. (1971). *Ecology and social behaviour of baboons (*Papio anubis*) in the Gombe National Park.* PhD thesis, University of California, Berkeley.

Rao, M., & van Schaik, C. P. (1997). The behavioral ecology of Sumatran orangutans in logged and unlogged forest. *Tropical Biodiversity, 4*, 173–185.

Redford, K. (1990). The ecologically noble savage. *Cultural Survival Quarterly, 15*, 46–48.

Redmond, I. (1983). Summary of parasitology research, November 1976 to April 1978. In D. Fossey (ed.), *Gorillas in the mist.* Houghton Mifflin, New York.

Rehmann, P., Gröne, A., Lawrenz, A., Pagan, O., Gottstein, B., & Bacciarine, L. N. (2003). *Echinococcus multilocularis* in two lowland gorillas (*Gorilla g. gorilla*). *Journal of Comparative Pathology, 129*, 85–88.

Remis, M. J. (1994). Feeding ecology and positional behaviour of Western lowland gorillas (*Gorilla gorilla gorilla*) in the Centrak African Republic. Ph. D Dissertation. Yale University.

Remis, M. J. (1997). Western lowland gorillas (*Gorilla gorilla gorilla*) as seasonal frugivores: Use of variable resources. *American Journal of Primatology, 43*, 87–109.

Remis, M. J. (2000). Initial studies on the contributions of body size and gastrointestinal passage rates to dietary flexibility among gorillas. *American Journal of Physical Anthropology, 112*, 171–180.

Remis, M. J. (2003). Are gorillas vacuum cleaners of the forest floor? The roles of body size, habitat, and food preferences on dietary flexibility and nutrition. In A. B. Taylor & M. L. Goldsmith (eds.), *Gorilla biology: A multidisciplinary perspective.* Cambridge University Press, Cambridge, pp. 385–404.

Remis, M. J., Dierenfeld, E. S., Mowry, C. B., & Carroll, R. W. (2001). Nutritional aspects of western lowland gorilla (*Gorilla gorilla gorilla*) diet during seasons of fruit scarcity at Bai Hokou, Central African Republic. *International Journal of Primatology, 22*, 807–836.

Remis, M. J., & Dierenfeld, E. S. (2004). Digesta passage, digestibility and behavior in captive gorillas under two dietary regimes. *International Journal of Primatology, 25*, 825–845.

Reynolds, V. (1965). *Budongo: A forest and its chimpanzees.* Methuen, London.

Reynolds, V. (1992). Chimpanzees in the Budongo Forest, 1962–1992. *Journal of Zoology, 228*, 695–699.

Reynolds, V. (1993). Sustainable forestry: The case of Budongo forest, Uganda. *SWARA, July–Aug*, 13–16.

Reynolds, V., & Reynolds, F. (1965). Chimpanzees of the Budongo Forest. In I. DeVore (ed.), *Primate behavior.* Holt, Rinehart and Winston, New York, pp. 368–424.

Reynolds, V., Burch, D., Knight, J., Smith, R., & Waller, J. (1996). The nature, causes and consequences of injuries sustained by wild chimpanzees in the Budongo Forest, Uganda. In *The XVIth Congress of the International Primatological Society*, Madison, WI.

Reynolds, V., Plumptre, A. J., Greenham, J., & Harborne, J., (1998). Condensed tannins and sugars in the diet of chimpanzees (*Pan troglodytes schweinfurthii*) in the Budongo Forest, Uganda. *Oecologia, 115*, 331–336.

Richards, P. W. (1996). *The tropical rain forest* (2nd ed.). Cambridge University Press, Cambridge, England.

Ristau, C. A., & Robbins, D. (1982). Language in the great apes: A critical review. *Advances in the Study of Behavior, 12*, 141–255.

Robbins, M. M., & Czekala, N. M. (1997). A preliminary investigation of urinary testosterone and cortisol levels in wild male mountain gorillas. *American Journal of Primatology, 43*, 51–64.

Robbins, M. M., & McNeilage, A. (2003). Home range and frugivory patterns of mountain gorillas in Bwindi-Impenetrable National Park, Uganda. *International Journal of Primatology, 24*, 467–491.

Roberts, M. F., & Wink, M. (eds.). (1998). *Alkaloids biochemistry, ecology, and medicinal applications.* Plenum Press, New York.

Robertson, J. B., & van Soest, P. J. (1980). The detergent system of analysis and its application to human foods. In W. P. T. James & O. Theander (eds.), *The analysis of dietary fiber in foods.* Marcel Dekker, New York, pp. 123–158.

Rode, K. D., Chapman, C. A., Chapman, L. J., & McDowell, L. R. (2003). Mineral resource availability and consumption by colobus in Kibale National Park, Uganda. *International Journal of Primatology, 24*, 541–573.

Rodriguez E., & Wrangham R. W. (1993). Zoopharmacognosy: the use of medicinal plants by animals. In K. R. Downum, J. T. Romeo, & H. Stafford (eds.), *Recent advances in phytochemistry, Vol. 27: Phytochemical potential of tropic plants.* New York, Plenum Press, pp. 89–105.

Rodriguez, E., Aregullin, M., Nishida, T., Uehara, S., Wrangham, R., Abramowski, Z., Finlayson, A., & Towers, G. H. N. (1985). Thiarubrin A, a bioactive constituent of *Aspilia* (Asteraceae) consumed by wild chimpanzees. *Experientia, 41*, 419–420.

Rogers, A. (1991). Conserving resources for children. *Human Nature, 2*, 73–82.

Rogers, M. E., Maisels, F., Williamson, E. A., Fernandez, M., & Tutin, C. E. G. (1990). Gorilla diet in the Lopé Reserve, Gabon: A nutritional analysis. *Oecologia, 84*, 326–339.

Rogers, M. E., Abernathy, K., Bermejo, M., Cipolletta, C., Doran, D., McFarland, K., Nishihara, T., Remis, M., & Tutin, C. E. G. (2004). Western gorilla diet: A synthesis from six sites. *American Journal of Primatology, 64*, 173–192.

Romero, L. M., & Wingfield, J. C. (1998). Seasonal changes in adrenal sensitivity alter corticosterone levels in Gambel's white-crowned sparrows (*Zonotrichia leucophrys gambelii*). *Comparative Biochemistry & Physiology—C: Comparative Pharmacology & Toxicology, 119*(1), 31–36.

Romero, L. M., Reed, J. M., & Wingfield, J. C. (2000). Effects of weather on corticosterone responses in wild free-living passerine birds. *General & Comparative Endocrinology, 118*, 113–122.

Rose, M. D. (1978). Feeding and associated positional behaviour of black and white colobus monkeys (*Colobus guereza*). In G. G. Montgomery (ed.), *The ecology of arboreal folivores.* Smithsonian Institution Press, New York, pp. 71–96.

Rosenbaum B., O'Brien, T. G., Kinnard, M., & Supriatna, J. (1998). Population densities of Sulawesi crested black macaques (*Macaca nigra*) on Bacan and Sulawesi, Indonesia: Effects of habitat disturbance and hunting. *American Journal of Primatology, 44*, 89–106.

Ross, G. T., Cargile, C. M., Lipsett, M. B., Rayford, P. L., Marshall, J. R., Strott, C. A., & Rodbard, D. (1970). Pituitary and gonadal hormones in women during spontaneous and induced ovulatory cycles. *Recent Progress in Hormone Research, 26*, 1–62.

Rothman, J. M., Bowman, D. D., Eberhard, M. L., & Pell, A. N. (2002). Intestinal parasites in the research group of mountain gorillas in Bwindi Impenetrable National Park, Uganda: Preliminary results. *Annals of the New York Academy of Sciences, 969*, 346–349.

Rothman, J. M., Dierenfeld, E. S., Molina, D. O., Shaw, A. V., Hintz, H. F., & Pell, A. V. (in press). Nutritional chemistry of foods eaten by gorillas in Bwindi Impenterable National Park, Uganda. American Journal of Primatology.

Rousselot, R., & Pellissier, A. (1952). III. Esophagostomose nodulaire a *Oesophagostomum stephanostomum* Pathologie du gorille et du chimpanzee. *Bulletin of the Exotic Pathology Society, 9*, 569–574.

Rowell, T. (1966). Forest-living baboons in Uganda. *Journal of Zoology, 149*, 344–364.

Rudran, R. (1978a). Socioecology of the blue monkeys (*Cercopithecus mitis stuhlmanni*) of the Kibale Forest, Uganda. *Smithsonian Contributions to Zoology, 249*, 1–88.

Rudran, R. (1978b). Intergroup dietary comparisons and folivorous tendencies of two groups of blue monkeys (*Cercopithecus mitis stuhlmanni*). In G. G. Montgomery (ed.), *The ecology of arboreal folivores.* Smithsonian Institution Press, Washington, DC, pp. 483–504.

Ruhiyat, Y. (1983). Socio-ecological study of *Presbytis aygula* in West Java. *Primates, 42*, 344–359.

Rukundo, T. (1996). *The long term effect of canopy treatment on tree diversityin Budongo Forest: An evaluation of a 25 hectare permanent plot study.* Unpublished MSc thesis, Makerere University, Uganda.

Rumbaugh, D. M. (1977). *Language learning by a chimpanzee.* Academic Press, New York.

Russon, A. E. (1998). The nature and evolution of intelligence in orangutans (*Pongo pygmaeus*). *Primates, 39*, 485–503.

Ruttan, L., & Borgerhoff-Mulder, M. (1999). Are east African pastoralists truly conservationsists? *Current Anthropology, 40*, 621–652.

Ruvolo, M. (1988). Genetic evolution in the African guenons. In A. Gautier-Hion, F. Bourliere, & J-P. Gautier (eds.), *A primate radiation: Evolutionary biology of the African guenons.* Cambridge University Press, Cambridge, pp. 127–139.

Sakura, O. (1994). Factors affecting party size and composition of chimpanzees (*Pan troglodytes verus*) at Bossou, Guinea. *International Journal of Primatology, 15*, 167–183.

Sandosham, A. A. (1950). On *Enterobius vermicularis* (Linnaeus, 1758) and some related species from primates and rodents. *Journal of Helminthology, 24*, 171–204.

Sarmiento, E. E., Butynski, T. M., & Kalina, J. (1996). Gorillas of Bwindi-Impenetrable Forest and the Virunga Volcanoes: Taxonomic implications of morphological and ecological differences. *American Journal of Physical Anthropology, 40*, 1–21.

Savage-Rumbaugh, S., Shanker, S. G., & Taylor, T. J. (1998). *Apes, language and the human mind.* Oxford University Press, New York.

Schaller, G. B. (1963). *The mountain gorilla: Ecology and Behaviour.* University of Chicago Press, Chicago.

Schaller, G. B. (1964). *The year of the gorilla.* University of Chicago Press, Chicago.

Schlichte, H. J. (1978). The ecology of two groups of blue monkeys in an isolated habitat of poor vegetation. In G. G. Montgomery (ed.), *The Ecology of Arboreal Folivores.* Smithsonian Institution Press, Washington, DC, pp. 505–517.

Schmidt, J. (1999). Soldiers in the gorilla war. International Wildlife. January/February Edition. Available at: http://www.nwf.org/internationalwildlife/1998/gorilla.html

Schmitt, T. M. (1997). Close encounter with gorillas at Bwindi. *J. Berggorilla and Regenwald Direkthilfe, 14*, 12–13.

Schofield, P., Mbugua, D. M., & Pell, A. N. (2001). Analysis of condensed tannins: A review. *Animal Feed Science and Technology, 91*, 21–40.

Schuett, G. W., Harlow, H. J., Rose, J. D., Van Kirk, E. A., & Murdoch, W. J. (1996). Levels of plasma corticosterone and testosterone in male copperheads (*Agkistrodon contortrix*) following staged fights. *Hormones & Behavior, 30*(1), 60–68.

Scott, P. (1998). *From conflict to collaboration: People and forests at Mount Elgon, Uganda.* Nairobi Report: IUCN Eastern Africa Regional Office.

Seehausen, O., Koetsier, E., Scneider, M. V., Chapman, L. J., Chapman, C. A., Knight, M. E., Turner, G. F., van Alphen, J. J. M., & Bills, R. (2002). Nuclear markers reveal unexpected genetic variation and a Congolese-Nilotic origin of the Lake Victoria cichlid species flock. *Proceedings of the Royal Society of London, B, 270*, 129–137.

Seigler, D. A. (1991). Cyanide and cyanogenic glycosides. In G. A. Rosenthal & M. R. Berenbaum (eds.), *Herbivores: Their interactions with secondary plant metabolites, Vol. 1: The chemical participants.* Academic Press, New York, pp. 35–77.

Seraphin, S. B. (2000). *The reproductive ecology and stress physiology of free-ranging male chimpanzees in Budongo Forest, Uganda.* MSc thesis, Oxford University, Oxford, England.

Seyfarth, R. M., & Cheney, D. L. (2003). Signalers and recievers in animal communication. *Annual Review of Psycology, 54*, 145–173.

Seyfarth, R. M, Cheney, D. L., & Marler, P. (1980). Monkey responses to three different alarm calls: Evidence for predator classification and semantic communication. *Science, 210*, 801–803.

Sheppard, D. J. (2000). *Ecology of the Budongo Forest redtail: Patterns of habitat use and population density in primary and regenerating forest sites.* MA thesis, University of Calgary, Canada.

Shuttleworth, S. J. (1998). *Cognition, evolution and behavior.* Oxford University Press, New York.

Shideler, S. E., Munro, C. J., Tell, L., Owitt, G., Laughlin, L. S., Chatterton, R., Jr., & Lasley, B. L. (1990). The relationship of serum estradiol and progesterone concentrations to the enzyme immunoassay measurements of urinary estrone conjugates and immunoreactive pregnanediol-3-glucuronide in *Macaca mulatta. American Journal of Primatology, 22*, 113–122.

Shimizu, K., Udono, T., Mitsunaga, F., & Hayashi, M. (1999). Assesment of reproductive status by using urinary hormone evaluation in great apes. In *COE International Symposium: Evolution of the Apes and the Origins of Human Beings, Kyoto*, pp. 60.

Shimizu, K., Douke, C., Fujita, S., Matsuzawa, T., Tomonaga, M., Tanaka, M., Matsubayashi, K., & Hayashi, M. (2003a). Urinary steroids, FSH and CG measurements for monitoring the ovarian cycle and pregnancy in the chimpanzee. *Journal of Medical Primatology, 32*, 15–22.

Shimizu, K., Udono, T., Tanaka, C., Narushima, E., Yoshihara, M., Takeda, M., Tanahashi, A., van Elsacker, L., Hayashi, M., & Takenaka, O. (2003b). Comparative study of urinary reproductive hormones in great apes. *Primates, 44*, 183–190.

Sholley, C. R. (1991). Conserving gorillas in the midst of guerrillas. *AAZPA Annual Conference, San Diego*, pp. 30–37.

Sholley, C. R., & Hastings, B. (1989). Outbreak of illness among Rwanda's gorillas. *Gorilla Conservation News, 3*, 7.

Shuttleworth, S. J. (1988). *Cognition evolution and behaviour.* Oxford, Oxford University Press.

Simmen, B., & Sabatier, D. (1996). Diets of some French Guianan primates: Food composition and food choices. *International Journal of Primatology, 17*, 661–693.

Skorupa, J. P. (1986). Responses of rainforest primates to selective logging in Kibale Forest, Uganda: a summary report. In K. Benirschke (ed.), *Primates: The road to self-sustaining populations.* Springer Verlag, New York, pp. 55–70.

Skorupa, J. P. (1988). *The effect of selective timber harvesting on rain-forest primates in Kibale Forest, Uganda.* PhD dissertation, University of California, Davis.

Sleeman, J. M., Meader, L. L., Mudakikwa, A. B., Foster, J. W., & Patton, S. (2000). Gastro-intestinal parasites of mountain gorillas (*Gorilla gorilla beringei*) from the

Parc National des Volcans, Rwanda. *Journal of Zoo and Wildlife Management, 31*, 322–328.

Smith, E. (1992). Human behavioral ecology. *Evolutionary Anthropology, 10*, 15–25.

Snyder, D. L., Cowan, R. L., Hagen, D. R., & Schanbacher, B. D. (1983). Effect of pinealectomy on seasonal changes in antler growth and concentrations of testosterone and prolactin in white-tailed deer. *Biology of Reproduction, 29*, 63–71.

Songorwa, A. (1999). Community-based wildlife management (CWM) in Tanzania: Are the communities interested? *World Development, 27*, 2061–2079.

Sokal, R. R., & Rohlf, F. J. (1981). *Biometry: The principles and practice of statistics in biological research.* Freeman, San Francisco, 859 pp.

Sourd, C., & Gautier-Hion, A. (1986). Fruit selection by a forest guenon. *Journal of Animal Ecology, 55*, 235–244.

Stambolic-Robb, A. (1997). *Geophagy amongst free ranging Sumatran orangutans* (Pongo pygmaeus abelii) *of Gunung Leuser National Park and ex-captive Borean orangutans (*Pongo pygmaeus pygmaeus*) of Sungai Wain Forest, Indonesia.* MSc thesis, York University, Toronto.

Stanford, C. B. (1990). Colobine socioecology and female-bonded models of primate social structure. *Kroeber Anthropology Society Papers, 71–72*, 21–28.

Stanford, C. B. (1991). The diet of the capped langur (*Presbytis pileata*) in a moist forest in Bangladesh. *International Journal of Primatology, 12*, 199–216.

Stanford, C. B. (2001). The subspecies concept in primatology: the case of mountain gorillas. *Primates, 42*, 309–318.

Stanford, C. B., & Nkurunungi, J. B. (2003). Behavioral ecology of sympatric chimpanzees and gorillas in Bwindi Impenetrable National Park, Uganda: Diet. *International Journal of Primatology, 24*, 901–918.

Stanford, C. B., Wallis, J., Mpongo, E., & Goodall, J. (1994). Hunting decisions in wild chimpanzees. *Behaviour, 131*, 1–18.

Steiner, P. E. (1954). Anatomical observations in a gorilla. *American Journal of Physical Anthropology, 12*, 145–170.

Steklis, H. D., Hodgkinson, C., Fawcett, K., Gerald-Steklis, N., Czekala, N., Lilly, A., & Mehlman, P. T. (2004). The impact of tourism on mountain gorillas. *Folia Primatologica, 75*, 40–41.

Stem C., Lassoie, J., Lee, R., Deshler, D., & Schelhas, J. (2003). Community participation in ecotourism benefits: The link to conservation practices and perspectives. *Society and Natural Resources, 16*, 387–413.

Stevens, C. E., & Hume, I. D. (1995). Contributions of microbes in vertebrate gastrointestinal tract to production and conservation of nutrients. *Physiology Review, 78*, 393–427.

Stewart, K. J. (1991). Editorial. *Gorilla Conservation News, 5*, 1–2.

Stewart, K. J. (1992). Gorilla tourism: Problems of control. *Gorilla Conservation News 6*, 15–16.

Stewart, K. J., Sicotte, P., & Robbins, M. M. (2001). Mountain gorillas of the Virungas: a short history. In M. M. Robbins, P. Sicotte, K. J. Stewart (eds.), *Mountain gorillas: Three decades of research at Karisoke*. Cambridge University Press, Cambridge, pp. 1–26.

Stokes, E. J. (1999). *Feeding skills and the effect of injury on wild chimpanzees*. PhD dissertation, University of St Andrews, Fife.

Stokes, E. J., & Byrne, R. W. (2001). Cognitive capacities for behavioural flexibility in wild chimpanzees (*Pan troglodytes*): The effect of snare injury on complex manual food processing. *Animal Cognition, 4*, 11–28.

Stokes, E., & Byrne, R. (in review). Feeding competition and food accessibility: Everyday use of complex behaviour by chimpanzees (*Pan troglodytes*) during natural foraging. *Behaviour*

Stossich, M. (1904). Sopra alcuni nematodi. *Annuario del Museo Zoologico Della R. Universita Di Napoli, 1*, 1–5.

Ströbele-Gregor, J. (2000). Visit to the mountain gorillas in Rwanda. J. Berggorilla and Regenwald Direkthilfe Archives, Dec. Available at: http://www.kilimanjaro.com/gorilla/brd/12-00.htm

Struhsaker, T. T. (1975). *The red colobus monkey*. University of Chicago Press, Chicago.

Struhsaker, T. T. (1978). Food habits of five monkey species in the Kibale Forest, Uganda. In D. J. Chivers & J. Herbert (eds.), *Recent advances in primatology*. Academic Press, London.

Struhsaker, T. T. (1981). Polyspecific associations among tropical rain-forest primates. *Zeitschrift Fur Tierpsychologie, 57*, 268–304.

Struhsaker, T. T. (1997). *Ecology of an African rain forest: Logging in Kibale and the conflict between conservation and exploitation*. University Press of Florida, Gainsville, FL.

Struhsaker, T. T., & Leland, L. (1979). Socioecology of five sympatric monkey species in the Kibale Forest, Uganda. In J. Rosenblatt, R. A. Hinde, C. Beer, & M. C. Busnel (eds.), *Advances in the Study of Behavior* (Vol. 9). Academic Press, New York, pp. 158–228.

Struhsaker, T. T., & Oates, J. F. (1975). Comparison of the behavior and ecology of red colobus and black-and-white colobus monkeys in Uganda: A summary. In R. H. Tuttle (ed.), *Socio-ecology and psychology of primates*. The Hague, pp. 130–123.

Struhsaker, T. T., & Oates, J. F. (1979). Comparison of the behaviour and ecology of red colobus and black and white colobus monkeys in Uganda: A summary. In R. W. Sussman (ed.), *Primate ecology: Problem-oriented field studies*. Wiley, New York, pp. 165–183.

Struhsaker, T. T., Lwanga, J. S., & Kasenene, J. M. (1996). Elephants, selective logging, and forest regeneration in the Kibale Forest, Uganda. *Journal of Tropical Ecology, 12*, 45–64.

Sugiyama, Y. (1968). Social organization of chimpanzees in the Budongo Forest, Uganda. *Primates, 9*, 225–258.

Sugiyama, Y. (1969). Social behavior of chimpanzees in the Budongo Forest, Uganda. *Primates, 10*, 197–225.

Sugiyama, Y. (1989). Population dynamics of chimpanzees at Bossou, Guinea. In P. G. Heltne & L. A. Marquardt (eds.), *Understanding chimpanzees.* Harvard University Press, Cambridge, MA, pp. 134–145.

Sugiyama, Y. (1994). Age-specific birth rate and lifetime reproductive success of chimpanzees at Bossou, Guinea. *American Journal of Primatology, 32*, 311–318.

Sugiyama, Y., & Koman, J. (1992). The flora of Bossou: Its utilization by chimpanzees and humans. *African Study Monographs, 13*, 127–169.

Sukhija, S., & Palmquist, D. L. (1988). Rapid method for determination of total fatty acid composition of feedstuffs and feces. *Journal of Agricultural and Food Chemistry, 36*, 1202–1206.

Sussman, R. W. (1977). Feeding behaviour of *Lemur catta* and *Lemur fulvus.* In T. H. Clutton-Brock (ed.), *Primate ecology: Studies of feeding and ranging behaviour in lemurs, monkeys and apes.* Academic Press, London, pp. 1–36.

Suzuki, A. (1972). On the problems of conservation of the chimpanzees in East Africa and of the preservation of their environment. *Primates, 12*, 415–418.

Suzuki, A. (1979). The variation and adaptation of social groups of chimpanzees and black and white colobus monkeys. In I. S. Bernstein & E. O. Smith (eds.), *Primate ecology and human origins.* Garland STPM Press, New York, pp. 153–173.

Symington, M. (1990). Fission-fusion social organization in *Ateles* and *Pan. International Journal of Primatology, 11*, 47–61.

Synnott, T. J. (1985). *A checklist of the flora of the Budongo Forest Reserve, Uganda with notes on ecology and phenology.* CFI Occasional Papers No. 27. Department of Forestry Commonwealth Forestry Institute, Oxford, University of Oxford.

Takahata, Y., Ihobe, H., & Idani, G. (1996). Comparing copulations of chimpanzees and bonobos: do females exhibit proceptivity or receptivity? In W. C. McGrew, L. F. Marchant, & T. Nishida (eds.), *Great ape societies.* Cambrige University Press, Cambridge, pp. 146–155.

Taniguchi, M., Chapya, A., Kubo, I., & Nakanishi, K. (1978). Screening of East African plants for antimicrobial activity. *Chemical and Pharmaceutical Bulletin, 26*, 2910–2913.

Taussky, H. H. (1954). A microcolorimetric determination of creatinine in urine by the Jaffe reaction. *Journal of Biological Chemistry, 208*, 853–861.

Teare, A., & Loomis, M. R. (1982). Epizootic balantidiasis in lowland gorillas. *Journal of the American Veterinary Medical Association, 181*, 1345–1347.

Teelen, S. (1994). *Group size and group structure of guereza,* Colobus guereza occidentalis *(Rochebrune 1886), in the Kibale Forest, Uganda.* Diplomarbeit, Universitat Braunschweig.

Teleki, G. (1973). *The predatory behaviour of wild chimpanzees.* Bucknell University Press, E. Brunswick, NJ.

Terashima, H., Kalala, S., & Malasi, N. (1991). Ethnobotany of the Lega in the tropical rain forest of Eastern Zaire. *African Study Monographs, 15.* The Centre for African Studies, Kyoto University.

Terashima, H., Kalala, S., & Malasi, N. (1992). Ethnobotany of the Lega in the tropical rain forest of Eastern Zaire, Part II: Zone de Walikale. *African Study Monographs, 19.* The Centre for African Studies, Kyoto University.

Terborgh, J. (1992). *Diversity and the tropical rain forest.* Scientific American Library, New York.

Terrace, H. S., Pettito, L. A., Sanders, R. J., & Bever, T. G. (1979). Can an ape create a sentence? *Science, 206*, 891–902.

The American Heritage College Dictionary. (3rd ed.). (1993). Habituate. Houghton Mifflin Co, Boston, MA.

Thomas, S. C. (1991). Population densities and patterns of habitat use among anthropoid primates of the Ituri forest, Zaire. *Biotropica, 23*, 68–83.

Thomas, D. W., & Mbenkum, F. T. (1987). *Medicinal and food plants from Cameroon's forests. Development and Conservation.* UNDP/FAO.

Thompson, R. C. A. (2000). Giardiasis as a re-emerging infectious disease and its zoonotic potential. *International Journal for Parasitology, 30*, 1259–1267.

Toft, C. A., Aeschlimann, A., & Bolis, L. (1991). *Parasite–host associations: Coexistence or conflict?* Oxford Science Publications, Oxford.

Tomasello, M., & Call, J. (1997). *Primate cognition.* Oxford University Press, New York.

Tosi, A. J., Disotell, T. R., Morales, J. C., & Melnick, D. J. (2003). Cercopithecine Y-chromosome data provide a test of competing morphological evolutionary hypotheses. *Molecular Phylogenetics and evolution, 27*, 510–521.

Tosi, A. J., Melnick, D. J., & Disotell, T. R. (2004). Sex chromosome phylogenetics indicate a single transition to terrestriality in the guenons (tribe Cercopithecini). *Journal of Human Evolution, 46*, 223–237.

Toubiana, R., & Gaudemer, A. (1967). Structure du vernolide, nouvel ester sesquiterpique isole de Vernonia colorata. *Tetrahedron Letters, 14*, 1333–1336.

Touitou, Y., Sulon, J., Bogdan, A., Reinberg, A., Sodoyez, J. C., & Demey-Ponsart, E. (1983). Adrenocortical hormones, ageing and mental condition: seasonal and circadian rhythms of plasma 18-hydroxy-11-deoxycorticosterone, total and free cortisol and urinary corticosteroids. *Journal of Endocrinology, 96*, 53–64.

Treves, A. (1999). Has predation shaped the social systems of arboreal primates? *International Journal of Primatology, 20*, 35–67.

Tripp, A. (2004). Women's movements, customary law, and land rights in Africa: The case of Uganda. *African Studies Quarterly, 7*, 1–19.

Tsukahara, T. (1993). Lions eat chimpanzees: The first evidence of predation by lions on wild chimpanzees. *American Journal of Primatology, 29*, 1–11.

Turke, P., & Betzig, L. (1985). Those who can do: Wealth, status and reproductive success on Ifaluk. *Ethnology and Sociobiology, 6*, 79–87.

Turkkan, J. S. (1994). Biobehavioral effects of extended salt loading and conflict stress in intact baboons. *Journal of the Experimental Analysis of Behavior, 61*, 263–272.

Turton, C. (2000). *Enhancing livelihood through participatory watershed development in India.* Sustainable Livelihoods Working Paper 131. UNDP, 2000. Development Institute, London.

Tutin, C. E. G. (1979). Mating patterns and reproductive strategies in a community of wild chimpanzees (*Pan troglodytes schweinfurthii*). *Behavioral Ecology and Sociobiology, 6*, 29–38.

Tutin, C. E. G. (1980). Reproductive behavior of wild chimpanzees in the Gombe National Park, Tanzania. *Journal of Reproduction and Fertility, 28*(Suppl.), 43–57.

Tutin, C. E. G. (1994). Reproductive success story: variability among chimpanzees and comparisons with gorillas. In R. W. Wrangham, W. C. McGrew, F. B. M. de Waal, & P. G. Heltne (eds.), *Chimpanzee cultures.* Harvard University Press, Cambridge, MA, pp. 181–194.

Tutin, C. E. G. (1996). Ranging and social structure of lowland gorillas in the Lopé Reserve, Gabon. In W. C. McGrew, L. F. Marchant, & T. Nishida (eds.), *Great ape societies.* Cambridge University Press, Cambridge, pp. 58–70.

Tutin, C. E. G., & McGinnis, P. R. (1981). Sexuality of the chimpanzee in the wild. In C. E. Graham (ed.), *Reproductive biology of the great apes: Comparative and biomedical perspectives.* Academic Press, New York.

Tutin, C. E. G., & Fernandez, M. (1983). Gorillas feeding on termites in Gabon, West Africa. *Journal of Mammals, 64*, 511–513.

Tutin, C. E. G., & Fernandez, M. (1985). Foods consumed by sympatric populations of *Gorilla g. gorilla* and *Pan t. troglodytes* in Gabon: Some preliminary data. *International Journal of Primatology, 6*, 27–43.

Tutin, C. E. G., & Fernandez, M. (1991). Responses of wild chimpanzees and gorillas to the arrival of primatologists: behavior observed during habituation. In H. O. Box (ed.), *Primate responses to environmental change.* Chapman & Hall, London, pp. 187–197.

Tutin, C. E. G., & Fernandez, M. (1992). Insect-eating by sympatric lowland gorillas and chimpanzees (*Pan t. troglodytes*) in the Lopé Reserve, Gabon. *American Journal of Primatology, 28*, 29–40.

Tutin, C. E. G., & Fernandez, M. (1993). Composition of the diet of chimpanzees and comparisons with that of sympatric lowland gorillas in the Lopé Reserve, Gabon. *American Journal of Primatology, 30*, 195–211.

Tweheyo, M. (2003). *Abundance, distribution and phenology of chimpanzee food in the Budongo Forest Reserve, Uganda*. Doctoral Dissertation, Agricultural University of Norway. [ISBN 0802–3220]

Tweheyo, M., & Obua, J. (2001). Feeding habits of chimpanzees (*Pan troglodytes*), red-tail monkeys (*Cercopithecus ascanius schmidti*) and blue monkeys (*Cercopithecus mitis stuhlmanii*) on figs in Budongo Forest Reserve, Uganda. *African Journal of Ecology, 39*, 133–139.

Twinomugisha, D. (1999). *The status and ecology of the golden monkey (*Cercopithecus mitis kandti*) in Mgahinga Gorilla National Park, Uganda*. MSc thesis, Makerere University, Kampala, Uganda.

Twinomugisha, D., Basuta, G. I., & Chapman, C. A. (2003). Status and ecology of the golden monkey (*Cercopithecus mitis kandti*) in Mgahinga Gorilla National Park, Uganda. *African Journal of Ecology, 41*, 47–55.

Uehara, S. (1982). Seasonal changes in the techniques employed by wild chimpanzees in the Mahale Mountains, Tanzania, to feed on termites (*Pseudacanthotermes spiniger*). *Folia Primatologica, 37*, 44–76.

Uganda National Parks (1996). *Mgahinga Gorilla National Park Management Plan 1996–2000*. Unpublished report, Uganda Government, Kampala, Uganda.

Ugandan Wildlife Authority (1996). *Mgahinga Gorilla National Park Management Plan 1996–2000*. Unpublished Government Report, Kampala, Uganda.

Uphof, J. C. Th. (1968). *Dictionary of economic plants* (2nd ed.). Verlag von J. Cramer.

Uphoff, N., & Langholz, J. (1998). Incentives for avoiding the tragedy of the commons. *Environmental Conservation, 25*, 251–261.

Valeggia, C., & Ellison, P. T. (2001). Lactation, energetics, and postpartum fecundity. In P. T. Ellison (ed.), *Reproductive ecology and human evolution*, Aldine de Gruyter, Chicago.

van den Berghe, L., Chardome, M., & Peel, E. (1964). The filarial parasites of the eastern gorilla in the Congo. *Journal of Helmithology, 38*, 349–368.

van Puyvelde, L. D. , Geysen, F. X., & Ayobangira, E. H. (1985). Screening of medical plants of Rwanda for acaricidal activity. *Journal of Ethnopharmacology, 13*, 209–215.

van Schaik, C. P. (1983). Why are diurnal primates living in groups? *Behaviour, 8*, 120–143.

van Schaik, C. P. (1989). The ecology of social relationships amongst female primates. In V. Standen & R. A. Foley (eds.), *Comparative socioecology: The behavioural ecology of humans and other mammals*. Blackwell, Oxford, pp. 195–218.

van Schaik, C., Mancrenaz, M., Borgen, G., Galdikas, B., Knott, C., Singleton, I., Suzuki, A. Utami S. S., & Merrit, M. (2003). Orangutan cultures and the evolution of material culture. *Science, 299*, 102–105.

van Soest, P. J. (1963). Use of detergents in the analysis of fibrous feeds, II: A rapid method for the determination of fiber and lignin. *Journal of the Association of Official Agricultural Chemists, 46*, 829–835.

van Soest, P. J. (1982). *Nutritional ecology of the ruminant*. O & E Books, Corvallis, OR.

van Soest, P. J. (1994). *Nutritional ecology of the ruminant* (2nd ed.). Cornell University Press, Ithaca.

van Waerebeke, D., Chabaud, A. G., & Collet, J. Y. (1988). New atractid nematode parasites of chimpanzee and gorilla in Gabon. *Annales de Parasitologie Humaine et Comparee, 63*, 37–47.

Vaughan, M. K., Menendez-Pelaez, A., Buzzell, G. R., Vaughan, G. M., Little, J. C., & Reiter, R. J. (1994). Circadian rhythms in reproductive and thyroid hormones in gonadally regressed male hamsters exposed to natural autumn photoperiod and temperature conditions. *Neuroendocrinology, 60*, 96–104.

Vedder, A. L. (1984). Movement patterns of a group of free-ranging mountain gorillas (*Gorilla gorilla beringei*) and their relation to food availability. *American Journal of Primatology, 7*, 73–88.

Vedder, A., & Weber, W. (1990). The mountain gorilla project (Volcanoes National Park). In A. Kiss (ed.), *Living with wildlife: Wildlife resource management with local participation in Africa*. World Bank Technical Paper, No. 130.

Verheyen, W. (1962). Contribution à la craniologie comparée des primates. *Annales du Musée Royale de l'Afrique Centrale, Series 80, Zoologie, 105*, 1–256.

Vitzthum, V. (2001). Why not so great is still good enough. In P. T. Ellison (ed.), *Reproductive ecology and human evolution*. Aldine de Gruyter, New York, pp. 179–202.

Vitzthum, V. J. (1997). Flexibility and paradox: the nature of adaptation in human reproduction. In M. E. Morbeck, A. Galloway, & A. L. Zihlman (eds.), *The evolving female: A life-history perspective*. Princeton University Press, Princeton, pp. 242–258.

Wade, G. N., & Jones, J. E. (2004). Neuroendocrinology of nutritional infertility. *American Journal of Physiology Regulatery Integrative and Comparative Physiology, 287*, R1277–96.

Wakibara, J. V., Huffman, M. A, Wink, M., Reich, S., Aufreiter, S., Hancock, R. G. V., Sodhi, R., Mahaney, W. C., & Russell, S. (2001). The adaptive significance of geophagy for japanese macaques (*Macaca fuscata*) at Arashiyama, Japan. *International Journal of Primatology, 22*, 495–520.

Waller, J. C. W. (1995). *The aetiologies of major limb injuries amongst chimpanzees in the Sonso area of the Budongo Forest*, Uganda. MSc thesis, University of Oxford, Oxford.

Waller, J. C. W., & Reynolds, V. (2001). Limb injuries resulting from snares and traps in chimpanzees (*Pan troglodytes schweinfurthii*) of the Budongo Forest, Uganda. *Primates, 42*, 135–139.

Wallis, J. (1985). Synchrony of estrous swelling in captive group-living chimpanzees (*Pan troglodytes*). *International Journal of Primatology, 6*, 335–350.

Wallis, J. (1995). Seasonal influence on reproduction in chimpanzees of Gombe National Park. *International Journal of Primatology, 16*, 435–451.

Wallis, J. (1997). A survey of reproductive parameters in the free-ranging chimpanzees of Gombe National Park. *Journal of Reproduction and Fertility, 109*, 297–307.

Wallis, J. (2002). Seasonal aspects of reproduction and sexual behavior in two chimpanzee populations: a comparison of Gombe (Tanzania) and Budongo (Uganda). In C. Boesch, G. Hohmann, & L. F. Marchant (eds.), *Behavioural diversity in chimpanzees and bonobos*. Cambridge University Press, Cambridge, pp. 181–191.

Wallis, J., & Lee, D. R. (1999). Primate conservation: The prevention of disease transmission. *International Journal of Primatology, 20*, 803–826.

Wallis, J., & Reynolds, V. (1999). Seasonal aspects of sociosexual behaviour in two chimpanzee populations: a comparison of Gombe (Tanzania) and Budongo (Uganda). *American Journal of Primatology, 49*, 111.

Washburn, S. L., & DeVore, I. (1961). The social life of baboons. *Scientific American, 204*, 62–71.

Waser, P., & Waser, M. (1977). Experimental studies of primate vocalization: Specializations for long-distance propagation. *Zeitschrift fur Tierpsychologie, 43*, 239–263.

Waser, P., & Brown, C. H. (1986). Habitat acoustics and primate communication. *American Journal of Primatology, 10*, 135–154.

Wasser, S. K. (1996). Reproductive control in wild baboons measured by fecal steroids. *Biology of Reproduction, 55*, 393–399.

Wasser, S. K., & Starling, A. K. (1988). Proximate and ultimate causes of reproductive suppression among female yellow baboons at Mikumi National Park, Tanzania. *American Journal of Primatology*, 16, 97–121.

Wasserman, M. D., & Chapman, C. A. (2003). Determinants of colobine monkey abundance: The importance of food energy, protein and fibre content. *Journal of Animal Ecology*, 72, 650–659.

Wasson, K., & Peper, R. L. (2000). Mammalian microsporidiosis. *Veterinary Pathology, 37*, 113–128.

Watkins, B. E., Ullrey, D. E., & Whetter, P. A. (1985). Digestibility of a high-fibre biscuit-based diet by black and white colobus (*Colobus guereza*). *American Journal of Primatology, 9*, 137–144.

Waterman, P. G. (ed.). (1993). *Methods in plant biochemistry* (Vol. 8). Academic Press, New York.

Waterman, P. G., & Choo, G. M. (1981). The effects of digestibility-reducing compounds in leaves on feed selection of some colobinae. *Malaysian Applied Biology, 10*, 147–162.

Waterman, P. G., & Mole, S. (1994). *Analysis of phenolic plant metabolites.* Blackwell Science Publications, Oxford, U.K.

Waterman, P. G., & Kool, K. M. (1994). Colobine food selection and plant chemistry. In A. G. Davies & J. F. Oates (eds.), *Colobine monkeys: Their ecology, behaviour and evolution.* Cambridge University Press, Cambridge, pp. 251–284.

Waterman, P. G., Choo, G. M., Vedder, A. L., & Watts, D. (1983). Digestibility, digestion-inhibitors and nutrients of herbaceous foliage and green stems from an African montane flora and comparison with other tropical flora. *Oecologia, 60*, 244–249.

Waterman, P. G., Ross, J. A. M., Bennett, E. L, & Davies, A. G. (1988). A comparison of the floristics and leaf chemistry of the tree flora in two Malaysian rain forests and the influence of leaf chemistry on populations of colobine monkeys in the old world. *Biological Journal of the Linnaean Society, 34*, 1–32.

Watterson, J. J., & Butler, L. G. (1983). Occurrence of an unusual leukoanthocyanidin and absence of proanthocyanidins in sorghum leaves. *Journal of Agricultural and Food Chemistry, 31*, 41–45.

Watson, E. E., Easteal, S., & Penny, S. (2001). *Homo* genus: A review of the classification of humans and the great apes. In P. V. Tobias, M. A. Raath, J. Moggi-Cecchi, & G. A. Doyle (eds.), *Humanity from African Naissance to coming Millenia.* Firenze University Press, Firenze and Witwatersrand University Press, Johannesburg, pp. 307–318.

Watt, J. M., & Breyer-Brandwijk, M. G. (1962). *The medicinal and poisonous plants of Southern and Eastern Africa, E&S.* Livingstone, Edinburgh, Scotland.

Watts, D. P. (1984). Composition and variability of mountain gorilla diets in the central Virungas. *American Journal of Primatology, 7*, 323–356.

Watts, D. P. (1989). Ant eating behaviour of mountain gorillas. *Primates,* 30, 121–125.

Watts, D. P. (1990). Ecology of gorillas and its relation to female transfer in mountain gorillas. *International Journal of Primatalogy,* 11, 21–45.

Watts, D. P. (1991). Strategies of habitat use by mountain gorillas. *Folia Primatologica, 56*, 1–16.

Watts, D. P. (1996). Comparative socio-ecology of gorillas. In W. C. McGrew, L. F. Marchant, & T. Nishida (eds.), *Great ape societies.* Cambridge University Press, Cambridge, pp. 16–28.

Watts, D. P. (1998a). Long-term habitat use by mountain gorillas (*Gorilla gorilla beringei*), I: Consistency, variation and home range size and stability. *International Journal of Primatology, 19*, 651–680.

Watts, D. P. (1998b) Long term habitat use by mountain gorillas (*Gorilla gorilla beringei*), II: Reuse of foraging areas in relation to resource abundance, quality and depletion. *International Journal of Primatology, 19*, 681–702.

Watts, D. P. (1998c). Coalitionary mate-guarding by male chimpanzees at Ngogo, Kibale National Park, Uganda. *Behavioral Ecology and Sociobiology, 44*, 43–55.

Watts, D. P., & Mitani, J. C. (2000). Infanticide and cannibalism by chimpanzees at Ngogo, Kibale National Park, Uganda. *Primates, 41*, 357–365.

Watts, D. P., & Mitani, J. C. (2001). Boundary patrols and intergroup encounters in wild chimpanzees. *Behaviour, 138*, 299–327.

Watts, D. P., & Mitani, J. C. (2002). Hunting and meat sharing by chimpanzees at Ngogo, Kibale National Park, Uganda. In C. Boesch, G. Hohmann, L. F. Marchant (eds.), *Behavioural Diversity in Chimpanzees and Bonobos.* Cambridge University Press, Cambridge, pp. 244–255.

Weber, W. (1993). Primate conservation and ecotourism in Africa. In C. A. Potter, J. I. Cohen, & D. Janczewski (eds.), *Perspectives on biodiversity: Case studies of genetic resource conservation and development.* AAAS Press, Washington, DC.

Weber, W., & Vedder, A. (2001). *In the kingdom of gorillas: Fragile species in a dangerous land.* Simon & Schuster, New York.

Weenen, H., Nkunya, M. H. H., Bray, D. H., Mwasumbi, L. B., Kinabo, S., & Kilimali, V. A. E. B. (1990). Antimalarial activity of Tanzanian medicinal plants. *Planta Medica, 56*, 368–370.

Weick, R. F., Dierschke, D. J., Karsch, F. J., Butler, W. R., Hotchkiss, J., & Knobil, E. (1973). Periovulatory time courses of circulating gonadotropic and ovarian hormones in the rhesus monkey. *Endocrinology, 93*, 1140–1147.

Weiss, L. M. (2001). Microsporidia: Emerging pathogenic protists. *Acta Tropica, 78*, 89–102.

Weisenseel, K., Chapman, C. A., & Chapman, L. J. (1993). Nocturnal primates of Kibale Forest: Effects of selective logging on prosimian densities. *Primates, 34*, 445–450.

Werikhe, S., Macfie, L., Rosen, N., & Miller, P. (1997). *Can the mountain gorilla survive? Population and habitat viability assessment for* Gorilla gorilla beringei, *Kampala, Uganda.* Conservation Breeding Specialist Group, Apple Valley, Minnesota.

White, F. J. (1992). Pygmy chimpanzee social organization: Variation with party size and between study sites. *American Journal of Primatology, 26*, 203–214.

Whiten, A., Byrne, R. W., Barton, R. A., Waterman, P. G., & Henzi, S. P. (1991). Dietary and foraging strategies of baboons. *Philosophical Transactions of the Royal Society of London B, 334*, 187–197.

Whiten, A., Goodall, J., McGrew, W. C., Nishida, T., Reynolds, V., Sugiyama, Y., Tutin, C. E. G., Wrangham, R. W., & Boesch, C. (1999). Cultures in chimpanzees. *Nature, 399*, 682–685.

Whitten, P. L., & Naftolin, F. (1998). Reproductive actions of phytoestrogens. *Baillieres Clinical Endocrinology and Metabolism, 12*, 667–690.

Whitten, P. L., & Patisaul, H. B. (2001). Cross-species and interassay comparisons of phytoestrogen action. *Environmental Health Perspectives, 109*, 5–20.

Whitten, P. L., Brockman, D. K., & Stavisky, R. C. (1998a). Recent Advances in Noninvasive Techniques to Monitor Hormone–Behavior Interactions. *Yearbook of Physical Anthropology, 41*, 1–23.

Whitten, P. L., Stavisky, R. C., Aureli, F., & Russell, E. (1998b). Response of fecal cortisol to stress in captive chimpanzees (*Pan troglodytes*). *American Journal of Primatology, 44*, 57–69.

Wickings, J. E., Ambrose, L., & Bearder, S. K. (1998). Sympatric populations of Galagoides demidoff and G. thomasi in the Haut Ogooué region of Gabon. *Folia Primatologica, 69*(suppl. 1), 389–393.

Wild, R., & Mutebi, J. (1997). Bwindi Impenetrable Forest, Uganda: Conservation through collaborative management. *Nature and Resources, 33*, 33–51.

Wildman, D. E., Uddin, M., Liu, G-Z., Grossman, L. I., & Goodman, M. (2003). Implications of natural selection in shaping 99.4% nonsynonymous DNA identity between humans and chimpanzees: enlarging genus *Homo*. *Proceedings of the National Academy of Sciences, 100*, 7181–7188.

Wilkie, D. S., Sidle, J. G., & Boundzanga, G. C. (1992). Mechanized logging, market hunting, and a bank loan in Congo. *Conservation Biology, 6*, 570–580.

Wilkie, D. A., Curran, B., Tshombe, R., & Morelli, G. A. (1998). Modeling sustainability of subsistence farming and hunting in the Ituri Forest of Zaire. *Conservation Biology, 12*, 137–147.

Williams, J. (1999). *Female strategies and the reasons for territoriality in chimpanzees: Lessons from three decades of research at Gombe*. Unpublished PhD thesis, University of Minnesota.

Williams, J. M., Liu, H-Y., & Pusey, A. E. (2002a). Costs and benefits of grouping for female chimpanzees at Gombe. In C. Boesch, G. Hohmann, & L. F. Marchant (eds.), *Behavioural diversity in chimpanzees and bonobos*. Cambridge University Press, Cambridge, pp. 192–203.

Williams, J. M., Pusey, A. E., Carlis, J. V., Farm, B. P., & Goodall, J. (2002b). Female competition and male territorial behaviour influence female chimpanzees' ranging patterns. *Animal Behaviour, 63*, 347–360.

Williamson, E. A. (1988). *Behavioural ecology of western lowland gorillas in Gabon*. PhD Dissertation, University of Stirling, Scotland.

Williamson, E. A., Tutin, C. E. G., Rogers, M. E., & Fernandez, M. (1990). Composition of the diet of lowland gorillas at Lopé in Gabon. *American Journal of Primatology, 21*, 265–277.

Wilson, M. L. (2001). *Imbalances of power: How chimpanzees respond to the threat of intergroup aggression.* PhD dissertation, Department of Anthropology, Harvard University, Cambridge, MA.

Wilson, M., Daly, M., Gordon, S., & Pratt, A. (1996). Sex differences in valuations of the environment. *Population and Environment, 18*, 143–159.

Wilson, M. I., Daly, M., & Gordon, S. (1998). The evolved psychological apparatus of human decision-making is one source of environmental problems. In T. Caro (ed.), *Behavioral ecology and conservation.* Oxford University Press, Oxford, pp. 501–523.

Wilson, M. L., Hauser, M. D., & Wrangham, R. W. (2001). Does participation in intergroup conflict depend on numerical assessment, range location, or rank for wild chimpanzees? *Animal Behaviour, 61*, 1203–1216.

Wilson, M. L., Britton, N. F., & Franks, N. R. (2002). Chimpanzees and the mathematics of battle. *Proceedings of the Royal Society of London B, 269*, 1107–1112.

Wingfield, J. C., Hahn, T. P., Wada, M., Astheimer, L. B., & Schoech S. (1996). Interrelationship of day length and temperature on the control of gonadal development, body mass, and fat score in white-crowned sparrows, *Zonotrichia leucophrys gambelii. General & Comparative Endocrinology, 101*, 242–255.

Winslow, J. T., & Miczek, K. A. (1988). Androgen dependency of alcohol effects on aggressive behavior: A seasonal rhythm in high-ranking squirrel monkeys. *Psychopharmacology, 95*, 92–98.

Wittig, R. M., & Boesch, C. (2003). Food competition and linear dominance hierarchy among female chimpanzees of the Tai National Park. *International Journal of Primatology, 24*, 847–867.

Whitehead, J. M. (1986). Developement of feeding selectivity in mantled howler monkeys, *Alouatta palliata.* In Else, J. G. & Lee, P. C. (eds.), *Primate ontogeny, cognition and social behaviour.* Proceedings of the 10th Congress of the International Primatological Society, Vol. 3, Cambridge University Press, Cambridge, pp. 105–117.

Whitesides, G. H., Oates, J. F., Green, S. M., & Kluberdanz, R. P. (1988). Estimating primate densities from transects in a West African Rain Forest: A comparison of techniques. *Journal of Animal Ecology, 57*, 345–367.

Wong, C. C., Dohler, K. D., Geerlings, H., & von zur Muhlen, A. (1983). Influence of age, strain and season on circadian periodicity of pituitary, gonadal and adrenal hormones in the serum of male laboratory rats. *Hormone Research, 17*, 202–215.

Wood, J. W. (1994). *Dynamics of human reproduction: Biology, biometry, demography.* Aldine de Gruyter, New York.

Woodford, M. H., Butynski, T. M., & Karesh, W. B. (2002). Habituating the great apes: The disease risks. *Oryx, 36*, 153–160.

Woodroffe, R. (1999). Managing disease threats to wild animals. *Animal Conservation, 2*, 185–193.

Wrangham, R. W. (1977). Feeding behaviour of chimpanzees in Gombe National Park, Tanzania. In T. H. Clutton-Brock (ed.), *Primate ecology: Studies of feeding and ranging behaviour in lemurs, monkeys and apes*. Academic Press, New York, pp. 503–537.

Wrangham, R. W. (1980). An ecological model of female-bonded primate groups. *Behaviour, 75*, 262–300.

Wrangham, R. W. (1986). Ecology and social relationships in two species of chimpanzee. In D. I. Rubenstein & R. W. Wrangham (eds.), *Ecology and social evolution: Birds and mammals*. Princeton University Press, Princeton, NJ.

Wrangham, R. W. (1995). Relationship of chimpanzee leaf-swallowing to a tapeworm infection. *American Journl of Primatology, 37*, 297–303.

Wrangham, R. W. (2000). Why are male chimpanzees more gregarious than mothers? A scramble competition hypothesis. In P. M. Kappeler (ed.), *Primate males: Causes and consequences of variation in group composition*. Cambridge University Press, Cambridge, pp. 248–258.

Wrangham, R. W. (2002). The cost of sexual attraction: Is there a trade-off in female *Pan* between sex appeal and received coercion? In C. Boesch, G. Hohmann, & L. F. Marchant (eds.), *Behavioural diversity in chimpanzees and bonobos*. Cambridge University Press, Cambridge, pp. 204–215.

Wrangham, R. W., & Smuts, B. B. (1980). Sex differences in the behavioural ecology of chimpanzees in the Gombe National Park, Tanzania. *Journal of Reproductive Fertility, 28*(Suppl.), 13–31.

Wrangham, R. W., & Waterman, P. G. (1981a). Feeding behaviour of vervet monkeys on *Acacia totilis* and *Acacia xanthophloea*: With special reference to reproductive strategies and tannin production. *Journal of Animal Ecology, 50*, 715–731.

Wrangham, R. W., & Waterman, P. G. (1981b). Condensed tannins in fruits eaten by chimpanzees. *Biotropica, 15*, 217–222.

Wrangham, R. W., & Nishida, T. (1983). *Aspilia* spp. leaves: A puzzle in the feeding behavior of wild chimpanzees. *Primates, 24*, 276–282.

Wrangham, R. W., & Goodall, J. (1989). Chimpanzee use of medicinal leaves. In P. G. Heltne & L. A. Marquardt (eds.), *Understanding chimpanzees*. Harvard University Press, Cambridge, MA, pp. 22–37.

Wrangham, R. W., & Goldberg, T. (1997). An overview of chimpanzee conservation and management strategies. In E. Edroma, N. Rosen, & P. S. Miller (eds.), *Conserving the chimpanzees of Uganda: A population and habitat viability assessment for* Pan troglodytes schweinfurthii. IUCN/SSC, Apple Valley, MN.

Wrangham, R. W., & Mugume, S. (2000). Snare removal program in Kibale National Park: A preliminary report. *Pan Africa News, 7*, 18–20.

Wrangham, R. W., Clark, A. P., & Isabirye-Basuta, G. (1992). Female social relationships and social organization of Kibale Forest chimpanzees. In T. Nishida, W. C. McGrew, P. Marler, M. Pickford, & F. B. M. de Waal (eds.), *Topics in primatology, Vol I: Human origins.* University of Tokyo Press, Tokyo, pp. 81–98.

Wrangham, R. W., Conklin, N. L., Etot, G., Obua, J., Hunt, K. D., Hauser, M. D., & Clark, A. P. (1993). The value of figs to chimpanzees. *International Journal of Primatology, 14*, 243–256.

Wrangham, R. W., Chapman, C. A., Clark-Arcadi, A. P., & Isabirye-Basuta, G. (1996). Social ecology of Kanyawara chimpanzees: implications for understanding the costs of great ape groups. In W. C. McGrew, L. F. Marchant, & T. Nishida (eds.), *Great ape societies.* Cambridge University Press, Cambridge, pp. 45–57.

Wrangham, R. W., Conklin-Brittain, N. L., & Hunt, K. D. (1998). Dietary response of chimpanzees and Cercopithecines to seasonal variation in fruit abundance, I: Antifeedants. *International Journal of Primatology, 19*, 949–970.

Wright, B. E., Abadie, J., Svec, F., & Porter, J. R. (1994). Does taste aversion play a role in the effect of dehydroepiandrosterone in Zucker rats? *Physiology & Behavior, 55*, 225–229.

Wunder, S. (2000). Ecotourism and economic incentives: An empirical approach. *Ecological Economics, 32*, 465–479.

Yamagiwa, J., Mwanza, N., Yumoto, T., & Maruhashi, T. (1991). Ant eating by eastern lowland gorillas. *Primates, 32*, 247–253.

Yamagiwa, J., Mwanza, N., Yumoto, T., & Maruhashi, T. (1994). Seasonal change in the composition of the diet of eastern lowland gorillas. *Primates, 35*, 1–14.

Yamagiwa, J., Maruhashi, T., Yumoto, T., & Mwanza, N. (1996). Dietary and ranging overlap in sympatric gorillas and chimpanzees in Kahuzi-Biega National Park, Zaïre. In W. C. McGrew, L. F. Marchant, & T. Nishida (eds.), *Great ape societies.* Cambridge University Press, Cambridge, pp. 82–98.

Yamagiwa, J., Kahekwa, J., & Basabose, A. K. (2003). Intra-specific variation in social organization of gorillas: Implications for their social evolution. *Primates, 44*, 359–369.

Yamagiwa, J., & Mwanza, N. (1994). Day-journey length and daily diet of solitary male gorillas in lowland and highland habitats. *International Journal of Primatology, 15*, 207–224.

Yamashita, J. Y. (1963). Ecological relationships between parasites and primates. *Primates, 4*, 1–96.

Yamauchi, A., Takei, I., Kasuga, A., Kitamura, Y., Ohashi, N., Nakano, S., Takayama, S., Nakamoto, S., Katsukawa, F., & Saruta, T. (1996). Depression of dehydroepiandrosterone in Japanese diabetic men: Comparison between non-insulin-dependent diabetes mellitus and impaired glucose tolerance. *European Journal of Endocrinology, 135*, 101–104.

Yeager, C. P., & Kool, K. (2000). The behavioural ecology of Asian colobines. In P. F. Whitehead & C. J. Jolly (eds.), *Old World monkeys*. Cambridge University Press, Cambridge, pp. 498–521.

Yoshimura, Y., & Wallach, E. E. (1987). Studies of the mechanism(s) of mammalian ovulation. *Fertility & Sterility, 47*, 22–34.

Zuberbühler, K. (2000). Referential labelling in wild Diana monkeys. *Animal Behaviour*, 63, 293–299.

INDEX

A
Acalypha sp., (plant), 318
Acanthus pubescens (plant), 118, 120, 129
acid detergent fiber, 51
ADF, *See* acid detergent fiber
African forest , division of, 15
Afromomum (plant), 68, 80–81, 118, 120
Albizia coriaria (tree), 88
Albizia grandibracteata (tree), 80, 118, 124
Albizia zygia (tree), 67–68, 80
Alcornea floribunda (tree), 203
alkaloids, 31
Allochrocebus lhoesti, 16
Allouatta palliata, 40
Alstonia boonei (plant), 118, 120
Amaranthus sp. (herb), 64
Aneiloma aquietiale (plant), 68, 80
Aningeria altissima (tree), 49
Annona chrysophylla (tree), 80
Antiaris toxicaria (plant), 118
Arundinaria alpina (tree), 49, 53, 55
Aspilia mossambicensis (plant), 107
Aspilia pluriseta (plant), 107
Aspilia rudis (plant), 107
Ateles geoffroyi, 375

B
baboons, 11
  olive, *See* feeding behavior, of primate species
  olive savannah, *See Papio cynocephalus anubis*
  sonso, 63
Balsamocitrus dawei (plant), 118
bamboo, *See Arundinaria alpina*
Bertiella studeri (parasite), 116
black-and-white colobus, *See Colobus guereza*
blood testosterone, 96, 98
blue monkeys, 9, 46–47, 378
Brachiaria brizantha (seed), 67, 87
Broussonetia papyrifera (plant), 64, 69, 71–72, 80–81, 88, 101, 118, 294, 318, 320
Budongo forest reserve, 24, 110, 120–124
  animal vocalizations, interpretation of, *See* chimpanzee vocalizations
  conservation project
    critical issues related to, 432–436
    fate of natural resources, 431
    motivating incentives and desired trade-offs, 432
    perception of chimpanzees, 427–430
    project site, 425–427
    stakeholder responsibility and ownership, 430–431
  dietary abundance and habitat preferences
    dietary overlap, 358–363
    diets at dawn and dusk, 348–349, 352–358
    diets from transect scans, 348, 351–352
    effects of logging on diets, 368–370
    fruit availability, 349–351, 364–368
    tree species enumeration and fruit production, 347
  fig feeding by chimpanzees, study of, *See* fig feeding by chimpanzees, study of
  fission-fusion grouping, of chimpanzees, *See* fission–fusion grouping, study of
  leaf-assisted drinking behavior, of chimpanzees, *See* sonso community chimpanzees, leaf-assisted drinking behavior
  physicochemistry and mineralogy of soils eaten by chimpanzees, *See Pan troglodytes schweinfurthii*
  protective measures, to chimpanzees
    buffer zone, 397–398
    live-trap project, 400–401
    logging effects, 401–402
    snare removal project, 399–400
    tree corridors, 399
  species comsumed, by primates, 80
  threats to chimpanzees
    contagious diseases, 396–397
    snares and traps, 392–396
Bwindi gorillas, *See* gorillas
Bwindi national park, 6, 10, 154, 171, 174–175, 186–189, 191, 193, 195–197, 203, 330, 333, 335, 337, 341, 397, 405, 407, 410

C
Calconcoba schweinfurthi (tree), 80, 83
Canna indica (plant), 80–82

Cassia mannii (tree), 80
CCT, *See* corticosterone
Celtis africana (plant), 118, 383
Celtis durandii (tree), 64, 71–72, 80–81, 118, 125, 369, 383
Celtis gomphophylla (tree), 24, 35, 39
Celtis mildraedii (plant), 118, 284, 369
Celtis wightii (plant), 118
Celtis zenkeri (plant), 118, 284
Central African red colobus, *See* colobus monkeys
Centrosema pubescens (plant), 80
Cercocebus, 12
  agilis, 16
cercopithecinae
  cercopithecini
    Allenopithecus, 8–9
    Cercopithecus, 7, 9
    Chlorocebus, 7–8
    Erythrocebus, 7, 9
  papionini
    Colobus, 13–14
    Lophocebus, 12
    Papio, 11
    Piliocolobus, 12–13
    Theropithecus, 12
Cercopithecus
  albogularis, 9
  ascanius, 59, 346, 351, 369, 378
    schmidti, 9, 16
  denti, 16, 18
  doggetti, 10–11, 17–19
  guereza, 348, 369
  hamlyni, 16
  kandti, 18
  lhoesti, 9
  linnaeus, 9
mitis, 16, 59, 346, 369, 375, 378
    doggetti, 54
    erythrarchus, 54, 56–57
    kandti, 54, 57, *See also* feeding behavior, of primate species
    labiatus, 54, 56–57
    mitis, 10
    nyasae, 54
    stuhlmanni, 10, 16–19, 54, 56–57
  neglectus, 11
  nictitans, 16
  pogonias, 16
chadonneret traps, 331
Chaetacme aristata (plant), 118, 124
Chelonia mydas, *See* green turtles
chimpanzee ovarian cycles, 229
chimpanzee vocalizations
  deciphering junglespeak
    implictions for referential communication, 306–308
    informational cue processing, 308–310
  implications for human language, 310–312
  lexicon in, 302–305
  referential communications, 301–302
Chlorocebus
  pygerythrus rufoviridis, 8
  tantalus budgetti, 8
chromium, 150
Chrysophyllum albidum (plant), 284, 370
churr call, 332
Cleistopholis patens (tree), 64, 80–82
Colobus
  angolensis, 17–18, 26, 30
    cottoni, 14, 17, 19
    ruwenzorii, 14, 18
  guereza, *See Colobus guereza*
  polycomos, 26, 29, 31
  satanas, 26, 29
Colobus angolensis ruwenzori, 330
Colobus guereza, 17, 19
  behavior pattern of, in logged and unlogged areas, 376
  dietary composition, 26–27
  feeding ecology and plant chemistry, 28, 31–32
  feeding strategies, 25, 28
  foraging behavior of, in logged and unlogged areas
    behavioral observations, 378–379
    food tree density, 379
    statistical evaluations, 379–387
  habitat preferences, 25, 28
  and low energy high fiber diet, 38–40
  methodological analysis of densities of, 32–34
    in correlation to forest structural attributes, 33
    intrasite variation of, 33–34
  physiology and mixed-fruit leaf diet, 36–38
  responses to habitat disturbances, 29–30
  response to habitat modification, 24–25
  and significance of liana leaves, 35–36
Combretum molle (tree), 80
Commelinaceae leaves, 126
Commelina erecta (plant), 80
community-based natural resource management, 425
coprophagy, 155
copulation behavior, *See* estrous cycles, of female chimpanzees
Cordia ovalis (tree), 80
CORT, *See* urinary free cortisol
corticosterone, 95, 97–98
crescendo callers, 5

Croton macrostachys (tree), 80, 83
Cubitermes speciosus (termite mound soil), 140, 146, 151
cyanogenic glycosides, 52, 55
Cynometra alexandri (plant), 318, 346, 351

D
De Brazza's monkey, 11
dehydroepiandrosterone, 95
 -sulfate, 95
Demidoff's bushbaby, *See* galagoides
Dendrohyrax dorsalis (tree), 343
Dent's mona, 11
Desmodium canum (plant), 80–81
Desmodium intotueme (plant), 80–81
Desmodium intotuene (creeper), 64
Desmodium ovalifolium (plant), 159
Desmodium repandum (plant), 80–81
Desplatsia dewevrei (plant), 318
DHEA, *See* dehydroepiandrosterone
DHEA-S, *See* dehydroepiandrosterone, -sulfate
Dioscorea dumeforum (plant), 80
Dragendorff's reagent, 52

E
eagle alarm calls, 301
eastern-central forest area, 15
Eastern gorilla, *See Gorilla beringei*
eastern needle-clawed bushbaby, *See* galago
ecotourism, 406
Eleusine jaegeri (grass), 67, 80
Erica arborea (tree), 49
Erythrina abysinica (tree), 80
Erythrocebus, *See* cercopithecidae, cercopithecinae
estrogens, 230
estrous cycles, of female chimpanzees, 250–253
estrous parties, *See* fission–fusion grouping, study of
Eucalyptus (tree), 338
eyes
 of *Chlorocebus tantalus*, 8
 of *Galagoides demidoff*, 336
 of *Galagoides thomasi*, 335
 of *Galago matschiei*, 5, 332
 of *Pthirus pubis*, 188

F
family, definition of, 4
feeding behavior, of primate species
 of *Colobus guereza*, 31, 35–40
 nutritional ecology of golden monkey
  group observations, 49–50
  plant collections and nutritional analyses, 50–53
  source of diet of, 53–58
  study areas, 47–49
Papio anubis, *See also* primate diets
  dietary profiles of, 71
  food items, in diet of, 65–66
Papio cynocephalus anubis
  diet diversity, 79–81
  pith, root and bark eating, 87–90
  seed predation, 81–87
Ficus barteri (plant), 102, 118
Ficus capensis (tree), 80
Ficus exasperata (tree), 55, 64, 66, 68, 80, 87–88, 117–118, 125–126
Ficus igens (tree), 80
Ficus mucuso (tree), 83, 118, 284, 288, 292–294
Ficus natalensis (plant), 117–118
Ficus polita (tree), 80, 83, 118
Ficus sur (tree), 64, 66, 80, 101, 116, 119–120, 122, 126, 275, 284, 288, 290–294
Ficus vallis-choudae (tree), 80
Ficus varifolia (tree), 68, 80, 87
fig feeding by chimpanzees, study of
 data analysis, 289
 data collection: feeding behavior
  feeding efficiency, 288
  feeding technique, 287–288
  position, 286
  posture, 286–287
 feeding tasks, 284, 286
 injuries classification and description, 285
 result analysis
  feeding efficiency, 293–294
  feeding position and posture, 289–290
  feeding techniques, 290–293
 study community, injury characteristics of, 284
 study site, 283–284
Fiscus sur (tree), 318
fission–fusion grouping, study of, 209–210
 definitions, 214
 observation methods, 213–214
 parameters of, 210–212
 result analysis
  estrous parties, 221–222
  intercommunity differences, 215–221
  sex difference in gregariousness, 215, 222–223
 statistics, 215
 study sites, 212
follicle-stimulating hormone, 94
follicular estradiol levels, 230
froth test, 52
FSH, *See* follicle-stimulating hormone
Funtumia elastica (plant), 284

G

Galago sotikae, 5
Galagoides demidoff, 6, 16, 18, 330
behavior and habitat use, 337
conservation of, 342–343
distribution and density, 335–336
morphological description and loud call repertoire, 336–337
patterns of distribution, 340–341
Galagoides thomasi, 6, 16, 330
behavior and habitat use, 335
distribution and density, 334–335
morphological description and loud call repertoire, 335
Galago matschiei, 5, 16, 329–330
behavior and habitat use, 332
conservation of, 342–343
distribution and density of, 331
morphological description and loud call repertoire, 332
patterns of distribution, 339–340
Galago senegalensis, 5, 19
behavior and habitat use, 334
conservation of, 342–343
distribution and density, 332–333
morphological description and loud call repertoire, 333
patterns of distribution, 341
Galago thomasi
conservation of, 342–343
patterns of distribution, 340
genus, definition of, 4
geophagy, in chimpanzees, *See Pan troglodytes schweinfurthii*
gibbsite, 86
goethite, 86
golden monkey, 10, *See also* feeding behavior, of primate species
goodness of fit test, 289
Gorilla beringei, 14, 16, 154, 172, 330
beringei, 18, 154
graueri, 17, 155
Gorilla gorilla, 14
gorilla, 155
gorillas, *See also* primate ranging patterns, study of
captive zoo diet, 157
diet and foraging strategies of, 154–157
digestive anatomy and physiology of, 157–158
nutritional characteristics of foods
Bwindi gorillas, 165–167
eastern gorillas, 165
western gorillas, 161, 164
nutritional composition, environmental effects on, 158–160
nutritional ecology for conservation of, 168–169
parasite transmission in
through large intestine, 184–188
through skin, 188–190
through small intestine, 176–184
through stomach, 174–176
gray-cheeked mangabey, 12
green turtles, 42
Grewia mollis (tree), 80
grey-olive monkeys, 11
grunt yaps, 335
gut microflora, 42, 47

H

halloysite, *See* physicochemistry, of termite mound soil ingested by chimpanzees
haplorrhini, *See* simiiformes
homininae
Gorilla, 14
Homo, 14
Pan, 14
hominoidae
homininae, 14–15
ponginae, 14
Hylobates muelleri, 375
Hyparrhenia cymbarica (grass), 80
Hyparrhenia folipendula (grass), 80
Hypericum revolutum (plant), 49, 59
hypothalamic–pituitary–gonadal axis, 94
Hypsignathus monstrosus (tree), 343

I

incremental callers, 5
Indigoflora emarginella (plant), 80
infant howler monkeys, *See Allouatta palliata*

J

Jacaranda ovalifolia (tree), 80

K

Kagombe–Matiri forest complex, 13
Kalinzu forest reserve, 247–249
kanyawara community chimpanzees, *See also* physicochemistry, of termite mound soil ingested by chimpanzees; sexual behavior of chimpanzees, study of
fission-fusion grouping of, *See* fission–fusion grouping, study of
reproductive parameters of, *See* urinary estrone conjugates in wild chimpanzees, study of
self-medicative behavior of
fecal analysis, 114–116
plant consumption, 113–114, 117–120, 124–127

soil consumption, 127–128, 136
urine analysis, 116–117
snare related injuries to, 395–396
kanzi, 310
kaolinite, *See* physicochemistry, of termite mound soil ingested by chimpanzees
Kaplan–Meier survival analysis, 232, 235–236
Khaya anthotheca (tree), 68, 80, 88, 119, 124, 284, 369
Khaya grandifoliola (tree), 88
Kibale national park, 9, 12, 24, 26, 40, 46, 49, 51, 55–56, 107, 109–110, 117, 136, 158, 212, 227, 230, 232, 239, 261, 330–332, 339–341, 364–365, 375–377, 384, 395, 401
Kinyara sugar works, 397–398

L
lactating, 233
Lannea kerstingii (tree), 80
Lantana camara (tree), 64, 80–81
Lasiodiscus mildbraedii (plant), 318
leaf sponging, 322
leopard alarm calls, 301
Lesser or Senegal bushbaby, *See* galago
LH, *See* luteinizing hormone
L'Hoest's monkey, 9
lianas leaves, 24, 31, 33, 35–36
lipids, 160
locomotive behavior, in chimpanzees
of banura, 266
of kalema, 266–267
of kewaya, 267
of kigere, 267–268
of zana, 267
Lophocebus albigena, 16, 59, *See also* gray-cheeked mangabey
johnstoni, 12
lorisiformes, *See* strepsirrhini
luteinizing hormone, 94
Lychnodiscus cerospermus (tree), 80

M
Macaca arctoides, 99
Macaca mulatta, 98
Maesopsis eminii (tree), 80, 318, 369
mantled guereza, 13
Marantochloa leucantha (plant), 68, 80, 318
Margaritaria discoideus (plant), 64, 83
Markhamia platycalyx (tree), 24
maximum sexual swelling, 233–234
medicinal plants, 106–108
consumed, by chimpanzees, 117–124
sites, 110–111
metahalloysite, *See* physicochemistry, of termite mound soil ingested by chimpanzees
Mgahinga national park, 10, 14, 47–49, 55, 60, 330
Microcebus murinus, 98
Miocene–Pliocene boundary, 4
montane forest belt, 49
mountain gorillas, *See Gorilla beringei*
Myrianthus arboreus (plant), 127
Myrianthus holstii (plant), 102

N
Newtonia buchananii (tree), 49
Nkuringo gorilla group, behavioral ecology of
daily path length of, 408–409
habitat use by Nkuringo gorillas, 409–413
monthly summaries, 415–416
other behaviors, 419–420
recommendations for conservation, 420–422
time spent outside park, 416–417
dietary behavior, 413–415
influences on daily path length, 414–415
nest cohesion, 413–414
Nuxia congesta (tree), 58
Nyakafunjo villagers, *See* Budongo forest reserve, conservation project

O
Oesophagostomum stephanostomum (parasite), 107–108, 111, 116, 120, 130–131, 144, *See also* parasites, attacking gorillas
Olea welwitschii (tree), 49, 383
Oligocene–Miocene boundary, 4
olive baboons, *See* feeding behavior, of primate species, *Papio anubis*
olive savannah baboon, *See* feeding behavior, of primate species, *Papio cynocephalus anubis*
Otolemur argentatus, 6
Otolemur monteiri, 342
Oxytenanthera abysinica (plant), 80

P
Pannicum maximum (grass), 64, 80, 87
pant–grunt behavior, 255
pant hoots, *See* chimpanzee vocalizations
Pan troglodytes, 59, 62, 351–352, 354, 363, 369, 376, 389, *See also* fig feeding by chimpanzees, study of; primate ranging patterns, study of; sexual behavior of chimpanzees, study of hormones, role of
and climate, 96–99, 102
and diets, 94–96, 101–102
marungensis, 15
schweinfurthii, *See Pan troglodytes schweinfurthii*

Pan troglodytes, (*cont.*)
seasonality changes, role of, 99–100
self-medicative behavior
kanyawara community, 112–113
sonso community, 111–112
vocalizations, *See* chimpanzee vocalizations
Pan troglodytes schweinfurthii, *See also* primate ranging patterns, study of; sexual behavior of chimpanzees, study of
locomotive behavior and injury effects, study of
data analysis, 264–266
data collection, 264
definitions of terms, 263
result analysis: congenital deformities, 278
result analysis: mother carrying dependent during locomotion, 270–273, 277–278
result analysis: moving pattern, 266–268, 274
result analysis: snare injuries, 278
result analysis: time spent moving, 268–270, 274
study site, 261–263
physicochemistry and mineralogy of soils eaten
data analysis, 139–140
field protocols, 138–139
laboratory protocols, 139
result analysis:diet and health, 143–145
result analysis:general behavior, 140–143
result analysis:physicochemistry, 145–149
study site, 137–138
paper mulberry, *See Broussonetia papyrifera*
Papio tesselatum, 11
Papio cynocephalus anubis
foraging behavior
diet diversity, 79–81
forest structure, influence of, 90–91
pith, root and bark eating, 87–90
seed predation, 81–87
unidentified species consumed, 91
species consumed, from Budongo forest reserve, 80
Paralibyostrongylus kalina, 175–176
parasites, attacking gorillas
Anoplocephala gorillae, 179–180
Ascaris lumbricoides, 177–178
Cryptosporidium parvum, 182–184
Encephalitozoon intestinalis, 182
of free-ranging gorillas, 191–192
Giardia lamblia, 181–182
Hyostrongylus kigeziensis, 174–175
Murshidia devians, 184
Oesophagostomum stephanostomum, 185–186
Probstmayria gorillae, 186–187
Pthirus gorillae, 188–189
Sarcoptes scabiei, 189–190
Trichuris trichiura, 187–188
Parinari excelsa forest, 5
Paspalum conjugatum (grass), 80
Patas monkey, 9
Pennisetum purpureum (grass), 68, 80, 87, 120
peripheral female chimpanzees, 217, 219, 222, 229–230, 234–235, 238–240, 243–244
Perodicticus potto ibeanus, 7, 16, 330
behavior and habitat use, 338
conservation of, 342–343
distribution and density, 337–338
morphological description, 338
patterns of distribution, 341–342
Philipa johnstonii (tree), 49
Physalis peruviana (plant), 64
physicochemistry, of termite mound soil ingested by chimpanzees
characteristics of ingested and control soil samples, 145
mineralogy of clay fraction, 147
mineralogy of sand and silt fraction, 146–147
particle size, 146
soil chemistry, 147–149
soil structure, 146
Phytolacca dodecandra (plant), 109, 124
Piliocolobus foai, 13, 16, 18–19
ellioti, 17
Piliocolobus tephrosceles, 18
Piper capense stems, 125
Pollia condensata (tree), 80, 83
postpartum infecundity, 230
Presbytis
aygula, 29
comata, 27, 29
hosei, 29
melalophos, 27, 29
obscura, 27, 29
rubicunda, 29
primate diets
bark, 68
crop raiding, 69
fruit, 64–66
garbage raiding, 69
leaves, 64
other food items, 69
roots, 68
and season changes, 70
seedlings, 69
seeds, 66–67
stems, 67–68
time spent on feeding, 67

primate ranging patterns, study of
chimpanzee and gorillas nesting sites, 198–203
GPS recordings of, 196–198
study area, 195–196
primates, of western Uganda
biogeography of, 16–20
definitions of, 3–4
haplorrhini
simiiformes, 7–15
north–south division in fauna, 17
strepsirrhini
lorisiformes, 4–7
vocalizations, 5
Procolobus badius, 26, 29–30
forging behavior of, in logged and unlogged areas, 376
behavioral observations, 378–379
food tree density, 379
statistical evaluations, 379–387
Procolobus badius tephrosceles, 330
Prunus africanus (tree), 332, 382
Psidium guajava (tree), 64, 80

R
Raphia farinifera (tree), 64, 69–71, 80–81
red colobus, *See Procolobus badius*
red-tailed monkey, 9, 51, *See Cercopithecus ascanius*
Ricinodendron arinifera (tree), 66
Ricinodendron heudelotii (tree), 66, 70–71, 80–81
rolling callers, 5
Rubia cordifolia (plant), 107
Rutshuru plains, 8
Rwenzoris, 18–19

S
saponins, 52
Sarcophrynium schweinfurthianium (herb), 67
schoutedeni intermediates, 11
Securinega virosa (plant), 80
Semnopithecus entellus, 27, 99
sexual behavior of chimpanzees, study of
methodology, 249–250
result analysis, 250–253
silver monkey, 10, 18
silvery Greater galago, *See* otolemur monteiri
simiiformes
cercopithecidae, 7–11
hominoidea, 14
SLOWPOKE-2 reactor, 139
snake alarm calls, 301
snare removal project, *See* Budongo forest reserve, protective measures, to chimpanzees
snaring, of chimpanzees, 260
avoidance of, 277
conservation implications of, 279–280
injuries, 278
sonso community chimpanzees, *See also* physicochemistry, of termite mound soil ingested by chimpanzees; sexual behavior of chimpanzees, study of
feeding behavior of, *See* fig feeding by chimpanzees, study of
fission-fusion grouping of, *See* fission–fusion grouping, study of
leaf-assisted drinking behavior, 314–315
cultural behavior in, 321–323
geographic distribution of, 319–320
leaf use within community, 316–317
leaves used in, 318
reproductive parameters of, *See* urinary estrone conjugates in wild chimpanzees, study of
self-medicative behavior
fecal analysis, 114
plant comsumption, 113, 117–120, 125–127
soil consumption, 127, 136
urine analysis, 116
snare related injuries to, 393–394
termite mound soil ingestion, 141–142
unassisted drinking behavior, 315–316
cultural behavior in, 321–323
geographic distribution of, 319–320
vocalizations of, *See* chimpanzee vocalizations
Spathodea campanulata (tree), 80
species, definition of, 3
spectacled galago, *See* galago
staccato yaps, 332
standardized analytical techniques, for assessing nutrient content
butanol–HCl assay method, 159
freeze-drying technique, 158
lipid content assessment, 160
in vitro digestibility method, 159–160
strepsirrhini
lorisiformes
galagidae, 4–5
lorisidae, 7
Strombosia scheffleri (tree), 49, 383
Strongyloides fulleborni (parasite), 116, 120, 144, 178–179
Strychnos innocua (tree), 80
subspecies, definition of, 3

T

tails
- of *Ascaris lumbricoides*, 177
- of *Cercopithecus ascanius*, 9
- of *Chlorocebus pygerythrus*, 8
- of *Chlorocebus tantalus*, 8
- of *Colobus angolensis ruwenzorii*, 14
- of *Colobus guereza occidentalis*, 13
- of *Galagoides demidoff*, 336
- of *Galagoides thomasi*, 335
- of *Galago matschiei*, 332
- of *Galago senegalensis*, 333
- of *Galago senegalensis senegalensis*, 5
- of *Macaca arctoides*, 99
- of *Otolemur*, 6
- of *Perodicticus potto ibeanus*, 338
- of *Piliocolobus tephrosceles*, 13

tannins, 31, 159
Tantalus monkey, 8
termite mound soil, 136
Ternidens deminutus, 120
TEST, *See* blood testosterone
Tetrapleura tetraptera (plant), 64
thin layer chromatography, 89
Thomas's bushbaby, *See* galagoides
TLC, *See* thin layer chromatography
Trachypithecus
- *johnii*, 27
- *pileatus*, 27
- *senex*, 27

Trichilia rubescens (plant), 109, 124–125
Trichostrongylus sp., 176–177
Trichuris trichiura (parasite), 116
Troglodytella abrassarti (parasite), 116, 120, 144

U

ugandae Matschie, 12
Ugandan red colobus, *See* colobus monkeys
urinary estrone conjugates in wild chimpanzees, study of
- birth interval analysis, 232
- data treatment, 233–234
- endocrine methods, 232–233
- result analysis: assay validation, 234–235
- result analysis: estrone conjugates, 238–240, 242
- result analysis: interbirth intervals, 235–238
- study population, 231–232

urinary free cortisol, 95, 97

V

Vernonia amygdalina (plant), 108
Vervet monkey, 8
Virunga gorillas, *See* gorillas
vulcanism, 19

W

wetern gorillas, *See* gorillas

Z

zoopharmacognosy, 106